D0891761

INTRODUCTION TO METALLURGICAL THERMODYNAMICS

INTRODUCTION TO METALLURGICAL THERMODYNAMICS

SECOND EDITION

DAVID R. GASKELL

Professor of Metallurgical Engineering
Purdue University

HEMISPHERE PUBLISHING CORPORATION
A member of the Taylor & Francis Group

New York Washington Philadelphia London

INTRODUCTION TO METALLURGICAL THERMODYNAMICS, Second Edition

10 11 12 13 14 15 B C B C 8 9 8 7 6 5 4 3 2 1 0 9 8

Library of Congress Cataloging in Publication Data

Gaskell, David R., date.
Introduction to metallurgical thermodynamics.

Bibliography: p.
Includes index.
1. Metallurgy. 2. Thermodynamics. I. Title.
II. Series.
TN673.G33 1981 536'.7'024669 81-668
ISBN 0-89116-486-3 AACR2

For Sheena

Sarah

Claire

and Andrew

CONTENTS

PREFACE

The major difference between the first and second editions is the adoption of S.I. units in the latter. In addition, Chapter 14 has been revised to include a discussion of the thermodynamics of aqueous solutions and Pourbaix diagrams, the numbers of worked examples and end-of-chapter problems have been increased, and a solutions manual has been prepared.

I wish to acknowledge my gratitute to Professor S. K. Tarby of Lehigh University who provided me with many valuable comments and suggestions, all of which have been incorporated in the later printings of the first edition and in this new edition.

David R. Gaskell

PREFACE TO THE FIRST EDITION

Thermodynamics tends to be a confusing subject to the beginning student. This confusion is caused by the initial conceptual difficulties experienced by the student rather than as a result of the subject being, in any way, inherently "difficult." Indeed the beauty of the subject stems from its simplicity, whereby, from the statement of a few initial laws, there evolves systematically a method of rigorously describing the behavior of matter in a manner which is devoid of temporal theories. The ease with which the average undergraduate begins to understand the thermodynamic method and its applications depends very much on the nature of his introduction to the subject. Although there exist many excellent standard treatises on thermodynamics, these standard works, being written for thermodynamicists by thermodynamicists, are, almost by definition, not suitable as introductory texts for beginning students. There thus exists a need for introductory texts, the prime purpose of which is to render the student more capable of fully utilizing the standard treatises.

At one extreme thermodynamics can be regarded as being a subject which is of sufficient beauty to be worth studying solely for its own sake, and at the other extreme it can be regarded as being simply a tool to be applied for the purposes of solving real problems. Although the ideal introduction to the subject should take some midway path between these extremes, the majority of undergraduates are introduced to the subject near one or the other of its extremes. Traditionally these are the "chemistry" students and the "engineering" students. On the one hand, students who are introduced to thermodynamics by chemists can find themselves rapidly immersed in a pseudoqualitative world inhabited by a countless number of strange equations, among which there appear to be few connections. The student rates the difficulty of the subject as being proportional to the number of equations which he is required to memorize, and the lack of any apparent raison d'etre or even application of the subject, as presented, leaves the student thinking that thermodynamics is a subject to be learned solely for reasons of acquiring a feeling of intellectual superiority. On the other hand, the students who are introduced to thermodynamics by the traditional engineers learn the subject purely as a mathematical tool. They learn the minimum possible number of equations which are then available for the "plugging in" of numbers. The subtleties of the subject are all but ignored, as a result of which the arrival of the student at the wrong numerical answer is caused by his plugging the wrong numbers into the wrong places in the wrong equation. A student simultaneously learning thermodynamics from the chemists and the engineers might have difficulty in realizing that the two subjects are, in fact, the same.

As metallurgy can be partially defined as being the meeting point of the physical sciences and engineering, in that a balance is struck between physical principles and practical applications, it is appropriate that a similar balance be obtained in the presentation of metallurgical thermodynamics. In this book an attempt has been made to simultaneously demonstrate the underlying principles and their applicability. Wherever possible an attempt has been made to accompany the normal thermodynamic derivation of an equation with a derivation which illustrates the thermodynamics of the situation, and the order in which the material is presented is such as to attempt to maintain, throughout, a continuity of the development and use of the principles. Also an attempt has been made to illustrate the principles, as they are developed, with quantitative examples of their applicability. It has been stated that thermodynamics is a subject which must be applied with intelligence. In this context the "intelligence" is acquired as a result of an understanding of the principles.

As the treatment of the subject in this book is introductory in nature, its scope is necessarily limited; but, once having read the book, it is hoped that the student of metallurgy will be in a better position than he otherwise would be both for using the thermodynamic method and for going more deeply into the subject with the aid of the standard reference texts.

The author gratefully acknowledges the aid of Dr. Dwarika P. Agarwal, who verified the answers to the problems.

David R. Gaskell

SYSTEM OF UNITS, SYMBOLS, NOTATION, AND SELECTED CONSTANTS

S.I. UNITS

S.I. units (Système International d'Unites) are arranged in three categories: (1) base units, which are dimensionally independent of one another, (2) derived units, which are obtained by suitable combination of the base units, and (3) supplementary units.* The six base units and the specially named derived units commonly used in chemical thermodynamics are listed in Tables 1 and 2, respectively.

Fractions and multiples of the basic units are as designated in Table 3.

*"The International System of Units (SI)," NBS Special Publication 330, U.S. Department of Commerce, 1974 edition.

Table 1. The S.I. Base Units

Quantity	Name	Symbol
Length	Meter	m
Mass	Kilogram	kg
Time	Second	s
Electric current	Ampere	A
Thermodynamic temperature	Kelvin	K
Amount of substance	Mole	mol

Table 2. S.I. Derived Units with Special Names

Quantity	Name	Symbol	Expression in terms of other units	Expression in terms of S.I. base units
Force	Newton	N		$m \cdot kg \cdot s^{-2}$
Energy	Joule	J	$N \cdot m$	$m^2 \cdot kg \cdot s^{-2}$
Pressure	Pascal	Pa	N/m^2	$m^{-1} \cdot kg \cdot s^{-2}$
Power	Watt	W	J/s	$m^2 \cdot kg \cdot s^{-3}$
Electric charge	Coulomb	C	$A \cdot s$	$s \cdot A$
Electromotive force	Volt	V	W/A	$m^2 \cdot kg \cdot s^{-3} \cdot A^{-1}$

An exponent attached to a symbol containing a prefix indicates that the multiple or submultiple of the unit is raised to the power expressed by the exponent. For example,

$$1 \text{ cm}^3 = (10^{-2} \text{ m})^3 = 10^{-6} \text{ m}^3$$
$$1 \text{ cm}^{-1} = (10^{-2} \text{ m})^{-1} = 10^2 \text{ m}^{-1}$$

The S.I. system specifically eliminates the use of such units as

$$\text{erg } (1 \text{ erg} = 10^{-7} \text{ J})$$

and

$$\text{dyne } (1 \text{ dyne} = 10^{-5} \text{ N})$$

and deprecates the use of

$$\text{torr} \left(1 \text{ torr} = 1 \text{ mmHg} = \frac{101{,}325}{760} \text{ Pa}\right)$$

and

$$\text{calorie } (1 \text{ thermochemical calorie} = 4.184 \text{ J})$$

Table 3. S.I. Prefixes

Factor	Prefix	Symbol	Factor	Prefix	Symbol
10^1	Deka	da	10^{-1}	Deci	d
10^2	Hecto	h	10^{-2}	Centi	c
10^3	Kilo	k	10^{-3}	Milli	m
10^6	Mega	M	10^{-6}	Micro	μ
10^9	Giga	G	10^{-9}	Nano	n
10^{12}	Tera	T	10^{-12}	Pico	p
			10^{-15}	Femto	f
			10^{-18}	Atto	a

However, the system tolerates, "for the time being," use of the

standard atmosphere (1 atm = 101,325 Pa)

and bar (1 bar = 10^5 Pa)

In view of the familiarity, on the part of the metallurgist, with the standard atmosphere as the unit of pressure and 1 atm pressure as the standard state for an ideal gas, this unit has been adopted in the present text.

LIST OF SYMBOLS

a	van der Waals constant
a_i	the activity of species i with reference to a specified standard state
A	Helmholtz free energy (or work function)
$\mathcal{A}$	Avogadro's number
b	van der Waals constant
C	the number of components
C	heat capacity
c_p	constant pressure molar heat capacity
c_v	constant volume molar heat capacity
$\mathcal{E}$	electromotive force
$\mathcal{E}^0_{M/M^{z+}}$	standard oxidation potential of species M
$\mathcal{E}^0_{A/A^{z-}}$	standard reduction potential of species A
E	electromotive force with sign according to the European convention
e^i_j	the interaction parameter of i on j
F	the number of degrees of freedom of an equilibrium
$\mathcal{F}$	Faraday's constant
f	fugacity
f_i	the Henrian activity coefficient of the species i
$f_{i(\mathrm{wt\%})}$	the activity coefficient of the species i with respect to the 1 weight percent standard state
f^i_j	the interaction coefficient of i on j
G	Gibbs free energy
H	enthalpy
h_i	the Henrian activity of species i

$h_{i(\text{wt\%})}$ the activity of the species i with respect to the 1 weight percent standard state

K the equilibrium constant
k Boltzmann's constant

m mass

n the number of moles
n_i the number of moles of the species i

P pressure
P partition function
P the number of phases occurring in a system
p_i the partial pressure of the species i
p_i^0 the saturated vapor pressure of the species i

q heat

R the Gas Constant

S entropy

T temperature
T_m melting temperature
T_b boiling temperature

U internal energy

V volume

w work

X_i the mole fraction of the species i

Z the compressibility factor

α coefficient of thermal expansion
α the regular solution constant

β coefficient of isothermal compressibility

γ	ratio of c_p to c_v
γ_i	the activity coefficient of the species i
γ_i^0	the Henry's law constant
ϵ_i	the energy of the ith energy level
ϵ_j^i	the interaction parameter of i on j
μ_i	the chemical potential of the species i
(s)	solid
(l)	liquid
(g)	gas

NOTATION FOR EXTENSIVE THERMODYNAMIC PROPERTIES (exemplified by G, the Gibbs free energy)

G'	the Gibbs free energy of the system containing n moles
G	the Gibbs free energy per mole of the system
ΔG	the change in G due to a specified change in the state of the system
ΔG^M	the integral molar Gibbs free energy change due to mixing of the components to form a solution
$\Delta G^{M,\text{id}}$	the integral molar Gibbs free energy change due to mixing of the components to form an ideal solution
G_i	the molar Gibbs free energy of the species i
G_i^0	the molar Gibbs free energy of the species i in its designated standard state
$\overline{G}_i$	the partial molar Gibbs free energy of i in some specified solution
$\Delta \overline{G}_i^M$	$= \overline{G}_i - G_i^0$, the partial molar Gibbs free energy of mixing of i
G^{xs}	$= \Delta G^M - \Delta G^{M,\text{id}}$, the integral excess molar Gibbs free energy of a solution
$\overline{G}_i^{xs}$	the partial molar excess Gibbs free energy of mixing of i
ΔG_m	the molar Gibbs free energy of melting
ΔG_b	the molar Gibbs free energy of boiling

VALUES OF SELECTED PHYSICAL CONSTANTS

Absolute temperature of the ice point (0°C) $= 273.15$ K

Absolute temperature of the triple point of H_2O (by definition) $= 273.16000$ K

Thermochemical calorie (by definition)	1 calorie = 4.184 joules
Faraday's constant	$\mathcal{F}$ = 96,487 coulomb/mole = 23,060 calories/volt·mole
Avogadro's number	$\mathcal{A}$ = 6.0232 × 10^{23} /gram·mole
Boltzmann's constant	k = 1.38054 × 10^{-23} joules/degree
Atmosphere	1 atm = 1.01325 × 10^{6} dyne/cm^2 = 1.01325 bar = 101.325 kPa = 760 mmHg
Gas Constant	R = 8.3144 joules/degree·mole = 1.987 calories/degree·mole = 82.06 cm^3·atm/degree·mole = .08206 l-atm/K·mole

$h = 6.6252 \times 10^{-34}$ J·s

INTRODUCTION TO METALLURGICAL THERMODYNAMICS

CHAPTER ONE

INTRODUCTION AND DEFINITION OF TERMS

1.1 INTRODUCTION

Classical Thermodynamics is concerned with the behavior of matter, where matter is anything which occupies space, and the matter which makes up the subject of a thermodynamic discussion is called a *system*. In metallurgy and chemistry the systems to which thermodynamic principles are applied are generally chemical reaction systems. The central aim of applied chemical thermodynamics is the determination of the effect of environment on the state of rest (equilibrium state), of a given system, where environment is generally determined as the pressure exerted on the system and the temperature of the system. The aim of chemical thermodynamics is thus the establishment of the relationships which exist between the equilibrium state of existence of a given system and the influences which are brought to bear on the system.

1.2 THE CONCEPT OF STATE

The most important concept in Classical Thermodynamics is that of *state*. If it were possible to know the masses, velocities, positions, and all modes of motion of all the constituent particles in a system, then this mass of knowledge would serve to describe the *microscopic state* of the system, which, in turn, would determine all of the properties of the system. In the absence of such detailed knowledge as is required to determine the microscopic state of the system, classical thermodynamics begins with a consideration of the properties of the system which, when determined, determine the *macroscopic state* of the system; i.e., when all the properties of the system are fixed, then the macroscopic state of the system is fixed. It might seem that, in order to uniquely fix the macroscopic, or thermodynamic, state of a system, an enormous amount of information might be required; i.e., all of the properties of the system might have to be known. In fact, it is found that when a very small number of properties are fixed, then all of the rest are automatically fixed. Indeed, when a simple system such as a given quantity of substance of fixed composition is being considered, the fixing of two of the properties of the system fixes the values of all the other properties. Thus only two properties are independent, and these, in thermodynamic parlance, are termed the independent variables, while the remaining multitude of properties are termed the dependent variables. The thermodynamic state of the simple system is thus uniquely fixed when the values of the two independent variables are fixed.

In the case of the simple system any two properties could be chosen as the independent variables, and the choice is purely a matter of convenience. Properties most amenable to experimental control are the pressure P and the temperature T of the system. When P and T are fixed, then the state of the simple system is fixed and all of the other properties have unique values corresponding to this state. Consider the volume V of the system as a property the value of which is dependent on the values of P and T. Thus we can write

$$V = V(P, T) \tag{1.1}$$

The relation between V, P, and T for a system is called an *equation of state* for that system and, in a three-dimensional diagram, the coordinates of which are

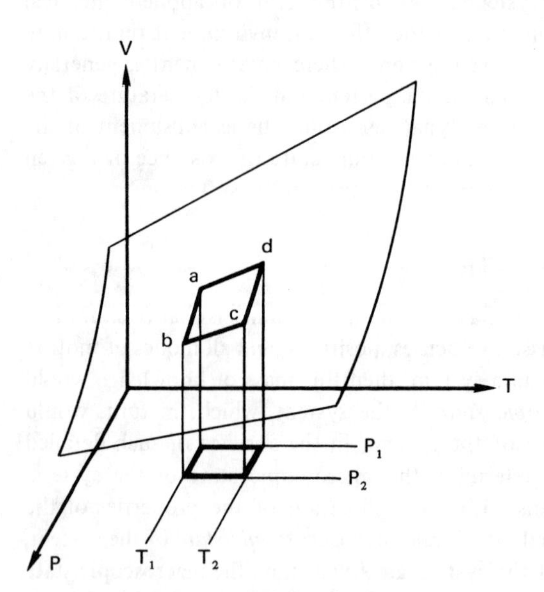

Fig. 1.1. Schematic representation of the states of existence of a fixed quantity of gas in V–P–T space.

volume, pressure, and temperature, the points in P-V-T space which represent the equilibrium states of existence of the system lie on a surface. This is shown in Fig. 1.1, from which it can be seen that the fixing of any two of the three variables fixes the value of the third variable. Consider a process which moves the system from the state a to the state c. For such a process the change in the volume of the system is

$$\Delta V = V_c - V_a$$

It is obvious that such a process could proceed along an infinite number of paths on the P-V-T surface, two of which, $a \to b \to c$ and $a \to d \to c$, are shown in Fig. 1.1. Consider the path $a \to b \to c$. The volume change

$$\begin{aligned} \Delta V &= V_c - V_a \\ &= (V_b - V_a) + (V_c - V_b) \end{aligned}$$

where $a \to b$ occurs at the constant temperature T_1 and $b \to c$ occurs at the constant pressure P_2.

$$(V_b - V_a) = \int_{P_1}^{P_2} \left(\frac{\partial V}{\partial P}\right)_{T_1} dP$$

and

$$(V_c - V_b) = \int_{T_1}^{T_2} \left(\frac{\partial V}{\partial T}\right)_{P_2} dT$$

Thus

$$\Delta V = \int_{P_1}^{P_2} \left(\frac{\partial V}{\partial P}\right)_{T_1} dP + \int_{T_1}^{T_2} \left(\frac{\partial V}{\partial T}\right)_{P_2} dT \tag{1.2}$$

Consider the path $a \to d \to c$.

$$(V_d - V_a) = \int_{T_1}^{T_2} \left(\frac{\partial V}{\partial T}\right)_{P_1} dT$$

and

$$(V_c - V_d) = \int_{P_1}^{P_2} \left(\frac{\partial V}{\partial P}\right)_{T_2} dP$$

and hence, again

$$\Delta V = \int_{T_1}^{T_2} \left(\frac{\partial V}{\partial T}\right)_{P_1} dT + \int_{P_1}^{P_2} \left(\frac{\partial V}{\partial P}\right)_{T_2} dP \tag{1.3}$$

Equations (1.2) and (1.3) are identical and are the physical representations of what is obtained when the complete differential of Eq. (1.1), i.e.,

$$dV = \left(\frac{\partial V}{\partial P}\right)_T dP + \left(\frac{\partial V}{\partial T}\right)_P dT$$

is integrated between the limits P_2, T_2 and P_1, T_1.

The value of the volume change from *a* to *c* thus depends only on values of the volume at *a* and *c* and is independent of the path taken by the system between the states *a* and *c*. This is a consequence of the facts that the volume is a *state function* and the above equation is an exact differential of the volume *V*.

1.3 SIMPLE EQUILIBRIUM

In Fig. 1.1 the state of existence of the system (or simply the state of the system) is such that it may only occur on the surface in *P-V-T* space; i.e., for any values of temperature and pressure the system is only at equilibrium when it has that unique volume which corresponds to the particular values of pressure and temperature. A particularly simple system is illustrated in Fig. 1.2. This is a fixed quantity of gas contained in a cylinder fitted with a movable piston. The system is at rest, i.e., is at equilibrium, when:

1. The pressure exerted on the gas by the piston equals the pressure exerted on the piston by the gas
2. The temperature of the gas equals the temperature of the surroundings (provided that the cylinder material is a conductor of heat)

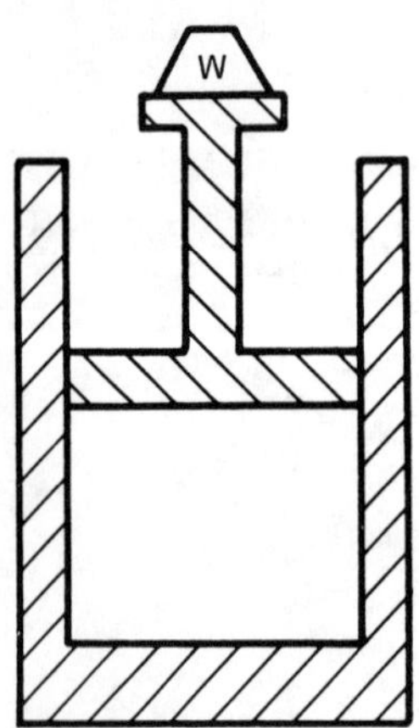

Fig. 1.2. A quantity of gas contained in a cylinder by a piston.

The state of the gas is thus fixed, and equilibrium occurs as the result of the establishment of a balance between the tendency of the external influences acting on the system to cause a change in the system and the tendency of the system to resist change (or vice versa). The fixing of the pressure at P_1 and the temperature at T_1 defines the state of the system and hence fixes the volume at the value V_1. If, by suitable decrease of the weight *w*, the pressure exerted on the gas by the piston is decreased to P_2, then physically, the resultant imbalance between the pressure exerted by the gas and the pressure exerted on the gas causes the piston to move out of the cylinder; this increases the volume occupied by the gas and decreases the pressure exerted by the gas, until pressure equalization is restored. When this process has occurred the volume is found to have increased from V_1 to V_2. Thermodynamically, the changing of the pressure from P_1 to P_2 changes the state of the system from the state 1 (characterized by P_1,T_1) to the state 2 (characterized by P_2,T_1). Hence the volume, as a dependent variable, changes from the value V_1 to the value V_2.

If the pressure exerted by the piston remains constant at P_2 and the temperature of the surroundings is raised from T_1 to T_2, then physically, the resultant temperature gradient across the cylinder walls causes a flow of heat into the gas, thus raising its temperature. This increase in the gas temperature at the constant pressure P_2 causes expansion of the gas which pushes the piston further out of the cylinder. When heat flow has occurred to the extent that the gas temperature has been increased to T_2, such that the temperature gradient has been eliminated, then it is found that the volume of the gas has increased to the value V_3. Again, thermodynamically, the changing of the temperature of the system from T_1 to T_2 at constant pressure P_2 changes the state of the system from state 2 (P_2,T_1) to state 3 (P_2,T_2); and hence the volume, again as the dependent variable, changes from V_2 in the state 2 to V_3 in the state 3. As volume is a state function, the final volume V_3 is independent of the order in which the above steps are carried out.

1.4 THE EQUATION OF STATE OF AN IDEAL GAS

The pressure-volume relationship of a gas at constant temperature was first determined experimentally in 1660 by Robert Boyle, who found that, at constant T,

$$P \propto \frac{1}{V}$$

This equation is known as Boyle's law. Similarly the volume-temperature relationship of a gas at constant pressure was first determined experimentally by Charles in 1787. This relationship, which is known as Charles' law, is

$$V \propto T$$

at constant P. Hence in Fig. 1.1, which is drawn for a fixed quantity of gas, a section of the *P-V-T* surface, drawn at constant *T*, produces a rectangular hyperbola which asymptotically approaches the *P* and *V* axes, and a section of the surface drawn at constant *P* produces a straight line. These two sections are shown in Fig. 1.3*a* and *b*.

In 1802 Joseph Gay-Lussac observed that the coefficient of thermal expansion of what he termed "permanent gases" was a constant. The coefficient of thermal expansion, α, is defined as the fractional increase, with temperature at constant pressure, of the volume of a gas at 0°C; i.e.,

$$\alpha = \frac{1}{V_0}\left(\frac{\partial V}{\partial T}\right)_P$$

where V_0 is the volume of the gas at 0°C. Gay-Lussac obtained a value of 1/267 for α, but more refined experimentation by Regnault in 1847 showed α to be equal to 1/273. Later it was found that the accuracy with which Boyle's and Charles' laws describe the behavior of different gases varies slightly from one gas to another and that, generally, gases with lower boiling points obey the laws more closely than do gases with higher boiling points. It was also found that the laws are more closely obeyed by all gases as the pressure of the gas is decreased. It was thus found convenient to invent a hypothetical gas which obeys Boyle's and Charles' laws exactly at all values of temperature and pressure. This hypothetical gas is termed the *ideal gas*, and for this ideal gas α equals 1/273.16.

The existence of a finite coefficient of thermal expansion sets a limit on the

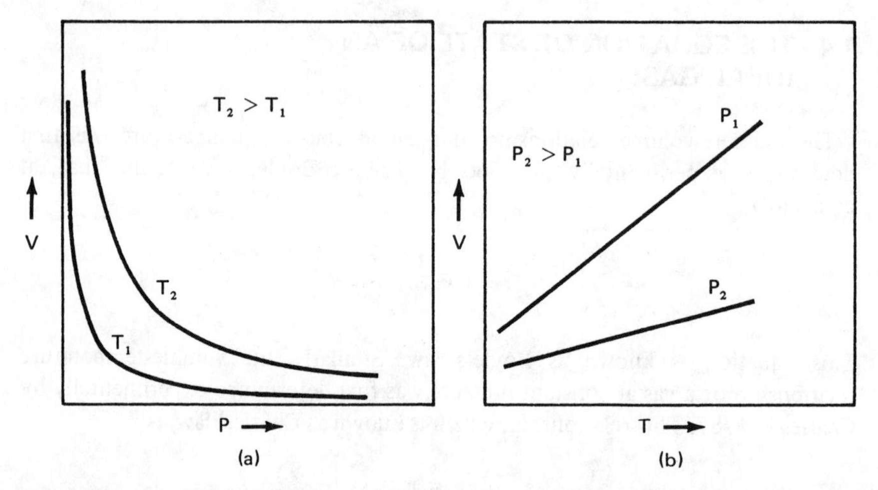

Fig. 1.3. (*a*) The variation with pressure of the volume of a fixed quantity of gas at constant temperature. (*b*) The variation with temperature of the volume of a fixed quantity of gas at constant pressure.

thermal contraction of the ideal gas; i.e., as α equals 1/273.16, then the fractional decrease in the volume of the gas, per degree decrease in temperature, is 1/273.16 of the volume at 0°C. Thus at −273.16°C the volume of the gas is zero, and hence the limit of temperature decrease, −273.16°C, is thus the absolute zero of temperature. This defines an absolute scale of temperature called the *ideal gas temperature scale*, which is related to the centigrade scale by the equation

$$T(\text{degrees absolute}) = T(\text{degrees centigrade}) + 273.16$$

Combination of Boyle's law

$$P_0 V(T, P_0) = PV(T, P)$$

and Charles' or Gay-Lussac's law

$$\frac{V(P_0, T_0)}{T_0} = \frac{V(P_0, T)}{T}$$

where P_0 = standard pressure (1 atm)
T_0 = standard temperature (0°C)
$V(T, P)$ = volume at temperature T and pressure P

gives

$$\frac{PV}{T} = \frac{P_0 V_0}{T_0} = \text{constant} \tag{1.4}$$

From Avogadro's hypothesis the volume per mole* of all gases at 0°C and 1 atm pressure (termed *standard temperature and pressure*–STP) is 22.414 liters. Thus the constant in Eq. (1.4) can be evaluated as

$$\frac{P_0 V_0}{T_0} = \frac{1 \text{ atm} \times 22.414 \text{ liters}}{273.16 \text{ degree} \cdot \text{mole}}$$

$$= 0.082057 \text{ liter} \cdot \text{atm/degree} \cdot \text{mole}$$

This constant is termed R, the *Gas Constant*, and, being applicable to all gases, it

*A mole of a substance is its molecular weight expressed in grams; e.g., a mole of oxygen weighs 32 grams, a mole of carbon weighs 12 grams, and a mole of carbon dioxide weighs 44 grams.

is a universal constant. Equation (1.4) can now be written as

$$PV = RT \tag{1.5}$$

which is thus the equation of state for 1 mole of ideal gas. Equation (1.5) is called the *ideal gas law*. Because of the simple form of its equation of state, the ideal gas is used extensively as a system in thermodynamics discussions.

1.5 THE UNITS OF WORK

It is to be realized that the unit "liter•atmosphere" occurring in the units of R is an energy term. Work is said to be done when a force moves through a distance, and work and energy have the dimensions force X distance. Pressure is defined as force divided by area, and hence work and energy can have the dimensions of pressure X distance X area, or pressure X volume. The unit of mechanical work is the joule, which is the work done when a force of 1 newton moves through a distance of 1 meter. Liter•atmospheres are converted to joules as follows.

$$\begin{aligned} 1 \text{ atm} &= 0.76 \text{ meters of mercury} \\ &= 0.76 \times \rho_{\text{Hg}} \times g \end{aligned}$$

where ρ_{Hg} = the density of mercury = 1.3595×10^4 kg/m^3

g = the standard gravitational acceleration = 9.80665 m/s^2

Hence

$$\begin{aligned} 1 \text{ atm} &= 0.76 \times 1.3595 \times 10^4 \times 9.80665 \text{ kg/m·s}^2 \\ &= 1.0132 \times 10^5 \text{ newtons/meter}^2 \end{aligned}$$

Multiplying both sides by liters (= 10^{-3} m^3)

$$\begin{aligned} 1 \text{ liter·atm} &= 1.0132 \times 10^2 \text{ newton·meters} \\ &= 1.0132 \times 10^2 \text{ joules} \end{aligned}$$

Hence

$$\begin{aligned} R &= 0.082057 \text{ liter·atm/degree·mole} \\ &= 8.3144 \text{ joules/degree·mole} \end{aligned}$$

The calorie, which is the energy unit traditionally used in thermodynamics and thermochemistry, will be introduced in Chap. 2.

1.6 EXTENSIVE AND INTENSIVE PROPERTIES, HETEROGENEOUS AND HOMOGENEOUS SYSTEMS

Properties (or state variables) can be categorized into two types—*extensive* and *intensive*. Extensive variables are properties, the magnitudes of which depend on the size of the system; and intensive properties are variables whose magnitudes do not depend on the size of the system. Volume is thus an extensive variable, and pressure and temperature are intensive variables. The values of extensive properties, expressed per unit quantity of the system or substance, have the characteristics of intensive variables; e.g., the volume per unit mass (specific volume) and the volume per mole (the molar volume) are properties the magnitudes of which are independent of the size of the system. Density, which is the reciprocal of specific volume, also occurs in this specific property category. For a system comprising n moles of an ideal gas the equation of state is

$$PV' = nRT$$

where V' is the volume of the system. Per mole of the system, the equation of state is

$$PV = RT$$

where V, the molar volume, equals V'/n.

1.7 PHASE DIAGRAMS AND THERMODYNAMIC COMPONENTS

Of the several ways of graphically representing the equilibrium states of existence of a system, the *constitution* or *phase diagram* is the most popular and convenient. The complexity of the phase diagram of a system is determined primarily by the number of *components* which occur in the system, where components are chemical species of fixed composition. The simplest components are chemical elements and stoichiometric compounds. Systems can be categorized by the number of components which they contain, e.g., one-component systems, two-component (binary) systems, three-component (ternary) systems, four-component (quaternary) systems, etc.

The phase diagram of a one-component system (i.e., a system of fixed composition) is a two-dimensional representation of the dependence of the equilibrium state of existence of the system on the two independent variables. Temperature and pressure are normally chosen as the two independent variables, and Fig. 1.4 shows a schematic representation of part of the H_2O phase diagram. The full lines in Fig. 1.4 divide the diagram into three areas designated liquid, solid, and vapor. If a quantity of pure H_2O is at some temperature and pressure which is represented by a point *within* the area *AOB*, the equilibrium state of the H_2O is a liquid. Similarly, within the areas *COA* and *COB* the equilibrium states are solid and vapor, respectively. If the state of existence lies on a line, e.g., on the line *AO*, then liquid and solid H_2O exist in equilibrium, and the equilibrium is said to be two-phase, in contrast to existence within any of the three areas, which is a one-phase equilibrium. A *phase* is defined as being a finite region in the physical system across which the properties are uniformly constant, i.e., do not experience any abrupt change in passing from one point in the region to another. Within any of the one-phase areas in the phase diagram, the system is said to be *homogeneous*. The system is *heterogeneous* when it contains two or more phases, e.g., coexisting ice and liquid water (on the line *AO*) is a heterogeneous system comprising two phases, and the phase boundary between the solid ice and the liquid water is that very thin region across which the density changes abruptly from the value for homogeneous ice to the higher value for liquid water.

The line *AO* represents the simultaneous variation of *P* and *T* required for maintenance of the equilibrium between solid and liquid H_2O. Similarly the lines *CO* and *OB* represent the simultaneous variations of *P* and *T* required for

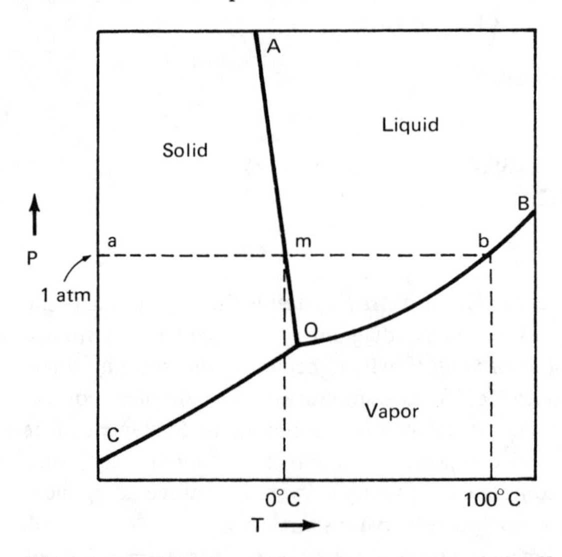

Fig. 1.4. A schematic representation of part of the H_2O phase diagram.

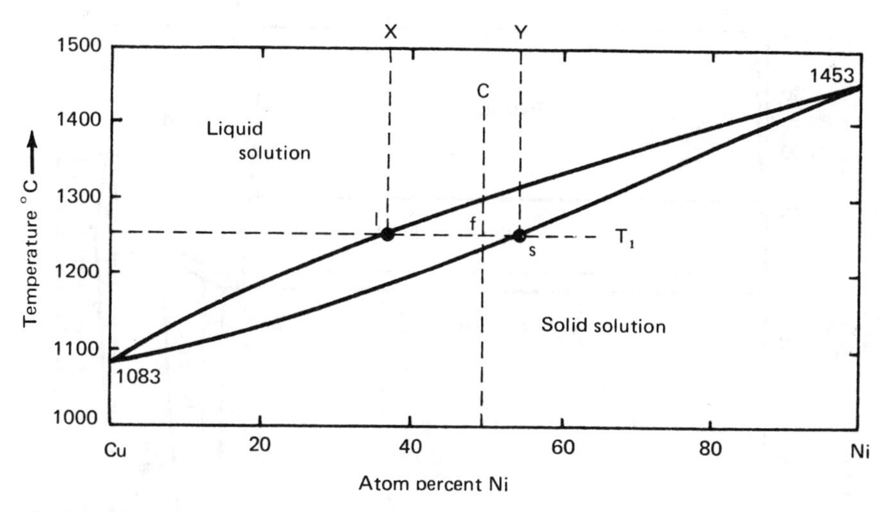

Fig. 1.5. The phase diagram of the binary copper-nickel system.

maintenance of the equilibrium between solid and vapor H_2O and between liquid and vapor H_2O, respectively. All three two-phase equilibrium lines meet at the point *O* (the triple point) which thus represents the unique values of *P* and *T* required for the establishment of the three-phase (solid + liquid + vapor) equilibrium. The path *amb* indicates that if a quantity of solid ice is heated at a constant pressure of 1 atm, melting occurs at the state *m* (the normal melting temperature of 0°C) and boiling occurs at the state *b* (the normal boiling temperature of 100°C).

If the system contains two components, then a composition axis must be included in the phase diagram, and consequently the complete phase diagram is three-dimensional with the coordinates composition, temperature, and pressure. As three-dimensional representation in two dimensions is often difficult to visualize, it is usual (and normally sufficient) to present a binary phase diagram as a constant-pressure section of the three-dimensional composition-pressure-temperature diagram. The constant pressure chosen is normally 1 atm, and the coordinates are composition and temperature. Figure 1.5, which is a typical simple binary phase diagram, shows the phase relationships occurring in the system Cu-Ni at 1 atm pressure. This phase diagram shows that below the melting temperature of Cu (1083°C) solid copper and solid nickel are completely miscible in all proportions, and that above the melting temperature of Ni (1453°C) liquid copper and liquid nickel are completely miscible in all proportions. The diagram thus contains areas of complete solid solubility and complete liquid solubility which are separated from one another by a two-phase area in which solid solutions and liquid solutions coexist in equilibrium with one

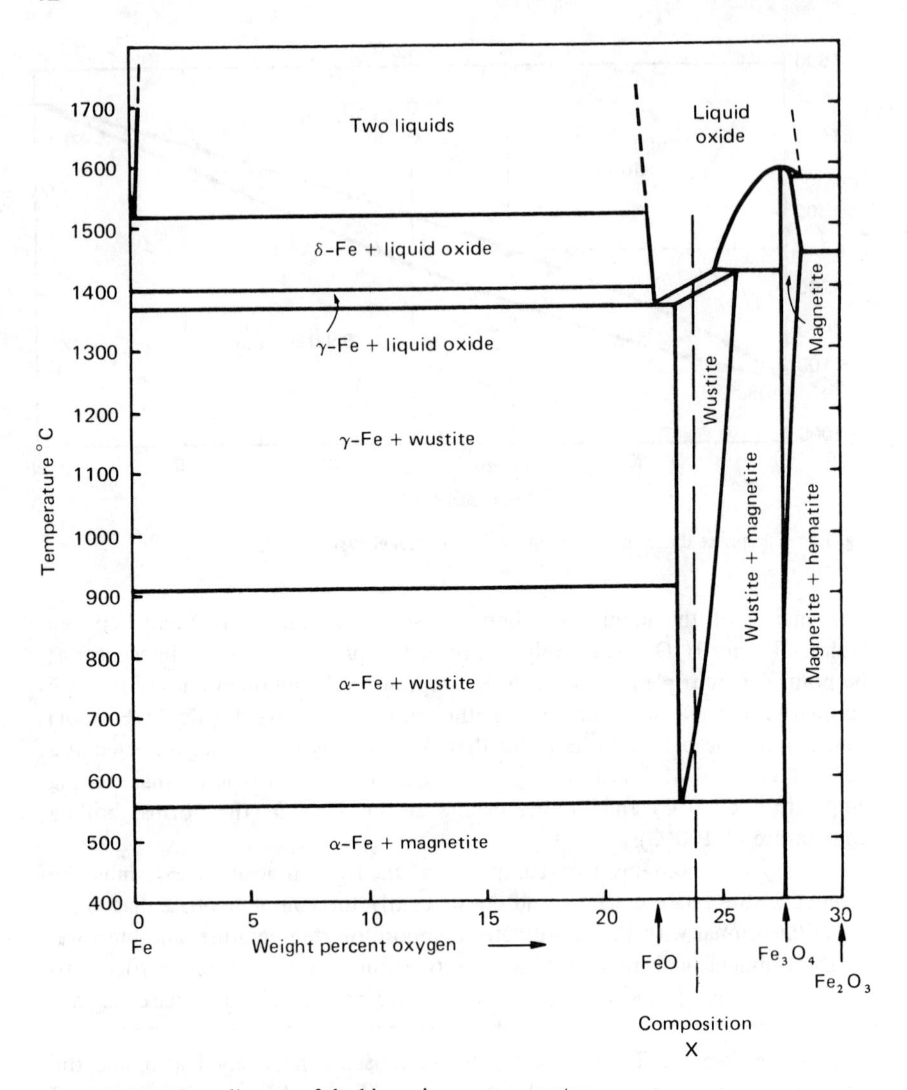

Fig. 1.6. The phase diagram of the binary iron-oxygen system.

another. For example, at the temperature T_1 a Cu-Ni system of composition between X and Y exists as a two-phase system comprising a liquid solution of composition l in equilibrium with a solid solution of composition s. The relative proportions of the two phases present depend only on the overall composition of the system in the range X-Y and are determined by the lever rule as follows. For the overall composition C at the temperature T_1 the lever rule states that if a fulcrum is placed at f on the lever ls, then the relative proportions of liquid and solid phases present are such that, placed on the ends of the lever at s and l respectively, the lever balances about the fulcrum; i.e., the ratio of liquid to solid present at T_1 is given by the ratio fs/lf.

Because the only requirement of a component is that it have a fixed composition, the designation of the components of a system is purely arbitrary. In the system Ni-Cu the obvious choice of the two components is Ni and Cu. However, the most convenient choice is not always as obvious, and the general arbitrariness in selecting the components of a system can be demonstrated by considering the iron-oxygen system, the phase diagram of which is shown in Fig. 1.6. This phase diagram shows that Fe and O form two stoichiometric compounds, Fe_3O_4 (magnetite) and Fe_2O_3 (hematite), and a limited range of solid solution (wustite). Of particular significance is the fact that neither a stoichiometric compound of the formula FeO nor a wustite solid solution in which the Fe/O atomic ratio is unity occurs. In spite of this it is often found convenient to consider the stoichiometric FeO composition as a thermodynamic component of the system. The available choice of the two components of the binary system can be demonstrated by considering the composition X in Fig. 1.6. This composition can equivalently be considered as being in any one of the following systems:

1. The system Fe-O (at 24 weight % O, 76 weight % Fe)
2. The system FeO-Fe_2O_3 (at 77.81 weight % FeO, 22.19 weight % Fe_2O_3)
3. The system FeO-Fe_3O_4 (at 67.83 weight % FeO, 32.17 weight % Fe_3O_4)
4. The system Fe-Fe_3O_4 (at 13.18 weight % Fe, 86.82 weight % Fe_3O_4)
5. The system Fe-Fe_2O_3 (at 20.16 weight % Fe, 79.84 weight % Fe_2O_3)
6. The system FeO-O (at 97.78 weight % FeO, 2.22 weight % O)

The actual choice of the two components for use in a thermodynamic discussion is thus purely a matter of convenience. The ability of the thermodynamic method to deal with descriptions of the compositions of systems in terms of arbitrarily chosen components, which need not correspond to physical reality, is a distinct advantage. The thermodynamic behavior of highly complex systems, such as steelmaking slags, can be completely described in spite of the fact that the actual constitution of such systems is, to a great extent, unknown.

CHAPTER TWO

THE FIRST LAW OF THERMODYNAMICS

2.1 INTRODUCTION

In a frictionless kinetic system of interacting rigid elastic bodies, kinetic energy is conserved. A collision between two of these bodies results in a transfer of kinetic energy from one to the other. The work done by the one equals the work done on the other, and the total kinetic energy of the system is unchanged as a result of the collision. If the kinetic system is in the influence of a gravitational field, then the sum of the kinetic and potential energies of the bodies is constant; changes of position of the bodies with respect to the gravitational field, in addition to changes in the velocities of the bodies, do not alter the total dynamic energy of the system. As the result of possible interactions, kinetic energy may be converted to potential energy and vice versa, but the sum of the two remains constant. If, however, frictional agencies are operative in the system, then with continuing collision and interaction among the bodies, the total dynamic energy of the system decreases and heat is produced. It would thus be reasonable to expect that a relationship exists between the dynamic energy lost and the heat produced as a result of the effects of friction.

The establishment of this relationship laid the foundations for the development of the thermodynamic method. As a subject, this has now gone far beyond simple considerations of the interchange of energy from one form to another, e.g., from dynamic energy to thermal energy. The development of thermodynamics from its early beginnings to its present state was achieved as the result of the invention of convenient thermodynamic functions of state. In this chapter the first two of these thermodynamic functions—the internal energy U and the enthalpy H—are introduced.

2.2 THE RELATIONSHIP BETWEEN HEAT AND WORK

The relation between heat and work was first suggested in 1798 by Count Rumford; during the boring of cannon at the Munich Arsenal he noticed that the heat produced during the boring was roughly proportional to the work performed during the boring. This suggestion was novel, as hitherto, heat had been regarded as being an invisible fluid called caloric which resided between the constituent particles of a substance. In the caloric theory of heat, the temperature of a substance was considered to be determined by the quantity of caloric gas which it contained; and two bodies of differing temperature, when placed in contact with one another, came to an intermediate common temperature as the result of caloric flowing between them. Thermal equilibrium was reached when the pressure of caloric gas in the one body equaled that in the other body. Rumford's observation that heat production accompanied the performance of work was accounted for by the caloric theory as being due to the fact that the amount of caloric which could be contained by a body, per unit mass of the body, depended on the mass of the body. Small pieces of metal (the metal turnings produced by the boring) contained less caloric per unit mass than did the original large mass of metal, and thus, in reducing the original large mass to a number of smaller pieces, caloric was evolved as sensible heat. Rumford then demonstrated that when a blunt borer was used (which produced very few metal turnings), the same heat production accompanied the same expenditure of work. The caloric theory "explained" the heat production in this case as being due to the action of air on the metal surfaces during the performance of work.

The caloric theory was finally discredited in 1799 when Humphrey Davy melted two blocks of ice by rubbing them together in a vacuum. In this experiment the latent heat necessary to melt the ice was provided by the mechanical work performed in rubbing the blocks together.

From 1840 onwards the relationship between heat and work was placed on a firm quantitative basis as the result of a series of experiments carried out by James Joule. Joule conducted experiments in which work was performed in a certain quantity of adiabatically* contained water and measured the resultant temperature rise of the water. He observed that a direct proportionality existed between the work done and the resultant temperature rise; he observed further

*An adiabatic vessel is one which is constructed in such a way as to prohibit, or at least minimize, the passage of heat through its walls. The most familiar example of an adiabatic vessel is the Dewar flask (known more popularly as a thermos flask). Heat transmission by conduction into or out of this vessel is minimized by using double glass walls separated by an evacuated space, and a rubber or cork stopper, and heat transmission by radiation is minimized by using highly polished mirror surfaces.

that the same proportionality existed no matter what means were employed in the work production. Methods of work production used by Joule included

1. Rotating a paddle wheel immersed in the water
2. An electric motor driving a current through a coil immersed in the water
3. Compressing a cylinder of gas immersed in the water
4. Rubbing together two metal blocks immersed in the water

This proportionality gave rise to the notion of a *mechanical equivalent of heat*, and for the purpose of defining this figure it was necessary to define a unit of heat. This unit is the *calorie* (or 15° calorie) and is the quantity of heat required to raise the temperature of 1 gram of water from 14.5°C to 15.5°C. On the basis of this definition Joule determined the value of the mechanical equivalent of heat to be 0.241 calories per joule. The presently accepted value is 0.2389 calories (15° calories) per joule. Rounding this to 0.239 calories per joule defines the *thermochemical calorie*, which, until the introduction, in 1960, of S.I. units, was the traditional energy unit used in thermochemistry. The gas constant R, in Eq. (1.5), thus equals $8.3144 \times 0.239 = 1.987$ calories/degree·mole.

2.3 INTERNAL ENERGY AND THE FIRST LAW OF THERMODYNAMICS

Joule's experiments resulted in the statement that "the change of a body inside an adiabatic enclosure from a given initial state to a given final state involves the same amount of work by whatever means the process is carried out." This statement is a preliminary formulation of the First Law of Thermodynamics, and in view of this statement, it is necessary to define some function which depends only on the internal state of a body or system. Such a function is U, the internal energy. This function is best introduced by means of comparison with more familiar concepts. When a body of mass m is lifted in a gravitational field from height h_1 to height h_2, the work w done on the body is given by

$$\begin{aligned} w &= \text{force} \times \text{distance} \\ &= mg \times (h_2 - h_1) \\ &= mgh_2 - mgh_1 \\ &= \text{potential energy at position } h_2 \text{ minus potential energy at position } h_1 \end{aligned}$$

As the potential energy of the body of given mass m depends only on the position of the body in the gravitational field, it is seen that the work done on the body is dependent only on its final and initial positions and is independent of the path taken by the body between the two positions, i.e., between the two states. Similarly the application of a force f to a body of mass m causes the body to accelerate according to Newton's Law

$$f = ma = m\frac{du}{dt}$$

where $a = du/dt$, the acceleration.

The work done on the body is thus obtained by integrating

$$dw = f\,dl$$

where l is distance.

$$\therefore dw = m\frac{du}{dt}dl = m\frac{dl}{dt}du = mu\,du$$

Integration gives

$$\begin{aligned} w &= \tfrac{1}{2}mu_2^2 - \tfrac{1}{2}mu_1^2 \\ &= \text{the kinetic energy of the body at velocity } u_2 \text{ (state 2)} \\ &\quad - \text{the kinetic energy of the body at velocity } u_1 \text{ (state 1)} \end{aligned}$$

Thus, again, the work done on the body is the difference between the values of a function of the state of the body and is independent of the path taken by the body between the states.

In the case of work being done on an adiabatically contained body of constant potential and kinetic energy, the pertinent function which describes the state of the body, or the change in the state of the body, is the internal energy U. Thus the work done on, or by, an adiabatically contained body equals the change in the internal energy of the body, i.e., equals the difference between the value of U in the final state and the value of U in the initial state. In describing work, it is conventional to assign a negative value to work done *on* a body and a positive value to work done *by* a body. Hence for an adiabatic process in which work w is done on a body, as a result of which its state moves from A to B,

$$w = -(U_B - U_A)$$

If work w is done on the body, then $U_B > U_A$; and if the body itself performs work, then $U_B < U_A$.

In Joule's experiments the change in the state of the adiabatically contained water was measured as an increase in the temperature of the water. The same increase in temperature, and hence the same change of state, could have been produced by placing the water in thermal contact with a source of heat and allowing heat q to flow into the water. In describing heat changes it is conventional to assign a negative value to heat which flows *out* of a body (an exothermic process) and a positive value to heat which flows into a body (an endothermic process). Hence

$$q = (U_B - U_A)$$

Thus, when heat flows into the body, q is a positive quantity and $U_B > U_A$; whereas if heat flowed out of the body; $U_B < U_A$ and q would be a negative quantity.

It is now of interest to consider the change in the internal energy of a body which simultaneously performs work and absorbs heat. Consider a body, initially in the state A, which performs work w, absorbs heat q and, as a consequence, moves to the state B. The absorption of heat q *increases* the internal energy of the body by the amount q and the performance of work w by the body *decreases* its internal energy by the amount w. Thus the total change in the internal energy of the body, ΔU, is

$$\Delta U = U_B - U_A = q - w \tag{2.1}$$

This is a statement of the *First Law of Thermodynamics*.

For an infinitesimal change of state, Eq. (2.1) can be written as a differential

$$dU = \delta q - \delta w \tag{2.2}$$

It is to be noticed that the left-hand side of Eq. (2.2) gives the value of the increment in an already existing property of the system, whereas the right-hand side has no corresponding interpretation. As U is a state function, the integration of dU between two states gives a value which is independent of the path taken by the system between the two states. Such is not the case when δq and δw are integrated. The heat and work effects depend on the path taken between the two states, as a result of which the integrals of δw and δq cannot be evaluated without a knowledge of the path. This is illustrated in Fig. 2.1. In Fig. 2.1 the value of $U_2 - U_1$ is independent of the path taken between state 1 ($P_1 V_1$) and state 2 ($P_2 V_2$). However the work done by the system, which is given by the integral $\int_1^2 \delta w = \int_1^2 P dV$ and hence is the area under the curve between V_2 and V_1, can vary greatly depending on the path. In Fig. 2.1 the work done in the process $1 \rightarrow 2$ via c is less than that done via b which, in turn, is less than that done via a. From Eq. (2.1) it is seen that the integral of δq must also depend on the path, and in the process $1 \rightarrow 2$ more heat is absorbed by the system via a than is absorbed via b which, again in turn, is greater than the heat absorbed via c. In Eq. (2.2) use of the symbol "d" indicates a differential element of a state function or state property, the integral of which is independent of the path, and use of the symbol "δ" indicates a differential element of some quantity which is not a state function. In Eq. (2.1) it is to be realized that the algebraic sum of two quantities, neither of which individually is independent of the path, gives a quantity which is independent of the path.

In the case of a cyclic process which returns the system to its initial state, e.g., the process $1 \to 2 \to 1$ in Fig. 2.1, the change in U as a result of this process is zero; i.e.,

$$\Delta U = \int_1^2 dU + \int_2^1 dU = (U_2 - U_1) + (U_1 - U_2) = 0$$

The vanishing of a cyclic integral $\oint dU = 0$ is a property of a state function.

In Joule's experiments, where $(U_2 - U_1) = -w$, it is to be noted that as the process was adiabatic ($q = 0$), then the path of the process was specified.

As U is a state function, then for a simple system consisting of a given amount of substance of fixed composition, the value of U is fixed once any two properties (the independent variables) are fixed. If temperature and volume are chosen as the independent variables, then

$$U = U(V, T)$$

The complete differential of U in terms of the partial derivatives gives

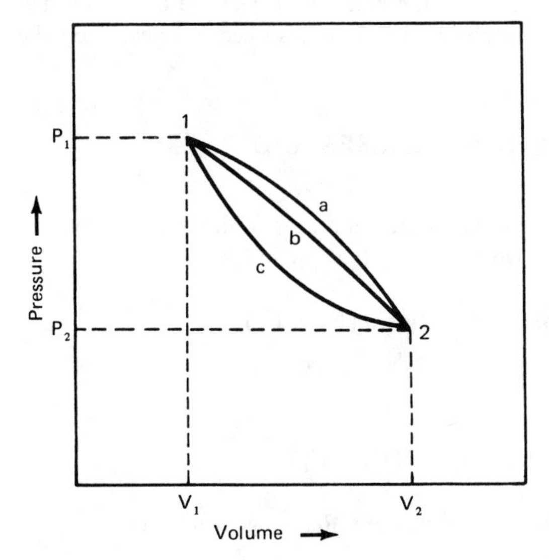

Fig. 2.1. Three process paths taken by a fixed quantity of gas in moving from the state 1 to the state 2.

$$dU = \left(\frac{\partial U}{\partial V}\right)_T dV + \left(\frac{\partial U}{\partial T}\right)_V dT$$

As the state of the system is fixed when the two independent variables are fixed, it is of interest to examine those processes which can occur when the value of one of the independent variables is maintained constant and the other is allowed to vary. Thus we can examine processes in which the temperature T is maintained constant (isothermal processes), or the pressure P is maintained constant (isobaric processes), or the volume V is maintained constant (isochore or isometric processes). We can also examine adiabatic processes in which $q = 0$.

2.4 CONSTANT-VOLUME PROCESSES

If the volume of a system is maintained constant during a process, then the system does no work ($\int PdV = 0$), and from the First Law, Eq. (2.2),

$$dU = \delta q_v \tag{2.3}$$

where the subscript v indicates constant volume. Integration of Eq. (2.3) gives

$$\Delta U = q_v$$

for such a process, which indicates that the increase or decrease in the internal energy of the system equals, respectively, the heat absorbed or rejected by the system during this process.

2.5 CONSTANT-PRESSURE PROCESSES AND THE ENTHALPY *H*

If the pressure is maintained constant during a process which takes the system from state 1 to state 2, then the work done by the system is given as

$$w = \int_1^2 PdV = P\int_1^2 dV = P(V_2 - V_1)$$

and the First Law gives

$$U_2 - U_1 = q_p - P(V_2 - V_1)$$

where the subscript p indicates constant pressure. Rearrangement gives

$$(U_2 + PV_2) - (U_1 + PV_1) = q_p$$

and as the expression $(U + PV)$ contains only state functions, then the expression itself is a state function. This is termed the *enthalpy*, H; i.e.,

$$H = U + PV \tag{2.4}$$

Hence for a constant-pressure process,

$$H_2 - H_1 = \Delta H = q_p \tag{2.5}$$

Thus the enthalpy change during a constant-pressure process simply equals the heat admitted to or withdrawn from the system during the process.

2.6 HEAT CAPACITY

Before discussing isothermal and adiabatic processes, it is convenient to introduce the concept of heat capacity. The heat capacity, C, of a system is the ratio of the heat added to or withdrawn from the system to the resultant change in the temperature of the system. Thus

$$C = \frac{q}{\Delta T}$$

or if the temperature change is made vanishingly small, then

$$C = \frac{\delta q}{dT}$$

The concept of heat capacity is only used when the addition of heat to or withdrawal of heat from the system produces a temperature change; the concept is not used when a phase change is involved. For example, if the system is a mixture of ice and water at 1 atm pressure and 0°C, then the addition of heat simply melts some of the ice and no temperature change results. In such a case the heat capacity, as defined, would be infinite.

It is to be noted that if a system is in a state 1 and the absorption of a certain quantity of heat by the system raises its temperature from T_1 to T_2, then the statement that the final temperature is T_2 is insufficient to determine the final state of the system. This is because the system has two independent variables and so one other variable, in addition to the temperature, must be specified in order to define the state of the system. This second independent variable could be varied in a specified manner or could be maintained constant during the change. The latter possibility is the more practical, and so the addition of heat to a system to produce a temperature change is normally considered at constant pressure or at constant volume. In this way the path of the process is specified and the final state of the system is known.

Thus we define a heat capacity at constant volume, C_v, and a heat capacity at constant pressure, C_p.

$$C_v = \left(\frac{\delta q}{dT}\right)_v \quad \text{the heat capacity at constant volume}$$

$$C_p = \left(\frac{\delta q}{dT}\right)_p \quad \text{the heat capacity at constant pressure}$$

Thus from Eqs. (2.3) and (2.5)

$$C_v = \left(\frac{\delta q}{dT}\right)_v = \left(\frac{dU}{dT}\right)_v \quad \text{or} \quad dU = C_v dT \tag{2.6}$$

$$C_p = \left(\frac{\delta U}{dT}\right)_p = \left(\frac{dH}{dT}\right)_p \quad \text{or} \quad dH = C_p dT \tag{2.7}$$

The heat capacity, being dependent on the size of the system, is an extensive property. However, in normal usage it is more convenient to use the heat capacity per unit quantity of the system. Thus the specific heat of the system is the heat capacity per gram at constant P, and the molar heat capacity is the heat capacity per mole at constant pressure or at constant volume. Thus, for a system containing n moles,

$$nc_p = C_p$$

and

$$nc_v = C_v$$

where c_p and c_v are the molar values.

It is to be expected that, for any substance, c_p will be of greater magnitude than c_v. If it is required that the temperature of a system be increased by a certain amount, then, if the process is carried out at constant volume, all of the heat added is used solely to raise the temperature of the system. However, if the process is carried out at constant pressure, then, in addition to raising the temperature by the required amount, the heat added is required to provide the work necessary to expand the system at the constant pressure. This work of expansion against the constant pressure per degree of temperature increase is calculated as

$$\frac{PdV}{dT} \quad \text{or} \quad P\left(\frac{\partial V}{\partial T}\right)_P$$

and hence it might be expected that

$$c_p - c_v = P\left(\frac{\partial V}{\partial T}\right)_P$$

The difference between c_p and c_v is calculated as follows.

$$c_p = \left(\frac{\partial H}{\partial T}\right)_P = \left(\frac{\partial U}{\partial T}\right)_P + P\left(\frac{\partial V}{\partial T}\right)_P$$

and

$$c_v = \left(\frac{\partial U}{\partial T}\right)_V$$

Hence

$$c_p - c_v = \left(\frac{\partial U}{\partial T}\right)_P + P\left(\frac{\partial V}{\partial T}\right)_P - \left(\frac{\partial U}{\partial T}\right)_V$$

but

$$dU = \left(\frac{\partial U}{\partial V}\right)_T dV + \left(\frac{\partial U}{\partial T}\right)_V dT$$

and therefore

$$\left(\frac{\partial U}{\partial T}\right)_P = \left(\frac{\partial U}{\partial T}\right)_V + \left(\frac{\partial U}{\partial V}\right)_T \left(\frac{\partial V}{\partial T}\right)_P$$

Hence

$$c_p - c_v = \left(\frac{\partial U}{\partial T}\right)_V + \left(\frac{\partial U}{\partial V}\right)_T \left(\frac{\partial V}{\partial T}\right)_P + P\left(\frac{\partial V}{\partial T}\right)_P - \left(\frac{\partial U}{\partial T}\right)_V$$

$$= \left(\frac{\partial V}{\partial T}\right)_P \left[P + \left(\frac{\partial U}{\partial V}\right)_T\right] \qquad (2.8)$$

The two expressions differ by the term $(\partial V/\partial T)_P\,(\partial U/\partial V)_T$, and in an attempt to evaluate the term $(\partial U/\partial V)_T$ for gases, Joule performed an experiment which involved filling a copper vessel with a gas at some pressure and connecting this vessel via a stopcock to a similar but evacuated vessel. The two-vessel system was immersed in a quantity of adiabatically contained water and the stopcock was opened, thus allowing free expansion of the gas into the evacuated vessel. After this expansion, Joule could not detect any change in the temperature of the system. As the system was adiabatically contained and no work was performed, then, from the First Law,

$$\Delta U = 0$$

and hence

$$dU = \left(\frac{\partial U}{\partial V}\right)_T dV + \left(\frac{\partial U}{\partial T}\right)_V dT = 0$$

Thus as $dT = 0$ (experimentally determined) and $dV \neq 0$ then the term $(\partial U/\partial V)_T$ must be zero. Joule concluded thus that the internal energy of a gas is a function only of temperature and is independent of the volume (and hence pressure). Thus for a gas

$$c_p - c_v = P\left(\frac{\partial V}{\partial T}\right)_P$$

However, in a more critical experiment performed by Joule and Thomson, in which an adiabatically contained gas of molar volume V_1 at the pressure P_1 was throttled through a porous diaphragm to the pressure P_2 and the molar volume V_2, a change in the temperature of the gas was observed, thus showing that, for real gases, $(\partial U/\partial V)_T \neq 0$.

Nevertheless, if

$$\left(\frac{\partial U}{\partial V}\right)_T = 0$$

then, from Eq. (2.8),

$$c_p - c_v = P\left(\frac{\partial V}{\partial T}\right)_P$$

and as, for one mole of ideal gas, $PV = RT$, then

$$c_p - c_v = \frac{R}{P} \times P = R$$

The reason for Joule's not observing a temperature rise in the original experiment was that the heat capacity of the copper vessels and the water was considerably greater than the heat capacity of the gas; thus the small heat changes which actually occurred in the gas were absorbed in the copper vessels and the water. This decreased the actual temperature change to below the limits of the then-available means of temperature measurement.

In Eq. (2.8) the term

$$P\left(\frac{\partial V}{\partial T}\right)_P$$

represents the work done by the system per degree rise in temperature in expanding against the constant external pressure P acting on the system. The other term in Eq. (2.8), namely,

$$\left(\frac{\partial U}{\partial V}\right)_T\left(\frac{\partial V}{\partial T}\right)_P$$

represents the work done per degree rise in temperature in expanding against the internal cohesive forces acting between the constituent particles of the substance. As will be seen in Chap. 8, an ideal gas is a gas consisting of noninteracting particles, and hence no work is done against the internal cohesive forces. Thus for an ideal gas the above term, and so the term

$$\left(\frac{\partial U}{\partial V}\right)_T$$

are zero.

In real gases the internal pressure contribution is very much smaller in magnitude than the external pressure contribution; but in liquids and solids, in which the interatomic forces are considerable, the work done in expanding the system against the external pressure is insignificant in comparison with the work done against the internal pressure. Thus for liquids and solids the term

$$\left(\frac{\partial U}{\partial V}\right)_T$$

is very large.

2.7 · REVERSIBLE ADIABATIC PROCESSES*

In an adiabatic process $q = 0$, and thus, from the First Law, $dU = -\delta w$. Consider a system comprising one mole of an ideal gas. From Eq. (2.6)

$$dU = c_v dT$$

*The significance of a reversible process is discussed in Chap. 3.

and

$$\delta w = PdV$$

Thus

$$c_V dT = -PdV$$

As the system is one mole of ideal gas, then $P = RT/V$ and hence

$$c_V dT = -\frac{RTdV}{V}$$

Integrating between states 1 and 2 gives

$$c_V \ln\left(\frac{T_2}{T_1}\right) = R \ln\left(\frac{V_1}{V_2}\right)$$

or

$$\left(\frac{T_2}{T_1}\right)^{c_V} = \left(\frac{V_1}{V_2}\right)^{R}$$

or

$$\left(\frac{T_2}{T_1}\right) = \left(\frac{V_1}{V_2}\right)^{R/c_V}$$

For an ideal gas it has been shown that $c_p - c_v = R$.

Thus $c_p/c_v - 1 = R/c_v$; and if $c_p/c_v = \gamma$, then $R/c_v = \gamma - 1$,

and hence

$$\frac{T_2}{T_1} = \left(\frac{V_1}{V_2}\right)^{\gamma-1}$$

From the ideal gas law,

$$\frac{T_2}{T_1} = \frac{P_2 V_2}{P_1 V_1} = \left(\frac{V_1}{V_2}\right)^{\gamma-1}$$

Thus

$$\frac{P_2}{P_1} = \left(\frac{V_1}{V_2}\right)^{\gamma}$$

and hence

$$P_2 V_2^\gamma = P_1 V_1^\gamma = PV^\gamma = \text{constant} \tag{2.9}$$

This is the relationship between the pressure and the volume of an ideal gas undergoing a reversible adiabatic process.

2.8 REVERSIBLE ISOTHERMAL PRESSURE OR VOLUME CHANGES OF AN IDEAL GAS

From the First Law

$$dU = \delta q - \delta w$$

and as $dT = 0$ (isothermal process), then $dU = 0$. Therefore $\delta w = \delta q = PdV = RTdV/V$ per mole of gas.

Integrating between the states 1 and 2 gives

$$w = q = RT \ln\left(\frac{V_2}{V_1}\right) = RT \ln\left(\frac{P_1}{P_2}\right) \tag{2.10}$$

Thus, for an ideal gas, an isothermal process is one of constant internal energy during which the work done by the system equals the heat absorbed by the system, both of which are given by Eq. (2.10).

A reversible isothermal process and a reversible adiabatic process are shown on a P-V diagram in Fig. 2.2 in which it is seen that, for a given pressure decrease, the work done by the reversible isothermal process exceeds that done by the reversible adiabatic process. This difference is due to the fact that during the isothermal process heat is absorbed by the system in order to maintain the temperature constant, whereas during the adiabatic process no heat is admitted to the system. During the isothermal expansion the internal energy of the gas remains constant, and during the adiabatic expansion the internal energy decreases by an amount equal to the work done.

2.9 SUMMARY

1. The establishment of the relationship between the work done on or by a system and the heat entering or leaving the system was facilitated by the introduction of the thermodynamic function U—the internal energy. U is a function of state, and thus the difference between the values of U in two states depends only on the states and is independent of the process path taken by the

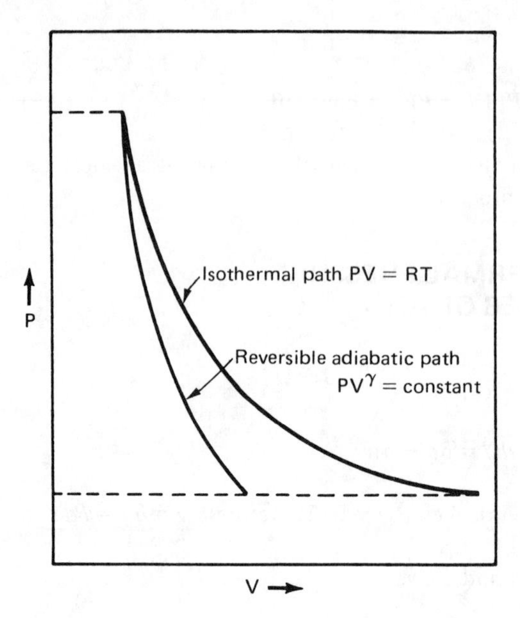

Fig. 2.2. Comparison of the process path taken by a reversible isothermal expansion of an ideal gas with the process path taken by a reversible adiabatic expansion of an ideal gas between P_1 and P_2.

system in moving between the states. The relationship between the internal energy change, the work done, and the heat absorbed per mole by a system of fixed composition in moving from one state to another is given as $\Delta U = q - w$, or, for an increment of this process, $dU = \delta q - \delta w$. This relationship is referred to as the First Law of Thermodynamics.

2. The integrals of δq and δw can only be obtained if the process path taken by the system in moving from one state to another is known. Process paths which are convenient for consideration include

a. Constant-volume processes in which $\int \delta w = \int P dV = 0$
b. Constant-pressure processes in which $\int \delta w = P \int dV = P\Delta V$
c. Constant-temperature processes
d. Adiabatic processes in which $q = 0$

3. For a constant-volume process, as $w = 0$, then $\Delta U = q_v$. The definition of the constant-volume molar heat capacity as $c_v = (\delta q/dT)_V = (\partial U/\partial T)_V$ (which is an experimentally measurable quantity) facilitates determination of the change in U resulting from a constant-volume process as $\Delta U = \int_1^2 c_v \, dT$.

4. Consideration of constant-pressure processes is facilitated by the introduction of the thermodynamic function H—the enthalpy—defined as $H = U + PV$. As the expression for H contains only functions of state, then H is a function of

state, and thus the difference between the values of H in two states depends only on the states and is independent of the path taken by the system in moving between them. For a constant-pressure process, $\Delta H = \Delta U + P\Delta V = (q_p - P\Delta V) + P\Delta V = q_p$. The definition of the constant-pressure molar heat capacity as $c_p = (\delta q/dT)_P = (\partial H/\partial T)_P$ (which is an experimentally measurable quantity) facilitates determination of the change in H as the result of a constant-pressure process as $\Delta H = \int_1^2 c_p \, dT$.

5. For an ideal gas, the internal energy U is a function only of temperature, and $c_p - c_v = R$.

6. The process path of an ideal gas undergoing a reversible adiabatic change of state is described by $PV^\gamma = \text{constant}$, where $\gamma = c_p/c_v$. During an adiabatic expansion, as $q = 0$, the decrease in the internal energy of the system equals the work done by the system.

7. As the internal energy of an ideal gas is a function only of temperature, the internal energy of an ideal gas remains constant during an isothermal change of state. Thus the heat which enters or leaves the gas as a result of the isothermal process equals the work done by or on the gas, with both quantities being given by

$$w = q = RT \ln\left(\frac{V_2}{V_1}\right) = RT \ln\left(\frac{P_1}{P_2}\right)$$

8. Only the differences in the values of U and H between two states, i.e., the values of ΔU and ΔH, can be measured. The absolute values of U and H in any given state cannot be determined.

2.10 NUMERICAL EXAMPLES

Ten liters of a monatomic ideal gas at 25°C and 10 atm pressure are expanded to a final pressure of 1 atm. The molar heat capacity of the gas at constant volume, c_v, is $3/2\,R$ and is independent of temperature. Calculate the work done, the heat absorbed, and the change in U and in H for the gas if the process is carried out (*a*) isothermally and reversibly, and (*b*) adiabatically and reversibly. Having determined the final state of the gas after the reversible adiabatic expansion, verify that the change in U for the process is independent of the path taken between the initial and final states by considering the process to be carried out as

(*i*) An isothermal process followed by a constant-volume process
(*ii*) A constant-volume process followed by an isothermal process
(*iii*) An isothermal process followed by a constant-pressure process
(*iv*) A constant-volume process followed by a constant-pressure process
(*v*) A constant-pressure process followed by a constant-volume process

The size of the system must first be calculated. From consideration of the initial state of the system (the point *a* in Fig. 2.3)

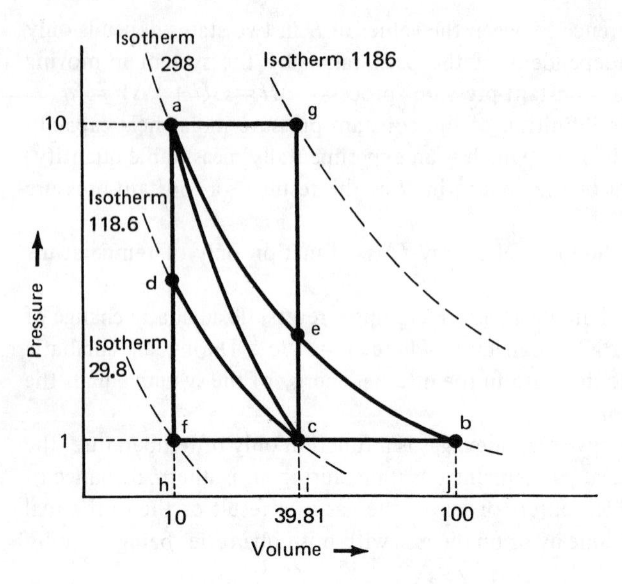

Fig. 2.3.

$$n = \text{the number of moles} = \frac{P_a V_a}{RT_a} = \frac{10 \times 10}{0.08206 \times 298} = 4.089$$

(*a*) *The isothermal reversible expansion*. The state of the gas moves from *a* to *b* along the isotherm 298. As, along any isotherm, the product *PV* is constant,

$$V_b = \frac{P_a V_a}{P_b} = \frac{10 \times 10}{1} = 100 \text{ liters}$$

For an ideal gas undergoing an isothermal process, $\Delta U = 0$ and hence, from the First Law,

$$q = w = \int_a^b P\,dV = nRT \int_a^b \frac{dV}{V} = 4.089 \times 8.3144 \times 298 \times \ln \frac{100}{10} \text{ joules}$$
$$= 23.33 \text{ kilojoules}$$

Thus in passing from the state *a* to the state *b* along the 298 K isotherm, the system performs 23.33 kilojoules of work and absorbs 23.33 kilojoules of heat from the constant-temperature surroundings.

As for an ideal gas *H* is a function only of temperature, then $\Delta H_{(a \to b)} = 0$; that is,

$$\Delta H_{(a\to b)} = \Delta U_{(a\to b)} + (P_b V_b - P_a V_a) = (P_b V_b - P_a V_a)$$
$$= nRT_b - nRT_a = nR(T_b - T_a) = 0$$

(*b*) *The reversible adiabatic expansion*. If the adiabatic expansion is carried out reversibly, then during the process the state of the system is, at all times, given by PV^{γ} = constant, and the final state is the point c in the diagram. The volume V_c is obtained from $P_a V_a^{\gamma} = P_c V_c^{\gamma}$ as

$$V_c = (10 \times 10^{5/3})^{3/5} = 39.81 \text{ liters}$$

and

$$T_c = \frac{P_c V_c}{nR} = \frac{1 \times 39.81}{4.089 \times 0.08206} = 118.6 \text{ K}$$

The point c thus lies on the 118.6 K isotherm. As the process is adiabatic, $q = 0$ and hence

$$\begin{aligned} \Delta U_{(a \to c)} &= -w = \int_a^c nc_v dT = nc_v(T_c - T_a) \\ &= 4.089 \times 1.5 \times 8.3144 \times (118.6 - 298) \text{ joules} \\ &= -9.149 \text{ kilojoules} \end{aligned}$$

The work done by the system as a result of the process equals the decrease in the internal energy of the system = 9.149 kilojoules.

(*i*) *An isothermal process followed by a constant-volume process* (the path $a \to e \to c$; that is, an isothermal change from a to e, followed by a constant-volume change from e to c).

$$\Delta U_{(a \to e)} = 0 \text{ as this is an isothermal change of state}$$

$$\Delta U_{(e \to c)} = q_v \qquad (\Delta V = 0 \text{ and hence } w = 0)$$

$$= \int_e^c nc_v \, dT \qquad \text{and as the state } e \text{ lies on the 298 K isotherm then}$$

$$\Delta U_{(e \to c)} = 4.089 \times 1.5 \times 8.3144 \times (118.6 - 298) \text{ joules} = -9.149 \text{ kilojoules}$$

Thus

$$\Delta U_{(a \to c)} = \Delta U_{(a \to e)} + \Delta U_{(e \to c)} = -9.149 \text{ kilojoules}$$

(*ii*) *A constant-volume process followed by an isothermal process* (the path $a \to d \to c$; that is, a constant-volume change from a to d, followed by an isothermal change from d to c).

$\Delta U_{(a\to d)} = q_v \quad (\Delta V = 0 \text{ and hence } w = 0)$

$= \int_a^d nc_v dT$ and as the state d lies on the 118.6 K isotherm then

$\Delta U_{(a\to d)} = 4.089 \times 1.5 \times 8.3144 \times (118.6 - 298)$ joules $= -9.149$ kilojoules

$\Delta U_{(d\to c)} = 0$ as this is an isothermal process and hence

$\Delta U_{(a\to c)} = \Delta U_{(a\to d)} + \Delta U_{(d\to c)} = -9.149$ kilojoules

(*iii*) *An isothermal process followed by a constant-pressure process* (the path $a \to b \to c$; that is, an isothermal change from a to b, followed by a constant-pressure change from b to c).

$$\Delta U_{(a\to b)} = 0 \text{ as this process is isothermal}$$

$$\Delta U_{(b\to c)} = q_p - w \text{ and as } P_b = P_c \text{ then } w = P_b(V_c - V_b)$$

$$= \int_b^c nc_p dT - P_b(V_c - V_b)$$

As $c_v = 1.5\,R$ and $c_p - c_v = R$, then $c_p = 2.5\,R$; and as 1 liter•atm equals 101.3 joules,

$$\Delta U_{(b\to c)} = [4.089 \times 2.5 \times 8.3144 \times (118.6 - 298)] - [1 \times (39.81 - 100) \times 101.3] \text{ joules}$$

$$= -15.247 + 6.097 = -9.150 \text{ kilojoules}$$

Thus

$$\Delta U_{(a\to c)} = U_{(a\to b)} + U_{(b\to c)} = -9.150 \text{ kilojoules}$$

(*iv*) *A constant-volume process followed by a constant-pressure process* (the path $a \to f \to c$; that is, a constant-volume change from a to f, followed by a constant-pressure change from f to c).

$$\Delta U_{(a\to f)} = q_v \quad (V_a = V_f \text{ and hence } w = 0)$$

$$= \int_a^f nc_v dT$$

From the ideal gas law

$$T_f = \frac{P_f V_f}{nR} = \frac{1 \times 10}{4.089 \times 0.08206} = 29.8 \text{ K}$$

i.e., the state f lies on the 29.8 K isotherm. Thus

$$\Delta U_{(a\to f)} = 4.089 \times 1.5 \times 8.3144 \times (29.8 - 298) \text{ joules} = -13.677 \text{ kilojoules}$$

$$\Delta U_{(f\to c)} = q_p - w$$

$$= \int_f^c nc_p\, dT - P_f(V_c - V_f)$$

$$= [4.089 \times 2.5 \times 8.3144 \times (118.6 - 29.8)] - [1 \times (39.8 - 10) \times 101.3] \text{ joules}$$

$$= +7.547 - 3.019 \text{ kilojoules}$$

Thus

$$\Delta U_{(a\to c)} = \Delta U_{(a\to f)} + \Delta U_{(f\to c)} = -13.677 + 7.547 - 3.019$$
$$= -9.149 \text{ kilojoules}$$

(v) *A constant-pressure process followed by a constant-volume process* (the path $a \to g \to c$; that is, a constant-pressure step from a to g, followed by a constant-volume step from g to c).

$$\Delta U_{(a\to g)} = q_p - w$$

From the ideal gas law

$$T_g = \frac{P_g V_g}{nR} = \frac{10 \times 39.81}{4.089 \times 0.08206} = 1186 \text{ K}$$

and hence the state g lies on the 1186 K isotherm. Thus

$$\Delta U_{(a\to g)} = [4.089 \times 2.5 \times 8.3144 \times (1186 - 298)] \text{ joules} - [10 \times (39.8 - 10) \times 101.3] \text{ joules}$$
$$= 75.475 - 30.187 \text{ kilojoules}$$

$$\Delta U_{(g\to c)} = q_v = 4.089 \times 1.5 \times 8.3144 \times (118.6 - 1186) \text{ joules}$$
$$= -54.433 \text{ kilojoules}$$

Thus

$$\Delta U_{(a\to c)} = \Delta U_{(a\to g)} + \Delta U_{(g\to c)} = 75.475 - 30.187 - 54.433$$
$$= -9.145 \text{ kilojoules}$$

The value of $\Delta U_{(a \to c)}$ is thus seen to be independent of the path taken by the process between the states a and c.

The change in enthalpy from a to c. The enthalpy change is most simply calculated from consideration of a path which involves an isothermal portion over which $\Delta H = 0$ and an isobaric portion over which $\Delta H = q_p = \int nc_p dT$. For example, consider the path $a \to b \to c$.

$$\Delta H_{(a \to b)} = 0$$

$$\begin{aligned} \Delta H_{(b \to c)} &= q_p = nc_p(T_c - T_b) \\ &= 4.089 \times 2.5 \times 8.3144 \times (118.6 - 298) \text{ joules} \\ &= -15.248 \text{ kilojoules} \end{aligned}$$

and hence

$$\Delta H_{(a \to c)} = -15.248 \text{ kilojoules}$$

or alternatively

$$\begin{aligned} \Delta H_{(a \to c)} &= \Delta U_{(a \to c)} + (P_c V_c - P_a V_a) \\ &= -9.149 \text{ kilojoules} + [(1 \times 39.81 - 10 \times 10) \times 101.3] \text{ joules} \\ &= -9.149 - 6.097 = -15.246 \text{ kilojoules} \end{aligned}$$

In each of the paths (*i*) to (*v*) the heat and work effects differ, although in each case the difference $q - w$ equals -9.149 kilojoules. In the case of the reversible adiabatic path, $q = 0$ and hence $w = +9.149$ kilojoules. If the processes (*i*) to (*v*) are carried out reversibly, then

for path (*i*) $q = -9.149 +$ the area *aeih*
for path (*ii*) $q = -9.149 +$ the area *dcih*
for path (*iii*) $q = -9.149 +$ the area *abjh* $-$ the area *cbji*
for path (*iv*) $q = -9.149 +$ the area *fcih*
for path (*v*) $q = -9.149 +$ the area *agih*

PROBLEMS

2.1 A quantity of an ideal gas occupies 10 liters at 10 atm and 100 K. Calculate (1) the final volume of the system, (2) the work done by the system, (3) the heat entering or leaving the system, and (4) the internal energy and enthalpy changes in the system if it undergoes

a. A reversible isothermal expansion to 1 atm
b. A reversible adiabatic expansion to 1 atm

For the gas, the molar heat capacity $c_v = 1.5\,R$.

2.2 A system comprises 7.14 grams of Ne gas at 0°C and 1 atm. When 2025 joules of heat are added to the system at constant pressure, the resultant expansion causes the system to perform 810 joules of work. Calculate (*a*) the initial state, (*b*) the final state, (*c*) ΔU and ΔH for the process, and (*d*) C_p and C_v. The molecular weight of Ne is 20 and it can be assumed that Ne behaves as an ideal gas.

2.3 A system comprises one mole of an ideal gas at 0°C and 1 atm. The system is subjected to the following processes, each of which is conducted reversibly:

a. A 10-fold increase in volume at constant temperature
b. Then a 100-fold adiabatic increase in pressure
c. Then a return to the initial state along a straight-line path in the *P-V* diagram

Calculate the work done by the system in step (*c*) and the total heat added to or withdrawn from the system as a result of the cyclic process.

2.4 One mole of an ideal gas at 25°C and 1 atm undergoes the following reversibly conducted process:

a. Isothermal expansion to 0.5 atm, followed by
b. Isobaric expansion to 100°C, followed by
c. Isothermal compression to 1 atm, followed by
d. Isobaric compression to 25°C

The system then undergoes the following cyclic process:

a. Isobaric expansion to 100°C, followed by
b. A decrease in pressure at constant volume to *P* atm, followed by
c. An isobaric compression at *P* atm to 24.5 liters, followed by
d. An increase in pressure at constant volume to 1 atm

Calculate the value of *P* which makes the work done on the gas in the first cycle equal to the work done by the gas in the second cycle.

2.5 Two moles of an ideal gas, in an initial state $P = 10$ atm, $V = 5$ liters, are taken, reversibly, in a clockwise direction, around a circular path given by $(V - 10)^2 + (P - 10)^2 = 25$. Calculate the amount of work done by the gas as a result of this process, and calculate the maximum and minimum temperatures attained by the gas during the cycle.

CHAPTER THREE

THE SECOND LAW OF THERMODYNAMICS

3.1 INTRODUCTION

In Chap. 2 it was seen that when a system undergoes a change of state, the consequent change in the internal energy of the system, which is dependent only on the initial and final states, is equal to the algebraic sum of the heat and work effects. The question thus arises—what magnitudes may the heat and work effects have, and what criteria govern these magnitudes? Two obvious cases occur, namely the extreme cases in which either $w = 0$ or $q = 0$, in which cases, respectively, $q = \Delta U$ and $w = -\Delta U$. But if $q \neq 0$ and $w \neq 0$, is there a definite maximum amount of work which the system can do during its change of state? The answers to these questions require an examination of the nature of processes. This examination, which will be made in this chapter, results in the identification of two classes of processes (reversible and irreversible processes) and in the introduction of a state function called the entropy, S. The concept of entropy will be introduced from two different starting points. Firstly, in Secs. 3.2 to 3.8 entropy will be introduced and discussed as the result of a need for quantification of the degree of irreversibility of a process; and secondly, in Secs. 3.10 to 3.14 it will be seen that, as a result of the examination of the properties of reversibly operated cyclic heat engines, there naturally develops a quantity which has all the properties of a state function. This function is found to be the entropy.

The examination leads to a statement known as the Second Law of Thermodynamics, which, in conjunction with the First Law, lays the foundation for the development of the thermodynamic method of describing the behavior of matter.

3.2 SPONTANEOUS, OR NATURAL, PROCESSES

A system, left to itself, will do one of two things: it will remain in the state in which it happens to be, or it will move, of its own accord, to some other state. That is, if the system is initially in equilibrium with its surroundings, then left to itself it will remain in this, its equilibrium, state; or, if it is not initially in its equilibrium state, it will spontaneously tend to move towards its equilibrium state. The equilibrium state is a state of rest, and hence a system, once at equilibrium, will only move away from equilibrium if it is acted on by some external agency. Even then, the combined system, comprising the original system plus the external agency, is simply moving towards the equilibrium state of the combined system. A process which involves the spontaneous movement of a system from a nonequilibrium state to an equilibrium state is called a *spontaneous* or *natural process*. As such a process cannot be reversed without the application of an external agency (a process which would leave a permanent change in the external agency), such a process is said to be *irreversible*. (The terms natural, spontaneous, and irreversible are synonymous in this context.)

Common examples of natural processes are (1) the mixing of two gases, and (2) the equalization of temperature. In (1), if the initial state of a system comprising two gases, A and B, is one in which gas A is contained in one vessel and gas B is contained in a separate vessel, then when the vessels are connected to one another, the system will spontaneously move to the equilibrium state in which the two gases are completely mixed. In (2), if the initial state of a system comprising two bodies is that in which one body is at one temperature and the other body is at a different temperature, then when the bodies are placed in contact with one another, the spontaneous process which occurs is the flow of heat from the hotter to the colder body, and the equilibrium state is reached when the temperatures of the bodies are equal. In both of these examples the reverse process (unmixing of the gases, and heat flow up a temperature gradient) will never spontaneously occur, and in both of these examples the simplicity of the systems, together with common experience, allows the equilibrium states of the systems to be predicted without any knowledge of the criteria for equilibrium. However, in less simple systems, the equilibrium state cannot be predicted from common experience, and the criteria governing equilibrium must be established before calculation of the equilibrium state can be made.

Determination of the equilibrium state is of prime importance in thermodynamics, as knowledge of this state for any chemical reaction system will allow determination to be made of the direction in which any spontaneous chemical reaction will proceed from any starting or initial state. For example, knowledge of the equilibrium state of a chemical reaction such as

$$A + B = C + D$$

will afford knowledge of whether, from an initial state (which would be some mixture of A, B, C, and D), the reaction will proceed from right to left or from left to right and, in either case, to what extent before equilibrium is reached.

If a system undergoes a spontaneous process involving the performance of work and the production of heat, then as the process continues, during which the system is being brought nearer and nearer to its equilibrium state, the capacity of the system for further spontaneous change decreases. Finally, once equilibrium is reached, the capacity of the system for doing further work is exhausted. In the initial nonequilibrium state of an isolated system (a system of constant internal energy), some of the energy of the system is available for the doing of useful work; and when the equilibrium state is reached, as the result of the completion of a spontaneous process, none of the energy of the system is available for the doing of useful work. Thus, as a result of the spontaneous process, the system has become degraded, in that energy, which was available for the doing of useful work, has been converted into thermal energy (or heat), in which form it is no longer available for external purposes.

3.3 ENTROPY AND THE QUANTIFICATION OF IRREVERSIBILITY

Two distinct types of spontaneous process are (1) the conversion of work into heat (i.e., the degradation of mechanical energy to thermal energy), and (2) the flow of heat down a temperature gradient.

If it is considered that an irreversible process is one in which the system undergoing the process is degraded, then the possibility that the extent of degradation can differ from one process to another suggests that there exists a quantitative measure of the extent of degradation, or degree of irreversibility, of a process. The existence of processes which exhibit differing degrees of irreversibility can be illustrated in the following manner. Consider the weight-heat-reservoir system shown schematically in Fig. 3.1. This system consists of a weight-pulley arrangement which is coupled to a constant-temperature heat reservoir, and the system is at equilibrium when an upward force acting on the weight exactly balances the downward force, W, of the weight. If the upward force is removed, then the equilibrium is upset and the weight spontaneously falls, thus performing work, which is converted, by means of a suitable system of paddle wheels, into heat which enters the constant-temperature heat reservoir. Equilibrium is reattained when the upward force acting on the weight is replaced, and the net effect of this process is that mechanical energy has been converted into thermal energy.

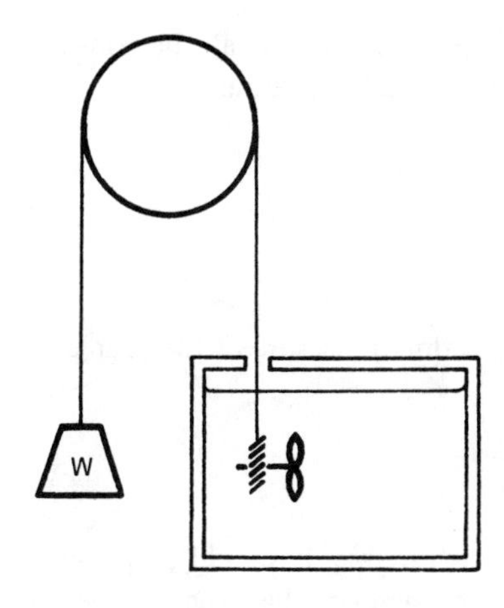

Fig. 3.1. A weight-pulley-heat-reservoir arrangement in which the work done by the falling weight is degraded to heat entering the heat reservoir.

Lewis and Randall* considered the following three processes:

1. The heat reservoir in the weight-heat reservoir system is at the temperature T_2. The weight is allowed to fall, performing work w, and the heat produced, q, enters the heat reservoir.
2. The heat reservoir at the temperature T_2 is placed in contact with a heat reservoir at a lower temperature T_1, and the same heat q is allowed to flow from the reservoir at T_2 to the reservoir at T_1.
3. The heat reservoir in the weight-heat reservoir system is at the temperature T_1. The weight is allowed to fall, performing work w, and the heat produced, q, enters the heat reservoir.

Each of these processes is spontaneous and hence irreversible, and degradation occurs in each of them. However, as process (3) is the sum of processes (1) and (2), the degradation occurring in process (3) must be greater than the degradation occurring in either process (1) or process (2). Thus it can be said that process (3) is more irreversible than either process (1) or process (2). Examination of the three processes indicates that both the magnitude of the heat production or the heat flow q and the temperature at which this heat is produced, or the temperatures between which the heat flows, are important in defining a quantitative scale of irreversibility. In the case of comparison between process (1) and process (3), the quantity q/T_2 is smaller than the quantity q/T_1, which agrees with the conclusion that process (1) is less irreversible than process (3). The quantity q/T is thus taken as a measure of the degree of irreversibility of the process, and the value of q/T is called the increase in *entropy*, S, occurring as a result of the process. Thus, when the weight-heat reservoir system undergoes

*G. N. Lewis and M. Randall, "Thermodynamics," rev. by K. S. Pitzer and L. Brewer, 2nd ed., p. 78, McGraw-Hill Book Company, New York, 1961.

a spontaneous process which causes the absorption of heat q at the constant temperature T, the entropy produced in the system, ΔS, is given as

$$\Delta S = \frac{q}{T} \tag{3.1}$$

The entropy increase as a result of the process is thus a measure of the degree of irreversibility of the process.

3.4 REVERSIBLE PROCESSES

As the degree of irreversibility of a process is variable, it should be possible for the process to be conducted in such a manner that the degree of irreversibility is minimized. The ultimate of this minimization is a process in which the degree of irreversibility is zero and in which no degradation occurs. This limit, toward which the behavior of actual systems can be made to approach, is the *reversible* process. If a process is reversible, then the concept of spontaneity is no longer applicable. It will be recalled that spontaneity occurred as a result of the system moving, of its own accord, from a nonequilibrium state to an equilibrium state. Thus if the spontaneity is removed, it is apparent that, at all times during the process, the system is at equilibrium. Thus a reversible process is one during which the system is never away from equilibrium, and a reversible process which takes a system from the state A to the state B is one in which the process path passes through a continuum of equilibrium states. Such a path is, of course, imaginary, but it is possible to conduct an actual process in such a manner that it is virtually reversible. Such an actual process is one which proceeds under the influence of an infinitesimally small driving force such that, during the process, the system is never more than an infinitesimal distance from equilibrium. If, at any point along the path, the minute external influence is removed, then the process ceases; or, if the direction of the minute external influence is reversed, then the direction of the process is reversed. Natural and reversible processes are illustrated in the following discussion.

3.5 AN ILLUSTRATION OF IRREVERSIBLE AND REVERSIBLE PROCESSES

Consider a system of water and water vapor at the uniform temperature T contained in a cylinder fitted with a frictionless piston, and let the cylinder be placed in thermal contact with a heat reservoir which is also at the constant temperature T. This system is shown in Fig. 3.2.

The water vapor in the cylinder exerts a certain pressure $P_{H_2O}(T)$, which is the saturated water vapor pressure at the temperature T. The system is exactly at equilibrium when the external pressure acting on the piston, P_{ext}, equals the internal pressure acting on the piston, $P_{H_2O}(T)$, and when the temperature of the water + water vapor in the cylinder equals the temperature T of the constant-temperature heat reservoir. If the external pressure acting on the piston, P_{ext}, is suddenly decreased by a finite amount ΔP, then the pressure imbalance causes the piston to accelerate rapidly out of the cylinder. The consequent rapid expansion of the water vapor in the cylinder decreases the water vapor pressure below its saturation (and hence equilibrium) value, and thus water spontaneously evaporates in an attempt to reestablish equilibrium. This spontaneous evaporation, being endothermic, decreases the temperature of the water and hence sets up a temperature gradient between the heat reservoir and the contents of the cylinder, such that heat spontaneously flows from the heat reservoir into the cylinder in an attempt to reestablish thermal equilibrium. If, when one mole of water has evaporated, the external pressure acting on the piston is instantaneously increased to its original value, P_{ext}, then evaporation of the water ceases, heat flow ceases, and complete equilibrium is reestablished. The work done by the system during this process equals $(P_{ext} - \Delta P)V$, where V is the molar volume of water vapor at $P_{H_2O}(T)$ and T. If the external pressure acting on the piston is suddenly increased by a finite amount ΔP, then the piston accelerates rapidly into the cylinder. The compression of the water vapor raises its pressure above the saturation value, and hence spontaneous condensation

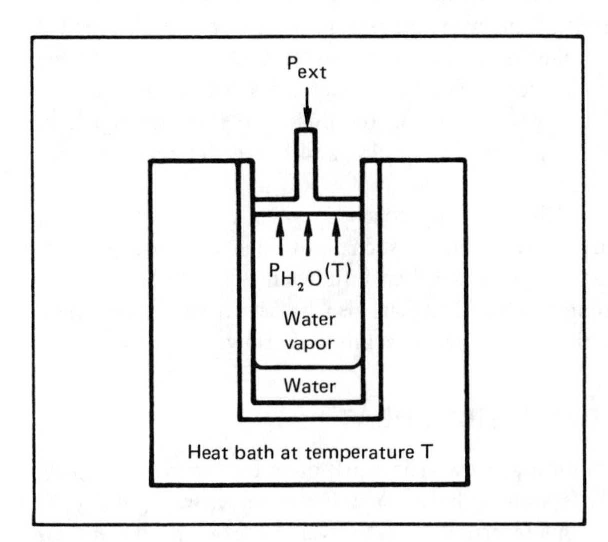

Fig. 3.2. A thermostatted piston and cylinder containing liquid water and water vapor.

occurs which, being exothermic, raises the temperature in the cylinder above T. The resultant temperature gradient between the cylinder and the heat reservoir causes the spontaneous flow of heat from the cylinder into the reservoir. If, when 1 mole of water vapor has condensed, the external pressure acting on the piston is instantaneously decreased to its original value, equilibrium is reestablished and the work done on the system equals $(P_{ext} + \Delta P)V$. The permanent change in the external agency as a result of the cyclic process is thus $2\Delta PV$.

Consider again the evaporation process. If the magnitude of P_{ext} is decreased by an infinitesimal amount δP, then the resulting minute imbalance between the pressures acting on the piston causes the piston to slowly move out of the cylinder. The slow expansion of the water vapor decreases its pressure. When this pressure has fallen by an infinitesimal amount below the saturation value, then evaporation of the water begins. The evaporation sets up an infinitesimal temperature gradient between the heat reservoir and the cylinder, down which flows the required latent heat of evaporation of the water. The smaller the value of δP, then the slower the process, the smaller the degree of undersaturation of the water vapor, and the smaller the temperature gradient. The more slowly the process is carried out, then the greater the opportunity afforded to the evaporation and heat flow processes to "keep up" with equilibrium. If, after the evaporation of 1 mole of water, the external pressure is instantaneously increased to its original value P_{ext}, then the work done by the system equals $(P_{ext} - \delta P)V$. If the external pressure is then increased by δP, then work, $(P_{ext} + \delta P)V$, is done on the system to condense 1 mole of water vapor; and the permanent change, as a result of the cyclic process, in the agency which supplied the external pressure equals the work done by the system minus the work done on the system. It can thus be seen that the smaller the value of δP, then the more nearly equal are the two work terms. In the limit that they are equal, no permanent change occurs in the external agency, and hence the cyclic process has been carried out reversibly.

It is thus seen that reversibility is approached when the evaporation or condensation processes are carried out in such a manner that the pressure exerted by the water vapor is never more than infinitesimally different from its saturation value at the temperature T. It can also be seen that, as complete reversibility is approached, the process becomes infinitely slow.

3.6 ENTROPY AND REVERSIBLE HEAT

Consider only the evaporation process. The work done by the system during the evaporation of 1 mole is seen to have a maximum value, $w_{max} = P_{ext}V$, when the process is carried out reversibly. Any irreversible evaporation process will perform less work, $w = (P_{ext} - \Delta P)V$. The change in the value of U for the system as the result of the evaporation of 1 mole of water vapor is independent of whether the process is carried out reversibly or not; and hence, from the First Law–Eq. (2.1)–it is seen that if the process is carried out reversibly, then the maximum amount of heat, q_{rev}, enters the cylinder from the heat reservoir, where $q_{rev} = \Delta U + w_{max}$. If the process is carried out irreversibly, then less heat q is transferred from the heat reservoir to the cylinder, where $q = \Delta U + w$. The difference between the work obtained in the reversible process and the work obtained in the irreversible process, $(w_{max} - w)$, is the mechanical energy which has been degraded to thermal energy (heat) in the cylinder as a result of the irreversible nature of the process. This heat produced by degradation, which is given as $(q_{rev} - q) = (w_{max} - w)$, accounts for the fact that less heat is absorbed by the cylinder from the heat reservoir during the irreversible process than is absorbed during the reversible process.

Thus, if the evaporation process is carried out reversibly, heat q_{rev} leaves the heat reservoir and enters the cylinder at the temperature T. The entropy change in the heat reservoir is thus given, from Eq. (3.1), as

$$\Delta S_{\text{heat reservoir}} = -\frac{q_{rev}}{T}$$

(The negative sign corresponds to heat leaving the reservoir, and hence the entropy of the reservoir decreases.)

and the entropy change of the water and water vapor in the cylinder is

$$\Delta S_{\text{water+vapor}} = \frac{q_{rev}}{T}$$

(The positive sign corresponds to heat entering the cylinder, and hence the entropy of the cylinder contents increases.)

The total entropy change in the combined heat-reservoir-water/water-vapor system, as a result of the reversible process, is thus

$$\Delta S_{\text{total}} = \Delta S_{\text{heat reservoir}} + \Delta S_{\text{water + water vapor}} = 0$$

This zero change in the total entropy is due to the fact that the process was carried out reversibly, i.e., no degradation occurred during the process.

If the evaporation process is carried out irreversibly, then heat q $(q < q_{rev})$ leaves the heat reservoir and enters the cylinder at the temperature T. The entropy change in the heat reservoir is thus,

$$\Delta S_{\text{heat reservoir}} = -\frac{q}{T}$$

However, the total heat appearing in the cylinder equals the heat q entering from the heat reservoir plus the heat which is produced by degradation of work due to the irreversible nature of the process. This degraded work equals $w_{max} - w$, which equals $(q_{rev} - q)$, and hence the entropy change in the contents of the cylinder is

$$\Delta S_{\text{water+water vapor}} = \frac{q}{T} + \frac{q_{rev} - q}{T}$$

which, it is seen, is simply q_{rev}/T. Thus the total entropy change in the system occurring as a result of the irreversible process is

$$\begin{aligned}\Delta S_{\text{total}} &= \Delta S_{\text{heat reservoir}} + \Delta S_{\text{water+vapor}} \\ &= -\frac{q}{T} + \frac{q_{rev}}{T} \\ &= \frac{q_{rev} - q}{T}\end{aligned}$$

As $q_{rev} > q$, this entropy change is positive, and hence entropy has been produced (or created) as a result of the occurrence of an irreversible process. The entropy produced, $(q_{rev} - q)/T$, is termed $\Delta S_{\text{irreversible}}$ (ΔS_{irr}) and is the measure of the degradation which has occurred as a result of the process. Thus, for the evaporation process, irrespective of the degree of irreversibility,

$$\Delta S_{\text{water+vapor}} = \frac{q}{T} + \Delta S_{\text{irr}} \tag{3.2}$$

Consideration of the condensation process shows that the work done *on* the system has a minimum value when the condensation is conducted reversibly, and correspondingly, the heat leaving the cylinder and entering the heat reservoir has a minimum value q_{rev}. If the process is conducted irreversibly, then a greater amount of work must be performed, and the excess of this work over the minimum required is the work which is degraded to heat in the irreversible process. This extra heat, in turn, is the difference between the heat leaving the cylinder, q, and the minimum heat q_{rev}.

Hence, for a reversible condensation,

$$\Delta S_{\text{water+vapor}} = -\frac{q_{rev}}{T}$$

$$\Delta S_{\text{heat reservoir}} = \frac{q_{rev}}{T}$$

and

$$\Delta S_{\text{total}} = 0$$

i.e., no entropy is created.

For an irreversible condensation,

$$\Delta S_{\text{water+vapor}} = -\frac{q}{T} + \frac{q - q_{\text{rev}}}{T}$$

$$= \frac{\text{heat leaving cylinder}}{T} + \frac{\text{heat produced in cylinder by degradation}}{T}$$

$$\Delta S_{\text{heat reservoir}} = \frac{q}{T}$$

and

$$\Delta S_{\text{total}} = \left[-\frac{q}{T} + \frac{q - q_{\text{rev}}}{T}\right] + \frac{q}{T}$$

$$= \frac{q - q_{\text{rev}}}{T}$$

And as $q > q_{\text{rev}}$, it is seen that entropy has been created as a result of the irreversible process. This created entropy is ΔS_{irr}, and hence again the entropy change of the water and water vapor is given as

$$\Delta S_{\text{water+vapor}} = -\frac{q}{T} + \Delta S_{\text{irr}} \tag{3.3}$$

The important feature to be noted from Eqs. (3.2) and (3.3) is that, in going from an initial to a final state (either the evaporation or the condensation of 1 mole of water at the pressure $P_{H_2O}(T)$ and the temperature T), the left-hand sides of Eqs. (3.2) and (3.3) are constants, being equal to q_{rev}/T and $-q_{\text{rev}}/T$, respectively. The entropy difference between the initial and final states is thus independent of whether the process is conducted reversibly or irreversibly and, being independent of the path of the process, can be considered as being the difference between the values of a state function. This state function is the entropy, and in going from state A to state B

$$\Delta S = S_B - S_A = \frac{q}{T} + \Delta S_{\text{irr}} \tag{3.4a}$$

$$= \frac{q_{\text{rev}}}{T} \tag{3.4b}$$

Equation (3.4*b*) indicates that, as the entropy change can only be determined via measurement of the heat flow at the temperature T, then entropy changes can only be measured for reversible processes, in which cases the measured heat flow is q_{rev}, and $\Delta S_{\mathrm{irr}} = 0$.

3.7 THE REVERSIBLE ISOTHERMAL COMPRESSION OF AN IDEAL GAS

Consider the reversible isothermal compression of one mole of an ideal gas from the state (V_A, T) to the state (V_B, T). The gas is placed in thermal contact with a heat reservoir at the temperature T and by application of a falling weight the gas is compressed slowly enough that, at all times during the compression, the pressure exerted on the gas is only infinitesimally greater than the instantaneous pressure of the gas, P_{inst}, where $P_{\mathrm{inst}} = RT/V_{\mathrm{inst}}$. The state of the gas thus lies, at all times, on a section at constant temperature T of the P-V-T surface (Figs. 1.1 and 1.3*a*), and hence the gas passes through a continuum of equilibrium states in going from the state (V_A, T) to the state (V_B, T). As the gas is never out of equilibrium, i.e., the process path is reversible, no degradation occurs, and hence no entropy is created. Entropy is simply transferred from the gas to the heat reservoir, where it is measured as the heat entering divided by the temperature T.*

As the process is isothermal and the gas is ideal, then $\Delta U = 0$, and hence from the First Law,

*The pertinent feature of a constant-temperature heat reservoir is that it only experiences heat effects and neither performs work nor has work performed upon it. A simple heat reservoir is the "ice calorimeter" which comprises a system of ice and water at 0°C and 1 atm pressure. The heat flowing into or out of this calorimeter at 0°C is measured as the change occurring in the ratio of ice to water present as a result of the heat flow; and as the molar volume of ice is greater than that of water, this change in ratio is measued as a change in the total volume of the ice + water in the calorimeter. Strictly speaking, if heat flows out of the calorimeter, thus freezing some of the water, the volume of the system increases and hence the calorimeter does in fact perform work of expansion against the atmospheric pressure. However, the ratio of the work done in expansion to the corresponding heat leaving the system is small enough that the work effects may be neglected, as is illustrated below.

At 0°C and 1 atm, the molar volume of ice = 19.8 cm^3. At 0°C and 1 atm, the molar volume of water = 18 cm^3. Thus the work done against the atmosphere in the freezing of 1 mole of water at 0°C $= 1 \times 1.8 \times 10^{-3} \times 101.3 = 0.182$ joules. The latent heat of freezing of 1 mole of water =6 kilojoules.

On the other hand, the falling weight which performs the work of compression experiences no heat effects, and, as entropy change is due to heat flow, then no entropy changes occur in the falling weight.

Work done on the gas = heat withdrawn from the gas, i.e.,

$$w_{max} = q_{rev}$$

and the heat entering the reservoir = $-q_{rev}$. But

$$w_{max} = \int_{V_A}^{V_B} PdV = \int_{V_A}^{V_B} \frac{RTdV}{V} = RT \ln\left(\frac{V_B}{V_A}\right)$$

(see Eq. 2.10) and

$$\Delta S_{\text{heat reservoir}} = \frac{\text{the heat entering the reservoir}}{T}$$

$$= \frac{-q_{rev}}{T} = \frac{-w_{max}}{T}$$

Thus

$$\Delta S_{\text{heat reservoir}} = -R \ln\left(\frac{V_B}{V_A}\right)$$

and hence, as the process is carried out reversibly,

$$-\Delta S_{\text{heat reservoir}} = \Delta S_{gas} = S_B - S_A = R \ln\left(\frac{V_B}{V_A}\right)$$

As $V_B < V_A$, it is seen that, as a result of the reversible compression, the entropy of the gas has decreased and the entropy of the heat reservoir has increased by an equal amount. The sign of the entropy change is determined by the direction of the heat flow.

3.8 THE REVERSIBLE ADIABATIC EXPANSION OF AN IDEAL GAS

Consider the reversible adiabatic expansion of one mole of an ideal gas from the state (P_A, T_A) to the state (P_B, T_B). For the process to be reversible, the expansion must be carried out slowly enough that the state of the gas, at all times, lies on the P-V-T surface; and, as has been shown in Chap. 2, this condition, together with the condition that $q = 0$ (adiabatic process), dictates that the process path across the P-V-T surface follows the line PV^γ = constant. As the process is reversible, then no degradation occurs; and as the process is adiabatic, then no heat flow occurs. The entropy change of the gas is thus zero, and hence all states of a gas lying on a PV^γ = constant line are states of equal entropy (cf. all states of an ideal gas lying on a $PV = RT$ line are states of equal internal energy). A reversible adiabatic process is thus an isentropic process. During a reversible adiabatic expansion, the work done by the gas, w_{max}, equals the decrease in the internal energy.

If the pressure exerted on the gas is suddenly decreased from P_A to P_B, then the state of the gas, which is initially (P_A, T_A), momentarily moves off the P-V-T surface and, being thus out of equilibrium, the expansion occurs irreversibly and degradation occurs. As the gas is adiabatically contained, the heat produced by the degradation remains in the gas, and hence the final temperature of the gas after the irreversible expansion is greater than the temperature T_B. Thus the final state of the gas after an irreversible adiabatic expansion differs from the final state after a reversible adiabatic expansion involving the same pressure change. (The irreversible adiabatic expansion does not follow the path PV^γ = constant.) The entropy produced in the gas due to the irreversible process is the difference in entropy between the final and initial states, and it is to be realized that the final state itself is determined by the degree of irreversibility of the process. That is, for a given pressure change $(P_A \rightarrow P_B)$, the more irreversible the process, the greater the heat produced in the gas due to degradation, the higher the final temperature and the internal energy, and the greater the entropy increase. Thus, during an irreversible expansion the work done by the gas still equals the decrease in the internal energy of the gas (as is required by the First Law); but the decrease in U is less than in the reversible expansion from P_A to P_B due to the heat appearing in the gas as the result of degradation.

3.9 SUMMARY STATEMENTS

From the discussion so far three points have emerged:

1. When a system undergoes a spontaneous or irreversible process, the entropy of the system increases.
2. When a system undergoes a reversible process, no entropy is created; entropy is simply transferred from one part of the system to another part.
3. Entropy is a state function.

3.10 THE PROPERTIES OF HEAT ENGINES

Traditionally the concept of entropy as a state function is introduced by considering the behavior and properties of heat engines. A heat engine is a device which converts heat into work; and it is interesting to note that the first steam engine, which was built in 1769, was operational for a considerable number of years before the reverse process, i.e., the conversion of work into heat, was investigated. In the operation of a heat engine, a quantity of heat is withdrawn from a high-temperature heat reservoir, and some of this heat is converted into work, with the remainder being transferred to a low-temperature heat reservoir. This process is shown schematically in Fig. 3.3.

A typical example of a heat engine is the familiar simple steam engine. In this device, superheated steam is passed from the boiler (the high-temperature heat reservoir) into the cylinders, where it performs work by expanding against the pistons (the engine). As a result of this expansion the temperature of the steam decreases, and at the completion of the piston stroke the spent steam is exhausted to the atmosphere (the low-temperature heat reservoir). A flywheel returns the piston to its original position, thus completing the cycle and preparing for the next working stroke.

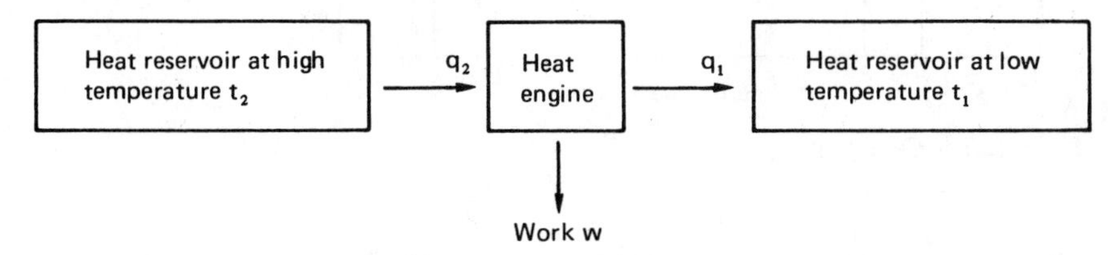

Fig. 3.3. Schematic representation of the working of a heat engine.

The efficiency of an engine is given as

$$\text{Efficiency} = \frac{\text{work obtained}}{\text{heat input}} = \frac{w}{q_2}$$

The factors governing the efficiency of this process were explained in 1824 by Sadi Carnot, who considered the cyclic process illustrated in Fig. 3.4.

In the step $A \to B$, heat q_2 is reversibly transferred from a heat reservoir at the temperature t_2 to a thermodynamic substance, as a result of which the thermodynamic substance isothermally and reversibly expands from the state A to the state B, and performs work w_1 equal to the area $ABba$.

In the step $B \to C$, the thermodynamic substance undergoes a reversible adiabatic expansion from the state B to the state C, as a result of which its temperature falls to t_1, and it performs work w_2 equal to the area $BCcb$.

In the step $C \to D$, heat q_1 is isothermally and reversibly transferred from the thermodynamic substance to a heat reservoir at the temperature t_1. Work w_3, equal to the area $DCcd$, is done on the substance, and its state moves from C to D.

In the step $D \to A$, the substance is reversibly and adiabatically compressed from the state D to the state A, during which change its temperature rises from t_1 to t_2, and work w_4, equal to the area $ADda$, is done on the substance.

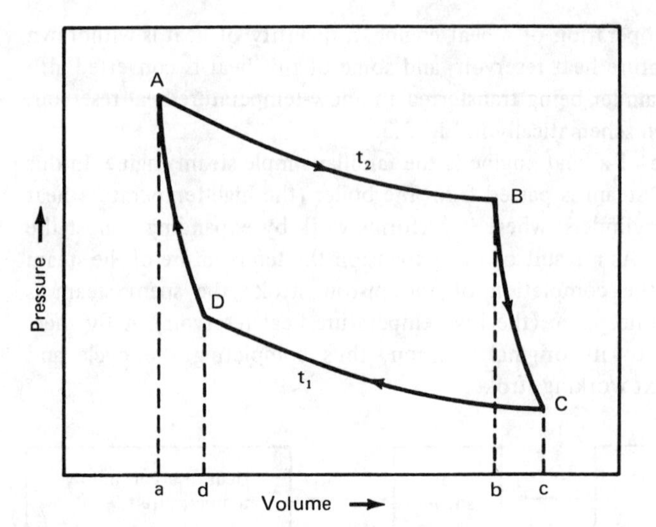

Fig. 3.4. The Carnot cycle.

During the cyclic process, which has returned the thermodynamic substance to its initial state, the substance has performed work $w = w_1 + w_2 - w_3 - w_4$ (equal to area *ABCD*) and has absorbed heat $q = q_2 - q_1$. For a cyclic process, $\Delta U = 0$; and hence, from the First Law,

$$q = w$$

Thus

$$q_2 - q_1 = w$$

and so the efficiency of the cyclic process (which is known as a Carnot cycle) is given as

$$\text{Efficiency} = \frac{w}{q_2} = \frac{q_2 - q_1}{q_2}$$

The consequence of all the steps in the cyclic process having been performed reversibly is illustrated in the following discussion. Consider a second engine working with a different substance, again between the temperatures t_1 and t_2, and let this second engine be more efficient than the first one. This greater efficiency could be obtained in either of two ways.

1. q_2 is withdrawn from the heat reservoir at t_2, and more work w' is obtained from it than was obtained from the first engine; that is, $w' > w$. Thus the second engine rejects less heat q'_1 to the cold reservoir at t_1 than does the first engine, that is, $q'_1 < q_1$.
2. The same work is obtained by withdrawing less heat q'_2 from the heat reservoir at t_2; that is, $q'_2 < q_2$. Thus less heat q'_1 is rejected into the heat reservoir at t_1; that is, $q'_1 < q_1$.

Consider, now, that the second engine is run in the forward direction and the first engine is run in the reverse direction, i.e., acts as a heat pump. Then, from (1), for the second engine running in the forward direction, $w' = q_2 - q'_1$; for the first engine running in the reverse direction, $-w = -q_2 + q_1$; and the sum of the two processes is

$$(w' - w) = (q_1 - q'_1)$$

i.e., an amount of work $(w' - w)$ has been obtained from a quantity of heat $(q_1 - q'_1)$ without any other change occurring. Although this conclusion does not contravene the First Law of Thermodynamics, it is contrary to human experience. Such a process corresponds to perpetual motion of the second kind; i.e., heat is converted into work without leaving a change in any other body. (Perpetual motion of the first kind is the creation of energy from nothing.) From (2), for the second engine run in the forward direction, $w = q'_2 - q'_1$; for the first engine run in the reverse direction, $-w = -q_2 + q_1$; and the sum of the two processes is

$$(q'_2 - q_2) = (q_1 - q'_1) = q$$

i.e., an amount of heat q at one temperature has been converted to heat at a higher temperature without any other change occurring. This corresponds to the spontaneous flow of heat up a temperature gradient, and is thus even more contrary to human experience than is perpetual motion of the second kind.

The above discussion gives rise to a preliminary formulation of the Second Law of Thermodynamics:

1. The principle of Thomsen states that it is impossible, by means of a cyclic process, to take heat from a reservoir and convert it into work without, in the same operation, transferring heat to a cold reservoir.
2. The principle of Clausius states that it is impossible to transfer heat from a cold to a hot reservoir without, in the same process, converting a certain amount of work into heat.

3.11 THE THERMODYNAMIC TEMPERATURE SCALE

The above suggests that all reversible Carnot cycles operating between the same upper and lower temperatures must have the same efficiency, namely, the maximum possible. This maximum efficiency is independent of the working substance and is a function only of the working temperatures t_1 and t_2. Thus,

$$\text{Efficiency} = \frac{q_2 - q_1}{q_2} = f'(t_1, t_2) = 1 - \frac{q_1}{q_2}$$

or

$$\frac{q_1}{q_2} = f(t_1, t_2)$$

Consider the Carnot cycles shown in Fig. 3.5. The two cycles operating between t_1 and t_2, and t_2 and t_3, are equivalent to a single cycle operating between t_1 and t_3. Thus

$$\frac{q_1}{q_2} = f(t_1, t_2)$$

$$\frac{q_2}{q_3} = f(t_2, t_3)$$

and

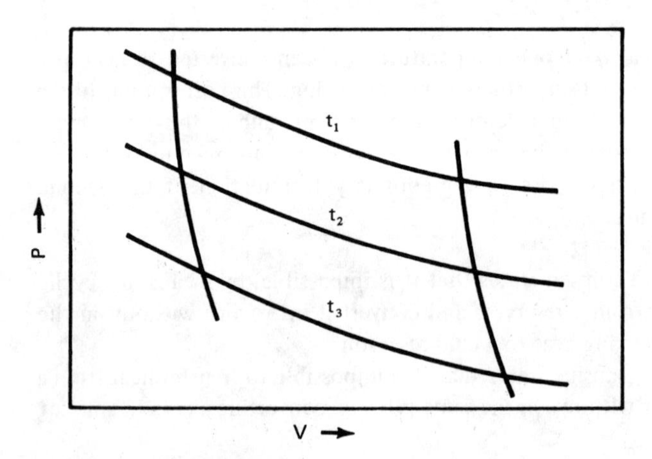

Fig. 3.5. Two Carnot cycles.

$$\frac{q_1}{q_3} = f(t_1, t_3)$$

so

$$\left(\frac{q_1}{q_3}\right) \times \left(\frac{q_3}{q_2}\right) = \frac{f(t_1, t_3)}{f(t_2, t_3)} = \frac{q_1}{q_2} = f(t_1, t_2)$$

As $f(t_1, t_2)$ is independent of t_3, then $f(t_1, t_3)$ and $f(t_2, t_3)$ must be of the form $f(t_1, t_3) = F(t_1)/F(t_3)$ and $f(t_2, t_3) = F(t_2)/F(t_3)$; i.e., the efficiency function $f(t_1, t_2)$ is the quotient of a function of t_1 alone and a function of t_2 alone.

Thus

$$\frac{q_1}{q_2} = \frac{F(t_1)}{F(t_2)}$$

Kelvin took these functions to have the simplest possible form, namely, T_1 and T_2. Thus

$$\frac{q_1}{q_2} = \frac{T_1}{T_2}$$

and hence the efficiency of a Carnot cycle is

$$\frac{q_2 - q_1}{q_2} = \frac{T_2 - T_1}{T_2} \tag{3.5}$$

This defines an absolute thermodynamic scale of temperature which is independent of the working substance. It is seen that the zero of this temperature scale is that temperature of the cold reservoir which makes the efficiency of a Carnot cycle equal to unity.

The absolute thermodynamic temperature scale (or Kelvin scale) is identical with the ideal gas temperature scale which was discussed in Chap. 1. This can be demonstrated by considering the working substance in a Carnot cycle to be 1 mole of an ideal gas. Referring to Fig. 3.4:

State A to state B. Reversible isothermal expansion at T_2:

$$\Delta U = 0$$

and from Eq. (2.10),

$$q_2 = w_1 = RT_2 \ln\left(\frac{V_B}{V_A}\right)$$

State B to state C. Reversible adiabatic expansion:

$$q = 0$$

and from Eq. (2.6),

$$w_2 = -\Delta U = -\int_{T_2}^{T_1} c_V dT$$

State C to state D. Reversible isothermal compression at T_1:

$$q_1 = w_3 = RT_1 \ln\left(\frac{V_D}{V_C}\right)$$

State D to state A. Reversible adiabatic compression:

$$w_4 = -\int_{T_1}^{T_2} c_V dT$$

Total work done by the gas $= w = w_1 + w_2 + w_3 + w_4$

$$= RT_2 \ln\left(\frac{V_B}{V_A}\right) - \int_{T_2}^{T_1} c_V dT + RT_1 \ln\left(\frac{V_D}{V_C}\right) - \int_{T_1}^{T_2} c_V dT$$

and

$$\text{Heat absorbed from the hot reservoir} = q_2 = RT_2 \ln\left(\frac{V_B}{V_A}\right)$$

It can easily be shown that

$$\frac{V_B}{V_A} = \frac{V_C}{V_D}$$

and hence $w = R(T_2 - T_1) \ln (V_B/V_A)$. Thus

$$\text{Efficiency} = \frac{w}{q_2} = \frac{(T_2 - T_1)}{T_2}$$

which is identical with Eq. (3.5).

3.12 THE SECOND LAW OF THERMODYNAMICS

The equation

$$\frac{q_2 - q_1}{q_2} = \frac{T_2 - T_1}{T_2}$$

can be written as

$$\frac{q_2}{T_2} - \frac{q_1}{T_1} = 0 \tag{3.6}$$

Now, any cyclic process can be broken down into a number of Carnot cycles as is shown in Fig. 3.6.

In going round the cycle *ABA*, the work done by the system equals the area enclosed by the path loop. This loop can be roughly approximated by a number of Carnot cycles as shown, and for the zigzag path of these cycles, from Eq. (3.6),

$$\sum \frac{q}{T} = 0$$

where the heat entering the system is positive and the heat leaving the system is

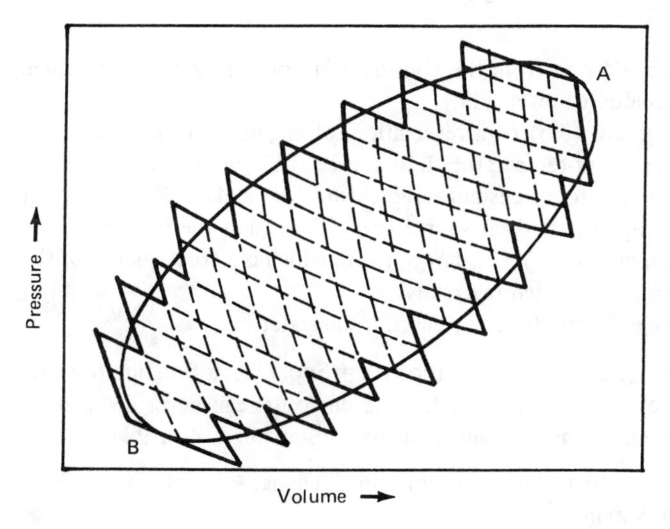

Fig. 3.6. A cyclic process broken into a large number of Carnot cycles.

negative. The zigzag path of the Carnot cycles can be made to coincide with the loop ABA by making the Carnot cycles smaller and smaller, and in the limit of coincidence the summation can be replaced by a cyclic integral; i.e.,

$$\oint \frac{\delta q}{T} = 0$$

The vanishing of the cyclic integral indicates that the integral is a perfect differential of some function of the state of the system. This function is called the entropy, S, and is defined as

$$dS = \frac{\delta q}{T} \tag{3.7}$$

Thus, for the loop ABA,

$$\oint dS = 0 = \int_A^B dS + \int_B^A dS = (S_B - S_A) + (S_A - S_B) = 0$$

It is to be emphasized that in Eq. (3.7) q is the reversible heat increment, and hence Eq. (3.7) should be properly written as

$$dS = \frac{\delta q_{\mathrm{rev}}}{T} \tag{3.8}$$

(This expression was derived from the consideration of Carnot cycles in which all operations are conducted reversibly.)

The application of Eq. (3.6) to a reversible heat engine, in which heat q_2 is withdrawn from a constant-temperature heat source at T_2, work w is performed, and heat q_1 is rejected into a constant-temperature heat sink at T_1, shows that the decrease in entropy of the heat source $(=q_2/T_2)$ equals the increase in entropy of the heat sink $(=q_1/T_1)$, i.e., $\Delta S_{\mathrm{total}} = 0$, which is a consequence of the fact that the process is conducted reversibly.

The Second Law of Thermodynamics can thus be stated as

1. The entropy S, defined by the equation $dS = \delta q_{\mathrm{rev}}/T$, is a function of state.
2. The entropy of a system in an adiabatic enclosure can never decrease; it increases in an irreversible process and remains constant in a reversible process.

From (2) it is seen that, for an infinitesimal change of state of an adiabatically contained system,

$$\sum dS_i \geqslant 0 \tag{3.9}$$

i.e., the sum of the incremental entropy changes of all i parts of the system which are in thermal contact is zero if the infinitesimal change of state occurs by a reversible process, and is greater than zero if the infinitesimal change of state occurs by an irreversible process. Equation (3.9) can be converted to an equality by writing

$$\sum_i dS_i = dS_{\text{irr}} \tag{3.10}$$

where dS_{irr} is the entropy created in the given incremental process. As has been seen, the magnitude of dS_{irr}, for the given change of state, is determined by, and is a measure of, the degree of irreversibility of the process. The value increases from zero for a reversible process to increasingly positive values for increasingly irreversible processes.

3.13 MAXIMUM WORK

For a change of state from A to B, the First Law gives

$$U_B - U_A = q - w$$

The First Law gives no indication of the allowed magnitudes of q and w in the given process. It has been seen, however, in the preceding discussion, that, although the values of q and w can vary depending on the degree of irreversibility of the path taken between states A and B, the Second Law of Thermodynamics sets a definite limit on the maximum amount of work which can be obtained from the system as a result of the given change of state, and hence sets a limit on the amount of heat which the system can absorb.

For an infinitesimal change of state, Eq. (3.4*a*) can be written as

$$dS_{\text{system}} = \frac{\delta q}{T} + dS_{\text{irr}}$$

and from the First Law,

$$\delta q = dU_{\text{system}} + \delta w$$

Thus

$$dS_{\text{system}} = \frac{dU_{\text{system}} + \delta w}{T} + dS_{\text{irr}}$$

or

$$\delta w = TdS_{\text{system}} - dU_{\text{system}} - TdS_{\text{irr}}$$

or

$$\delta w \leqslant TdS_{\text{system}} - dU_{\text{system}} \tag{3.11}$$

If the temperature T remains constant throughout the process (and equal to the temperature of the reservoir supplying the heat to the system), then Eq. (3.11), on integration from the state B to the state A, gives

$$w \leqslant T(S_B - S_A) - (U_B - U_A)$$

and as S and U are state functions, then w cannot be greater than a certain quantity w_{max}, the work which is obtained from the system when the process is carried out reversibly; i.e.,

$$w_{\text{max}} = T(S_B - S_A) - (U_B - U_A)$$

This work w_{max} corresponds to the absorption of the maximum heat q_{rev}.

As entropy is a state function, then in undergoing a specific change of state from A to B, *the entropy change is the same whether the process is carried out reversibly or irreversibly*. The above discussion indicates that *it is the heat effect that is different* in the two cases; i.e., if the process involves the absorption of heat and is conducted reversibly, then the heat which is absorbed, q_{rev}, is greater than the heat which would have been absorbed if the process had been conducted irreversibly. As has been seen, when 1 mole of an ideal gas is isothermally and reversibly expanded from state A to state B, heat $q = RT \ln (V_B/V_A)$ is reversibly transferred from the heat reservoir to the gas, and the entropy increase of the gas, $S_B - S_A$, equals $R \ln (V_B/V_A)$. The entropy of the heat reservoir decreases by an equal amount, and no entropy is created; that is, $\Delta S_{\text{irr}} = 0$. However, if the mole of ideal gas is allowed to expand freely from P_A to P_B (as in Joule's experiment discussed in Sec. 2.6), then, as the gas performs no work, no heat is transferred from the reservoir to the gas, and the entropy change in the reservoir is zero. As entropy is a state function, then the value of $S_B - S_A$ is independent of the path of the process, and hence the entropy created, ΔS_{irr}, equals $S_B - S_A = R \ln (V_B/V_A)$. This entropy is created as a result of the degradation of the work which would have been performed by the gas had the expansion not been carried out against a zero external pressure. This degraded work

$$= w_{max} - w$$

$$= w_{max} - 0$$

$$= w_{max}$$

$$= q_{rev}$$

Thus the free expansion represents the limit of irreversibility in which all the "potential" work is degraded to heat. The degraded heat appearing in the gas accounts for the increase in the entropy of the gas. Thus for the isothermal expansion of 1 mole of gas from state A to state B, the value of ΔS_{irr} can vary between zero and $R \ln (V_B/V_A)$, depending on the degree of irreversibility of the process.

3.14 ENTROPY AND THE CRITERION OF EQUILIBRIUM

At the beginning of this chapter it was stated that a system, left to itself, would either remain in the state in which it happened to be, or would spontaneously move to some other state; i.e., if the system is initially at equilibrium, then it will remain at equilibrium, and if it is initially in a nonequilibrium state, then it will spontaneously move to its equilibrium state. This spontaneous process is, by definition, irreversible, and during the movement of the system from its initial nonequilibrium to its final equilibrium state the entropy of the system increases. The attainment of the equilibrium state coincides with the entropy reaching a maximum value, and hence entropy can be used as a criterion for determination of the attainment of equilibrium.

For an isolated system of constant internal energy U and constant volume V, equilibrium is attained when the entropy of the system is a maximum, consistent with the fixed values of U and V. Consider the chemical reaction,

$$A + B = C + D$$

occurring in an adiabatic enclosure of constant volume. Starting with A and B, the reaction will proceed from left to right as long as the entropy of the system is thereby increased; or conversely, starting with C and D, the reaction will proceed from right to left again provided that the entropy of the system is thereby increased. Figure 3.7 shows a possible variation of entropy with extent of reaction. It is seen that a point is reached along the reaction ordinate at which the entropy of the system is a maximum. This is the equilibrium state of the

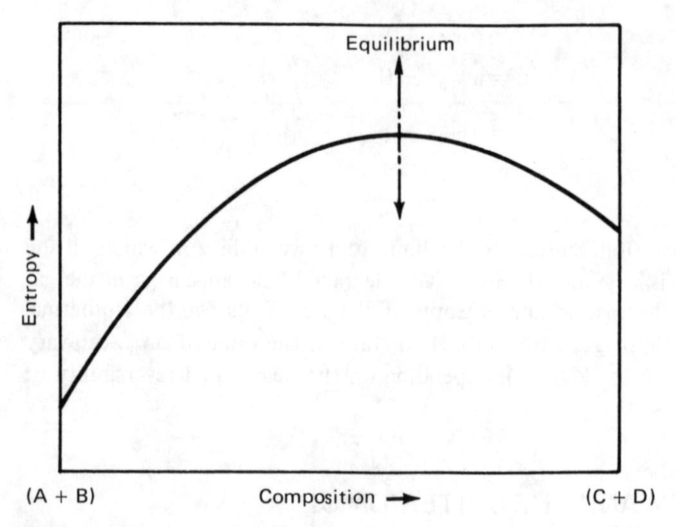

Fig. 3.7. Schematic representation of the entropy of a closed system containing A+B+C+D as a function of the extent of the reaction A+B = C+D at constant internal energy and volume.

system, as further reaction in either direction would decrease the entropy and hence will not occur spontaneously.

3.15 THE COMBINED STATEMENT OF THE FIRST AND SECOND LAWS OF THERMODYNAMICS

For an incremental change of state of a closed system, the First Law of Thermodynamics gives

$$dU = \delta q - \delta w$$

and if the process occurs reversibly, the Second Law of Thermodynamics gives

$$dS = \frac{\delta q}{T} \quad \text{or} \quad \delta q = TdS$$

and

$$\delta w = PdV$$

Combination of the two laws thus gives the equation

$$dU = TdS - PdV \tag{3.12}$$

Restrictions on the applicability of Eq. (3.12) are

1. That the system is closed; i.e., during the process the system does not exchange matter with its surroundings.
2. That work due to volume change is the only form of work performed by the system.

Equation (3.12) relates the dependent variable of the system, U to the independent variables S and V; i.e.,

$$U = U(S, V)$$

the total differential of which is

$$dU = \left(\frac{\partial U}{\partial S}\right)_V dS + \left(\frac{\partial U}{\partial V}\right)_S dV \tag{3.13}$$

Comparison of Eqs. (3.12) and (3.13) shows that

Temperature T is defined as $(\partial U/\partial S)_V$
Pressure P is defined as $-(\partial U/\partial V)_S$

The particularly simple form of Eq. (3.12) stems from the fact that in considering variations in U as the dependent variable, the "natural" choice of independent variables is S and V. Considering S to be the dependent variable and U and V to be the independent variables, i.e.,

$$S = S(U, V)$$

gives

$$dS = \left(\frac{\partial S}{\partial U}\right)_V dU + \left(\frac{\partial S}{\partial V}\right)_U dV \tag{3.14}$$

Rearranging Eq. (3.12) as

$$dS = \frac{dU}{T} + \frac{PdV}{T}$$

and comparing with Eq. (3.14) indicates that

$$(\partial S/\partial V)_U = P/T$$

Equilibrium, in a system of constant internal energy and constant volume, occurs when the entropy of the system is maximized and, in a system of constant entropy and volume, when the internal energy is minimized.

The further development of Classical Thermodynamics results from the fact that S and V are an inconvenient pair of independent variables; i.e., in considering a real system, considerable difficulty would be experienced in arranging the state of the system such that, simultaneously, it would have the required entropy and would occupy the required volume.

3.16 SUMMARY

1. The paths taken by a system undergoing a change of state can be classified into two types: reversible and irreversible. When the change of state of the system occurs as the result of the application of a finite driving force, the process proceeds irreversibly, and the degree of irreversibility of the process increases with increasing magnitude of the driving force. For a process to occur reversibly, the driving force must be infinitesimal, and hence a reversible process proceeds at an infinitesimal rate. During a reversible process the system moves through a continuum of equilibrium states.

2. When a system undergoes a change of state during which it performs work and absorbs heat, the magnitudes of the quantities w and q are maxima (w_{max} and q_{rev}, respectively) when the change of state occurs reversibly. For any irreversible path between the two states, less work is performed by the system, and correspondingly less heat is absorbed.

3. There exists a state function, entropy S, which is defined as $dS = \delta q_{rev}/T$. The entropy difference between the thermodynamic states A and B of a system is thus

$$\Delta S = S_B - S_A = \int_A^B \frac{\delta q_{rev}}{T}$$

If, in moving between the two states, the temperature T of the system remains constant, then the increase in entropy of the system is $\Delta S = q_{rev}/T$, where q_{rev} is the heat absorbed by the system in moving *reversibly* between the two states.

4. If the heat q_{rev} is provided by a constant-temperature heat reservoir at the temperature T, the decrease in the entropy of the heat reservoir as a result of the system moving from the state A to the state B is q_{rev}/T. The entropy of the combined system + heat reservoir is thus unchanged as a result of the reversible process; entropy has simply been transferred from the heat reservoir to the system.

5. If the change of state of the system from A to B was carried out irreversibly, then less heat q ($q < q_{rev}$) would be withdrawn from the heat

reservoir by the system. Thus the magnitude of the entropy decrease of the heat reservoir would be smaller (equal to q/T). However, as entropy is a state function, $S_B - S_A$ is independent of the process path, and hence $\Delta S_{system} + \Delta S_{heat\ reservoir} > 0$. Entropy has been created as the result of the occurrence of an irreversible process. The created entropy is termed ΔS_{irr}.

6. In the general case, $S_B - S_A = q/T + \Delta S_{irr}$; and as the degree of irreversibility increases, then the heat q withdrawn from the heat reservoir decreases, and the magnitude of ΔS_{irr} increases.

7. The increase in entropy, due to the occurrence of an irreversible process, arises from the degradation occurring in the system, wherein some of the internal energy, which is potentially available for the doing of useful work, is degraded to heat.

8. A process, occurring in an adiabatically contained system of constant volume (i.e., a system of constant U and V), will proceed irreversibly with a consequent production of entropy, until the entropy is maximized. The attainment of maximum entropy is the criterion that such a system has arrived at its equilibrium state. Thus the entropy of an adiabatically contained system can never decrease; it increases as the result of an irreversible process and remains constant during a reversible process.

9. Combination of the First and Second Laws of Thermodynamics gives, for a closed system which does no work other than work of volume expansion, $dU = T\,dS - P\,dV$. U is thus the natural choice of dependent variable for S and V as the independent variables.

3.17 NUMERICAL EXAMPLES

Example 1

Two moles of an ideal gas are contained adiabatically at 30 atm pressure and 298 K. The pressure is suddenly released to 10 atm, and the gas undergoes an irreversible adiabatic expansion as a result of which 2000 joules of work are performed. Show that the final temperature of the gas after the irreversible expansion is greater than that which the gas would attain if the expansion from 30 to 10 atm had been conducted reversibly. Calculate the entropy produced as a result of the irreversible expansion. c_v for the gas equals $1.5R$.

In the initial state 1,

$$V_1 = \frac{nRT_1}{P_1} = \frac{2 \times 0.08206 \times 298}{30} = 1.63 \text{ liters}$$

If the adiabatic expansion from 30 to 10 atm is carried out reversibly, then the process path follows PV^γ = constant, and in the final state 2,

$$V_2 = \left(\frac{P_1 V_1^\gamma}{P_2}\right)^{1/\gamma} = \left[\frac{30 \times 1.63^{5/3}}{10}\right]^{3/5} = 3.152 \text{ liters}$$

and

$$T_2 = \frac{P_2 V_2}{nR} = \frac{10 \times 3.152}{2 \times 0.08206} = 192 \text{ K}$$

For the irreversible process, which takes the gas from the state 1 to the state 3, as $q = 0$,

$$\Delta U = -w = -2000 = nc_v(T_3 - T_1) = 2 \times 1.5 \times 8.3144 \times (T_3 - 298)$$

and hence $T_3 = 217.8$ K, which is seen to be higher than T_2.

As the irreversible expansion from state 1 to state 3 was conducted adiabatically, no heat entered the system, and hence the entropy difference between states 3 and 1 equals the entropy created, ΔS_{irr}, as a result of the irreversible process. This entropy difference can be calculated by considering a reversible path from 1 to 3 as follows. Taking S to be a function of T and V gives

$$dS = \left(\frac{\partial S}{\partial T}\right)_V dT + \left(\frac{\partial S}{\partial V}\right)_T dV$$

For a reversible constant-volume process,

$$\delta q_v = nc_v dT = TdS$$

and from one of Maxwell's relations, given by Eq. (5.33),

$$(\partial S/\partial V)_T = (\partial P/\partial T)_V$$

Thus

$$dS = \frac{nc_v}{T} dT + \left(\frac{\partial P}{\partial T}\right)_V dV$$

or, as the system is an ideal gas, in which case $(\partial P/\partial T)_V = nR/V$,

$$dS = \frac{nc_v}{T} dT + \frac{nR}{V} dV$$

Integrating between the states 3 and 1 gives

$$S_3 - S_1 = nc_v \ln\left(\frac{T_3}{T_1}\right) + nR \ln\left(\frac{V_3}{V_1}\right) \tag{3.15}$$

As

$$V_3 = \frac{nRT_3}{P_3} = \frac{2 \times 0.08206 \times 217.8}{10} = 3.57 \text{ liters}$$

then

$$\begin{aligned} S_3 - S_1 &= 2 \times 8.3144 \times 1.5 \times \ln\left(\frac{217.8}{298}\right) \\ &\quad + 2 \times 8.3144 \times \ln\left(\frac{3.57}{1.63}\right) \\ &= -7.82 + 13.04 \\ &= +5.22 \text{ joules/degree} \end{aligned}$$

= the entropy created in the irreversible adiabatic expansion from state 1 to state 3

$= \Delta S_{irr}$

The process path in this calculation involved a reversible change of temperature at constant volume, followed by a reversible change of volume at constant temperature, i.e., the path $1 \to 4 \to 3$ in Fig. 3.8.

For the step $1 \to 4$,

$$\Delta S_{(1\to4)} = \int_1^4 \frac{\delta q_v}{T} = nc_v \ln\left(\frac{T_4}{T_1}\right) = nc_v \ln\left(\frac{T_3}{T_1}\right)$$

and for the step $4 \to 3$, as T is constant, $\Delta U = 0$, and hence

$$\delta w = \delta q = PdV = \frac{nRTdV}{V} = nRTd \ln V$$

Thus

$$\Delta S_{(4\to3)} = \int_4^3 \frac{\delta q}{T} = \int_4^3 nRd \ln V = nR \ln\left(\frac{V_3}{V_4}\right) = nR \ln\left(\frac{V_3}{V_1}\right)$$

So

$$\Delta S_{(1\to4)} + \Delta S_{(4\to3)} = \Delta S_{(1\to3)}$$

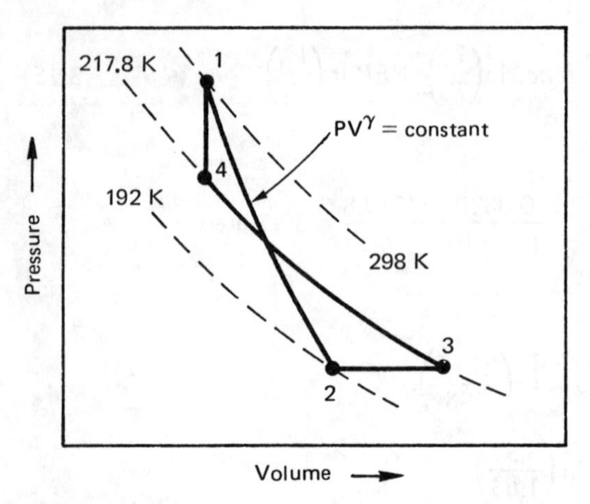

Fig. 3.8.

as given by Eq. (3.15). For any change of state $1 \rightarrow 2$, Eq. (3.15), per mole of system can be written as

$$S_2 - S_1 = c_v \ln\left(\frac{T_2}{T_1}\right) + R \ln\left(\frac{V_2}{V_1}\right)$$

or

$$e^{(S_2 - S_1)} = \left(\frac{T_2}{T_1}\right)^{c_v}\left(\frac{V_2}{V_1}\right)^R$$

or

$$\exp\left(\frac{S_2 - S_1}{c_v}\right) = \left(\frac{T_2}{T_1}\right)\left(\frac{V_2}{V_1}\right)^{R/c_v}$$

$$= \left(\frac{P_2 V_2}{P_1 V_1}\right)\left(\frac{V_2}{V_1}\right)^{\gamma - 1}$$

$$= \frac{P_2 V_2^\gamma}{P_1 V_1^\gamma}$$

or

$$P_1 V_1^\gamma e^{-S_1/c_v} = P_2 V_2^\gamma e^{-S_2/c_v} \tag{3.16}$$

In the special case of the change of state $1 \rightarrow 2$ being a reversible adiabatic process, i.e., $S_1 = S_2$, Eq. (3.16) becomes Eq. (2.9). Equation (3.16) indicates that if the process $1 \rightarrow 2$ is an irreversible adiabatic expansion from P_1 to P_2, then as $S_2 > S_1$, the final volume (and hence temperature) of the system is greater than would have been the case were the adiabatic expansion from P_1 to P_2 carried out reversibly.

The derivation of Eq. (3.15) required a knowledge of Maxwell's relations, which are introduced in Chap. 5. In view of this, $\Delta S_{(1 \to 3)} = \Delta S_{irr}$ could have been calculated by considering some other convenient reversible path. The most simple path is $1 \to 2 \to 3$; i.e., as $S_1 = S_2$, then

$$\Delta S_{(1\to3)} = \Delta S_{(2\to3)}$$

For the constant-pressure process $2 \to 3$,

$$\Delta S = \int_2^3 \frac{\delta q_p}{T} = \int_2^3 \frac{nc_p}{T}\, dT = nc_p \ln\left(\frac{T_3}{T_2}\right)$$

$$= 2 \times 2.5 \times 8.3144 \times \ln\left(\frac{217.8}{192}\right)$$

$$= 5.22 \text{ joules/degree}$$

$$= \Delta S_{irr}$$

Examination shows that the more irreversible the adiabatic expansion, the less the work performed, the higher the final temperature, and the greater the value of ΔS_{irr}. In a problem such as this, the final state of the gas after an irreversible adiabatic expansion from a known initial state to a final pressure cannot be determined without a knowledge of the work performed as a result of the process.

Example 2

At a pressure of 1 atm the equilibrium melting temperature of lead is 600 K, and at this temperature the latent heat of fusion of lead is 4810 joules per mole. If 1 mole of supercooled liquid lead spontaneously freezes at 590 K and 1 atm pressure, calculate the entropy produced. The constant-pressure molar heat capacity of liquid lead, as a function of temperature at 1 atm pressure, is given as

$$c_{p(l)} = 32.4 - 3.1 \times 10^{-3}\, T \text{ joules/degree}$$

and the corresponding expression for solid lead is given as

$$c_{p(s)} = 23.6 + 9.75 \times 10^{-3}\, T \text{ joules/degree}$$

The entropy created due to the irreversible freezing of the lead equals the difference between the entropy change in the lead as a result of the process, and the entropy change in the constant-temperature heat reservoir (at 590 K) as a result of the process.

First calculate the entropy difference between 1 mole of solid lead at 590 K and 1 atm, and 1 mole of liquid lead at 590 K and 1 atm.

Consider the process scheme illustrated in Fig. 3.9:

Step $a \to b$: 1 mole of supercooled liquid lead is heated reversibly from 590 to 600 K at 1 atm pressure.

Step $b \to c$: 1 mole of liquid lead is solidified reversibly at 600 K. (The equilibrium melting or freezing temperature is the only temperature at which the melting or freezing process can be conducted reversibly.)

Step $c \to d$: 1 mole of solid lead is reversibly cooled from 600 to 590 K at 1 atm pressure.

As entropy is a state function,

$$\Delta S_{(a \to d)} = \Delta S_{(a \to b)} + \Delta S_{(b \to c)} + \Delta S_{(c \to d)}$$

Step $a \to b$:

$$\Delta S_{(a \to b)} = \int_a^b \frac{\delta q_{\text{rev}}}{T} = \int_a^b \frac{\delta q_p}{T} = \int_{590}^{600} \frac{nc_{p(l)}dT}{T} = \int_{590}^{600} \left[\frac{32.4}{T} - 3.1 \times 10^{-3}\right] dT$$

$$= 32.4 \times \ln \frac{600}{590} - 3.1 \times 10^{-3} \times (600 - 590) = +0.514 \text{ joules/degree}$$

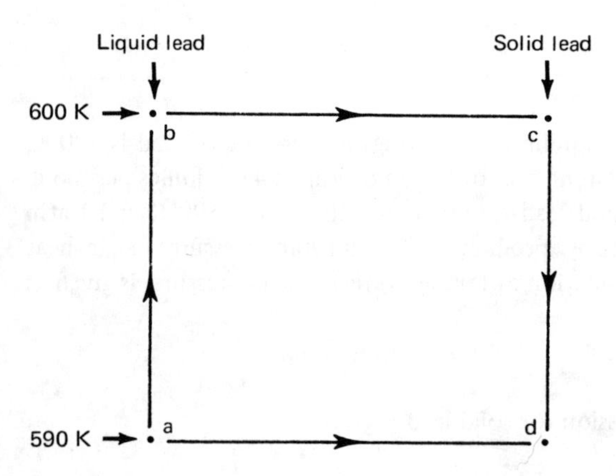

Fig. 3.9.

Step $b \to c$:

$$\Delta S_{(b\to c)} = \frac{q_{\text{rev}}}{T} = \frac{q_p}{T} = \frac{\text{latent heat of freezing}}{\text{freezing temperature}}$$

$$= -\frac{4810}{600} = -8.017 \text{ joules/degree}$$

Step $c \to d$:

$$\Delta S_{(c\to d)} = \int_c^d \frac{\delta q_{\text{rev}}}{T} = \int_c^d \frac{\delta q_p}{T} = \int_{600}^{590} \frac{nc_{p(s)}dT}{T}$$

$$= \int_{600}^{590} \left[\frac{23.6}{T} + 9.75 \times 10^{-3}\right] dT$$

$$= 23.6 \times \ln\frac{590}{600} + 9.75 \times 10^{-3}\,(590 - 600)$$

$$= -0.494 \text{ joules/degree}$$

Thus

$$\Delta S_{(a\to d)} = +0.514 - 8.017 - 0.494 = -7.997 \text{ joules/degree}$$

Consider the heat entering the constant-temperature heat reservoir at 590 K. As the heat transfer is at constant pressure, then $q_p = \Delta H$, where ΔH is the enthalpy difference between the state d and the state a. As H is a state function, then

$$\Delta H_{(a\to d)} = \Delta H_{(a\to b)} + \Delta H_{(b\to c)} + \Delta H_{(c\to d)}$$

$$\Delta H_{(a\to b)} = \int_a^b nc_p dT = \int_{590}^{600} [32.4 - 3.1 \times 10^{-3}T]\, dT$$

$$= 32.4 \times (600 - 590) - \frac{3.1 \times 10^{-3}}{2}(600^2 - 590^2)$$

$$= 306 \text{ joules}$$

$$\Delta H_{(b\to c)} = -4810 \text{ joules}$$

$$\Delta H_{(c\to d)} = \int_c^d nc_p dT = \int_{600}^{590} [23.6 + 9.75 \times 10^{-3}T]\, dT$$

$$= 23.6 \times (590 - 600) + \frac{9.75 \times 10^{-3}}{2}(590^2 - 600^2) = -294 \text{ joules}$$

Thus

$$\Delta H_{(a\to d)} = -4799 \text{ joules}$$

and so 4799 joules of heat enter the constant-temperature heat reservoir at 590 K. Then

$$\Delta S_{\text{heat reservoir}} = \frac{4799}{590} = 8.134 \text{ joules/degree}$$

and so the entropy created,

$$\Delta S_{\text{irr}} = -7.994 + 8.134$$
$$= 0.137 \text{ joules/degree}$$

Examination shows that the lower the temperature of irreversible freezing of the supercooled liquid, the more irreversible the process and the greater the value of ΔS_{irr}.

PROBLEMS

3.1 A steam engine operating between 150 and 30°C performs 1000 joules of work. What is the minimum quantity of heat which must be drawn from the heat source in order to obtain this amount of work? Which of the following would give the greater increase in the efficiency of the engine: (*a*) an increase of ΔT in the temperature of the heat source, or (*b*) a decrease of ΔT in the temperature of the heat sink?

3.2 The initial state of one mole of an ideal gas is $P = 10$ atm and $T = 300$ K. Calculate the entropy change in the gas for (*a*) an isothermal decrease in the pressure to 1 atm; (*b*) a reversible adiabatic decrease in the pressure to 1 atm; (*c*) a constant-volume decrease in the pressure to 1 atm.

3.3 One mole of an ideal gas is subjected to the following sequence of steps:

a. Starting at 25°C and 1 atm, the gas expands freely into a vacuum to double its volume.
b. The gas is next heated to 125°C at constant volume.
c. The gas is reversibly expanded at constant temperature until its volume is again doubled.
d. The gas is finally reversibly cooled to 25°C at constant pressure.

Calculate ΔU, ΔH, q, w, and ΔS in the gas.

3.4 Calculate the final temperature and the entropy produced when 1500 grams of lead at 100°C is placed in 100 grams of adiabatically contained water, the initial temperature of which is 25°C. Given: $c_{p,H_2O} = 75.44$ joules/degree and $c_{p,Pb} = 26.7$ joules/degree. The molecular weights of H_2O and Pb are 18 and 207, respectively.

3.5 From 298 K up to its melting temperature of 1048 K, the constant-pressure molar heat capacity of RbF is given as

$$c_p = 33.3 + 38.5 \times 10^{-3} T + 5.06 \times 10^5 T^{-2} \text{ joules/degree}$$

and from the melting temperature to 1200 K, the constant-pressure molar heat capacity of liquid RbF is given as

$$c_p = -47.3 + 3.49 \times 10^{-3} T + 1467 \times 10^5 T^{-2} \text{ joules/degree}$$

At its melting temperature the molar heat of fusion of RbF is 26,400 joules. Calculate the increase in the entropy of 1 mole of RbF when it is heated from 300 to 1200 K.

3.6 A reversible heat engine, operating in a cycle, withdraws heat from a high-temperature reservoir (the temperature of which consequently decreases), performs work w, and rejects heat into a low-temperature heat reservoir (the temperature of which consequently increases). If the two reservoirs are initially at the temperatures T_1 and T_2 and have constant heat capacities C_1 and C_2 respectively, calculate the final temperature of the system and the maximum amount of work which can be obtained from the engine.

CHAPTER FOUR

THE STATISTICAL INTERPRETATION OF ENTROPY

4.1 INTRODUCTION

The introduction, in Chap. 3, of entropy as a state function was achieved as the result of the realization that there exist possible and impossible processes, and from an examination of the relationships which occur between the heat and work effects of these processes. From the formal statement of the Second Law of Thermodynamics, as developed from Classical Thermodynamics arguments, it is difficult to assign a physical significance or physical quality to entropy. In this respect entropy differs from internal energy, in spite of the fact that, within the scope of Classical Thermodynamics, both properties are simply mathematical functions of the state of a system. That the physical significance of internal energy is readily understandable is evidenced by the rapid acceptance of the First Law of Thermodynamics after its enunciation, whereas the difficulty of assigning a corresponding physical significance to entropy is evidenced by the slow acceptance of the Second Law of Thermodynamics after its initial enunciation. From the classical viewpoint, the Second Law is only valid as a direct result of the fact that human ingenuity has, to date, failed to invent a perpetual motion machine. Within the scope of Classical Thermodynamics, the Second Law is thus a "law" only in that it has not yet been disproved.

The physical interpretation of entropy had to await the invention of quantum theory and the development of statistical mechanics.

4.2 ENTROPY AND DISORDER ON AN ATOMIC SCALE

Gibbs described the entropy of a system as being a measure of its "degree of mixed-up-ness," where this term is applied to an atomic-scale picture of the system; i.e., the more mixed up the constituent particles of the system, the larger the value of its entropy. Entropy can thus be correlated with the atomic-scale randomness or disorder of the system. For example, in the crystalline solid state the vast majority of the constituent particles are confined to vibrate about their regularly arranged lattice positions, whereas in the liquid state confinement of the particles to lattice positions is absent, and the particles are relatively free to wander through the liquid volume. The atomic arrangement in the solid state is thus more ordered than that of the liquid state or, alternatively, is less "mixed up" than that of the liquid state, and hence the entropy of the liquid state is greater than that of the solid state. Similarly the atomic disorder in the gaseous state greatly exceeds that of the liquid state, and so the entropy of the gaseous state exceeds that of the liquid state.

This correlates generally with macroscopic phenomena; e.g., a solid melts at its melting temperature T_m as the result of the addition of a quantity of heat q which is termed the latent heat of melting. The entropy of the substance being melted is thus increased by the amount q/T_m; and if the melting is carried out at constant pressure, then from Eq. (2.5), as $q = \Delta H$,

$$\Delta S_{\text{melting}} = \frac{\Delta H_{\text{melting}}}{T_m}$$

This increase in the entropy of the substance being melted correlates with an increase in the degree of disorder, or an increase in the degree of mixed-up-ness of the constituent particles of the system. The above correlation cannot, however, be universally applied, because if a supercooled liquid spontaneously freezes, then it would appear that a decrease in the degree of disorder accompanies an entropy increase (the entropy increase due to irreversible freezing). The anomaly in this latter example arises from the fact that although the degree of order in the freezing system is considered, the change in the degree of order of the heat bath absorbing the heat of freezing has not been considered. In fact, if spontaneous freezing of a supercooled liquid occurs, then the increase in the degree of order of the freezing system is less than the decrease in the degree of order in the heat reservoir; and hence the spontaneous freezing process produces an overall decrease in order and an overall increase in entropy. If the freezing occurs at the equilibrium melting point, then the increase in the degree of order of the freezing system exactly equals the decrease in the degree of order of the heat reservoir absorbing the heat of solidification, and hence the total degree of order in the system + reservoir is unchanged as a result of the process. Disorder has simply been transferred from the system to the heat reservoir, and consequently the total entropy of the system + reservoir is unchanged as a result of the freezing process; i.e., entropy has simply been transferred from the system to the heat reservoir. The equilibrium melting or freezing temperature of a substance can thus be defined as being that temperature at which no change in the degree of order of the system + heat bath occurs as a result of the phase change; i.e., only at this temperature is the solid phase in equilibrium with the liquid phase, and hence only at this temperature can the phase change occur reversibly.

4.3 THE CONCEPT OF MICROSTATE

The development of a quantitative relationship between entropy and "degree of mixed-up-ness" necessitates the quantification of the term "degree of mixed-up-ness." This can be obtained from a consideration of elementary statistical mechanics. Statistical mechanics is developed from the assumption that the equilibrium state of a system is simply the most probable of all its possible states, and hence the subject is concerned with the determination of, the criteria governing, and the properties of this most probable state.

One of the major developments in physical science which has led to a considerable increase in the understanding of the behavior of matter is the quantum theory. A postulate of the quantum theory is that, if a particle is confined to move within a given fixed volume, then its energy is quantized; i.e., the particle may only have certain discrete allowed values of energy which are separated by "forbidden energy bands." For any given particle the spacing between the quantized values of energy (normally called the allowed energy levels) decreases as the volume available to the movement of the particles increases. Thus energy only becomes continuous when no restriction is placed on the position of the particle. As an illustration, compare a particle in a volume of gas with a particle in a solid. As the gas particle can move about within the entire volume of the gas (the so-called communal volume) and the position of the particle in the solid is confined to a small volume surrounding the lattice point on which the particle is located (the movement thus being restricted by the presence of particles on the surrounding lattice points), then, from quantum theory, it is seen that the energy levels available to the particle in the solid are considerably more widely spaced than are the levels available to the particle in the gas.

The effect of this quantization of energy can be illustrated by examining the following hypothetical system. Following the discussion of Gurney,* consider a perfect crystal in which identical particles occupy all lattice sites. The characteristics of the particles and the particular crystal structure determine the quantization of the energy levels. The lowest energy level, or the ground state, is designated ϵ_0 and the succeeding levels of increasing energy are designated ϵ_1, ϵ_2, ϵ_3, etc. Let the crystal contain n particles and have the total energy U. Given

*R. W. Gurney, "Introduction to Statistical Mechanics," Dover Publications, New York, 1966.

this system, statistical mechanics asks the question–How can the n particles be distributed among the available energy levels such that the total energy of the system equals U; and of the possible distributions, which is the most probable?

Consider the crystal to contain three identical, and hence indistinguishable, particles which are located on three distinguishable lattice sites A, B, and C. Suppose, for simplicity, that the quantization is such that the energy levels are equally spaced, with the ground level ϵ_0 being taken as zero, the first level $\epsilon_1 = u$, the second level $\epsilon_2 = 2u$, etc. Let the total energy of the system, U, equal $3u$. This system can be realized in three different distributions, as shown in Fig. 4.1:

a. All three particles in level 1
b. One particle in level 3, and the other two particles in level 0
c. One particle in level 2, one particle in level 1, and one particle in level 0

The distributions must now be examined to determine how many distinguishable arrangements they individually contain.

Distribution a. There is only one arrangement of this distribution; i.e., interchange of the particles among the three lattice sites does not produce a different arrangement.

Distribution b. Any of the three distinguishable lattice sites can be occupied by the particle of energy $3u$, and the remaining two lattice sites are each occupied by a particle of zero energy. As interchange of the particles of zero energy between their lattice sites does not produce a different arrangement, there are thus three arrangements of distribution b.

Distribution c. Any of the three distinguishable lattice sites can be occupied by the particle of energy $2u$, either of the two remaining sites can be occupied by the particle of energy $1u$, and the single remaining site is occupied by the particle of zero energy. The number of distinguishable arrangements of distribution c is thus $3 \times 2 \times 1 = 3! = 6$.

These arrangements are shown in Fig. 4.2.

Thus, altogether, there are 10 distinguishable ways in which three particles can be arranged among the energy levels such that the total energy of the system, U,

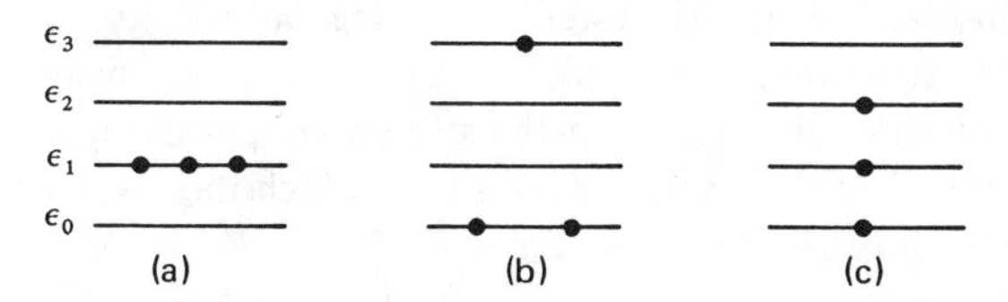

Fig. 4.1. Illustration of the distribution of particles among energy levels in a constant-energy system.

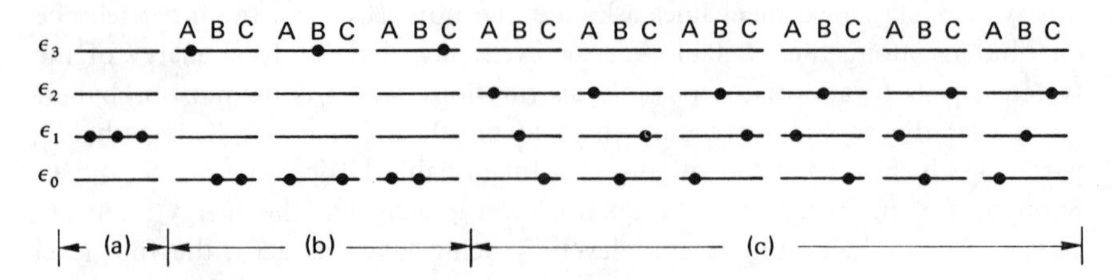

Fig. 4.2. Illustration of the complexions or microstates within distributions of particles among energy levels in a system of constant energy.

equals $3u$. These distinguishable arrangements are called *complexions* or *microstates*, and all of these 10 microstates correspond to a single *macrostate*.

4.4 DETERMINATION OF THE MOST PROBABLE MICROSTATE

The concept of macrostate lies within the domain of Classical Thermodynamics, and the macrostate of a system is fixed when the independent variables are fixed. As has been seen in Chap. 1, a system of fixed size and fixed composition has two independent variables. Thus a system of fixed composition has three independent variables–the original two plus a variable which describes the size of the system. In the above example the values of U, V, and n are fixed (constancy of volume being required in order that the quantization of the energy levels be determined), and hence the macrostate of the system is fixed. With respect to consideration of the microstates, in the absence of any reason to the contrary, it must be assumed that each of the microstates is equally probable. Thus we are as likely to find the system in any one microstate as in any other. The probability of observing the system in any microstate is thus 1/10. However, the 10 microstates are contained in three distributions, and hence the probability of the occurrence of the system in distribution a is 1/10, the probability of the occurrence of the system in distribution b is 3/10, and the probability of the occurrence of the system in distribution c is 6/10. Distribution c is thus seen to be the "most probable." The physical significance of these probabilities can be viewed in either of two ways: (1) if it were possible to make an instantaneous observation of the system, then the probability of observing an arrangement of distribution c is 6/10; or, (2) if the system were observed for a finite length of time, during which the system rapidly changes from one microstate to another, then the fraction of this time which the system spent in all of the arrangements of distribution c would be 6/10.

From the preceding discussion of a simple system containing three particles sharing an energy $3u$, it is obvious that as both the total energy of the system and the number of particles in the system increase, then the number of distinguishable arrangements (microstates) increases, and for fixed values of U, V, and n, these microstates still correspond to a single macrostate. Similarly the number of possible distributions increases, and with real systems—for example, 1 mole, which contains 6.023×10^{23} particles—it is found that the number of arrangements of the most probable distribution becomes very much larger than the number of arrangements in all other distributions. The number of arrangements within a given distribution Ω, is calculated as follows. If n particles are distributed among the energy levels such that there are n_0 in level ϵ_0, n_1 in level ϵ_1, n_2 in level ϵ_2, . . . , and n_r in the highest level of occupancy ϵ_r, then the number of arrangements, Ω, is given as

$$\Omega = \frac{n!}{n_0!n_1!n_2! \ldots n_r!} = \frac{n!}{\prod_{i=0}^{i=r} n_i!} \qquad (4.1)$$

e.g., consideration of the simple system discussed gives

$$\Omega(\text{distribution } a) = \frac{3!}{3!0!0!} = 1$$

$$\Omega(\text{distribution } b) = \frac{3!}{2!1!0!} = 3$$

$$\Omega(\text{distribution } c) = \frac{3!}{1!1!1!} = 6$$

The most probable distribution is obtained by determining the set of numbers $n_0, n_1, \ldots, n_r$ which maximizes the value of Ω. When the values for n_i are large, Stirling's approximation can be used (that is, $\ln X! = X \ln X - X$). Thus, taking the logarithms of Eq. (4.1),

$$\ln \Omega = n \ln n - n - \sum_{i=0}^{i=r} (n_i \ln n_i - n_i) \qquad (4.2)$$

As the macrostate of the system is determined by the fixed values of U, V, and n, any distribution of particles among the energy levels must conform with the conditions

$$U = \text{constant} = n_0\epsilon_0 + n_1\epsilon_1 + n_2\epsilon_2 + \cdots + n_r\epsilon_r = \sum_{i=0}^{i=r} n_i\epsilon_i \qquad (4.3)$$

and

$$n = \text{constant} = n_0 + n_1 + n_2 + \cdots n_r$$
$$= \sum_{i=0}^{i=r} n_i \tag{4.4}$$

From Eqs. (4.3) and (4.4), any interchange of particles among the energy levels must conform with the conditions

$$\delta U = \sum_i \epsilon_i \delta n_i = 0 \tag{4.5}$$

and

$$\delta n = \sum_i \delta n_i = 0 \tag{4.6}$$

Also, from Eq. (4.2), any interchange of particles among the energy levels gives

$$\delta \ln \Omega = -\Sigma\left(\delta n_i \ln n_i + \frac{n_i \delta n_i}{n_i} - \delta n_i\right)$$
$$= -\Sigma(\delta n_i \ln n_i) \tag{4.7}$$

If Ω has the maximum value possible, then a small rearrangement of particles among the energy levels will not alter the value of Ω or of $\ln \Omega$. Hence if the set of n_i's is such that Ω has its maximum value, then

$$\delta \ln \Omega = -\Sigma(\delta n_i \ln n_i) = 0 \tag{4.8}$$

The condition that Ω has its maximum value for the given macrostate is thus that Eqs. (4.5), (4.6), and (4.8) are simultaneously satisfied. Solution for the set of n_i values corresponding to the most probable distribution is most conveniently achieved by means of the method of undetermined multipliers. Multiply Eq. (4.5) by the constant β, where β has the units of reciprocal energy; i.e.,

$$\Sigma \beta \epsilon_i \delta n_i = 0 \tag{4.9}$$

Multiply Eq. (4.6) by the dimensionless constant α; i.e.,

$$\Sigma \alpha \delta n_i = 0 \tag{4.10}$$

and add Eqs. (4.8), (4.9), and (4.10)

$$\sum_{i=0}^{i=r} (\ln n_i + \alpha + \beta\epsilon_i)\delta n_i = 0 \tag{4.11}$$

i.e.,

$$(\ln n_0 + \alpha + \beta\epsilon_0)\delta n_0 + (\ln n_1 + \alpha + \beta\epsilon_1)\delta n_1 + (\ln n_2 + \alpha + \beta\epsilon_2)\delta n_2 + (\ln n_3 + \alpha + \beta\epsilon_3)\delta n_3 + \cdots + (\ln n_r + \alpha + \beta\epsilon_i)\delta n_i = 0$$

Solution of Eq. (4.11) requires that each of the bracketed terms be individually equal to zero; i.e.,

$$\ln n_i + \alpha + \beta\epsilon_i = 0$$

or

$$n_i = e^{-\alpha} e^{-\beta\epsilon_i} \tag{4.12}$$

Summing over all r levels gives

$$\sum_{i=0}^{i=r} n_i = n = e^{-\alpha} \sum_{i=0}^{i=r} e^{-\beta\epsilon_i}$$

or

$$e^{-\alpha} = \frac{n}{\Sigma e^{-\beta\epsilon_i}}$$

The summation

$$\Sigma e^{-\beta\epsilon_i} = e^{-\beta\epsilon_0} + e^{-\beta\epsilon_1} + e^{-\beta\epsilon_2} + \cdots + e^{-\beta\epsilon_r}$$

which is determined by the magnitude of β and by the quantization of the energy, is termed the partition function, P. Thus

$$e^{-\alpha} = \frac{n}{P}$$

and hence

$$n_i = \frac{ne^{-\beta\epsilon_i}}{P} \tag{4.13}$$

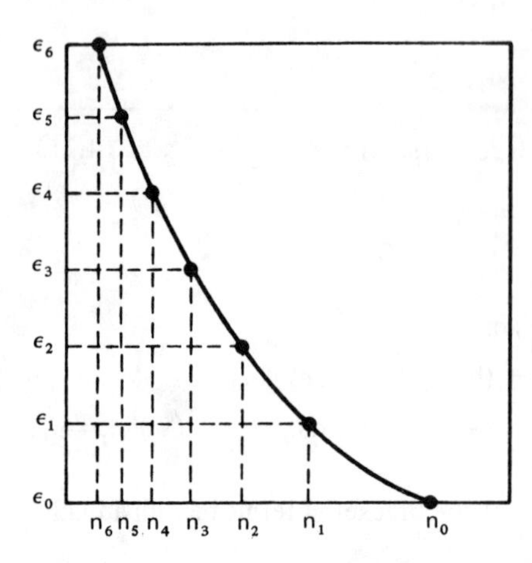

Fig. 4.3. Schematic representation of the most probable distribution of particles among energy states.

The distribution of particles in the energy levels which maximizes Ω (i.e., the most probable distribution) is thus one in which the occupancy of the levels decreases exponentially with increasing energy. The shape of this distribution is shown in Fig. 4.3. Examination of Eq. (4.13) indicates that β must be a positive quantity; otherwise the level of infinite energy would contain an infinite number of particles. The actual shape of the exponential curve in Fig. 4.3 (for a given system) is determined by the value of β, which, it can be shown, is inversely proportional to the absolute temperature; i.e.,

$$\beta \propto \frac{1}{T} \quad \text{or} \quad \beta = \frac{1}{kT} \tag{4.14}$$

where k is Boltzmann's constant, an expression of the Gas Constant per molecule; i.e.,

$$k = \frac{R}{\mathcal{A}} = \frac{8.3144}{6.0232 \times 10^{23}} = 1.38054 \times 10^{-23} \text{ joules/degree}$$

where $\mathcal{A}$ is Avogadro's number.

4.5 THE EFFECT OF TEMPERATURE

The nature of the exponential distribution of particles in Fig. 4.3 is determined by the temperature. However, as the macrostate of the system is fixed by fixing the values of U, V, and n, then T, as a dependent variable, is fixed. It can be seen that as T increases, then β decreases, and the shape of the exponential distribution changes as is shown in Fig. 4.4. As the temperature is increased, the upper levels become relatively more populated, and this corresponds to an increase in the average energy of the particles, (i.e., an increase in the value of U/n), which, for fixed values of V and n, corresponds to an increase in U.

As has been stated, when the number of particles in the system is very large, then the number of arrangements within the most probable distribution, Ω_{max}, is the only term which makes a significant contribution to the total number of arrangements, Ω_{total}, which the system may have; that is, Ω_{max} is significantly larger than the sum of all the other arrangements. Hence when the total number of particles is large, Ω_{max} can be equated with Ω_{total}.

Substituting $\beta = 1/kT$, Eq. (4.2) can be written as

$$\ln \Omega_{total} = \ln \Omega_{max} = n \ln n - \Sigma n_i \ln n_i$$

where the values of n_i are given by Eq. (4.13). Thus

$$\ln \Omega_{total} = n \ln n - \sum \frac{n}{P} e^{-\epsilon_i/kT} \ln\left(\frac{n}{P} e^{-\epsilon_i/kT}\right)$$

$$= n \ln n - \frac{n}{P} \sum \left[e^{-\epsilon_i/kT}\left(\ln n - \ln P - \frac{\epsilon_i}{kT}\right)\right]$$

$$= n \ln n - \frac{n}{P} (\ln n - \ln P) \sum e^{-\epsilon_i/kT} + \frac{n}{PkT} \sum \epsilon_i e^{-\epsilon_i/kT}$$

$$= n \ln n - n \ln n + n \ln P + \frac{n}{PkT} \sum \epsilon_i e^{-\epsilon_i/kT}$$

But

$$U = \sum n_i \epsilon_i = \sum \frac{n}{P} \epsilon_i e^{-\epsilon_i/kT} = \frac{n}{P} \sum \epsilon_i e^{-\epsilon_i/kT}$$

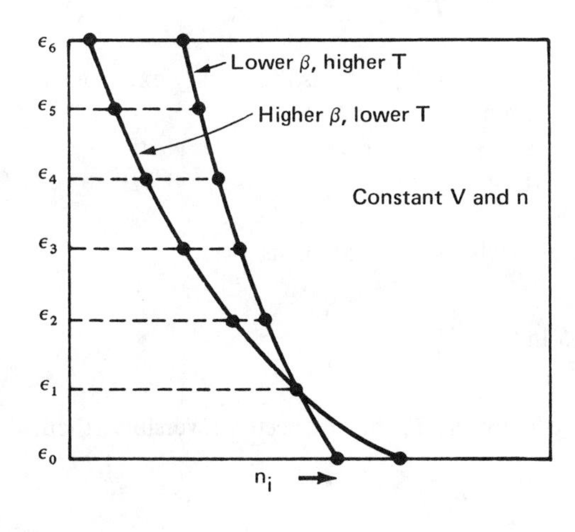

Fig. 4.4. The effect of temperature change on the most probable distribution of particles among energy levels in a closed system of constant volume.

Therefore

$$\sum \epsilon_i e^{-\epsilon_i/kT} = \frac{UP}{n}$$

and hence

$$\ln \Omega = n \ln P + \frac{U}{kT} \tag{4.15}$$

4.6 THERMAL EQUILIBRIUM WITHIN A SYSTEM AND THE BOLTZMANN EQUATION

Consider now a system of particles in thermal equilibrium with a heat bath, and let the state of the combined system (particles + heat bath) be fixed by fixing the values of U, V, and n, where

$$U = U_{\text{particles system}} + U_{\text{heat bath}}$$

$$V = V_{\text{particles system}} + V_{\text{heat bath}}$$

n = number of particles in the system + the heat bath of fixed size

As the particles system and the heat bath are in thermal equilibrium, then small exchanges of energy between the two can occur; and for such a small exchange at constant U, V, and n, Eq. (4.15), for the particles system, gives

$$\delta \ln \Omega = \frac{\delta U}{kT}$$

(P is dependent only on the values of ϵ_i and T.) And as this energy exchange is carried out at constant total volume, then,

$$\delta U = \delta q$$

i.e., the energy exchange occurs as an exchange of heat. Thus

$$\delta \ln \Omega = \frac{\delta q}{kT}$$

As the heat exchange is effected at constant T, that is, occurs reversibly, then, from Chap. 3

$$\frac{\delta q}{T} = \delta S$$

and hence

$$\delta S = k\,\delta \ln \Omega$$

As both S and Ω are state functions, the above equation can be written as a differential, appropriate integration of which gives*

$$S = k \ln \Omega \tag{4.16}$$

Equation (4.16) is known as Boltzmann's equation.

Equation (4.16) is thus the required quantitative relationship between the entropy of a system and its "degree of mixed-up-ness," where the "degree of mixed-up-ness," given as Ω, is the number of ways in which the available energy can be mixed or shared among the particles of the system. The "most probable" state of the system is that in which Ω is a maximum, consistent with the fixed values of U, V, and n; and hence the equilibrium state of the system is that state in which S is a maximum consistent with the fixed values of U, V, and n. The Boltzmann equation thus provides a physical quality to entropy.

4.7 HEAT FLOW AND THE PRODUCTION OF ENTROPY

Classical Thermodynamics arguments show that heat flow between two bodies at different temperatures is an irreversible process which gives rise to the production of entropy, and that the reverse process, i.e., the spontaneous flow of heat up a temperature gradient, is an impossible process. From an examination of microstates it can be shown that a microstate in which temperature differences occur within a system is less probable than a microstate in which the temperature of the system is uniformly constant.

Consider two closed systems A and B. Let the energy of A be U_A and the number of complexions of A be Ω_A. Similarly let the energy of the system B be U_B and the number of complexions of B be Ω_B. When thermal contact is made between A and B, the product $\Omega_A \Omega_B$ will generally not have its maximum possible value, and heat will flow either from A to B or from B to A. If heat flows from A to B, then this is because the increase in Ω_B, resulting from the increase in U_B, is greater than the simultaneous decrease in Ω_A, resulting from the decrease in U_A. Heat flow thus occurs as thereby the value of the product

*See Chap. 6.

$\Omega_A \Omega_B$ is increased, and the heat flow ceases when $\Omega_A \Omega_B$ reaches its maximum value, i.e., when the increase in Ω_B due to the transfer of an increment of heat from A, is exactly compensated by the decrease in Ω_A. The condition for thermal equilibrium between the two systems is thus that the transfer of a quantity of heat from A to B, or vice versa, does not produce a change in the product $\Omega_A \Omega_B$. That is,

$$\delta \ln \Omega_A \Omega_B = 0$$

Consider a rearrangement of the particles in the set of levels B such that the energy U_B is increased by a certain amount, and consider a simultaneous rearrangement of the particles in the set of levels A such that the energy U_A is diminished by the same amount; i.e., $(U_A + U_B)$ remains constant. If the levels of A are populated according to Eq. (4.13) with $T = T_A$, and if the levels of B are populated according to Eq. (4.13) with $T = T_B$, then

$$\delta \ln \Omega_A = \frac{\delta q_A}{kT_A}$$

$$\delta \ln \Omega_B = \frac{\delta q_B}{kT_B}$$

When an amount of heat is transferred from A to B at constant total energy, then

$$\delta q_A = -\delta q_B$$

Thus

$$\delta \ln \Omega_A \Omega_B = \delta \ln \Omega_A + \delta \ln \Omega_B = \left(\frac{1}{T_A} - \frac{1}{T_B}\right) \frac{\delta q}{k}$$

and hence the condition that $\delta \ln \Omega_A \Omega_B$ be zero is that $T_A = T_B$. The reversible transfer of heat between two bodies occurs, thus, only when the temperatures of the bodies are equal, as only in such a case does $\Omega_A \Omega_B$, and hence the total entropy of the combined system $(S_A + S_B)$, remain constant. An irreversible transfer of heat increases the product $\Omega_A \Omega_B$, and hence entropy is created. In microscopic terminology an irreversible process is one which takes the system from a less probable to the most probable state, and in the corresponding macroscopic terminology an irreversible process takes the system from a nonequilibrium to the equilibrium state. Thus what is considered in Classical Thermodynamics to be an impossible process, turns out, as the result of microstate examination, only to be an improbable process.

4.8 CONFIGURATIONAL ENTROPY AND THERMAL ENTROPY

In the preceding discussion entropy has been considered in the light of the number of ways in which energy can be mixed among identical particles, and the given example of a mixing process involved the redistribution of thermal energy among the particles of two closed systems placed in thermal contact. The entropy changes accompanying this redistribution are thus changes in the *thermal entropy*. Entropy can also be considered in light of the number of ways in which the particles themselves can be mixed or distributed over the available positions in space. Such consideration gives rise to the concept of *configurational entropy*.

Consider two crystals at the same temperature and pressure, one containing atoms of the element A and the other containing atoms of the element B. When the two crystals are placed in contact with one another, the spontaneous process which occurs is the diffusion of A atoms into the crystal B and the diffusion of B atoms into the crystal A. As this is a spontaneous process, the entropy of the system will increase; and intuitively, it might be predicted that equilibrium will be reached (i.e., the entropy of the system will reach a maximum value) when the diffusion processes have occurred to such an extent that the concentration gradients in the system have been eliminated. (Compare, in Sec. 4.7, equilibrium being reached when heat flow has occurred to the extent that the temperature gradients in the system have been eliminated.)

Following the discussion of Denbigh,* consider a crystal containing four atoms of A placed in contact with a crystal containing four atoms of B. The initial state of this system, in which all four atoms of A lie to the left of XY and all four of the B atoms lie to the right of XY, is shown in Fig. 4.5.

The number of distinguishable ways in which this arrangement can be realized is unity, as interchange among the identical A atoms on the left of XY and/or interchange among the identical B atoms on the right of XY does not produce a different configuration. Thus we can write

$$\Omega_{4:0} = 1$$

(The notation corresponds to four A atoms on the left of XY and none on the right.)

When one A atom is interchanged across XY with one B atom, the B atom can be located on any of four sites, and hence the left side can be realized in four different ways. Similarly the exchanged A atom can be located on any of four sites, and hence the right side can be realized in four different ways. As any of

*K. Denbigh, "The Principles of Chemical Equilibrium," 2nd ed., p. 48, Cambridge University Press, Cambridge, England, 1966.

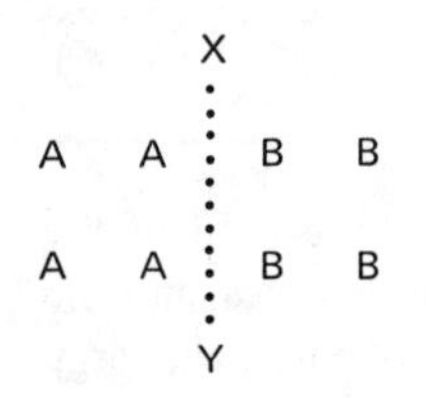

Fig. 4.5. Representation of a crystal of A in contact with a crystal of B.

the four former arrangements can be combined with any of the four latter arrangements, the total number of distinguishable configurations of the arrangement 3:1 is $4 \times 4 = 16$; that is,

$$\Omega_{3:1} = 16$$

When a second A is exchanged with a second B atom across XY, the first B atom on the left of XY can be located in any of four positions, and the second B atom can be located in any of the three remaining positions, thus giving, $4 \times 3 = 12$ configurations. However these 12 configurations include those which occur as the result of interchange between the two B atoms themselves. As such configurations are not distinguishable, they must be discounted; and so the number of distinguishable configurations on the left of XY is $4 \times 3/2! = 6$. Similarly six distinguishable arrangements occur on the right of XY, and hence the total number of configurations of the arrangement 2:2 is $6 \times 6 = 36$; that is,

$$\Omega_{2:2} = 36$$

When a third A atom is exchanged with a third B atom across XY, the first A atom can be located on any of four sites, the second on any of the three remaining sites, and the third on either of the remaining two sites. Factoring out the number of indistinguishable configurations due to interchange of the three B atoms among themselves gives the number of distinguishable configurations as $4 \times 3 \times 2/3! = 4$. Similarly four distinguishable configurations occur to the right of XY, and hence

$$\Omega_{1:3} = 16$$

Interchange of the final A and B atoms across XY gives

$$\Omega_{0:4} = \left(\frac{4 \times 3 \times 2}{4!}\right) \times \left(\frac{4 \times 3 \times 2}{4!}\right) = 1$$

Thus the total number of spatial configurations available to the system is $1 + 16 + 36 + 16 + 1 = 70$, which is the number of distinguishable ways in which four particles of one kind and four particles of another kind can be arranged on eight sites; i.e.

$$\text{number of ways} = \frac{8!}{4!4!} = 70$$

If, as before, it is assumed that, in the absence of any reason to the contrary, each of these 70 configurations is equally probable, then the probability of finding the system in the arrangement 4:0 is 1/70, in the arrangement 3:1 is 16/70, in the arrangement 2:2 is 36/70, in the arrangement 1:3 is 16/70, and in the arrangement 0:4 is 1/70. Arrangement 2:2 is thus seen to be the most probable and thus corresponds to the equilibrium state. This arrangement also corresponds to the elimination of the concentration gradients, as was to be expected.

Again as

$$S = k \ln \Omega$$

it is seen that maximization of Ω maximizes the entropy. In this case the entropy increase occurs as the result of the increase in the number of spatial configurations which become available to the system when the crystals A and B are placed in contact with one another. The increase in entropy of the system results from an increase in the configurational entropy, S_{conf}. The mixing process can be written as

$$A + B \text{ (unmixed)} \longrightarrow A + B \text{ (mixed) at constant } U, V \text{ and } n$$

i.e.,

$$\text{state (1)} \longrightarrow \text{state (2)}$$

$$\Delta S_{\text{conf}} = S_{\text{conf(2)}} - S_{\text{conf(1)}} = k \ln \Omega_{\text{conf(2)}} - k \ln \Omega_{\text{conf(1)}}$$

$$= k \ln \left(\frac{\Omega_{\text{conf(2)}}}{\Omega_{\text{conf(1)}}} \right)$$

and if n_a atoms of A are mixed with n_b atoms of B, then

$$\Omega_{\text{conf(2)}} = \frac{(n_a + n_b)!}{n_a! n_b!} \quad \text{and} \quad \Omega_{\text{conf(1)}} = 1$$

Thus

$$\Delta S_{\text{conf}} = k \ln \frac{(n_a + n_b)!}{n_a! n_b!} \tag{4.17}$$

The total entropy of a system can thus be considered as comprising two contributions, the thermal entropy S_{th}, which arises from the number of ways in which the energy of the system can be shared among the particles, and the configurational entropy S_{conf}, which arises from the number of distinguishable ways in which the particles of the system can be mixed over positions in space. Thus

$$\begin{aligned} S_{\text{total}} &= S_{\text{th}} + S_{\text{conf}} \\ &= k \ln \Omega_{\text{th}} + k \ln \Omega_{\text{conf}} \\ &= k \ln \Omega_{\text{th}} \Omega_{\text{conf}} \end{aligned}$$

The number of spatial configurations available to two closed systems placed in thermal contact, or to two open chemically identical systems placed in thermal contact, is unity. Thus in the case of heat flow down a temperature gradient between two such systems, as only Ω_{th} changes, the entropy increase resulting from heat flow which takes the system from state 1 to state 2 is

$$\Delta S_{\text{total}} = k \ln \frac{\Omega_{\text{th}(2)}\Omega_{\text{conf}(2)}}{\Omega_{\text{th}(1)}\Omega_{\text{conf}(1)}} = k \ln \frac{\Omega_{\text{th}(2)}}{\Omega_{\text{th}(1)}} = \Delta S_{\text{th}}$$

Similarly in the mixing of particles of A with particles of B, ΔS_{total} only equals ΔS_{conf} if this mixing process does not effect a redistribution of the particles among the energy levels, i.e., if $\Omega_{\text{th}(1)}$ equals $\Omega_{\text{th}(2)}$.

Such a situation corresponds to "ideal mixing" and requires that the quantization of energy within the two initial crystals be identical. Such a situation is the exception rather than the rule, and generally when two or more pure elements are mixed at constant U, V, and n, $\Omega_{\text{th}(2)} \neq \Omega_{\text{th}(1)}$, in which case complete spatial randomization of the constituent particles does not occur. In such cases, either clustering of like particles (indicating difficulty in mixing) or ordering (indicating a tendency towards compound formation) occurs. In all cases, however, the equilibrium state of the mixed system is that which, at constant U, V, and n, maximizes the product $\Omega_{\text{th}} \Omega_{\text{conf}}$.

4.9 SUMMARY

1. A single macrostate of a system, which is determined when the independent variables of the system are fixed, contains an exceedingly large number of microstates. Each microstate is characterized by the particular manner in which the thermal energy of the system is distributed among the constituent particles and the particular manner in which the particles themselves are distributed over the space available to them.

2. Although the occurrence of a system in any one of its microstates is equally probable, greatly differing numbers of microstates occur in differing

distributions. That single distribution which contains the maximum number of microstates is the most probable distribution; and in real systems the number of microstates occurring within this most probable distribution is considerably greater than the sum of all of the other microstates occurring in all the other distributions. This most probable distribution, in corresponding to the most probable state of existence of the system, is the equilibrium thermodynamic state of the system.

3. The relationship between the number of microstates available to the system, Ω, and the entropy of the system is given by the Boltzmann equation as $S = k \ln \Omega$, where k is Boltzmann's constant. Thus if a situation arises whereby the number of microstates available to the system is increased, then spontaneous redistribution of the energy among the particles (or particles over the available space) occurs until the "newly available" most probable distribution occurs. From the Boltzmann equation it is seen that an increase in the number of microstates made available to the system results in an increase in the entropy of the system.

4. The total entropy of a system comprises two contributions, namely, the thermal entropy S_{th} and the configurational entropy S_{conf}. The former arises from the number of ways in which the available thermal energy of the system can be shared among the constituent particles, Ω_{th}, and the latter arises from the number of ways in which the particles can be distributed over the space available to them, Ω_{conf}. As any of the former configurations can be combined with any of the latter, the total number of microstates available to the system is the product $\Omega_{th}\Omega_{conf}$; and hence, from the logarithmic form of Boltzmann's equation, the total entropy is the sum of S_{th} and S_{conf}.

5. As a result of the use of the postulates of the quantum theory and the methods of statistical mechanics, it is possible to obtain a correspondence between the physical existence of a system (considered to be an assemblage of individual particles) and the entropy of the system. This correspondence provides a physical quality to the thermodynamic function entropy.

PROBLEMS

4.1 A rigid container is divided into two compartments of equal volume by a partition. One compartment contains 1 mole of ideal gas A at 1 atm, and the other compartment contains 1 mole of ideal gas B at 1 atm. Calculate the entropy increase in the container if the partition between the two compartments is removed. If the first compartment had contained 2 moles of ideal gas A, what would have been the entropy increase due to gas mixing when the partition was removed? Calculate the corresponding entropy changes in each of the above two situations if both compartments had contained ideal gas A.

4.2 If a silver-gold alloy is a random mixture of gold and silver atoms, calculate the entropy increase when 10 grams of gold are mixed with 20 grams of silver to form an ideal homogeneous alloy. The atomic weights of Au and Ag are 197 and 107.88 respectively.

CHAPTER FIVE

AUXILIARY FUNCTIONS

5.1 INTRODUCTION

The power of thermodynamics lies in its provision of the criteria for equilibrium within a system and its ability to facilitate determination of the effect, on the equilibrium state, of a change in the external influences which can be brought to bear on the system. The practical usefulness of this power is consequently determined by the practicality of the equations of state of the system, i.e., the relationships among the state functions, which can be established.

The combination of the First and Second Laws of Thermodynamics leads to the derivation of Eq. (3.12),

$$dU = TdS - PdV$$

This equation of state gives the relationship between the dependent variable U and the independent variables S and V for a closed system of fixed composition which is in a state of equilibrium and is undergoing a process involving volume change against the external pressure as the only form of work performed on, or by, the system. Combination of the First and Second Laws also provides the criteria for equilibrium that (1) for a system of constant internal energy and constant volume, the entropy is a maximum, and (2) for a system of constant entropy and constant volume, the internal energy is a minimum.

The further development of thermodynamics stems, in part, from the fact that S and V are an inconvenient choice of independent variable from the point of view of experimental measurement or control. Although the volume of a system can be measured with relative ease and, with sufficient ingenuity on the part of the experimenter, can be controlled, entropy can be neither simply measured nor simply controlled. It is thus, firstly, desirable to develop a simple equation, similar in form to Eq. (3.12) which contains a more convenient choice of independent variables. From the experimentalist's point of view the most convenient pair of independent variables would be temperature and pressure, as these two variables are most easily measured and controlled in a practical experiment. The derivation of an equation of state of the same simple form as Eq. (3.12), but involving P and T as the independent variables, and a criterion for equilibrium in a constant-temperature, constant-pressure system is thus desirable. Alternatively, from the theoretician's viewpoint the choice of V and T as the independent variables would be most convenient, as constant T constant V processes are most amenable to theoretical calculation using the methods of statistical mechanics. This is because the fixing of the volume of a closed system fixes the quantization of its energy (i.e., fixes the ϵ_i value); hence the Boltzmann factor, exp $(-\epsilon_i/kT)$, and the partition function, both of which occur in Eq. (4.13), have constant values in constant-volume, constant-temperature processes. The derivation of an equation of state involving T and V as the independent variables, and the establishment of the criterion for equilibrium in a system of fixed volume and fixed temperature are thus desirable.

Secondly, Eq. (3.12) cannot be applied to systems which undergo compositional changes arising from the occurrence of chemical reactions within the system, nor can it be applied to systems which do work other than simple work of volume change against the external pressure (so-called P-V work). As systems which undergo compositional change, e.g., the transfer of an impurity from a metal being refined to the refining slag, or the precipitation of a second phase in an initially homogeneous alloy, are of prime importance to the metallurgist, it is necessary to include composition variables in any equation of state and in any criterion of equilibrium. Similarly it is necessary to establish a means of dealing with forms of work other than P-V work which a system may perform, e.g., electrical work performed by a galvanic cell.

Thus, although Eq. (3.12) represents the foundation of thermodynamics, in order that the practical usefulness of this equation be fully exercised it is necessary to invent auxiliary thermodynamic functions of state which, as dependent variables, are related in simple form to more convenient choices of independent variables. Also, with this increase in the number of thermodynamic functions it is necessary to establish the relationships which exist among them; it often occurs that some required thermodynamics expression which itself is not amenable to experimental measurement is related simply to an expression which is easily measurable. Examples of this have been seen already in Chap. 3, where it was found that $(\partial U/\partial S)_V = T$, $-(\partial U/\partial V)_S = P$, and $(\partial S/\partial V)_U = P/T$.

In this chapter the thermodynamic functions A (the Helmholtz free energy, or the work function), G (the Gibbs free energy), and μ_i (the chemical potential of the species i) are introduced and their particular properties and interrelationships are examined. It is interesting to note, in passing, that Slater* has calculated that the relationships among the thermodynamic functions P, V, T, S, U, H, A, and G occur in 521,631,180 separate formulas. It should be stated immediately, however, that only the more useful of these formulas will be discussed.

The functions A and G are defined as

$$A = U - TS \tag{5.1}$$

and

$$\begin{aligned} G &= U + PV - TS \\ &= H - TS \end{aligned} \tag{5.2}$$

As A and G are combinations of state properties, they are themselves state properties.

5.2 THE ENTHALPY *H*

As has been seen in Chap. 2, for a closed system undergoing a change of state, at constant pressure P, from the state 1 to the state 2, the First Law gives

$$U_2 - U_1 = q_p - P(V_2 - V_1)$$

from which

$$(U_2 + PV_2) - (U_1 + PV_1) = q_p$$

and hence

$$\Delta H = H_2 - H_1 = q_p \tag{2.5}$$

Thus, in a constant-pressure process, the enthalpy change of the system equals the heat entering or leaving the system. It is to be emphasized that Eq. (2.5) is only applicable to a system which does only P-V work.

The properties of H are examined in detail in Chap. 6.

5.3 THE HELMHOLTZ FREE ENERGY *A*

For a system undergoing a process from state 1 to state 2, Eq. (5.1) gives

$$(A_2 - A_1) = (U_2 - U_1) - (T_2S_2 - T_1S_1)$$

*J. C. Slater, "Introduction to Chemical Physics," p. 24, McGraw-Hill Book Company, New York, 1939.

If the system is closed, then

$$(U_2 - U_1) = q - w$$

and hence

$$(A_2 - A_1) = q - w - (T_2S_2 - T_1S_1)$$

If the process is isothermal–that is, $T_1 = T_2 = T$, the temperature of the reservoir which supplies or withdraws heat during the process–then from the Second Law

$$q \leqslant T(S_2 - S_1)$$

Hence

$$(A_2 - A_1) \leqslant -w$$

Comparison with Eq. (3.11) indicates that the equality can be written as

$$(A_2 - A_1) + T\Delta S_{irr} = -w \tag{5.3}$$

and thus, for a reversible isothermal process (which necessarily involves no entropy production) the amount of work done by the system, w_{max}, equals the decrease in the work function A. Further, for an isothermal process at constant volume, which necessarily performs no P-V work, Eq. (5.3) gives

$$(A_2 - A_1) + T\Delta S_{irr} = 0 \tag{5.4}$$

or, for an increment of such a process,

$$dA + TdS_{irr} = 0$$

Such a process could only occur spontaneously if dA were a negative quantity; i.e., entropy could be produced only if A decreased. And as the condition for thermodynamic equilibrium is that $dS_{irr} = 0$, then with respect to the described process, equilibrium is defined by the condition that

$$dA = 0 \tag{5.5}$$

Thus in a closed system, held at constant T and V, the Helmholtz free energy can only decrease or remain constant, and the attainment of equilibrium in the system coincides with the system having a minimum value of A, consistent with the fixed values of T and V. Consideration of A thus provides a criterion of equilibrium for a system at constant temperature and volume.

This criterion for equilibrium can be illustrated by examination of the following system. Consider n atoms of some element occurring in both a solid crystalline phase and a vapor phase contained in a constant-volume vessel which, in turn, is immersed in a constant-temperature heat reservoir. The problem involves the determination of the equilibrium state of this system, i.e., the equilibrium distribution of the n atoms between the solid phase and the vapor phase. At constant volume and constant temperature this distribution must be such that the Helmholtz free energy of the system is a minimum. From Eq. (5.1),

$$A = U - TS$$

and hence low values of A are obtained with low values of U and high values of S. The two extreme states of existence which are available to the system are (1) that in which all n atoms occur in the solid crystalline phase and none occur in the vapor phase, and (2) that in which all n atoms occur in the vapor phase, and the solid phase is absent. Consider the system occurring in the first of these two states. In the solid crystalline phase the atoms are held together by interatomic forces; if an atom is to be removed from the crystal surface and placed in the vacuum above the crystal (in which case what initially was a vacuum becomes the vapor phase), then work must be done against the attractive force existing between the crystal and the atom being removed. For this separation to be conducted isothermally, the energy required for the separation must be supplied as heat which flows from the heat reservoir into the system. This flow of heat into the system, in turn, increases both the internal energy and the entropy of the system. (As a system of [n-1] atoms in the solid phase and 1 atom in the vapor phase is more random than a system of n atoms in the solid phase, it is seen physically that the entropy of the former is greater than that of the latter.)

As the evaporation process continues, i.e., as more atoms are removed from the solid phase and placed in the vapor phase, heat continues to flow from the heat reservoir into the system, and the internal energy and entropy of the system continue to increase. Eventually, when all n atoms occur in the vapor phase, both the internal energy and the entropy of the constant-volume, constant-temperature system have their maximum values, in comparison with having minimum values when all n atoms occur in the solid phase. The variation of the entropy of the system, S, and the internal energy of the system, U, with n_v, the number of atoms in the vapor phase, can be calculated by the methods of statistical mechanics.* These relationships are shown in Fig. 5.1*a* and *b*,

*See K. G. Denbigh, "The Principles of Chemical Equilibrium," 2nd ed., chap. 13, Cambridge University Press, Cambridge, England, 1966.

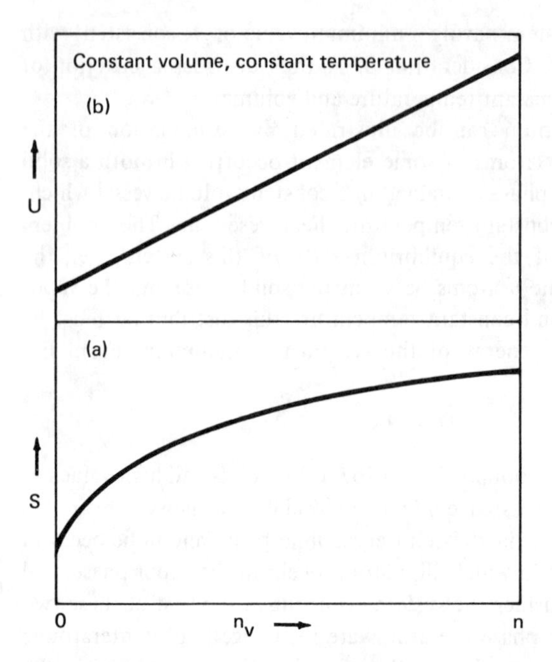

Fig. 5.1. The variations of internal energy and entropy with the number of atoms in the vapor phase of a closed solid-vapor system at constant volume and temperature.

respectively. As the transfer of an atom from the solid phase to the vapor results in a fixed increment in the internal energy of the system, U increases linearly with n_v as shown in Fig. 5.1*b*, but Fig. 5.1*a* shows that the corresponding entropy increase is nonlinear; the rate of increase of S with n_v decreases as n_v increases. The variation of A, which is obtained as the sum of the terms U and $-TS$, with n_v is shown in Fig. 5.2. This figure shows that A has a minimum value at a unique value of n_v, designated $n_{v(\text{eq},T)}$. This state represents a compromise between minimization of U and maximization of S, and in this state the solid exerts its equilibrium vapor pressure at the temperature T. If the vapor behaves ideally, then the vapor pressure, which is termed the saturated vapor pressure, is calculated as

$$p = \frac{n_{v(\text{eq},\,T)}kT}{(V - V_s)}$$

where V is the volume of the containing vessel, V_s is the volume of solid phase present, and k is Boltzmann's constant.

As the magnitude of the entropy contribution, $-TS$, is temperature-dependent and the internal energy contribution is independent of temperature, the entropy contribution becomes increasingly predominant as the temperature is increased, and the compromise between U and $-TS$ which minimizes A occurs at larger values of n_v. This is illustrated in Fig. 5.3, which is drawn for the temperatures T_1 and T_2 where $T_1 < T_2$. An increase in the temperature from T_1 to T_2 increases the saturated vapor pressure from

$$p_{(\text{at } T_1)} = \frac{n_{v(\text{eq}, T_1)} k T_1}{[V - V_{s(\text{at } T_1)}]}$$

to

$$p_{(\text{at } T_2)} = \frac{n_{v(\text{eq}, T_2)} k T_2}{[V - V_{s(\text{at } T_2)}]}$$

As will be seen in Chap. 7, the saturated vapor pressure increases exponentially with increasing temperature.

For the constant-volume system the maximum temperature at which both solid and vapor phases occur is that temperature at which minimization of A occurs at $n_v = n$. Above this temperature the entropy contribution overwhelms the internal energy contribution, and hence all n atoms occur in the vapor phase.

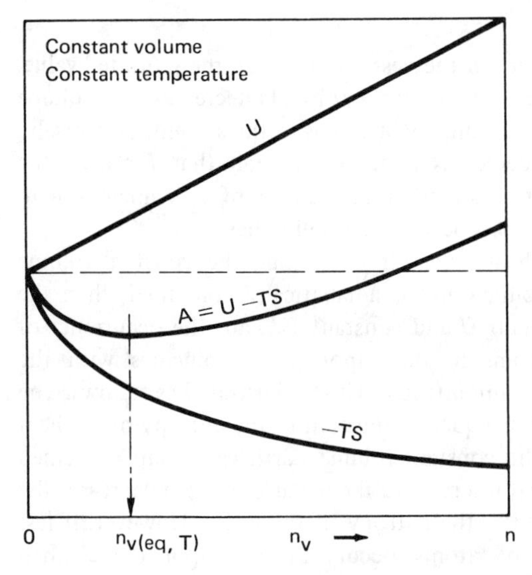

Fig. 5.2. Illustration of the criterion of equilibrium in a closed solid-vapor system of constant volume at constant temperature.

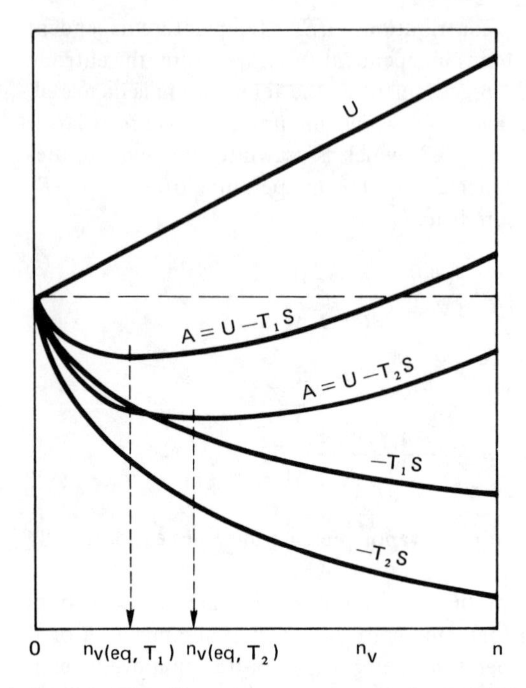

Fig. 5.3. The effect of temperature on the équilibrium state of a closed solid-vapor system of constant volume.

At such a temperature the pressure in the vessel is less than the saturated value, and the solid phase could only be made to reappear by (1) decreasing the volume of the system or (2) increasing the number of atoms in the system. Conversely, as T decreases, then $n_{v(\text{eq},T)}$ decreases and, in the limit that $T = 0$ K, the entropy contribution to A vanishes and minimization of A coincides with minimization of U, that is, all n atoms occur in the solid phase.

If the constant-temperature heat reservoir containing the constant-volume system is, itself, of constant volume and is adiabatically contained, then the combined system is one of constant U and constant V. Thus the occurrence of the equilibrium number of atoms in the vapor phase coincides with the combined system having a maximum entropy. This is illustrated as follows. The entropy of the combined system is equal to the sum of the entropy of the heat reservoir and the entropy of the constant-volume particles system contained within it. Heat flow from the heat reservoir to the particles system decreases the entropy of the former and increases the entropy of the latter. However, if less than the equilibrium number of atoms occurs in the vapor phase, then spontaneous evaporation of the solid occurs until the saturated vapor pressure is reached. During this process heat q flows spontaneously from the heat reservoir, the entropy of the reservoir decreases by the amount q/T, the entropy of the particles system increases by the amount $q/T + \Delta S_{irr}$, and the increase in the entropy of the combined system is ΔS_{irr}. From Eq. (5.4), the corresponding decrease in the Helmholtz free energy is

$$\Delta A = -T\Delta S_{irr}$$

and hence minimization of A in the constant-temperature, constant-volume particles system corresponds to maximization of S in the constant U, constant V combined system. Similarly if more than the equilibrium number of atoms occurs in the vapor phase, then spontaneous condensation occurs until equilibrium is reached. During this process, heat flows from the particles system into the heat reservoir, the entropy of the latter is increased by q/T, the entropy of the former is decreased by $q/T - \Delta S_{irr}$, and the entropy of the combined system is again increased by ΔS_{irr}.

It must be pointed out that as Classical Thermodynamics is only concerned with equilibrium states, only that single state, at the given T and V, in which A is a minimum, is of interest from the point of view of thermodynamic consideration. Although from the viewpoint of statistical mechanics, all values of n_v in the range $0 \leqslant n_v \leqslant n$ are possible at the given T and V, the probability that n_v deviates by even the smallest amount from the value $n_{v(eq,T)}$ is exceedingly small. This probability is small enough that, in practical terms, it corresponds to the thermodynamic statement that spontaneous deviation of a system from its equilibrium state is impossible.

5.4 THE GIBBS FREE ENERGY *G*

For a system undergoing a change of state from 1 to 2, Eq. (5.2) gives

$$\begin{aligned}(G_2 - G_1) &= (H_2 - H_1) - (T_2S_2 - T_1S_1)\\ &= (U_2 - U_1) + (P_2V_2 - P_1V_1) - (T_2S_2 - T_1S_1)\end{aligned}$$

For a closed system, the First Law gives

$$(U_2 - U_1) = q - w$$

and thus,

$$(G_2 - G_1) = q - w + (P_2V_2 - P_1V_1) - (T_2S_2 - T_1S_1)$$

If the process is carried out such that $T_1 = T_2 = T$, where T is the temperature of the reservoir which supplies or withdraws heat from the system, and also if

$P_1 = P_2 = P$, where P is the constant pressure at which the surroundings have undergone a volume change, then

$$(G_2 - G_1) = q - w + P(V_2 - V_1) - T(S_2 - S_1) \tag{5.6}$$

In the expression of the First Law, the work w is the *total* work done by the system during the process; i.e., if the system performs chemical or electrical work in addition to the work of expansion against the external pressure, then these work terms are included in w. Hence w can be written as

$$w = w' + P(V_2 - V_1)$$

where $P(V_2 - V_1)$ is the work done in volume change against the constant external pressure P, and w' is the sum of all the forms of work other than the work of expansion.

Substituting in Eq. (5.6) gives

$$(G_2 - G_1) = q - w' - T(S_2 - S_1)$$

and again, as

$$q \leqslant T(S_2 - S_1)$$

then

$$w' \leqslant -(G_2 - G_1) \tag{5.7}$$

Again the equality can be written

$$-w' = (G_2 - G_1) + T\Delta S_{irr} \tag{5.8}$$

In the case of an isothermal, isobaric process, during which no work other than that of expansion is performed, that is, $w' = 0$, then,

$$(G_2 - G_1) + T\Delta S_{irr} = 0$$

Such a process can only occur spontaneously (with resultant entropy production) if the free energy decreases. As the condition for thermodynamic equilibrium is that $dS_{irr} = 0$, then with respect to an increment of the isothermal, isobaric process, equilibrium is defined by the condition that

$$dG = 0 \tag{5.9}$$

Thus, for a system undergoing a process at constant T and P, the Gibbs free energy G can only decrease or remain constant, and hence the attainment of equilibrium in the system coincides with the system having the minimum value of G consistent with the fixed values of P and T. Consideration of G thus provides a criterion of equilibrium which is of considerable practical use. This criterion of equilibrium will be used extensively in the subsequent chapters.

5.5 SUMMARY OF EQUATIONS FOR A CLOSED SYSTEM

Equation (3.12) gives

$$dU = TdS - PdV$$

Now,

$$H = U + PV \text{ and thus } dH = TdS + VdP \tag{5.10}$$

$$A = U - TS \text{ and thus } dA = -SdT - PdV \tag{5.11}$$

$$G = H - TS \text{ and thus } dG = -SdT + VdP \tag{5.12}$$

5.6 VARIATION OF THE COMPOSITION AND SIZE OF THE SYSTEM

Thus far discussion has been restricted to closed systems of fixed size and composition, i.e., to systems containing a fixed number of moles of one component. In such cases it was found that the system had two independent variables which, when fixed, uniquely fixed the state of the system. However, if the size and/or composition can vary during a process, then specification of only two variables is no longer sufficient to fix the state of the system. For example it has been shown that, for a constant temperature and pressure process, equilibrium is attained when G is a minimum. If the composition of the system is variable, i.e., the numbers of moles of the various species present can vary as the result of the occurrence of a chemical reaction, then minimization of G at constant T and P occurs only when the system has a unique composition. For example, if the system contained the gaseous species CO, CO_2, and O_2, then at constant T and P minimization of G would occur when the reaction equilibrium $CO + \frac{1}{2}O_2 = CO_2$ was established. Similarly as G is an extensive property, i.e., is dependent on the size of the system, it is necessary that the number of moles within the system be specified.

G is thus a function of T, P, and the numbers of moles of all the species present in the system; i.e.,

$$G = G(T, P, n_i, n_j, n_k, \ldots) \tag{5.13}$$

where $n_i, n_j, n_k, \ldots$ are the numbers of moles of the species $i, j, k, \ldots$ present in the system, and the state of the system is fixed only when all of the independent variables are fixed. Differentiation of Eq. (5.13) gives

$$dG = \left(\frac{\partial G}{\partial T}\right)_{P, n_i, n_j, \ldots} dT + \left(\frac{\partial G}{\partial P}\right)_{T, n_i, n_j, \ldots} dP + \left(\frac{\partial G}{\partial n_i}\right)_{T, P, n_j, n_k, \ldots} dn_i$$

$$+ \left(\frac{\partial G}{\partial n_j}\right)_{T, P, n_i, n_k, \ldots} dn_j + \text{etc.} \qquad (5.14)$$

If the mole numbers of the various species remain constant during the process, then Eq. (5.14) simplifies to Eq. (5.12); i.e.,

$$dG = -SdT + VdP$$

from which it is seen that

$$\left(\frac{\partial G}{\partial T}\right)_{P, n_i, n_j, \ldots} = -S$$

and

$$\left(\frac{\partial G}{\partial P}\right)_{T, n_i, n_j, \ldots} = V$$

Substitution into Eq. (5.14) gives,

$$dG = -SdT + VdP + \sum_{i=1}^{i=k} \left(\frac{\partial G}{\partial n_i}\right)_{T, P, n_j, \ldots} dn_i \qquad (5.15)$$

where

$$\sum_{i=1}^{i=k} \left(\frac{\partial G}{\partial n_i}\right)_{T, P, n_j, \ldots} dn_i$$

is the sum of k terms (one for each of the k species) each of which is obtained by differentiating G with respect to the number of moles of the ith species at constant T, P, and n_j, where n_j represents the numbers of moles of every species other than the ith species.

5.7 THE CHEMICAL POTENTIAL

The term $(\partial G/\partial n_i)_{T, P, n_j, \ldots}$ is called the *chemical potential of the species i*, and is designated as μ_i; that is,

$$\left(\frac{\partial G}{\partial n_i}\right)_{T, P, n_j, \ldots} \equiv \mu_i \tag{5.16}$$

μ_i, the chemical potential of the species i in a homogeneous phase, is formally defined as the increase in the Gibbs free energy of the system (the homogeneous phase) for an infinitesimal addition of the species i, per mole of i added, with the addition being made at constant T and P and numbers of moles of all the other species present. Alternatively, if the system is large enough that the addition of 1 mole of i, at constant T and P, does not measurably change the composition of the system, then μ_i is the increase in G for the system accompanying the addition of 1 mole of i. Thus μ_i is the amount by which the capacity of the system for doing work, other than the work of expansion, is increased, per mole of i added at constant T, P, and composition.

Equation (5.15) can thus be written as

$$dG = -SdT + VdP + \sum_1^k \mu_i dn_i \tag{5.17}$$

in which form G is expressed as a function of T, P, and composition. Equation (5.17) can thus be applied to open systems which exchange matter as well as heat with their surroundings and to closed systems which undergo composition changes.

Similarly Eqs. (3.12), (5.10), and (5.11) can be made applicable to open systems by including the terms describing the composition dependences of the values of U, H, and A, respectively.

$$dU = TdS - PdV + \sum_1^k \left(\frac{\partial U}{\partial n_i}\right)_{S, V, n_j, \ldots} dn_i \tag{5.18}$$

$$dH = TdS + VdP + \sum_1^k \left(\frac{\partial H}{\partial n_i}\right)_{S, P, n_j, \ldots} dn_i \tag{5.19}$$

$$dA = -SdT - PdV + \sum_1^k \left(\frac{\partial A}{\partial n_i}\right)_{T, V, n_j, \ldots} dn_i \tag{5.20}$$

Inspection of Eqs. (5.16), (5.18), (5.19), and (5.20) indicates that

$$\left(\frac{\partial G}{\partial n_i}\right)_{T, P, n_j} \equiv \mu_i = \left(\frac{\partial U}{\partial n_i}\right)_{S, V, n_j} = \left(\frac{\partial H}{\partial n_i}\right)_{S, P, n_j} = \left(\frac{\partial A}{\partial n_i}\right)_{T, V, n_j} \tag{5.21}$$

and hence the complete set of equations is

$$dU = TdS - PdV + \Sigma\mu_i dn_i \tag{5.22}$$

$$dH = TdS + VdP + \Sigma\mu_i dn_i \tag{5.23}$$

$$dA = -SdT - PdV + \Sigma\mu_i dn_i \tag{5.24}$$

$$dG = -SdT + VdP + \Sigma\mu_i dn_i \tag{5.25}$$

U is thus the "characteristic function" of the independent variables S, V, and composition; H is the characteristic function of the independent variables S, P, and composition; A is the characteristic function of the independent variables T, V, and composition; and G is the characteristic function of the independent variables T, P, and composition. Although all four of the above equations are basic in nature, Eq. (5.25), as a result of its particular usefulness, for reference purposes, is termed the *fundamental equation*.

The First Law gives

$$dU = \delta q - \delta w$$

and comparison with Eq. (5.22) indicates that, for a closed system which undergoes a process involving a reversible change of composition (e.g., a reversible chemical reaction),

$$\delta q = TdS$$
$$\delta w = PdV + \Sigma\mu_i dn_i$$

Thus the term $\Sigma\mu_i dn_i$ is the chemical work done by the system which was denoted as w' in Eq. (5.8), and the total work w is the sum of the work of volume change and the chemical work.

5.8 THERMODYNAMIC RELATIONS

From Eqs. (5.22) to (5.25), the following relationships are obtained:

$$T = \left(\frac{\partial U}{\partial S}\right)_{V,\,\text{comp}} = \left(\frac{\partial H}{\partial S}\right)_{P,\,\text{comp}} \tag{5.26}$$

$$P = -\left(\frac{\partial U}{\partial V}\right)_{S,\,\text{comp}} = -\left(\frac{\partial A}{\partial V}\right)_{T,\,\text{comp}} \tag{5.27}$$

$$V = \left(\frac{\partial H}{\partial P}\right)_{S,\,\text{comp}} = \left(\frac{\partial G}{\partial P}\right)_{T,\,\text{comp}} \tag{5.28}$$

$$S = -\left(\frac{\partial A}{\partial T}\right)_{V,\,\mathrm{comp}} = -\left(\frac{\partial G}{\partial T}\right)_{P,\,\mathrm{comp}} \tag{5.29}$$

5.9 MAXWELL'S RELATIONS

If Z is a state function and x and y are the chosen independent variables for a closed system of fixed composition, then

$$Z = Z(x, y)$$

differentiation of which gives

$$dZ = \left(\frac{\partial Z}{\partial x}\right)_y dx + \left(\frac{\partial Z}{\partial y}\right)_x dy$$

If the partial derivative $(\partial Z/\partial x)_y$ is itself a function of x and y, being given as $(\partial Z/\partial x)_y = L(x, y)$ and similarly the partial derivative $(\partial Z/\partial y)_x = M(x, y)$, then

$$dZ = Ldx + Mdy$$

Thus

$$\left[\frac{\partial}{\partial y}\left(\frac{\partial Z}{\partial x}\right)_y\right]_x = \left(\frac{\partial L}{\partial y}\right)_x$$

and

$$\left[\frac{\partial}{\partial x}\left(\frac{\partial Z}{\partial y}\right)_x\right]_y = \left(\frac{\partial M}{\partial x}\right)_y$$

But as Z is a state function, the change in Z is independent of the order of differentiation, i.e.,

$$\left[\frac{\partial}{\partial y}\left(\frac{\partial Z}{\partial x}\right)_y\right]_x = \left[\frac{\partial}{\partial x}\left(\frac{\partial Z}{\partial y}\right)_x\right]_y = \frac{\partial^2 Z}{\partial x \partial y}$$

and hence

$$\left(\frac{\partial L}{\partial y}\right)_x = \left(\frac{\partial M}{\partial x}\right)_y \tag{5.30}$$

Application of Eq. (5.30) to Eqs. (3.12) and (5.10) to (5.12) gives a set of equations which are known as Maxwell's relations. These are:

$$\left(\frac{\partial T}{\partial V}\right)_S = -\left(\frac{\partial P}{\partial S}\right)_V \tag{5.31}$$

$$\left(\frac{\partial T}{\partial P}\right)_S = \left(\frac{\partial V}{\partial S}\right)_P \tag{5.32}$$

$$\left(\frac{\partial S}{\partial V}\right)_T = \left(\frac{\partial P}{\partial T}\right)_V \tag{5.33}$$

$$\left(\frac{\partial S}{\partial P}\right)_T = -\left(\frac{\partial V}{\partial T}\right)_P \tag{5.34}$$

Similarly, equations can be obtained by considering compositional variation. The value of the above relations lies in the fact that they contain many experimentally determinable quantities. An example of the use of Maxwell's equations was given in the first numerical example in Sec. 3.17, where, by Eq. (5.33), the variation of S with V at constant T was determined from a knowledge of the variation of P with T at constant V. A similar example of the use of this equation is as follows. For a closed system of fixed composition, Eq. (3.12) gives

$$dU = TdS - PdV$$

Thus

$$\left(\frac{\partial U}{\partial V}\right)_T = T\left(\frac{\partial S}{\partial V}\right)_T - P$$

Use of Maxwell's relation, Eq. (5.33) allows this to be written as

$$\left(\frac{\partial U}{\partial V}\right)_T = T\left(\frac{\partial P}{\partial T}\right)_V - P \tag{5.34b}$$

which is an equation of state relating the internal energy U of a closed one-component system to the experimentally measurable quantities T, P, and V. If the system is an ideal gas, i.e., one which obeys the equation of state $PV = RT$, then as $(\partial P/\partial T)_V = R/V$ and $P = RT/V$, it is seen that $(\partial U/\partial V)_T = 0$, indicating thus that the internal energy of an ideal gas is independent of the volume of the gas.

Similarly, for a closed system of fixed composition, Eq. (5.10) gives $dH = T\,dS + V\,dP$, in which case

$$\left(\frac{\partial H}{\partial P}\right)_T = T\left(\frac{\partial S}{\partial P}\right)_T + V$$

or, as from Eq. (5.34), $(\partial S/\partial P)_T = -(\partial V/\partial T)_P$,

$$\left(\frac{\partial H}{\partial P}\right)_T = -T\left(\frac{\partial V}{\partial T}\right)_P + V$$

This is an equation of state which relates the variation of enthalpy to the variation of the experimentally measurable quantities T, P, and V. If again the system is a fixed quantity of ideal gas, this equation of state indicates that the enthalpy of an ideal gas is independent of its pressure.

5.10 THE TRANSFORMATION FORMULA

Given three state properties x, y, and z, and a closed system of fixed composition, then, for the system,

$$x = x(y, z)$$

or

$$dx = \left(\frac{\partial x}{\partial y}\right)_z dy + \left(\frac{\partial x}{\partial z}\right)_y dz$$

For an incremental change of state at constant x,

$$\left(\frac{\partial x}{\partial y}\right)_z dy = -\left(\frac{\partial x}{\partial z}\right)_y dz$$

or

$$\left(\frac{\partial x}{\partial y}\right)_z \left(\frac{\partial y}{\partial z}\right)_x = -\left(\frac{\partial x}{\partial z}\right)_y$$

This can be written as

$$\left(\frac{\partial x}{\partial y}\right)_z \left(\frac{\partial y}{\partial z}\right)_x \left(\frac{\partial z}{\partial x}\right)_y = -1 \qquad (5.35)$$

and is known as the transformation formula. Equation (5.35) can be used with any three state functions with each of the functions appearing once in the denominator, once in the numerator, and once outside the bracket.

5.11 EXAMPLE OF THE USE OF THE THERMODYNAMIC RELATIONS

Equation (2.8) gives the relationship between c_p and c_v:

$$c_p - c_v = \left(\frac{\partial V}{\partial T}\right)_P \left[P + \left(\frac{\partial U}{\partial V}\right)_T\right] \tag{2.8}$$

The use of the thermodynamic relations allows the difference between c_p and c_v to be expressed in terms of experimentally determinable quantities. Equation (5.27) gives

$$P = -\left(\frac{\partial A}{\partial V}\right)_T$$

Thus

$$c_p - c_v = \left(\frac{\partial V}{\partial T}\right)_P \left[\left(\frac{\partial U}{\partial V}\right)_T - \left(\frac{\partial A}{\partial V}\right)_T\right]$$

By definition

$$A = U - TS$$

Thus

$$c_p - c_v = \left(\frac{\partial V}{\partial T}\right)_P \left[\left(\frac{\partial U}{\partial V}\right)_T - \left(\frac{\partial U}{\partial V}\right)_T + T\left(\frac{\partial S}{\partial V}\right)_T\right]$$

$$= \left(\frac{\partial V}{\partial T}\right)_P \left[T\left(\frac{\partial S}{\partial V}\right)_T\right]$$

The Maxwell's relation (5.33) gives

$$\left(\frac{\partial S}{\partial V}\right)_T = \left(\frac{\partial P}{\partial T}\right)_V$$

and the transformation formula, with respect to P, V, and T gives

$$\left(\frac{\partial P}{\partial T}\right)_V = -\left(\frac{\partial P}{\partial V}\right)_T \left(\frac{\partial V}{\partial T}\right)_P$$

Thus

$$c_p - c_v = -T\left(\frac{\partial V}{\partial T}\right)_P\left(\frac{\partial P}{\partial V}\right)_T\left(\frac{\partial V}{\partial T}\right)_P$$

In Chap. 1 the isobaric thermal expansivity was defined as

$$\alpha = \frac{1}{V}\left(\frac{\partial V}{\partial T}\right)_P$$

Similarly the isothermal compressibility of a system is defined as

$$\beta = -\frac{1}{V}\left(\frac{\partial V}{\partial P}\right)_T$$

(β is the fractional decrease in the volume of the system for unit increase of pressure acting on the system at constant temperature. The negative sign is used in order to make β a positive number.) Thus

$$c_p - c_v = \frac{VT\alpha^2}{\beta} \tag{5.36}$$

and the right side of this equation contains only experimentally measurable quantities. Equation (5.36) is particularly useful for comparing the theoretical, calculated values of the heat capacity of a system (which are normally calculated for constant-volume processes) with experimentally determined values (which are normally measured under conditions of constant pressure).

5.12 THE GIBBS-HELMHOLTZ EQUATION

Equation (5.2) gives,

$$G = H - TS$$

and from Eq. (5.12),

$$\left(\frac{\partial G}{\partial T}\right)_P = -S$$

Therefore, at constant pressure,

$$G = H + T\left(\frac{dG}{dT}\right)$$

or

$$GdT = HdT + TdG$$

Dividing through by T^2 and rearranging gives

$$\frac{TdG - GdT}{T^2} = -\frac{HdT}{T^2}$$

which on comparison with the identity $d(x/y) = (y\,dx - x\,dy)/y^2$ indicates that

$$\frac{d(G/T)}{dT} = -\frac{H}{T^2} \tag{5.37}$$

Equation (5.37) is known as the Gibbs-Helmholtz equation and is applicable to closed systems of fixed composition undergoing constant pressure processes.

For any isobaric change of state of a closed system of fixed composition, Eq. (5.37) gives

$$\frac{d(\Delta G/T)}{dT} = -\frac{\Delta H}{T^2} \tag{5.37a}$$

This equation is particularly useful in experimental thermodynamics, as it allows ΔH, the heat of a reaction, to be determined from the experimentally determined variation of ΔG, the free energy change of the reaction, with temperature; or, conversely, it allows ΔG to be determined from an experimentally measured ΔH. The full usefulness of this equation will be developed in Chap. 10.

The corresponding relationship between A and U is obtained as follows. Equation (5.1) gives

$$A = U - TS$$

$$= U + T\left(\frac{\partial A}{\partial T}\right)_V$$

and manipulation similar to the above gives

$$\frac{d(A/T)}{dT} = -\frac{U}{T^2} \tag{5.38}$$

This equation is only applicable to closed systems of fixed composition undergoing processes at constant volume. Again, for a given change of state under these conditions,

$$\frac{d(\Delta A/T)}{dT} = -\frac{\Delta U}{T^2} \tag{5.38a}$$

PROBLEMS

5.1 Show that for an isothermal change of state of an ideal gas, the rate of change of entropy with volume is inversely proportional to the volume of the gas.

5.2 Show that $\left(\frac{\partial^2 G}{\partial P^2}\right)_T = -\frac{1}{\left(\frac{\partial^2 A}{\partial V^2}\right)_T}$

5.3 Show that $\left(\frac{\partial c_p}{\partial P}\right)_T = -T\left(\frac{\partial^2 V}{\partial T^2}\right)_P$

5.4 Show that $\left(\frac{\partial T}{\partial P}\right)_S = \frac{T}{c_p}\left(\frac{\partial V}{\partial T}\right)_P$

5.5 Show that
$$c_p - c_v = V\left(\frac{\partial P}{\partial T}\right)_V + \left(\frac{\partial H}{\partial V}\right)_T\left(\frac{\partial V}{\partial T}\right)_P$$
$$= \left[V - \left(\frac{\partial H}{\partial P}\right)_T\right]\left(\frac{\partial P}{\partial T}\right)_V$$
$$= T\left(\frac{\partial V}{\partial T}\right)_P\left(\frac{\partial P}{\partial T}\right)_V$$

5.6 Show that $\left(\frac{\partial P}{\partial V}\right)_S = -\frac{c_p}{c_v V\beta}$

5.7 Joule and Thomson showed that when a steady stream of a real (i.e., nonideal) gas is passed through a thermally insulated tube in which is inserted a throttle valve, a temperature change occurs in the gas; i.e., the throttling causes a change of state from P_1, T_1 to P_2, T_2. Show that this process is isenthalpic. The change in T is described in terms of the Joule-Thomson coefficient, $\mu_{J\text{-}T}$, as

$$\mu_{J\text{-}T} = \left(\frac{\partial T}{\partial P}\right)_H$$

Show that

$$\mu_{J\text{-}T} = -\frac{V}{c_p}(1 - \alpha T)$$

and that, for an ideal gas, the Joule-Thomson coefficient is zero.

5.8 Determine the values of ΔU, ΔH, ΔS, ΔG, and ΔA for the following processes. (In (*c*), (*d*), and (*e*), show that the absolute value of the entropy is required.)

a. The four processes in Prob. 4.1.

b. One mole of ideal gas at the pressure P and temperature T expands into a vacuum to double its volume.

c. The reversible adiabatic expansion of 1 mole of ideal gas from P_1, T_1 to P_2, T_2.

d. A constant-pressure expansion of 1 mole of an ideal gas from V_1, T_1 to V_2, T_2.

e. A constant-volume change of state of 1 mole of ideal gas from P_1, T_1 to P_2, T_2.

CHAPTER SIX

HEAT CAPACITY, ENTHALPY, ENTROPY, AND THE THIRD LAW OF THERMODYNAMICS

6.1 INTRODUCTION

Equations (2.6) and (2.7) defined two heat capacities, the heat capacity at constant volume

$$C_v = \left(\frac{\partial U}{\partial T}\right)_v \tag{2.6}$$

and the heat capacity at constant pressure

$$C_p = \left(\frac{\partial H}{\partial T}\right)_p \tag{2.7}$$

At this point it is convenient to introduce a notation which will allow distinction to be made between the values of extensive properties per mole of a system and the values of extensive properties for the entire system. If E is an extensive property, then E' will be taken as denoting the value of the property for the system containing n moles, and E will be taken as denoting the value of the property per mole of the system. Thus, for a system containing n moles,

$$E' = nE$$

Equations (2.6) and (2.7) can thus be written equivalently as,

$$dU' = C_V dT = nc_V dT \quad \text{or} \quad dU = c_V dT \tag{2.6a}$$

$$dH' = C_p dT = nc_p dT \quad \text{or} \quad dH = c_p dT \tag{2.7a}$$

where c_p and c_V are, respectively, the constant-pressure, and constant-volume molar heat capacities.

Integration of Eq. (2.7*a*) between the states (T_2, P) and (T_1, P) gives,

$$\Delta H = H(T_2, P) - H(T_1, P) = \int_{T_1}^{T_2} c_p dT \tag{6.1}$$

from which it is seen that a knowledge of the temperature dependence of c_p is required for the determination of the temperature dependence of the molar enthalpy, and, as will be seen, for the determination of the temperature dependence of entropy. Similarly, integration of Eq. (2.6*a*) between T_2 and T_1 at constant volume indicates that the determination of the temperature dependence of internal energy requires a knowledge of the temperature dependence of c_V.

6.2 THEORETICAL CALCULATION OF THE HEAT CAPACITY

In 1819 Dulong and Petit introduced the empirical rule that the molar heat capacities of all solid elements equalled $3R$ (= 24.9 joules/degree). Subsequent experimental determination of values of the heat capacities of various elements showed that the heat capacities always increase with increasing temperature and that the values tend to be much lower than $3R$ at low temperatures. Also the heat capacities can be higher than $3R$ at temperatures greater than room temperature. Figure 6.1 shows the variation of c_V with temperature for several elements.

Calculation of the heat capacity of a solid element, as a function of temperature, was one of the early triumphs of the quantum theory. The first such calculation is due to Einstein, who considered the properties of a crystal comprising n atoms, each of which, in classical terms, is considered to behave as a harmonic oscillator vibrating independently about its lattice point. As each oscillator is considered to be independent, i.e., as its behavior is unaffected by the behavior of its neighbors, then each oscillator vibrates with a single fixed frequency ν, and a system of such oscillators is known as an Einstein crystal.

Quantum theory gives the energy of the ith level of a harmonic oscillator as

$$\epsilon_i = (i + \tfrac{1}{2})\, h\nu \tag{6.2}$$

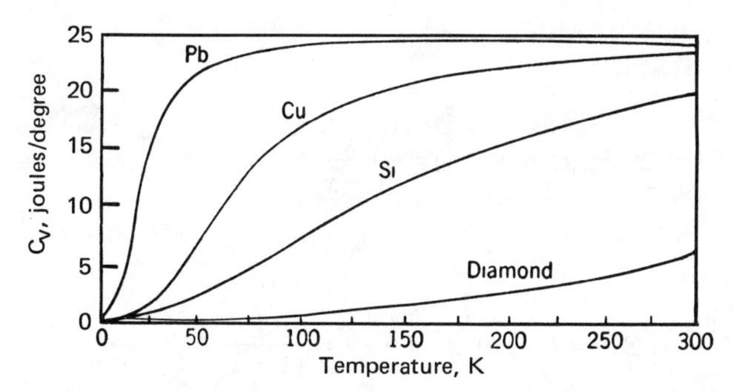

Fig. 6.1. The constant molar volume heat capacities of Pb, Cu, Si, and diamond as a function of temperature. *(From E. D. Eastman and G. K. Rollefson, "Physical Chemistry," McGraw-Hill Book Company, New York, 1947.)*

where i is an integer which can have values in the range zero to infinity, and h is Planck's constant of action ($= 6.6252 \times 10^{-34}$ joule·s). As each oscillator has three degrees of freedom, i.e., can vibrate in the x, y, and z directions, then the energy U' of such a system (which can be regarded as being a system of $3n$ linear harmonic oscillators) is given as

$$U' = 3\Sigma n_i \epsilon_i \tag{6.3}$$

where, as before, n_i is the number of atoms in the ith energy level. Substituting Eqs. (6.2) and (4.13) into Eq. (6.3) gives

$$U' = 3\Sigma \left(i + \tfrac{1}{2}\right) h\nu \left[\frac{ne^{-h\nu(i + \frac{1}{2})/kT}}{\Sigma e^{-h\nu(i + \frac{1}{2})/kT}}\right]$$

$$= 3nh\nu \left[\frac{\Sigma ie^{-h\nu(i + \frac{1}{2})/kT}}{\Sigma e^{-h\nu(i + \frac{1}{2})/kT}} + \frac{\frac{1}{2}\,\Sigma e^{-h\nu(i + \frac{1}{2})/kT}}{\Sigma e^{-h\nu(i + \frac{1}{2})/kT}}\right]$$

$$= 3nh\nu \left[\frac{\Sigma ie^{-h\nu i/kT}}{\Sigma e^{-h\nu i/kT}} + \frac{1}{2}\right]$$

$$= \tfrac{3}{2}\, nh\nu \left[1 + \frac{2\Sigma ie^{-h\nu i/kT}}{\Sigma e^{-h\nu i/kT}}\right]$$

Taking
$$\Sigma ie^{-h\nu i/kT} = \Sigma ix^i$$

where $x = e^{-h\nu/kT}$

gives

$$x(1 + 2x + 3x^2 + \cdots) = \frac{x}{(1-x)^2}$$

and

$$\Sigma e^{-h\nu i/kT} = \Sigma x^i = 1 + x + x^2 + \cdots = \frac{1}{(1-x)}$$

in which case

$$U' = \tfrac{3}{2}\, nh\nu \left[1 + \frac{2x}{1-x}\right]$$

$$= \tfrac{3}{2}\, nh\nu \left[1 + \frac{2e^{-h\nu/kT}}{1-e^{-h\nu/kT}}\right]$$

$$= \tfrac{3}{2}\, nh\nu + \frac{3nh\nu}{(e^{h\nu/kT}-1)} \tag{6.4}$$

Equation (6.4) then expresses the relationship between the energy of the system and the temperature. Differentiation of Eq. (6.4) with respect to temperature at constant volume gives, by definition, the constant-volume heat capacity C_v (as the volume remains constant, then so the quantization of the energy levels remains constant). Thus

$$C_v = \left(\frac{\partial U'}{\partial T}\right)_v = 3nh\nu(e^{h\nu/kT}-1)^{-2}\frac{h\nu}{kT^2}e^{h\nu/kT}$$

$$= 3nk\left(\frac{h\nu}{kT}\right)^2 \frac{e^{h\nu/kT}}{(e^{h\nu/kT}-1)^2}$$

Defining $h\nu/k = \theta_E$, where θ_E is the Einstein characteristic temperature, and taking n equal to Avogadro's number gives the constant-volume molar heat capacity of the substance as

$$c_v = 3R\left(\frac{\theta_E}{T}\right)^2 \frac{e^{\theta_E/T}}{(e^{\theta_E/T}-1)^2} \tag{6.5}$$

The variation of c_v with T/θ_E is shown in Fig. 6.2, which shows that as T/θ_E (and hence T) increases, $c_v \rightarrow 3R$ in agreement with the Dulong and Petit law, and as $T \rightarrow 0$, $c_v \rightarrow 0$ in agreement with experimental observation. The actual value of θ_E for any element, which is termed the Einstein characteristic temperature for that element, and its vibration frequency ν, are obtained by curve fitting Eq. (6.5) to experimentally measured heat capacity data. Such curve fitting (see Fig. 6.2) shows that, although the Einstein equation adequately represents the actual heat capacities at higher temperatures, the theoretical values approach zero significantly more rapidly than do the actual values. This discrepancy between theory and experiment is due to the fact that the oscillators do not all vibrate with a single frequency.

The next step in the development of the theory was made by Debye, who assumed that the range, or spectrum, of frequencies available to the oscillators is the same as that available to the elastic vibrations of a continuous solid. With respect to the wavelengths of these vibrations, the lower limit is fixed by the interatomic distances in the solid; i.e., if the wavelength was equal to the interatomic distance, then successive atoms would be in the same phase of vibration; hence vibration of one atom with respect to another, as such, would not occur. Theoretically the shortest allowable wavelength is twice the interatomic distance, in which case neighboring atoms vibrate in opposition to each other. Taking this minimum wavelength λ_{min} to be in the order of 5×10^{-8} cm and

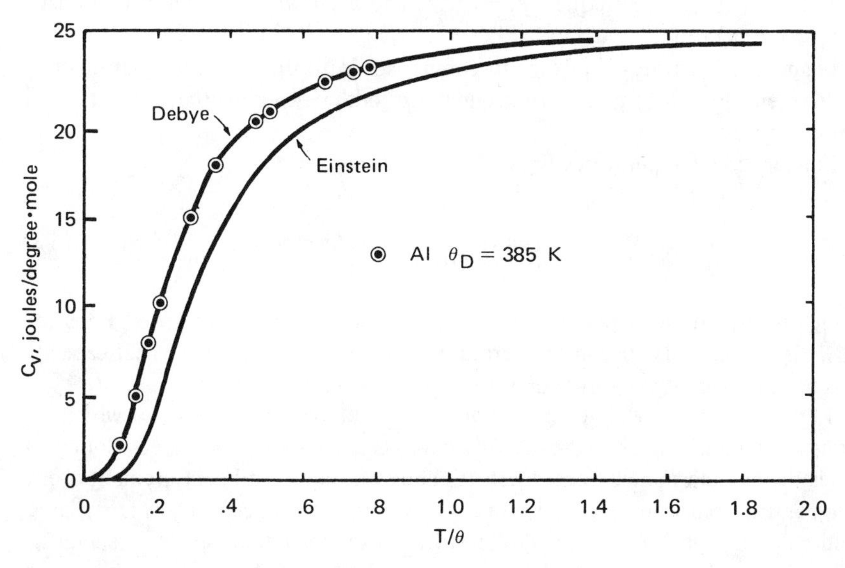

Fig. 6.2. Comparison among the Debye heat capacity, the Einstein heat capacity, and the actual heat capacity of aluminum.

the wave velocity v in the solid to be 5×10^5 cm/sec gives the maximum frequency of vibration of an oscillator to be in the order of

$$\nu_{max} = \frac{v}{\lambda_{min}} = \frac{5 \times 10^5}{5 \times 10^{-8}} = 10^{13} \text{ sec}^{-1}$$

Debye assumed the frequency distribution to be one in which the number of vibrations per unit volume per unit frequency range increases parabolically with increasing frequency in the allowed frequency range $0 \leqslant \nu \leqslant \nu_{max}$; and by integrating the Einstein expression over this frequency distribution, he obtained the heat capacity of the solid as

$$C_v = \frac{9nh^3}{k^2 \theta_D^3} \int_0^{\nu_D} \nu^2 \left(\frac{h\nu}{kT}\right)^2 \frac{e^{h\nu/kT}}{(1 - e^{h\nu/kT})^2} d\nu$$

which, with $x = h\nu/kT$, gives

$$c_v = 9R \left(\frac{T}{\theta_D}\right)^3 \int_0^{\theta_D/T} \frac{x^4 e^{-x}}{(1 - e^{-x})^2} dx \tag{6.6}$$

where ν_D (the Debye frequency) $= \nu_{max}$, and $\theta_D = h\nu_D/k$ is the characteristic Debye temperature.

Figure 6.2 illustrates Eq. (6.6) in comparison with the Einstein equation, and, as is seen, Eq. (6.6) gives an excellent fit to the experimental data at lower temperatures.

For very low temperatures Eq. (6.6) gives

$$c_v = 1943 \left(\frac{T}{\theta_D}\right)^3 \text{ joules/degree} \tag{6.7}$$

which is termed the Debye T^3 law for low-temperature heat capacities. Equation (6.7) is of particular use at temperatures below those at which the heat capacity has been measured experimentally for a given material.

Table 6.1 lists the Debye temperatures for a number of elements for which the constant-volume molar heat capacities have been accurately measured. Column 1 gives θ_D as determined from "best fit" between theory and measurement over a temperature range in the vicinity of $\theta_D/2$, where the heat capacity is fairly large; and column 2 gives θ_D from the "best fit" of the low-temperature data and Eq. (6.7). The Debye frequencies, ν_D, vary from 3.84×10^{13} sec^{-1} for diamond to 1.83×10^{12} sec^{-1} for lead, which shows good agreement with the theoretical maximum frequencies of elastic vibration in a solid.

Table 6.1

Substance	θ_D(high T)	θ_D(from T^3)
Pb	90 K	
K	99	
Na	159	
Sn	160	127 K
Cd	160	129
Au	180	162
Ag	213	
Pt	225	
Zn	235	205
Cu	315	321
Mo	379	379
Al	389	385
Fe	420	428
C(diamond)	1890	2230

Figures 6.3 and 6.4 illustrate the curve fitting of experimental data for Pb, Ag, Al, and C(diamond) with the Debye equation. Figure 6.3 shows that when the experimental data is plotted as a function of log T, the lines produced are nearly identical except for a horizontal displacement. The relative horizontal displacement is a measure of log θ_D for the particular element, and suitable choice of a characteristic Debye temperature for each element gives coincidence of the c_v — log T/θ_D plots shown in Fig. 6.4.

Debye's theory does not consider the contribution made to the heat capacity by the uptake of energy by the electrons; and since $c_v = (\partial U/\partial T)_v$, it follows that, in any temperature range where the energy associated with the electrons

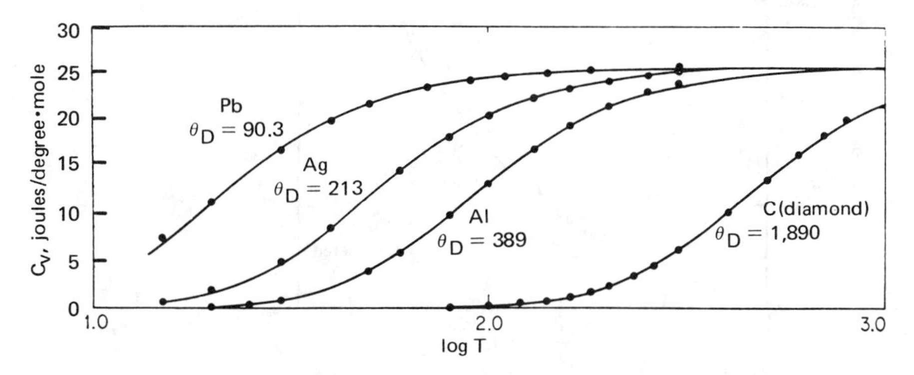

Fig. 6.3. The constant-volume molar heat capacities of several solid elements. The curves are from the Debye equation with the θ_D values given. *(From E. D. Eastman and G. K. Rollefson, "Physical Chemistry," McGraw-Hill Book Company, New York, 1947.)*

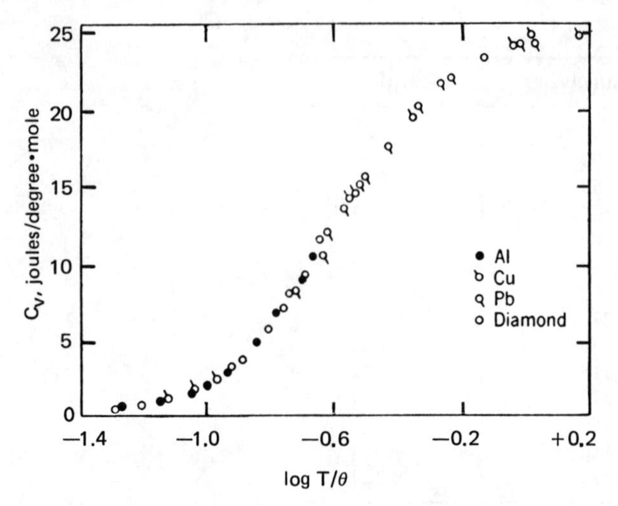

Fig. 6.4. The variation of c_v with log (T/θ_D) for several solid elements. *(From E. D. Eastman and G. K. Rollefson, "Physical Chemistry," McGraw-Hill Book Company, New York, 1947.)*

changes with temperature, a contribution to the heat capacity will result. Consideration of the electron gas theory of metals indicates that the electronic contribution to the heat capacity is proportional to the absolute temperature. Hence the electronic contribution becomes large in absolute value at elevated

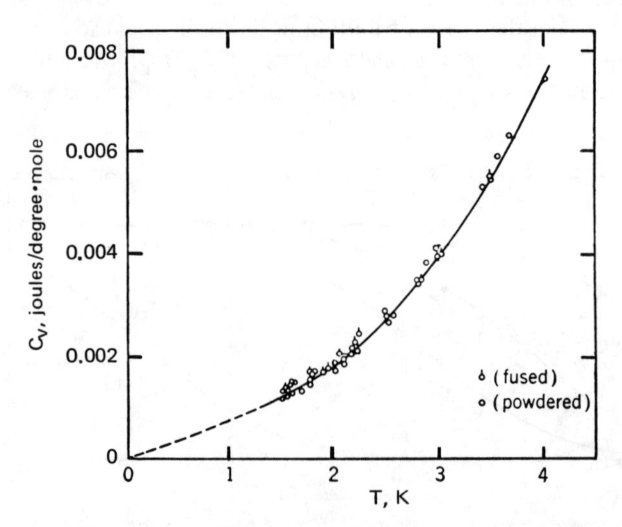

Fig. 6.5*a*. The low-temperature constant-volume molar heat capacity of zinc. *(From A. A. Silvidi and J. G. Daunt, Electronic Specific Heats in Tungsten and Zinc, Phys. Rev., 77:125 [1950].)*

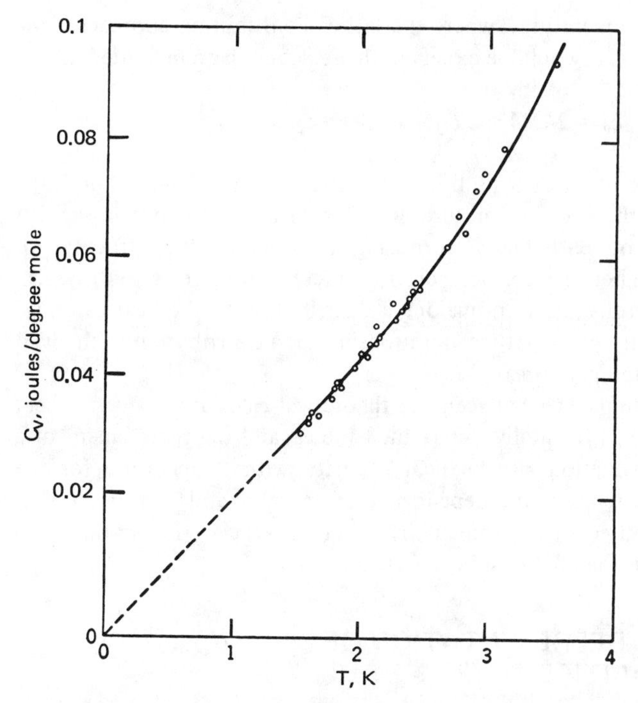

Fig. 6.5*b*. The low-temperature constant-volume molar heat capacity of tungsten. *(From A. A. Silvidi and J. G. Daunt, Phys. Rev., 77:125 [1950].)*

temperatures and also becomes large compared with the atomic vibration contribution (which is proportional to T^3) in the temperature range 0 to 1 K. This is illustrated by the low-temperature c_v data for Zn and W shown in Fig. 6.5*a* and *b*, respectively, where it is seen that the curves approach straight lines with a nonzero slope at 0 K. The curves in Fig. 6.5*a* and *b* are represented by

$$c_v = 1943\left(\frac{T}{291}\right)^3 + 6.28 \times 10^{-4}\,T \text{ joules/degree·mole for Zn}$$

and

$$c_v = 1943\left(\frac{T}{169}\right)^3 + 214 \times 10^{-4}\,T \text{ joules/degree·mole for W}$$

In both equations the first term on the right is the lattice contribution and the second term is the electronic contribution. If this interpretation is taken at face value, then at high temperatures, where the lattice contribution approaches the Dulong and Petit value, it would be expected that c_v could be represented as

$$c_v = 24.94 + bT \text{ joules/degree·mole}$$

where bT is the electronic contribution, and b has the same value as does the coefficient of T in the low-temperature equation. That this is not borne out experimentally is not particularly surprising in view of the difficulty of determining the number of electrons per atom present in the electron gas, i.e., the number of electrons per atom making a contribution to the heat capacity. Also the anharmonicity of the lattice vibrations makes a contribution to the heat capacity at elevated temperatures.

In view of the discrepancies between the theoretical calculated values of heat capacities and the experimentally determined values, and the requirement of a knowledge of the variation of the heat capacity with temperature for the determination of the temperature dependence of enthalpy and entropy, it is, at present, normal practice to experimentally determine the c_p-T variation of a substance and express the relationship analytically.

6.3 EMPIRICAL REPRESENTATION OF HEAT CAPACITIES

The experimentally measured variation of c_p for a substance with temperature is normally fitted to an expression of the form

$$c_p = a + bT + cT^{-2}$$

and it is to be noted that the analytical expression is only applicable in that stated temperature range over which the c_p values were measured; e.g., Ca has a high-temperature allotrope (the β-phase) which is stable above 737 K and a low-temperature allotrope (the α-phase) which exists below 737 K. Hence, for Ca, two equations are used:

$$\alpha\text{-Ca} \qquad c_p = 22.2 + 13.9 + 10^{-3}T \text{ joules/degree·mole}$$

over the temperature range 273 to 737 K, and

$$\beta\text{-Ca} \qquad c_p = 6.3 + 32.4 \times 10^{-3}T + 10.5 \times 10^{5}T^{-2} \text{ joules/degree·mole}$$

from 737 K to the melting temperature.

Often, as for example with α-Ca, the accuracy of measurement does not warrant inclusion of the third term, cT^{-2}, in which case c_p is expressed as a

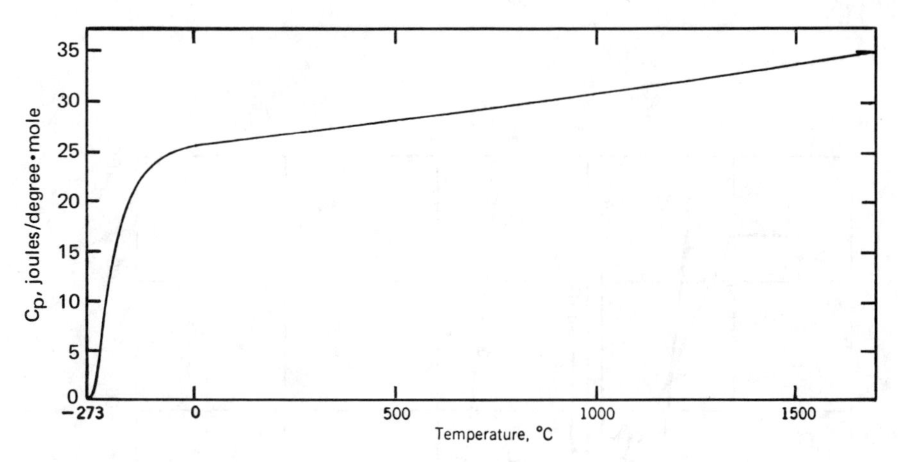

Fig. 6.6. The constant-pressure molar heat capacity of platinum.

linear function of temperature over the stated temperature range. Figure 6.6 shows the c_p-T variation for platinum, and from 298 K to the melting temperature the straight-line variation is given as $c_p = 24.25 + 5.376 \times 10^{-3} T$ joules/degree•mole. Where the experimentally measured data is insufficient to allow determination of the temperature dependence of c_p, as is often the case with liquid metals, c_p is assumed to be constant; i.e., the second term, bT, is omitted from the analytical equation. The variation of c_p with T for several elements at temperatures above room temperature is shown in Fig. 6.7, and the corresponding analytical representations of these data are listed in Table 6.2.

Table 6.2

Element	c_p (joules/degree•mole)	Temperature range, K
Al (solid)	$20.7 + 12.4 \times 10^{-3} T$	298 – T_m
Al (liquid)	29.3	T_m – 1273
Au (solid)	$23.7 + 5.19 \times 10^{-3} T$	298 – T_m
Au (liquid)	29.3	T_m – 1600
Cu (solid)	$22.6 + 6.28 \times 10^{-3} T$	298 – T_m
Cu (liquid)	31.4	T_m – 1600
Fe (α)	$17.5 + 24.8 \times 10^{-3} T$	273 – 1033
Fe (β)	38	1033 – 1181
Fe (γ)	$7.70 + 19.5 \times 10^{-3} T$	1181 – 1674
Fe (δ)	43.9	1674 – T_m
Fe (liquid)	41.8	T_m – 1873
C(diamond)	$9.12 + 13.2 \times 10^{-3} T - 6.19 \times 10^{5} T^{-2}$	298 – 1200
C(graphite)	$17.2 + 4.27 \times 10^{-3} T - 8.79 \times 10^{5} T^{-2}$	298 – 2300
O_2 (gas)	$30.0 + 4.18 \times 10^{-3} T - 1.7 \times 10^{5} T^{-2}$	298 – 3000

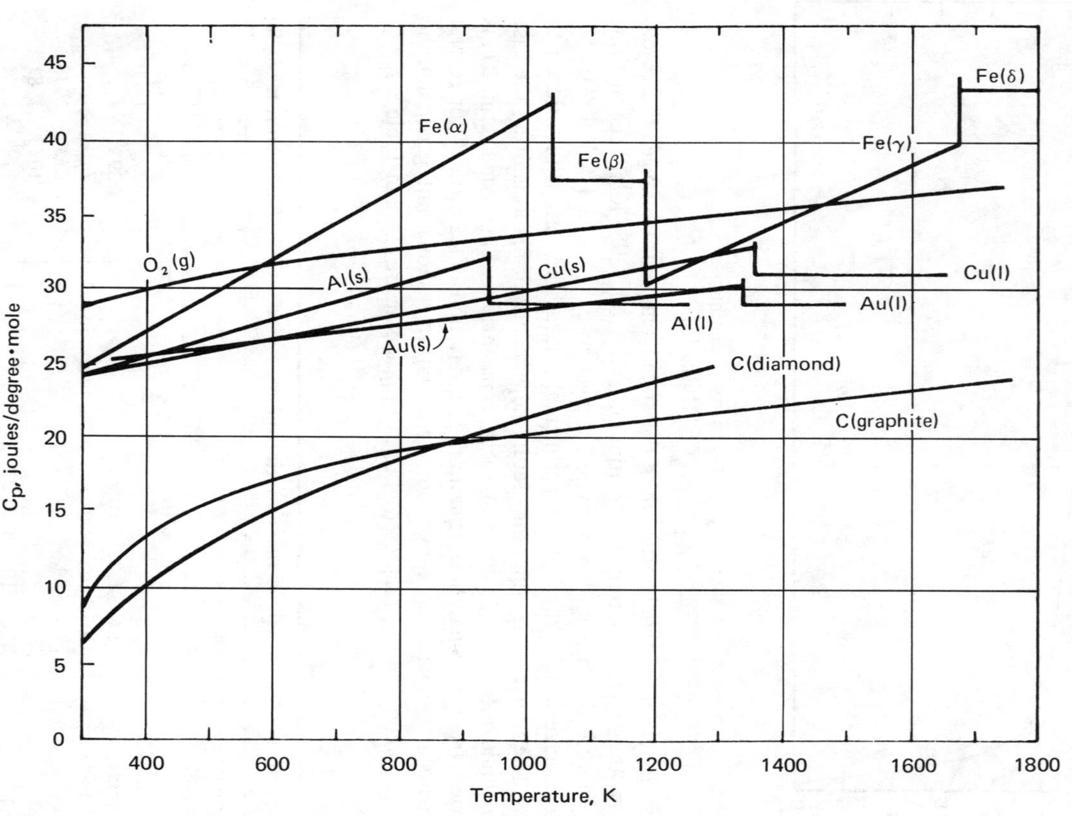

Fig. 6.7. The constant-pressure molar heat capacities of several elements.

In 1864 Kopp proposed that the heat capacity of a solid compound is equal to the sum of the heat capacities of its constituent elements. This rule, which is known as Kopp's rule, is reasonably reliable for ionic compounds at temperatures where the constituent elements of the compound obey the Dulong and Petit law. Thus if there are n atoms in the formula of the compound, the heat capacity per mole is $3nR$.

6.4 ENTHALPY AS A FUNCTION OF TEMPERATURE AND COMPOSITION

For a closed system of fixed composition undergoing a temperature change from T_1 to T_2 at the constant pressure P, integration of Eq. (2.7*a*) gives Eq. (6.1).

$$\Delta H = H(T_2, P) - H(T_1, P) = \int_{T_1}^{T_2} c_p dT \tag{6.1}$$

ΔH is thus the area under the c_p-T curve between the limits T_1 and T_2, and from Eq. (2.7) $\Delta H = q_p$ which is simply the amount of heat required to raise the temperature of 1 mole of the system from T_1 to T_2 at the constant pressure P.

For a system undergoing a chemical reaction or phase change at constant temperature and pressure, e.g., the reaction A + B = AB,

$$\Delta H(T, P) = H_{AB}(T, P) - H_A(T, P) - H_B(T, P) \tag{6.8}$$

ΔH is the difference between the enthalpy of the products of the process (state 2) and the enthalpy of the reactants (state 1), and Eq. (6.8) is a statement of Hess's law. If $\Delta H > 0$, the reaction occurs with an absorption of heat from the constant-temperature heat bath surrounding the system and hence is an *endothermic* process or reaction; and if $\Delta H < 0$, the reaction occurs with an evolution of heat and is thus an *exothermic* process. This convention corresponds to that used in connection with the First Law for the sign of q, the heat entering or leaving the system.

Enthalpy changes resulting from temperature and/or composition changes can be represented on an enthalpy-temperature diagram. Consider the process, or change of state, involving the phase change $A_{(s)} = A_{(l)}$, i.e., the melting of pure A. ΔH_{T_1} for this process equals the difference between the molar enthalpies of $A_{(l)}$ and $A_{(s)}$ at the temperature T_1.

$$\Delta H_{T_1} = H_{A_{(l)}}(T_1) - H_{A_{(s)}}(T_1)$$

In Fig. 6.8 this enthalpy change is represented by the line *ba*.

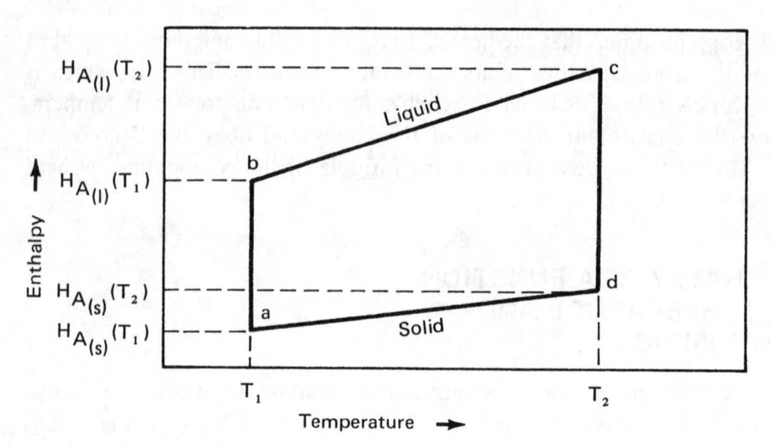

Fig. 6.8. The variation with temperature of the enthalpies of the solid and liquid phases of a substance.

For the phase change occurring at the temperature T_2,

$$\Delta H_{T_2} = H_{A(l)}(T_2) - H_{A(s)}(T_2)$$

which is represented in Fig. 6.8 by the line *cd*. As *H* is a state function, then

$$\Delta H(a \to b) = \Delta H(a \to d) + \Delta H(d \to c) + \Delta H(c \to b)$$

where $\Delta H(a \to d)$ = the heat required to raise the temperature of 1 mole of solid A from T_1 to T_2

$$= \int_{T_1}^{T_2} c_{p_{A(s)}} dT$$

$c_{p_{A(s)}}$ = the molar heat capacity of solid A

$\Delta H(d \to c) = \Delta H_{T_2}$

$\Delta H(c \to b)$ = the negative of the heat required to raise the temperature of 1 mole of liquid A from T_1 to T_2.

$$= -\int_{T_1}^{T_2} c_{p_{A(l)}} dT$$

$$= \int_{T_2}^{T_1} c_{p_{A(l)}} dT$$

$c_{p_{A(l)}}$ = the molar heat capacity of liquid A

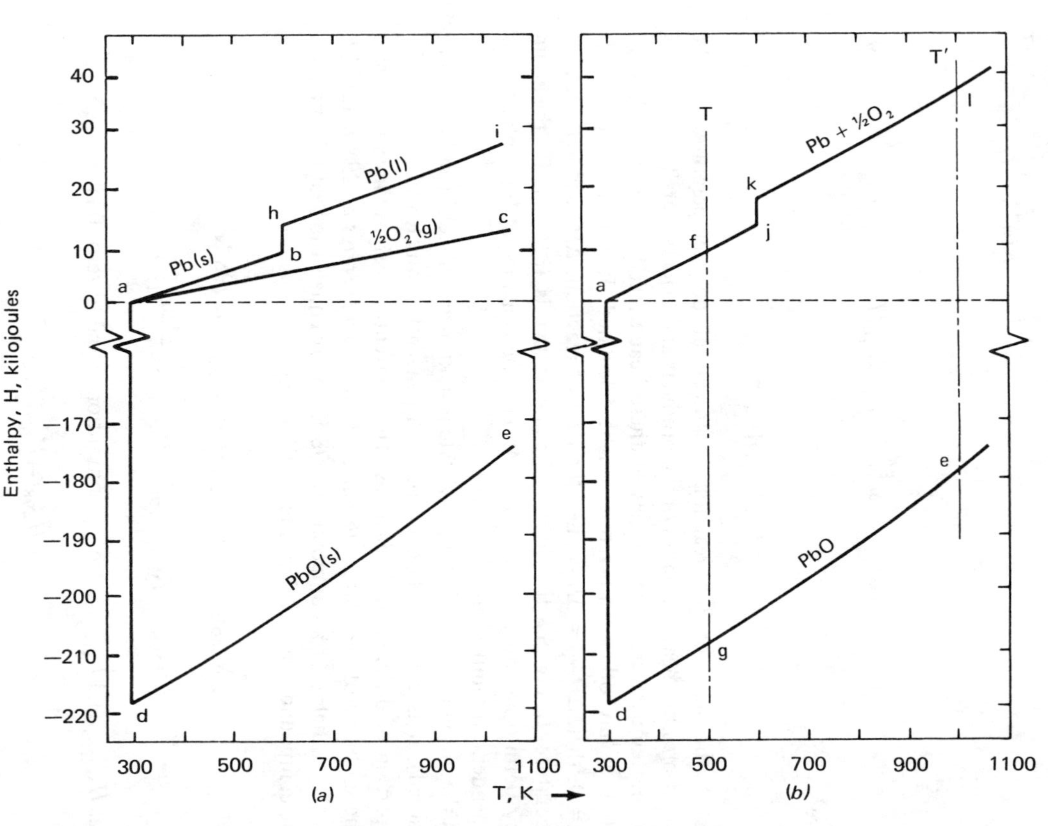

Fig. 6.9. (*a*) The variation with temperature of the enthalpies of $Pb_{(s)}$, $Pb_{(l)}$, $\frac{1}{2}O_2$, and $PbO_{(s)}$. (*b*) The variation with temperature of the enthalpies of ($Pb + \frac{1}{2}O_2$) and PbO.

Thus

$$\Delta H(A_{(s)} \rightarrow A_{(l)} \text{ at } T_1) = \Delta H(A_{(s)} \rightarrow A_{(l)} \text{ at } T_2) + \int_{T_1}^{T_2} c_{p_{A(s)}} dT - \int_{T_1}^{T_2} c_{p_{A(l)}} dT$$

or

$$\Delta H_{T_2} = \Delta H_{T_1} + \int_{T_1}^{T_2} \Delta c_p dT \tag{6.9}$$

where

$$\Delta c_p = c_{p_{A(l)}} - c_{p_{A(s)}}$$

Thus if the heat of the reaction is known at one temperature and the constant-pressure heat capacities of the products and reactants are known (along with the temperature dependencies of these heat capacities), then the heat of reaction at any other temperature can be calculated. It is to be noted that if $\Delta c_p = 0$, then $\Delta H_{T_2} = \Delta H_{T_1}$; i.e., the heat of reaction, ΔH, is independent of temperature. In Fig. 6.8 the slope of the line *bc*, that is, $(\partial H/\partial T)_p$ for the liquid A, by definition, equals c_p for liquid A. Thus *bc* is a straight line only if c_p is independent of temperature.

As the absolute value of H for any state is unknown (only changes in H can be measured), it is convenient to introduce a convention which will facilitate the comparison of different H-T diagrams. This convention assigns the value of zero to the enthalpy of *elemental substances in their stable states at 298 K* (25°C). Thus the enthalpy of a compound at 298 K simply equals the heat of formation of the compound from its elements at 298 K; e.g.,

$$M_{(solid)} + \tfrac{1}{2}O_{2(gas)} = MO_{(solid)} \text{ at } 298 \text{ K}$$

$$\Delta H_{298} = H_{MO_{(s)}298} - H_{M_{(s)}298} - \tfrac{1}{2}H_{O_2(g)298}$$

and as $H_{M,298}$ and $H_{O_2,298}$ are, by convention, equal to zero, then

$$\Delta H_{298} = H_{MO,298}$$

The variation of heats of chemical reaction (or heats of formation) with temperature at constant pressure can be represented on an enthalpy-temperature diagram such as Fig. 6.9, which is drawn for the oxidation,

$$Pb + \tfrac{1}{2}O_2 = PbO$$

Table 6.3

$H_{PbO(298)} = -219{,}000$ joules/mole
$c_{p,\,Pb(s)} = 23.6 + 9.75 \times 10^{-3}T$ joules/degree·mole from 298 K to $T_{m,\,Pb}$
$c_{p,\,Pb(l)} = 32.4 - 3.1 \times 10^{-3}T$ joules/degree·mole from $T_{m,\,Pb}$ to 1200 K
$c_{p,PbO(s)} = 37.9 + 26.8 \times 10^{-3}T$ joules/degree· mole from 298 K to $T_{m,\,PbO}$
$c_{p,\,O_2(g)}$ see Table 6.2

The latent heat of melting of Pb, $\Delta H_{m,\,Pb} = 4810$ joules/mole at the melting temperature, $T_{m,\,Pb}$, of 600 K

$T_{m,\,PbO} = 1159$ K

The pertinent thermochemical data for this system are listed in Table 6.3.

In Fig. 6.9*a*,

a represents the enthalpy of ½ mole of oxygen gas and 1 mole of $Pb_{(s)}$ at 298 K = (by convention).

ab represents the variation of $H_{Pb(s)}$ with temperature in the range $298 \leqslant T \leqslant 600$, where $H_{Pb(s),T}$ is given as $\int_{298}^{T} c_{p,\,Pb(s)}dT$.

ac represents the variation of $H_{\frac{1}{2}O_2(g)}$ with temperature in the range $298 \leqslant T \leqslant 3000$ where $H_{\frac{1}{2}O_2(g)}$ is given as $\frac{1}{2}\int_{298}^{T} c_{p,O_2(g)}dT$.

d represents the enthalpy of 1 mole of $PbO_{(s)}$ at 298 K $= H_{PbO(s),298} = \Delta H_{PbO(s),298} = -219{,}000$ joules.

de represents the variation of $H_{PbO(s)}$ with temperature in the range $298 \leqslant T \leqslant 1159$ where

$$H_{PbO(s),\,T} = -219{,}000 + \int_{298}^{T} c_{p,\,PbO(s)}\,dT \text{ joules/mole}$$

In Fig. 6.9*b*,

a represents the enthalpy of ½ mole of oxygen gas and 1 mole of $Pb_{(s)}$ at 298 K.

f represents the enthalpy of ½ mole of $O_2(g)$ and 1 mole of $Pb_{(s)}$ at the temperature T.

g represents the enthalpy of 1 mole of $PbO_{(s)}$ at the temperature T.

Thus

$$\Delta H_{PbO,298} = \Delta H(a \to f) + \Delta H(f \to g) + \Delta H(g \to d)$$

$$= \int_{298}^{T} \left(\tfrac{1}{2} c_{p,\,O_2(g)} + c_{p,\,Pb(s)}\right) dT + \Delta H_{PbO,T}$$

$$+ \int_{T}^{298} c_{p,\,PbO(s)}dT$$

and hence

$$\Delta H_T = \Delta H_{298} + \int_{298}^{T} \Delta c_p dT$$

where

$$\Delta c_p = c_{p,\,\mathrm{PbO}(s)} - c_{p,\,\mathrm{Pb}(s)} - \tfrac{1}{2}\, c_{p,\,\mathrm{O_2}(g)}$$

From the data in Table 6.3,

$$\Delta c_p = -0.7 + 14.96 + 10^{-3}T + 0.85 \times 10^5 T^{-2}$$

and hence, in the temperature range 298 to 600 K ($T_{m,\,\mathrm{Pb}}$),

$$\Delta H_T = -219{,}000 + \int_{298}^{T} (-0.7 + 14.96 \times 10^{-3}T + 0.85 \times 10^5 T^{-2})\, dT$$

$$= -219{,}000 - 0.7(T - 298) + 7.48 \times 10^{-3}(T^2 - 298^2)$$

$$-0.85 \times 10^5 \left(\frac{1}{T} - \frac{1}{298}\right)$$

For $T = 500$ K this gives $\Delta H_{500} = -217{,}800$ joules, as can be seen in Figs. 6.9*b* and 6.10.

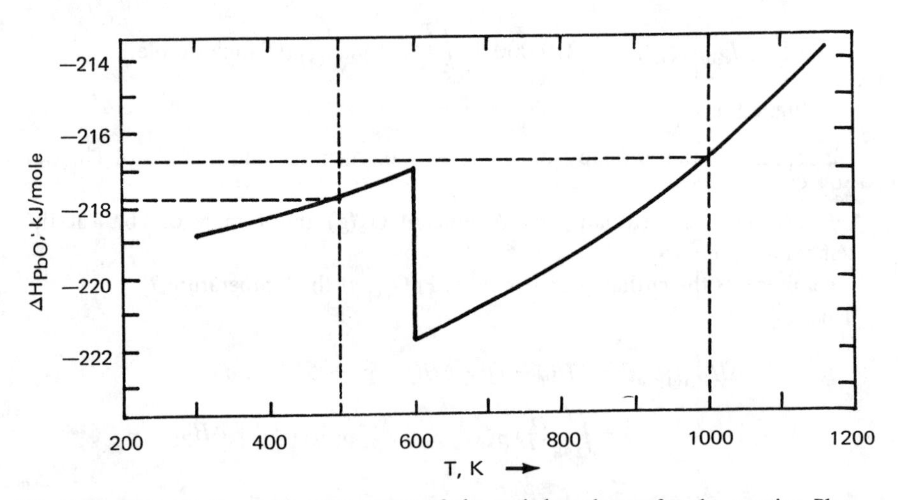

Fig. 6.10. The variation with temperature of the enthalpy change for the reaction Pb + $\frac{1}{2}O_2$ = PbO.

If a phase change occurs in one or more of the reactants or products, between the two temperatures at which the reaction is being considered, then the latent heats of the phase changes must be considered.

In Fig. 6.9*a*,

h represents the enthalpy of 1 mole of the Pb(*l*) at the melting temperature of 600°K, which is given as

$$H_{\text{Pb}(l),\,600} = \int_{298}^{600} c_{p,\,\text{Pb}(s)} dT + \Delta H_{m,\,\text{Pb}}$$

hb is the latent heat of melting of Pb at the melting temperature of 600 K = 4810 joules.

hi represents the variation of the enthalpy of 1 mole of Pb(*l*) with temperature in the range of 600 K to 1200 K.

$$H_{\text{Pb}(l),\,T} = \int_{298}^{600} c_{p,\,\text{Pb}(s)} dT + \Delta H_{m,\,\text{Pb}} + \int_{600}^{T} c_{p,\,\text{Pb}(l)} dT$$

In Fig. 6.9*b*, *ajkl* represents the variation of the enthalpy of 1 mole of Pb and ½ mole of O_2 (*g*), and hence $\Delta H_{T'}$ is calculated from the cycle,

$$\Delta H_{298} = \Delta H(a \to d) = \Delta H(a \to j) + \Delta H(j \to k) + \Delta H(k \to l) + \Delta H(l \to e) + \Delta H(e \to g) + \Delta H(g \to d)$$

where

$$\Delta H(a \to j) = \int_{298}^{T_{m,\,\text{Pb}}} (c_{p,\text{Pb}(s)} + \tfrac{1}{2} c_{p,\,\text{O}_2(g)}) dT$$

$\Delta H(j \to k)$ = the latent heat of melting of Pb at $T_{m,\,\text{Pb}}$ = 4810 joules

$$\Delta H(k \to l) = \int_{T_{m,\,\text{Pb}}}^{T'} (c_{p,\,\text{Pb}(l)} + \tfrac{1}{2} c_{p,\,\text{O}_2(g)}) dT$$

$$\Delta H(l \to e) = \Delta H_{T'}$$

$$\Delta H(e \to g) = \int_{T'}^{T_{m,\,\text{Pb}}} c_{p,\,\text{PbO}(s)} dT$$

$$\Delta H(g \to d) = \int_{T_{m,\,\text{Pb}}}^{298} c_{p,\,\text{PbO}(s)} dT$$

Thus,

$$\Delta H_{T'} = \Delta H_{298} + \int_{298}^{T_{m,\,\mathrm{Pb}}} (c_{p,\,\mathrm{PbO}(s)} - c_{p,\,\mathrm{Pb}(s)} - \tfrac{1}{2}\, c_{p,\,\mathrm{O_2}(g)})dT$$

$$- \Delta H_{m,\,\mathrm{Pb}} + \int_{T_{m,\,\mathrm{Pb}}}^{T'} (c_{p,\,\mathrm{PbO}(s)} - c_{p,\,\mathrm{Pb}(l)} - \tfrac{1}{2}\, c_{p,\,\mathrm{O_2}(g)})dT$$

$$= -219{,}000 + \int_{298}^{600} (-0.7 + 14.96 \times 10^{-3}T + 0.85 \times 10^{5}T^{-2})dT$$

$$-4810 + \int_{600}^{T'} (-9.5 + 27.8 \times 10^{-3}T + 0.85 \times 10^{5}T^{-2})dT$$

For $T' = 1000$ K, this gives $\Delta H_{1000} = -216{,}700$ joules as can be seen from Figs. 6.9*b* and 6.10. Figure 6.10 shows the variation of $\Delta H_{\mathrm{PbO},T}$ with temperature in the range 298 to 1100 K.

If the temperature of interest is greater than both the melting temperature of the metal and the oxide, then both latent heats of melting must be considered. For example, with respect to Fig. 6.11, which is drawn for the general oxidation reaction $\mathrm{M} + \frac{1}{2}\mathrm{O_2} = \mathrm{MO}$,

$$\Delta H_{T} = \Delta H_{298} + \int_{298}^{T_{m,\,\mathrm{M}}} (c_{p,\,\mathrm{MO}(s)} - c_{p,\,\mathrm{M}(s)} - \tfrac{1}{2}\, c_{p,\,\mathrm{O_2}(g)})dT$$

$$- \Delta H_{m,\,\mathrm{M}} + \int_{T_{m,\,\mathrm{M}}}^{T_{m,\,\mathrm{MO}}} (c_{p,\,\mathrm{MO}(s)} - c_{p,\,\mathrm{M}(l)} - \tfrac{1}{2}\, c_{p,\,\mathrm{O_2}(g)})dT$$

$$+ \Delta H_{m,\,\mathrm{MO}} + \int_{T_{m,\,\mathrm{MO}}}^{T} (c_{p,\,\mathrm{MO}(l)} - c_{p,\,\mathrm{M}(l)} - \tfrac{1}{2}\, c_{p,\mathrm{O_2}(g)})dT$$

When phase changes are to be considered, care must be exercised with the signs of the heats of the phase changes. The signs can be obtained from a consideration of Le Chatelier's principle, which states that "when a system, which is at equilibrium, is subjected to the effects of an external influence, the system moves in that direction which tends to nullify the effects of the external influence." Thus if the system is a low-temperature phase in equilibrium with a high-temperature phase at the equilibrium phase-transition temperature, e.g., solid and liquid coexisting at the melting point, then the introduction of heat into this system (the external influence) results in the system's undergoing a reaction which tends to nullify the effect of the introduction of heat. This reaction is the solid melting with an absorption of heat, and the absorption of heat (the latent heat of the phase change) thus nullifies the effect of the addition of heat (which is the raising of the temperature) and the phase change occurs isothermally. A phase change from a low- to a high-temperature phase is always endothermic, and hence ΔH for the change is always a positive quantity; for

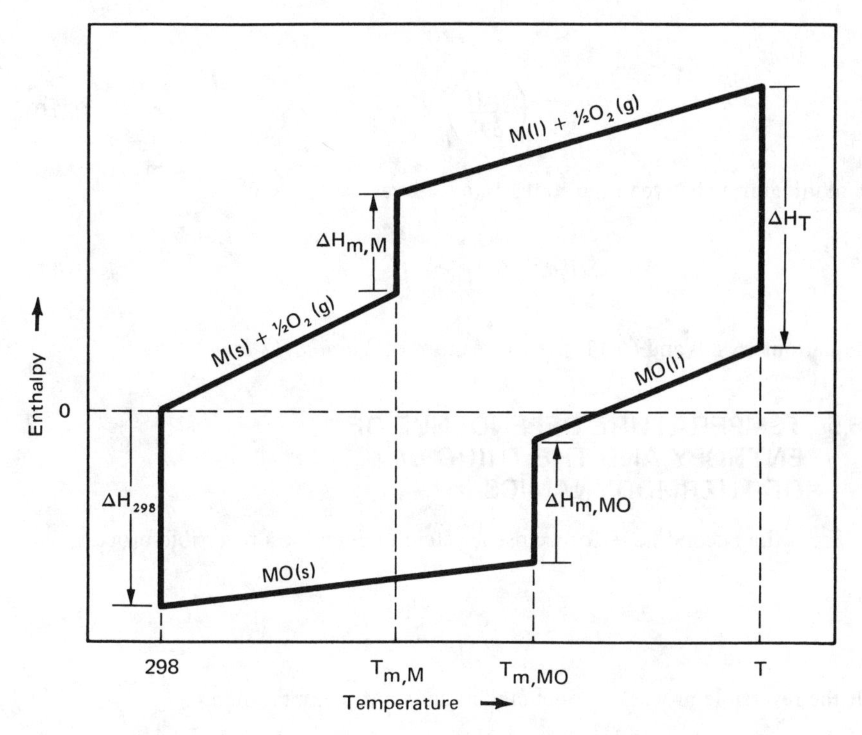

Fig. 6.11. The effect of phase changes on the enthalpy change of a reaction.

example, ΔH_m is always positive. The general Eq. (6.9) can be obtained as follows.

$$\text{For the state 1} \quad \left(\frac{\partial H_1}{\partial T}\right)_P = c_{p(1)}$$

$$\text{For the state 2} \quad \left(\frac{\partial H_2}{\partial T}\right)_P = c_{p(2)}$$

Subtraction of these equations gives, for the constant-pressure change of state, $1 \rightarrow 2$,

$$\left(\frac{\partial H_2}{\partial T}\right)_P - \left(\frac{\partial H_1}{\partial T}\right)_P = c_{p(2)} - c_{p(1)}$$

$$\left[\frac{\partial (H_2 - H_1)}{\partial T}\right]_P = \Delta c_p$$

or

$$\left(\frac{\partial \Delta H}{\partial T}\right)_P = \Delta c_p \tag{6.10}$$

and integrating between the states 1 and 2 gives

$$\Delta H_{T_2} - \Delta H_{T_1} = \int_{T_1}^{T_2} \Delta c_p \, dT \tag{6.11}$$

Equations (6.10) and (6.11) are expressions of *Kirchhoff's law*.

6.5 TEMPERATURE DEPENDENCE OF ENTROPY AND THE THIRD LAW OF THERMODYNAMICS

From the Second Law, for a closed system undergoing a reversible process,

$$dS = \frac{\delta q}{T} \tag{3.8}$$

If the reversible process is conducted at constant pressure, then

$$dS = \left(\frac{\delta q}{T}\right)_P = \left(\frac{dH}{T}\right)_P = c_p \frac{dT}{T}$$

and thus if the temperature of a closed system of fixed composition is increased from T_1 to T_2 at constant pressure, then the increase in the entropy per mole of the system, ΔS, is obtained as

$$\Delta S = S(T_2, P) - S(T_1, P) = \int_{T_1}^{T_2} c_p \frac{dT}{T} = \int_{T_1}^{T_2} c_p d \ln T \tag{6.12}$$

For this change of state, ΔS is obtained as the area under the curve of either of the following:

1. c_p/T versus T between the limits T_2 and T_1
2. c_p versus $\ln T$ between the limits $\ln T_2$ and $\ln T_1$

Generally, S_T, the entropy per mole of the system at any temperature T, is given as

$$S_T = S_0 + \int_0^T c_p d \ln T \tag{6.13}$$

where S_0 is the entropy of the system per mole of the system at 0 K. Consideration of the numerical value of S_0 leads to the enunciation of what is commonly termed the Third Law of Thermodynamics.

In 1906 Nernst postulated that, for chemical reactions between pure solids or liquids, both the terms

$$\left(\frac{\partial \Delta G}{\partial T}\right)_P \quad \text{and} \quad \left(\frac{\partial \Delta H}{\partial T}\right)_P$$

approach zero as the temperature approaches absolute zero. For a change of state of a system, e.g., a chemical reaction, at the constant temperature T, Eq. (5.2) gives

$$\Delta G_T = \Delta H_T - T\Delta S_T$$

and thus Nernst's postulate was that ΔG for the reaction varies with T as is shown in Fig. 6.12.

In Fig. 6.12 at any temperature T, the slope of the line equals $-\Delta S_T$, and the intercept of the tangent to the line at T with the $T = 0$ axis equals ΔH_T. As the temperature approaches zero, the slope of the line approaches zero and the variation of the tangential intercept with temperature approaches zero, the consequences of which are that, as $T \rightarrow 0$, then $\Delta S \rightarrow 0$ and $\Delta c_p \rightarrow 0$. This can be seen analytically by differentiation of Eq. (5.2) with respect to T at constant P.

$$\left(\frac{\partial \Delta G}{\partial T}\right)_P = \left(\frac{\partial \Delta H}{\partial T}\right)_P - T\left(\frac{\partial \Delta S}{\partial T}\right)_P - \Delta S$$

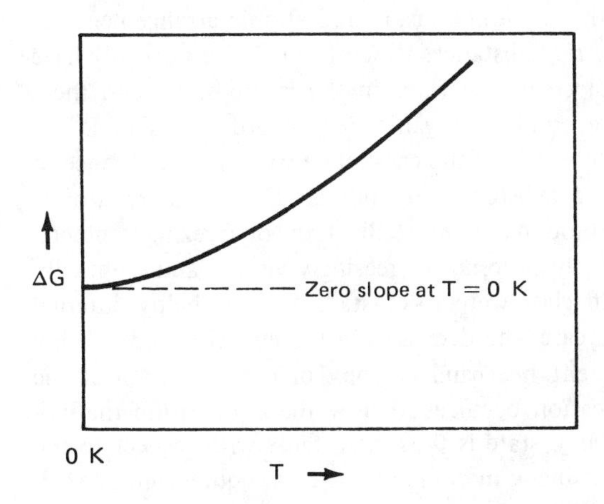

Fig. 6.12. The variation of ΔG with temperature as the temperature approaches absolute zero.

From Eq. (5.12)

$$\left(\frac{\partial \Delta G}{\partial T}\right)_P = -\Delta S$$

and hence

$$\left(\frac{\partial \Delta H}{\partial T}\right)_P = T\left(\frac{\partial \Delta S}{\partial T}\right)_P = \Delta c_p$$

Thus if $(\partial \Delta G/\partial T)_P$ and $(\partial \Delta H/\partial T)_P$ tend to zero as $T \to 0$, then ΔS and Δc_p tend to zero as $T \to 0$ [provided that $(\partial \Delta S/\partial T)_P$ is not infinite at $T = 0$].

The Nernst heat theorem states that "for all reactions involving substances in the condensed state ΔS is zero at the absolute zero of temperature." Thus for the general reaction

$$\text{A} + \text{B} = \text{AB}$$

$\Delta S = S_{\text{AB}} - S_{\text{A}} - S_{\text{B}} = 0$ when $T = 0$ K; and if S_{A} and S_{B} are assigned the value of zero at 0 K, then the compound AB also has zero entropy at 0 K.

The incompleteness of Nernst's statement was pointed out by Planck, who enunciated that "the entropy of any homogeneous substance, which is in complete internal equilibrium, may be taken at zero at 0 K." The requirement that the condensed phase be in complete internal equilibrium can be illustrated as follows.

1. Glasses are noncrystalline solids which are properly regarded as being supercooled liquids in which the liquidlike disordered atomic arrangement has been "frozen into" the solid state. Substances which form glasses normally have complex atomic, ionic, or molecular structures in the liquid state, and these liquids would require extensive atomic reorganization in order to assume the regular periodic structure characteristic of the crystalline state. In the absence of the ability of the glass-forming substance to undergo the necessary atomic rearrangement at a unique thermodynamic crystallization, or freezing, temperature, the liquid, on cooling, simply becomes increasingly viscous and eventually forms a solid glass. If the solid glass were to crystallize, its enthalpy, internal energy, and entropy would decrease; the decreases in the enthalpy and entropy would be, respectively, the latent heat and entropy of crystallization at the temperature at which devitrification occurred. Below the equilibrium thermodynamic freezing point, the glassy state is thus metastable with respect to the crystalline state; and a glass, not being in complete internal equilibrium, has an entropy at 0 K which is greater than zero by an amount which is dependent on the degree of atomic disorder in the glass.

2. Solutions are mixtures of atoms, ions, or molecules, and a contribution is made to their entropy by the fact that they are mixtures [see Eq. (4.17)]. This contribution is called the entropy of mixing and is determined by the randomness with which the component particles are mixed in the solution. The atomic randomness of a mixture determines its degree of order; e.g., in the case of an alloy containing 50 atomic percent of A and 50 atomic percent of B, complete order would correspond to every A atom being surrounded only by B atoms and vice versa, and complete randomness would correspond to each atom having, on the average, 50 percent of its neighbors as A atoms and 50 percent as B atoms. Respectively, the degrees of order in these two configurations would be unity and zero. The equilibrium degree of order is temperature-dependent and increases as the temperature is decreased. However, the maintenance of the equilibrium degree of order is dependent on the ability of the particles to change their positions in the solution; and, with ever-decreasing temperature, as atomic mobility decreases exponentially with decreasing temperature, the maintenance of internal equilibrium becomes increasingly difficult. Consequently a nonequilibrium degree of order can be frozen into the solid solution, in which case the entropy will not fall to zero at 0 K.

3. Even chemically pure elements are, in fact, mixtures of isotopes, and because of the chemical similarity between isotopes it would be expected that completely random mixing of the isotopes would occur. Thus an entropy of mixing results and so, in a strict sense, the entropy of such an element will not fall to zero at 0 K. For example, solid chlorine at 0 K is a solid solution of $Cl^{35}-Cl^{35}$, $Cl^{35}-Cl^{37}$, and $Cl^{37}-Cl^{37}$ molecules. However as this entropy of mixing is present in any other substance which contains the element, it is customary to ignore it.

4. At any temperature a pure crystalline solid contains an equilibrium number of point defects, e.g., vacant lattice sites which, as a result of their random positioning in the crystal, give rise to an entropy of mixing which is exactly the same as the entropy of mixing in a chemical solution. For internal equilibrium to be maintained, this number of defects must decrease as the temperature is decreased. This decrease is effected by the vacancies diffusing to the surface of the crystal where they disappear. Again, as diffusivity decreases with decreasing temperature, a nonequilibrium concentration of defects can be frozen into the crystal, giving rise, thus, to a nonzero entropy at 0 K. Random crystallographic orientation of molecules in the crystalline state can also give rise to a nonzero entropy at 0 K. Such is the case with solid CO, where a solid structure such as

```
CO      CO      CO      OC      CO      OC
    OC      OC      CO      CO      OC
OC      OC      CO      OC      OC      OC
```

can occur. The maximum entropy would occur if equal numbers of molecules were oriented in opposite directions and random mixing of the two orientations occurred. Per mole, the configurational entropy of mixing would then, from Eq. (4.17), be

$$\Delta S_{\text{conf}} = k \ln \frac{(\mathcal{A})!}{(\frac{1}{2}\mathcal{A})!(\frac{1}{2}\mathcal{A})!}$$

where $\mathcal{A}$ is Avogadro's Number, 6.0232×10^{23} per mole. Thus, via Stirling's approximation,

$$\begin{aligned} \Delta S_{\text{conf}} &= k\,[\mathcal{A} \ln \mathcal{A} - \tfrac{1}{2}\mathcal{A} \ln (\tfrac{1}{2}\mathcal{A}) - \tfrac{1}{2}\mathcal{A} \ln (\tfrac{1}{2}\mathcal{A})] \\ &= k \times 4.175 \times 10^{23} \\ &= 1.38054 \times 20^{-23} \times 4.175 \times 10^{23} \\ &= 5.76 \text{ joules/degree·mole} \end{aligned}$$

In comparison with the measured entropy of 4.2 joules/degree·mole, this indicates that the actual molecular orientations are not fully random.

As a result of the above considerations, the statement of the Third Law of Thermodynamics requires the inclusion of the qualification that the homogeneous phase be in complete internal equilibrium.

6.6 EXPERIMENTAL VERIFICATION OF THE THIRD LAW

The Third Law can be experimentally tested by examining the behavior of some simple phase transition of an element, e.g.,

$$\alpha = \beta$$

where α and β are solid allotropes of the element. In Fig. 6.13 T_{trans} is the temperature, at unit pressure, at which the α and β phases are in equilibrium with each other; and, as entropy is a state property, then for the cycle shown in Fig. 6.13,

$$\Delta S_{\text{IV}} = \Delta S_{\text{I}} + \Delta S_{\text{II}} + \Delta S_{\text{III}}$$

For the Third Law to be obeyed, $\Delta S_{\text{IV}} = 0$, and hence

$$\Delta S_{\text{II}} = -(\Delta S_{\text{I}} + \Delta S_{\text{III}})$$

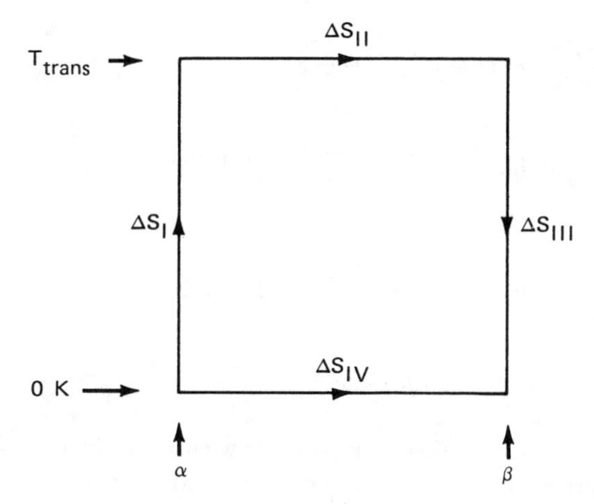

Fig. 6.13. Illustration of the experimental verification of the Third Law of Thermodynamics.

where

$$\Delta S_{\text{I}} = \int_0^{T_{\text{trans}}} c_{p(\alpha)} d \ln T$$

$$\Delta S_{\text{II}} = \frac{\Delta H_{\text{trans}}}{T_{\text{trans}}}$$

and

$$\Delta S_{\text{III}} = \int_{T_{\text{trans}}}^{0} c_{p(\beta)} d \ln T$$

ΔS_{II} is termed the "experimental entropy change," and $-(\Delta S_{\text{I}} + \Delta S_{\text{III}})$ is termed the "Third Law entropy change." These are equal if the Third Law is obeyed.

The cycle shown in Fig. 6.13 has been examined for the case of sulfur, which has two solid allotropes, a monoclinic form which is stable above 368.5 K and a rhombohedral form which is stable below 368.5 K, with the latent heat of transformation being 400 joules/mole at 368.5 K. As monoclinic sulfur can be supercooled with relative ease, the temperature dependence of the heat capacities of both allotropes can be determined experimentally. From these dependencies it was found that

$$\Delta S_{\text{I}} = \int_0^{368.5} c_{p(\text{rhombic})} d \ln T = 36.86 \text{ joules/degree}$$

$$\Delta S_{\text{II}} = \frac{\Delta H_{\text{trans}}}{T_{\text{trans}}} = \frac{400}{368.5} = 1.09 \text{ joules/degree}$$

$$\Delta S_{\text{III}} = \int_{368.5}^{0} c_{p\,(\text{monoclinic})} d \ln T = -37.8 \text{ joules/degree}$$

Thus

$$-(\Delta S_{\text{I}} + \Delta S_{\text{III}}) = -(36.86 - 37.8) = +0.94 \text{ joules/degree}$$

and

$$\Delta S_{\text{II}} = +1.09 \text{ joules/degree}$$

As the difference between the experimental and the Third Law entropy changes is less than the experimental error, this is taken as experimental verification of the Third Law.

Assigning a zero value to S_0 allows the absolute entropy of any pure substance to be determined as

$$S_T = \int_0^T c_p d \ln T \text{ joules/degree}$$

and molar entropies are normally tabulated at 298 K, where

$$S_{298} = \int_0^{298} c_p d \ln T \text{ joules/degree}$$

The values of S_{298} for a number of elements are listed in Table 6.4 along with the heats and temperatures of phase transition of the elements. The variations of S_T with temperature are illustrated in Fig. 6.14. Over the temperature range between 298 K and the temperature of the lowest phase change, e.g., the

Table 6.4

Element	S_{298} joules/degree	ΔH_m joules	T_m K	ΔS_m joules/degree
Al	28.3	10,500	932	11.3
Au	47.36	12,800	1336	9.58
Cu	33.3	13,000	1356	9.59
Fe	27.2	13,800	1808	7.63
Pb	64.9	4,810	600	8.02
C(gr)	5.694			
C(diam)	2.44			
O_2	205			

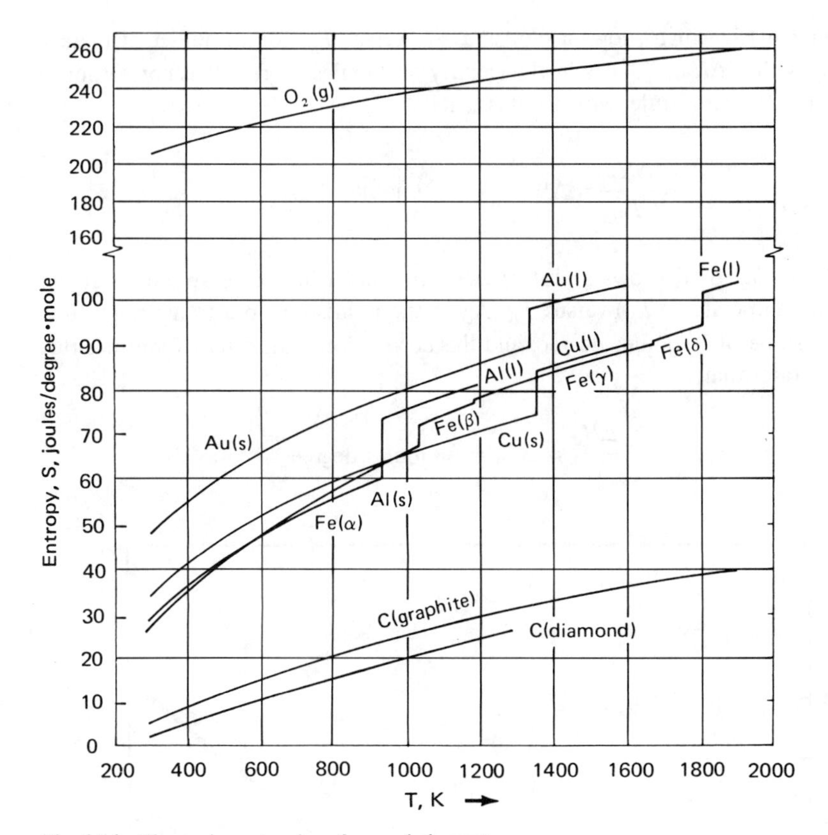

Fig. 6.14. The molar entropies of several elements.

melting temperature in the cases of Au, Al, and Cu, and the $\alpha \rightarrow \beta$ transition in the case of Fe, S_T is obtained as

$$S_T = S_{298} + \int_{298}^{T} c_p d \ln T$$

which, with $c_p = a + bT + cT^{-2}$, gives

$$S_T = S_{298} + a \ln\left(\frac{T}{298}\right) + b(T - 298) - \tfrac{1}{2}\, c\left(\frac{1}{T^2} - \frac{1}{298^2}\right)$$

In the temperature range over which the liquid state exists,

$$S_T = S_{298} + \int_{298}^{T_m} c_{p(s)} d \ln T + \Delta S_m + \int_{T_m}^{T} c_{p(l)} d \ln T$$

where ΔS_m, the entropy of melting, is obtained as $\Delta H_m / T_m$.

Figure 6.15, which shows a plot of ΔH_m versus T_m for a number of metals, indicates that $\Delta H_m/T_m$ lies in the vicinity of 9 joules/degree. This correlation is known as Richards' rule, which states that

$$\frac{\Delta H_m}{T_m} = \Delta S_m \sim 8 \text{ to } 16 \text{ joules/degree}$$

Similarly Fig. 6.16 shows a plot of ΔH_v (the molar heat of evaporation at the boiling temperature, T_b) versus T_b. Again the points lie close to a straight line with a slope of 88 joules/degree, and this correlation is known as Trouton's rule which states that

$$\frac{\Delta H_v}{T_b} = \Delta S_b = 88 \text{ joules/degree}$$

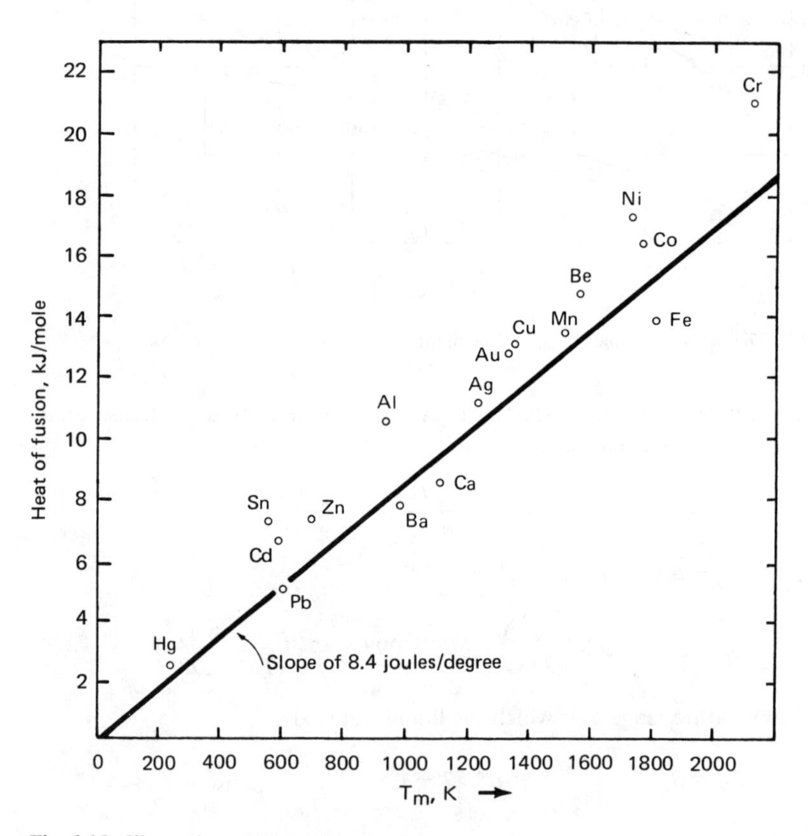

Fig. 6.15. Illustration of Richards' rule.

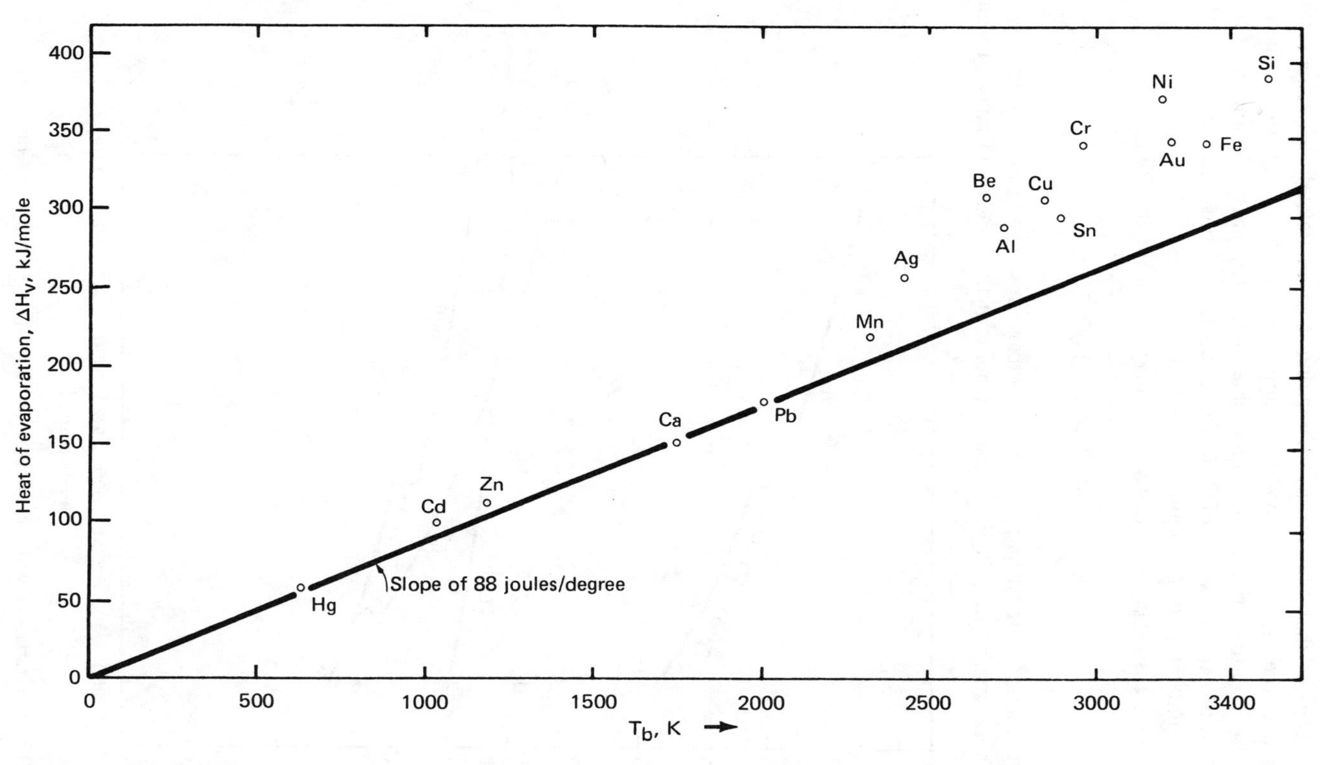

Fig. 6.16. Illustration of Trouton's rule.

It thus appears that the entropy of melting is approximately the same for all metals and the entropy of boiling is approximately the same for all metals. Generally Trouton's rule is more applicable than is Richards' rule, and it is often used to estimate the value of ΔH_v for a substance from a knowledge of the boiling temperature of the liquid.

Figure 6.17 is the entropy-temperature diagram for the reaction

$$Pb + \tfrac{1}{2}O_2 = PbO$$

corresponding to the enthalpy-temperature diagram shown in Fig. 6.9. In view of the similar magnitudes of the molar entropies of the condensed phases Pb and

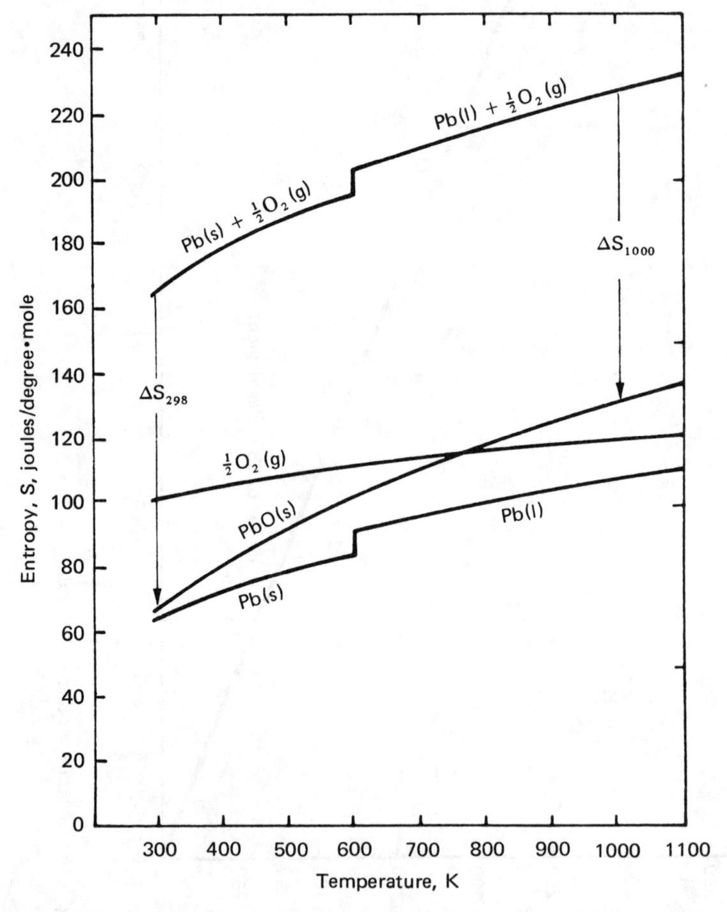

Fig. 6.17. The variation with temperature of the entropy change for the reaction $Pb + \frac{1}{2}O_2 = PbO$.

PbO, it is seen that the entropy change for the oxidation, being given as

$$\Delta S_T = S_{T,\,\mathrm{PbO}} - S_{T,\,\mathrm{Pb}} - \tfrac{1}{2} S_{T,\mathrm{O}_2}$$

is very nearly equal to $-\frac{1}{2} S_{T,\mathrm{O}_2}$; for example, at 298 K,

$$\begin{aligned}\Delta S_{298} &= S_{298,\mathrm{PbO}} - S_{298,\mathrm{Pb}} - \tfrac{1}{2} S_{298,\mathrm{O}_2} \\ &= 67.4 - 64.9 - \tfrac{1}{2} \times 205 \\ &= -100 \text{ joules/degree}\end{aligned}$$

which is very nearly equal to the entropy decrease resulting from the disappearance of half a mole of oxygen gas. This approximation is generally valid; i.e., in reactions in which a gas reacts with a condensed phase to produce a condensed phase, the entropy change is essentially that corresponding to the disappearance of the gas.

6.7 ENTHALPY AND ENTROPY AS FUNCTIONS OF PRESSURE

For a closed system of fixed composition undergoing a change of pressure at constant temperature,

$$dH = \left(\frac{\partial H}{\partial P}\right)_T dP$$

Equation (5.10) gives $dH = T\,dS + V\,dP$, and thus

$$\left(\frac{\partial H}{\partial P}\right)_T = T\left(\frac{\partial S}{\partial P}\right)_T + V$$

Maxwell's relation (5.34) gives $(\partial S/\partial P)_T = -(\partial V/\partial T)_P$, in which case

$$\left(\frac{\partial H}{\partial P}\right)_T = -T\left(\frac{\partial V}{\partial T}\right)_P + V$$

As the isobaric coefficient of thermal expansion α, is defined as $\alpha = 1/V(\partial V/\partial T)_P$, then

$$\begin{aligned}\left(\frac{\partial H}{\partial P}\right)_T &= -T\alpha V + V \\ &= V(1 - \alpha T)\end{aligned}$$

and hence for the change of state (P_1, T) to (P_2, T),

$$\Delta H = H(P_2, T) - H(P_1, T) = \int_{P_1}^{P_2} V(1 - \alpha T) dP \tag{6.14}$$

For an ideal gas, $\alpha = 1/T$, and hence Eq. (6.14) again demonstrates that the enthalpy of an ideal gas is independent of pressure. Taking the molar volume of iron to be 7.1 cm^3 and the expansivity to be 0.3×10^{-4} $(K)^{-1}$, and assuming both of these to be virtually independent of pressure, gives 71 joules as the increase in the molar enthalpy of iron resulting from an increase in pressure from 1 to 100 atm at 298 K. The same enthalpy increase would be obtained by heating iron from 298 to 301 K at 1 atm pressure. Similarly for aluminum, with $V_{Al} = 1$ cm^3 and $\alpha_{Al} = 0.69 \times 10^{-4}$ $(K)^{-1}$, the same pressure increase at 298 K raises the molar enthalpy by 99.2 joules and this equals the increase in H_{Al} obtained by heating from 298 to 302 K at 1 atm pressure.

For a closed system of fixed composition undergoing a change of pressure at constant temperature,

$$dS = \left(\frac{\partial S}{\partial P}\right)_T dP$$

From Maxwell's relation, Eq. (5.34),

$$\left(\frac{\partial S}{\partial P}\right)_T = -\left(\frac{\partial V}{\partial T}\right)_P \quad \text{and as} \quad \alpha = \frac{1}{V}\left(\frac{\partial V}{\partial T}\right)_P$$

then

$$\left(\frac{\partial S}{\partial P}\right)_T = -\alpha V$$

Hence, for the change of state (P_1, T) to (P_2, T),

$$\Delta S = S(P_2, T) - S(P_1, T) = -\int_{P_1}^{P_2} \alpha V dP \tag{6.15}$$

For an ideal gas, with $\alpha = 1/T$, Eq. (6.15) simplifies to

$$\Delta S = -\int_{P_1}^{P_2} R \ln P = -R \ln\left(\frac{P_2}{P_1}\right) = -R \ln\left(\frac{V_1}{V_2}\right)$$

as was obtained in Sec. 3.7.

In the case of iron at 298 K, an increase in the pressure from 1 to 100 atm decreases the molar entropy by 0.0022 joules/degree, and for aluminum the corresponding decrease is 0.007 joules/degree. These decreases would be obtained by lowering the temperatures of Fe and Al by 0.27 and 0.09 degrees, respectively, from 298 K at 1 atm pressure. It is seen, thus, that the molar enthalpies and entropies of condensed phases are relatively insensitive to pressure change. In metallurgical applications, where the range of pressure is normally from 0 to 1 atm, this is particularly so.

For a closed system of fixed composition undergoing both temperature and pressure changes, combination of Eqs. (6.1) and (6.14) gives

$$\Delta H = H(P_2, T_2) - H(P_1, T_1) = \int_{T_1}^{T_2} c_p dT + \int_{P_1}^{P_2} V(1 - \alpha T) dP \qquad (6.16)$$

and combination of Eqs. (6.12) and (6.15) gives

$$\Delta S = S(P_2, T_2) - S(P_1, T_1) = \int_{T_1}^{T_2} c_p d \ln T - \int_{P_1}^{P_2} \alpha V dP \qquad (6.17)$$

Just as it was required that the temperature dependence of c_p be known for integration of Eqs. (6.1) and (6.13), strictly, it is required that the pressure dependence of V and α be known for integration of Eqs. (6.14) and (6.15). However, for condensed phases being considered over small ranges of pressure, both of these pressure dependencies can usually be ignored.

6.8 SUMMARY

In this chapter it has been seen that knowledge of the heat capacities of substances and the heats of formation of compounds allows enthalpy and entropy changes to be evaluated for any process, i.e., for phase changes and chemical reactions. In practice the heat-capacity and heat-of-formation data used in such calculations are experimentally determined by calorimetric methods. In dealing with enthalpy it is conventional to assign the value of zero to the enthalpy of elements in their stable state of existence at 298 K; but as the entropy of all substances (which are in complete internal equilibrium) at 0 K is zero, then any substance in any thermodynamic state has a calculable absolute entropy.

Although both enthalpy and entropy are dependent on pressure and temperature, the pressure dependency of H and S of substances occurring in condensed phases is normally small enough to be neglected in most applications (which involve pressure variation in the range of 0 to 1 atm).

The determination of ΔH_T and ΔS_T for any reaction at any temperature (and pressure) allows the all-important Gibbs free energy change for the reaction to be calculated as

$$\Delta G_T = \Delta H_T - T\Delta S_T$$

As consideration of the Gibbs free energy in any isothermal, isobaric process provides the criterion of reaction equilibrium; it is seen that the equilibrium state of a reaction system can be determined from a knowledge of the thermochemical properties of the components of the system. An examination of this is started in Chap. 7.

6.9 NUMERICAL EXAMPLES

Example 1

A mixture of Fe_2O_3 and Al, present in the molar ratio 1:2, is adiabatically contained along with some Fe at 25°C. If the Thermit reaction ($Fe_2O_3 + 2Al = Al_2O_3 + 2Fe$) is allowed to proceed to completion, calculate the molar ratio of Fe to Fe_2O_3 in the initial mixture which, on completion of the reaction, will give liquid Fe and solid Al_2O_3 at 1600°C. Given:

$$H_{298(Fe_2O_3)} = -821{,}300 \text{ joules/mole}$$

$$H_{298(Al_2O_3)} = -1{,}674{,}000 \text{ joules/mole}$$

$$c_{p(Al_2O_3)} = 106.6 + 17.8 \times 10^{-3}T - 28.5 \times 10^5 T^{-2} \text{ joules/degree·mole}$$

in the temperature range 298–1873 K

$$c_{p(Fe\alpha)} = 17.5 + 24.8 \times 10^{-3}T \text{ joules/degree·mole from 273 to 1033 K}$$

$$c_{p(Fe\beta)} = 38 \text{ joules/degree·mole from 1033 to 1181 K}$$

$$c_{p(Fe\gamma)} = 7.7 + 19.5 \times 10^{-3}T \text{ joules/degree·mole from 1181 to 1674 K}$$

$$c_{p(Fe\delta)} = 43.9 \text{ joules/degree·mole from 1674 to 1808 K}$$

$$c_{p(Fel)} = 41.8 \text{ joules/degree·mole from 1808 to 1873 K}$$

For Feα → Feβ ΔH_{trans} = 5020 joules/mole at T_{trans} = 1033 K
For Feβ → Feγ ΔH_{trans} = 920 joules/mole at T_{trans} = 1181 K
For Feγ → Feδ ΔH_{trans} = 880 joules/mole at T_{trans} = 1674 K
For Feδ → Fel ΔH_m = 13,800 joules/mole at T_m = 1808 K

Consider the system per mole of Fe_2O_3 and consider the process path,

I. $Fe_2O_3 + 2Al \Rightarrow 2Fe + Al_2O_3$ at 298 K, followed by
II. $(2+n)Fe + Al_2O_3$ at 298°K $\Rightarrow (2+n)Fe + Al_2O_3$ at 1873 K

where n is the number of moles of Fe present in the initial mixture per mole of Fe_2O_3. As the system is adiabatically contained, $\Delta H_I + \Delta H_{II} = 0$. For the Thermit reaction at 298 K

$$Fe_2O_3 + 2Al = Al_2O_3 + 2Fe$$

$$\Delta H_{298} = -1{,}674{,}000 + 821{,}300 = -852{,}700 \text{ joules}$$
$$= \Delta H_I = H_b - H_a$$

ΔH_{II} = the heat to raise the temperature of $(n+2)$ moles of Fe from 298 to 1873 K (ΔH_1) + the heat to raise the temperature of 1 mole of Al_2O_3 from 298 to 1873 K (ΔH_2).

In Fig. 6.18 the slope of each individual segment of the line bc is $(n+2)c_{p(Fe)} + c_{p(Al_2O_3)}$, and the length of each vertical segment equals $(n+2)\Delta H_{Fe(trans)}$. Thus the temperature of the point c is determined by the value of n.

$$\Delta H_2 = \int_{298}^{1873} c_{p(Al_2O_3)}dT$$
$$= 106.6(1873-298) + \frac{17.8}{2} \times 10^{-3}(1873^2 - 298^2)$$
$$+ 28.5 \times 10^5 \left(\frac{1}{1873} - \frac{1}{298}\right) = 190{,}300 \text{ joules}$$

$$\frac{\Delta H_1}{(n+2)} = 17.5(1033-298) + \frac{24.8}{2} \times 10^{-3}(1033^2 - 298^2)$$
$$+ 5020 + 38 \times (1181 - 1033) + 920$$
$$+ 7.7(1674 - 1181) + \frac{19.5}{2} \times 10^{-3}(1674^2 - 1181^2)$$
$$+ 880 + 43.9 \times (1808 - 1674) + 13{,}800$$
$$+ 41.8 \times (1873 - 1808)$$
$$= 77{,}360 \text{ joules}$$

Thus

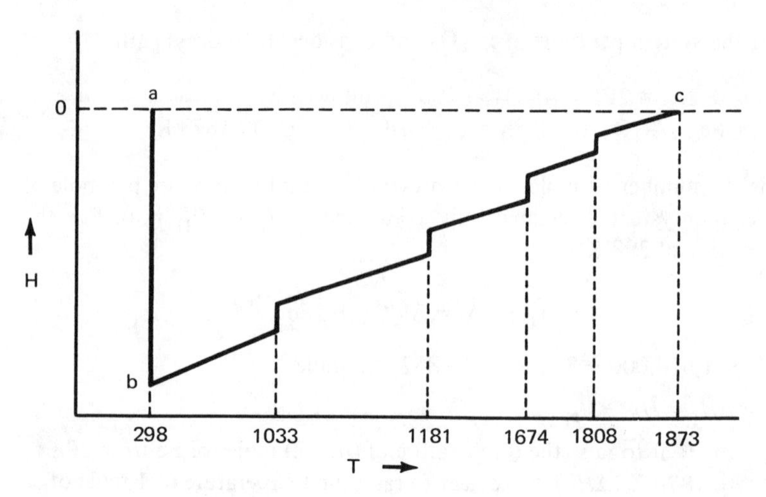

Fig. 6.18.

$$852{,}700 = 77{,}360n + 154{,}720 + 190{,}300$$

or $n = 6.56$ moles of Fe per mole of Fe_2O_3 initially present.

Example 2

A quantity of supercooled liquid tin is adiabatically contained at 495 K. Calculate the fraction of tin which spontaneously freezes. Given:

$$\Delta H_{m\,(\text{Sn})} = 7070 \text{ joules/mole at } T_m = 505 \text{ K}$$

$$c_{p,\text{Sn}\,(l)} = 34.7 - 9.2 \times 10^{-3}\,T \text{ joules/degree·mole}$$

$$c_{p,\text{Sn}\,(s)} = 18.5 + 26 \times 10^{-3}\,T \text{ joules/degree·mole}$$

The equilibrium state of the adiabatically contained system is that in which solid, which has spontaneously formed, and the remaining liquid coexist at 505 K. Thus the fraction of liquid which freezes is of such magnitude as to release the heat necessary to raise the temperature of the system from 495 to 505 K.

Let the molar fraction which freezes equal x and consider 1 mole of the system. In Fig. 6.19 the process is represented by the change of state from a to c, and as the process is adiabatic the enthalpy of the system remains constant; i.e.,

$$\Delta H = H_c - H_a = 0$$

Two paths can be conveniently considered:

Path I, $a \rightarrow b \rightarrow c$; i.e., all of the liquid is increased in temperature from 495 to 505 K, and then the fraction x freezes at 505 K. In this case,

$$\Delta H_{(a\rightarrow b)} = -\Delta H_{(b\rightarrow c)}$$

$$\Delta H_{(a\rightarrow b)} = \int_{495}^{505} c_{p,\,\mathrm{Sn}\,(l)}\,dT = 34.7(505-495) - \frac{9.2}{2} \times 10^{-3}(505^2 - 495^2)$$

$$= 301 \text{ joules}$$

$$\Delta H_{(b\rightarrow c)} = -7070x \text{ joules}$$

$$\text{and so } x = \frac{301}{7070} = 0.0426$$

i.e., on a molar basis 4.26 percent of the tin freezes.

Path II, $a \rightarrow d \rightarrow c$; i.e., the fraction x freezes at 495 K, and then the temperature of the solid formed and remaining liquid is increased from 495 to 505 K. In this case

$$\Delta H_{(a\rightarrow d)} = -\Delta H_{(d\rightarrow c)}$$

$$\Delta H_{(a\rightarrow d)} = \text{the heat of freezing of } x \text{ moles of tin at 495 K}$$

$$= -x\Delta H_m(495\ \mathrm{K})$$

$$\text{But}\quad \Delta H_m(495\ \mathrm{K}) = \Delta H_m(505\ \mathrm{K}) + \int_{505}^{495} \Delta c_{p(s\rightarrow l)}\,dT$$

$$= 7070 + 16.2(495-505) - \frac{35.2}{2} \times 10^{-3}(495^2 - 505^2)$$

$$= 7084 \text{ joules/mole}$$

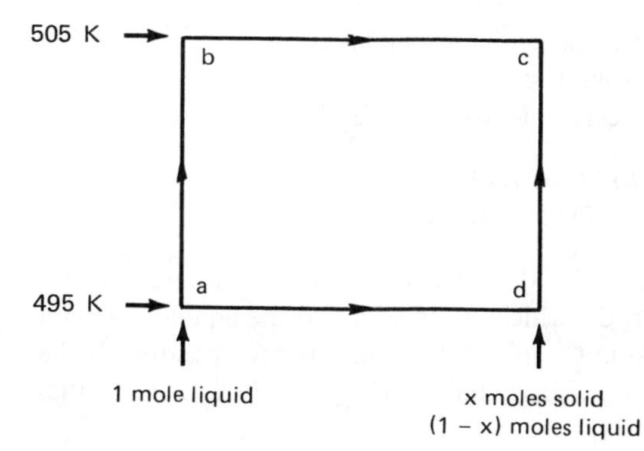

Fig. 6.19.

Thus $\Delta H_{(a \to d)} = -7084x$ joules

$$\Delta H_{(d \to c)} = x \int_{495}^{505} c_{p(s)} dT + (1-x) \int_{495}^{505} c_{p(l)}\, dT$$

$$= x\left[18.5(505-495) + \frac{26}{2} \times 10^{-3}(505^2 - 495^2)\right]$$

$$+ (1-x)\left[34.7(505-495) - \frac{9.2}{2} \times 10^{-3}(505^2 - 495^2)\right]$$

$$= 301 + 14x$$

Thus $$-7084x = -14x - 301$$

and so $$x = \frac{301}{7070} = 0.0426$$

The actual path followed by the process will be intermediate between paths I and II; i.e., the process of freezing and temperature increase will occur simultaneously.

PROBLEMS

Thermodynamic Data, Necessary for the Solution of These Problems, is Tabulated in the Appendix

6.1 Calculate ΔH_{1000} and ΔS_{1000} for the reaction

$$Pb_{(l)} + \tfrac{1}{2}O_2(g) = PbO_{(s)}$$

6.2 Calculate the difference in molar enthalpy, molar entropy, and molar free energy between liquid and solid bismuth at 800 K.

6.3 Which of the following processes releases more heat?

(*i*) Oxidation of graphite to CO at 1000 K
(*ii*) Oxidation of diamond to CO at 1000 K

6.4 An adiabatic vessel contains 1000 grams of liquid aluminum at 700°C. Calculate the weight of Cr_2O_3 which, when added to the liquid aluminum (with which it reacts to form Cr and Al_2O_3) raises the temperature of the resulting mixture of $Al_2O_{3(s)}$, $Cr_{(s)}$, and $Cr_2O_{3(s)}$ to 1000°C. The initial temperature of the added Cr_2O_3 is 25°C.

6.5 Calculate the value of ΔG_{600} for the reaction

$$NiO_{(s)} + Co_{(s)} = CoO_{(s)} + Ni_{(s)}$$

How much error is involved if it is assumed that Δc_p for this reaction is zero?

6.6 One mole of gold is taken from state 1 ($P = 1$ atm, $T = 293$ K) to state 2 ($P = 1$ atm, $T = 273$ K). What pressure must be applied to the gold at 273 K in order to raise its enthalpy back to that of state 1 (call this state 3)? Calculate the entropy difference between states 1 and 3. Given: the density of Au at 20°C is 19.30 grams/cm^3; the coefficient of cubical expansion of Au is 4.32×10^{-5} K^{-1}, which is independent of pressure; and the atomic weight of Au is 197.

6.7 Calculate ΔH_{298} and ΔG_{298} for the reaction

$$4CaO + Si = 2Ca + Ca_2SiO_4$$

6.8 A calorimeter of water-equivalent 20 grams contains 100 grams of water at 25°C. When a 100-gram mass of aluminum at 121°C is placed in the calorimeter, the final temperature at 40°C. When the initial temperature of the aluminum is 243°C, the final temperature in the calorimeter is 60°C; and when the initial temperature of the aluminum is 359°C, the final temperature is 80°C. From this data determine the molar heat capacity of aluminum in the form

$$c_p = a + bT$$

The atomic weight of Al is 26.98 and $c_{p,H_2O} = 75.44$ joules/degree in the temperature range 273 to 373 K.

6.9 Calculate ΔH_{1800} for the reaction

$$Mn_{(l)} + \tfrac{1}{2}O_2(g) = MnO_{(s)}$$

6.10 The molar heats of formation and entropies of formation at 298 K, of various vanadium oxides from vanadium metal and oxygen gas, are listed below. From this information calculate ΔG_{298}, ΔH_{298}, and ΔS_{298} for the reactions:

(*a*) $4VO(s) + O_2(g) = 2V_2O_3(s)$
(*b*) $2V_2O_3(s) + O_2(g) = 4VO_2(s)$
(*c*) $4VO_2(s) + O_2(g) = 2V_2O_5(s)$

Oxide	ΔH_{298} kJ/mole of oxide	ΔS_{298} joules/degree·mole of oxide
VO	−431.8	−92.5
V_2O_3	−1219	−358
VO_2	−713.8	−184
V_2O_5	−1550	−441

CHAPTER SEVEN

PHASE EQUILIBRIA IN A ONE-COMPONENT SYSTEM

7.1 INTRODUCTION

The intensive properties of a system are temperature, pressure, and the chemical potentials of the various species present; these intensive properties are all measures of potentials of one kind or another.

The temperature of a system is a measure of the potential or intensity of heat in the system. Temperature is thus a measure of the tendency of the heat to leave the system; i.e., if two parts of a system are at different temperatures, then a heat-potential gradient exists which produces a driving force for the flow of heat down this gradient from the higher temperature part to the lower temperature part. Spontaneous heat flow thus occurs until the potential gradient has been eliminated, in which state the heat is distributed at uniform intensity throughout the system; thermal equilibrium is thus established when the temperatures of all parts of the system are equal to one another.

The pressure of a system is a measure of its tendency towards massive movement. If the pressure exerted by one phase in a system of fixed volume is greater than the pressure exerted by another phase, then the tendency of the first phase to expand exceeds that of the second phase. The resultant pressure gradient causes the expansion of the first phase, thus decreasing its pressure and hence its tendency towards further expansion, and the contraction of the other phase, thus increasing its pressure and hence its tendency to oppose further contraction. Equilibrium occurs when massive movement of the two phases has occurred to the extent that the pressure gradient has been eliminated, in which state the pressures exerted by the various parts of the system are equal.

The chemical potential of the species i in a phase is a measure of the tendency of the species i to leave the phase. It is thus a measure of the "chemical pressure" exerted by i in the phase. If the chemical potential of i is different in different phases of the system which are at the same temperature and pressure, then as the escaping tendencies differ, the species i will tend to move from the phases in which it occurs at the higher chemical potential into the phases in which it occurs at the lower chemical potentials. The existence of a chemical-potential gradient is the driving force for chemical diffusion, and equilibrium occurs when the species i is distributed throughout the various phases of the system such that its chemical potential is the same in all phases.

In a closed system of fixed composition, e.g., a one-component system, equilibrium, at the temperature T and the pressure P, occurs when the system has a minimum value of G', consistent with the temperature and pressure of the system. Equilibrium can thus be discussed via examination of the G-T-P relationships. In the following discussion the H_2O system will be used as an example.

7.2 THE VARIATION OF GIBBS FREE ENERGY WITH TEMPERATURE AT CONSTANT PRESSURE

At 1 atm total pressure, ice and water are in equilibrium with one another at 0°C, and hence, for these values of temperature and pressure, the Gibbs free energy, G', of the system is a minimum. If heat is added to the system such that some of the ice melts at 0°C and 1 atm pressure, then provided that the water phase is still present, the equilibrium is not disturbed and the value of G' for the system remains unchanged. If, by the addition of heat, 1 mole of ice is melted, then for the reaction

$$H_2O_{(\mathrm{solid})} = H_2O_{(\mathrm{liquid})} \text{ at 1 atm and 273 K}$$

$$\Delta G = G_{H_2O_{(l)}} - G_{H_2O_{(s)}} = 0$$

Thus, at equilibrium,

$$G_{H_2O_{(l)}} = G_{H_2O_{(s)}} \tag{7.1}$$

where $G_{H_2O_{(s)}}$ is the molar free energy of H_2O in the solid (ice) phase, and $G_{H_2O_{(l)}}$ is the molar free energy of H_2O in the liquid (water) phase. For the system of ice + water containing n moles of H_2O, $n_{H_2O_{(s)}}$ of which are in the ice phase and $n_{H_2O_{(l)}}$ of which are in the water phase, the free energy of the system, G', is given as

$$G' = n_{H_2O_{(s)}} G_{H_2O_{(s)}} + n_{H_2O_{(l)}} G_{H_2O_{(l)}} \tag{7.2}$$

and from Eq. (7.1) it can be seen that, at 0°C and 1 atm pressure, the value of G' is independent of the proportions of the water phase and the ice phase present.

The equality of the molar free energies of H_2O in the solid and liquid phases at 0°C and 1 atm corresponds to the fact that, for equilibrium to occur, the escaping tendency of H_2O from the liquid phase equals the escaping tendency of H_2O from the solid phase. Hence it is to be expected that a relationship exists between the molar free energy and the chemical potential of a component in a phase. Integration of Eq. (5.25) at constant T and P gives

$$G' = \sum_i \mu_i n_i$$

which, for the ice + water system, is written as

$$G' = \mu_{H_2O_{(s)}} n_{H_2O_{(s)}} + \mu_{H_2O_{(l)}} n_{H_2O_{(l)}} \tag{7.3}$$

Comparison of Eqs. (7.2) and (7.3) indicates that $\mu_{H_2O} = G_{H_2O}$, or, generally, that $\mu_i = G_i$; that is, the chemical potential of a species in a particular state equals the Gibbs free energy per mole of the species in the particular state.

This result could have been obtained from a consideration of Eq. (5.16),

$$\left(\frac{\partial G'}{\partial n_i}\right)_{T,P} = \mu_i$$

In a one-component system, the chemical potential of the species i equals the increase in the value of G' for the system resulting from the addition of 1 mole of i at constant T and P. That is,

$$\Delta G' = \mu_i$$

and as the increase in the value of G' of the one-component system is simply the molar free energy of i, then

$$G_i = \mu_i$$

If the ice + water system is at 1 atm pressure and some temperature greater than 0°C, then the system is unstable and the ice spontaneously melts. This process decreases the free energy of the system, and equilibrium is attained when all the ice has melted. That is, for the reaction $H_2O_{(s)} = H_2O_{(l)}$ at $T > 273$ K, $P = 1$ atm,

$$\Delta G = G_{H_2O_{(l)}} - G_{H_2O_{(s)}} < 0$$

i.e.,

$$G_{H_2O(l)} < G_{H_2O(s)}$$

The escaping tendency of H_2O from the solid phase is greater than the escaping tendency of the H_2O from the liquid phase. Conversely, if the temperature is less than 0°C, then

$$G_{H_2O(l)} > G_{H_2O(s)}$$

The variations of $G_{H_2O(l)}$ and $G_{H_2O(s)}$ with temperature, at unit pressure, can be graphically represented as in Fig. 7.1, and the variation of $\Delta G_{(s \to l)}$ with temperature at unit pressure is as shown in Fig. 7.2.

Figures 7.1 and 7.2 show that, at unit pressure and temperatures greater than 0°C, minimum free energy occurs when all of the H_2O is in the liquid phase; and at unit pressure and temperatures lower than 0°C, minimum free energy occurs when all of the H_2O is in the solid phase. The slopes of the lines in Fig. 7.1 are obtained from Eq. (5.25) as

$$\left(\frac{\partial G}{\partial T}\right)_P = -S$$

and the curvatures are obtained, from Eq. (6.12), as

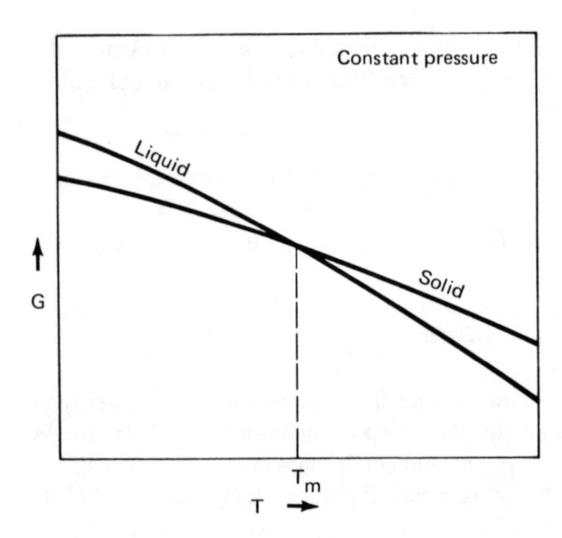

Fig. 7.1. Schematic representation of the molar Gibbs free energies of solid and liquid water as a function of temperature at constant pressure.

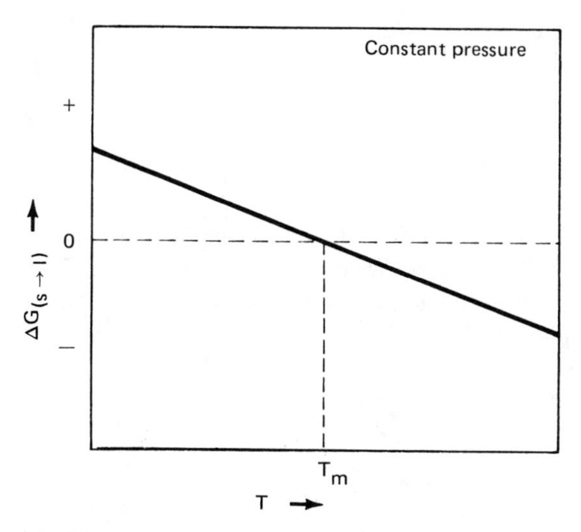

Fig. 7.2. Schematic representation of the molar Gibbs free energy of fusion of water as a function of temperature at constant pressure.

$$\left(\frac{\partial^2 G}{\partial T^2}\right)_P = -\left(\frac{\partial S}{\partial T}\right)_P = -\frac{c_P}{T}$$

Similarly, the slope of the line in Fig. 7.2 is given as

$$\left(\frac{\partial \Delta G}{\partial T}\right)_P = -\Delta S$$

where ΔS is the entropy change for the reaction. The slope in Fig. 7.2 is negative, indicating that at all temperatures

$$S_{H_2O_{(l)}} > S_{H_2O_{(s)}}$$

as is to be expected in view of the fact that at any temperature the liquid phase is more disordered than is the solid phase.

The state in which the solid and liquid phases of a one-component system are in equilibrium with one another can be determined from a consideration of the enthalpy H and the entropy S of the system. From Eq. (5.2),

$$G = H - TS$$

This can be written for both the solid and liquid phases,

$$G_{(l)} = H_{(l)} - TS_{(l)}$$

and
$$G_{(s)} = H_{(s)} - TS_{(s)}$$

and for the reaction solid → liquid, subtraction gives

$$\Delta G_{(s\to l)} = \Delta H_{(s\to l)} - T\Delta S_{(s\to l)}$$

where $\Delta H_{(s \to l)}$ and $\Delta S_{(s \to l)}$ are respectively the molar enthalpy and molar entropy changes accompanying melting at the temperature T. From Eq. (7.1), equilibrium between the solid and liquid phases occurs when $\Delta G_{(s \to l)} = 0$. This occurs at the temperature T_m at which

$$\Delta H_{(s\to 1)} = T_m \Delta S_{(s\to l)} \tag{7.4}$$

For H_2O,

$$\begin{aligned}
\Delta H_m = \Delta H_{(s\to l)} &= 6008 \text{ joules at } 273 \text{ K}\\
S_{H_2O(l),298} &= 70.08 \text{ joules/degree}\\
S_{H_2O(s),298} &= 44.77 \text{ joules/degree}\\
c_{p,H_2O(l)} &= 75.44 \text{ joules/degree}\\
c_{p,H_2O(s)} &= 38 \text{ joules/degree}
\end{aligned}$$

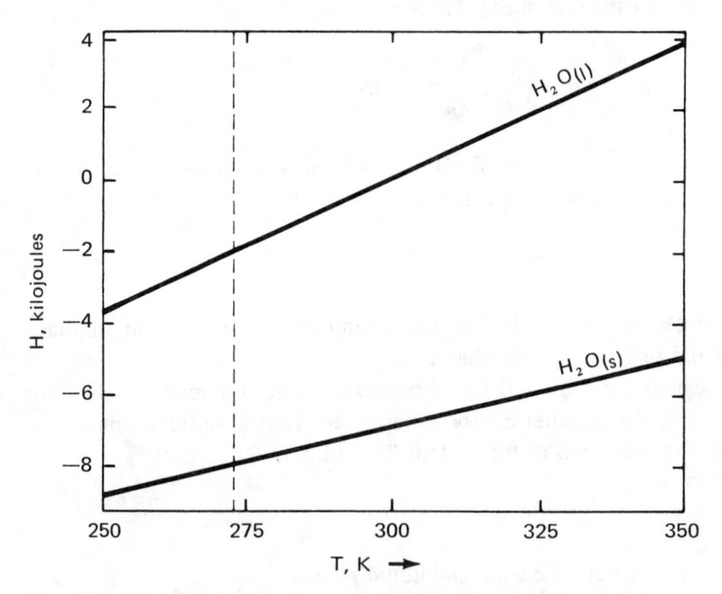

Fig. 7.3*a*. The variations with temperature of the molar enthalpies of solid and liquid water at 1 atm pressure ($H_{H_2O(l)} = 0$ at 298 K).

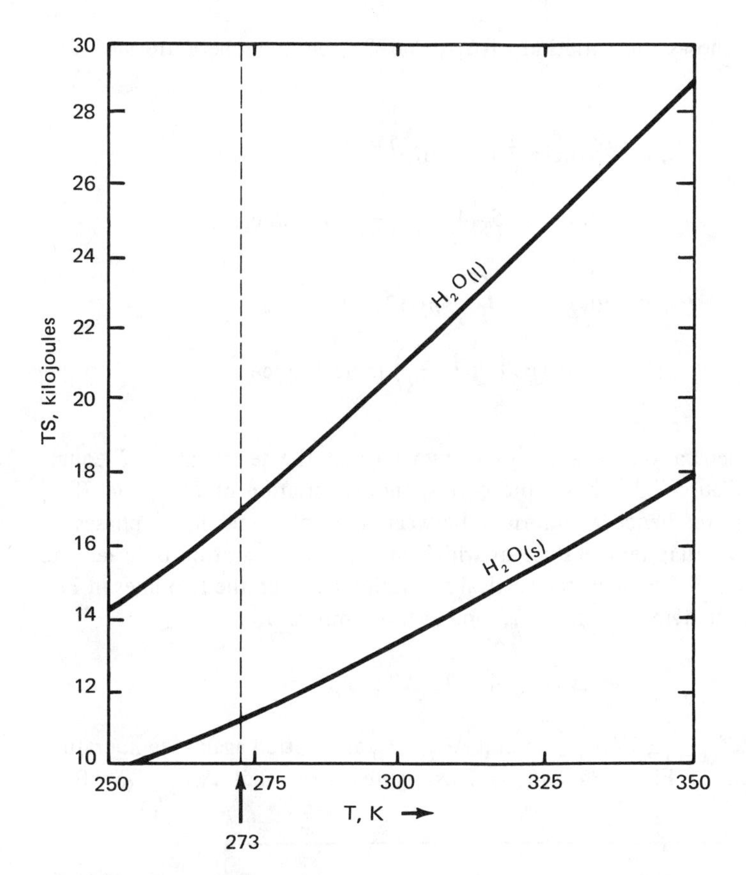

Fig. 7.3*b*. The variations with temperature of TS for solid and liquid water at 1 atm pressure.

Figure 7.3*a* shows the variation of $H_{(s)}$ and $H_{(l)}$ with T at 1 atm pressure where, for convenience, $H_{(l),298} = 0$, in which case,

$$H_{(l),\,T} = \int_{298}^{T} c_{p,(l)}dT = 77.44(T - 298) \text{ joules/degree}$$

$$H_{(s),\,T} = \int_{298}^{273} c_{p,(l)}dT - \Delta H_m + \int_{273}^{T} c_{p,(s)}dT$$

$$= 77.44(273 - 298) - 6008 + 38(T - 273) \text{ joules/degree}$$

The vertical separation between these two lines at the temperature T equals $\Delta H_{(s \to l),T}$.

Figure 7.4 shows the variation of $S_{(s)}$ and $S_{(l)}$ with temperature at 1 atm pressure where

$$S_{(l),\,T} = S_{(l),298} + \int_{298}^{T} c_{p\,(l)} d \ln T$$

$$= 70.08 + 75.44 \ln\left(\frac{T}{298}\right) \text{ joules/degree}$$

$$S_{(s),\,T} = S_{(s),298} + \int_{298}^{T} c_{p\,(s)} d \ln T$$

$$= 44.77 + 38 \ln\left(\frac{T}{298}\right) \text{ joules/degree}$$

The vertical separation between these two lines at the temperature T equals $\Delta S_{(s \to l),T}$. Figure 7.3*b* shows the corresponding variation of $TS_{(s)}$ and $TS_{(l)}$ with temperature; hence equilibrium between the solid and liquid phases of water occurs at that temperature at which the vertical separation between the two lines in Fig. 7.3*a* equals the vertical separation between the two lines in Fig. 7.3*b*. This unique temperature is T_m, and at this temperature,

$$\Delta H_{(s \to l)} = T_m \Delta S_{(s \to l)}$$

In Fig. 7.5, $\Delta H_{(s \to l)}$, $T\Delta S_{(s \to l)}$, and $\Delta G_{(s \to l)}$ are plotted against temperature using the data in Fig. 7.3*a* and *b*. This figure shows that $\Delta G_{(s \to l)} = 0$ at

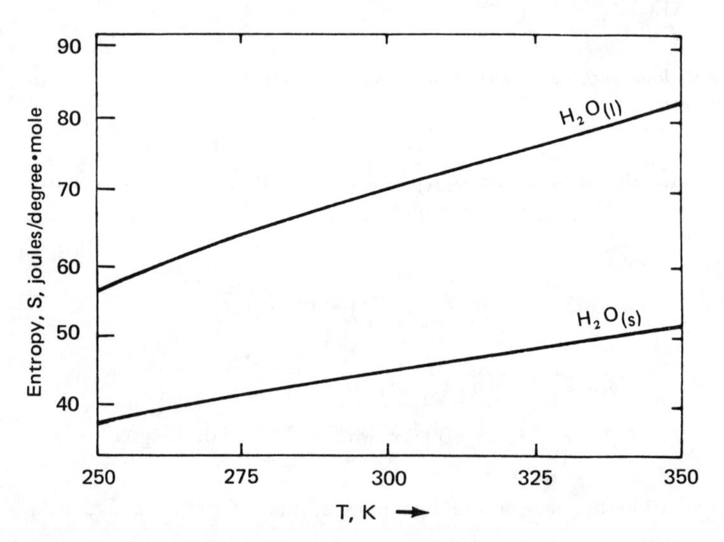

Fig. 7.4. The variations with temperature of the molar entropies of solid and liquid water at 1 atm pressure.

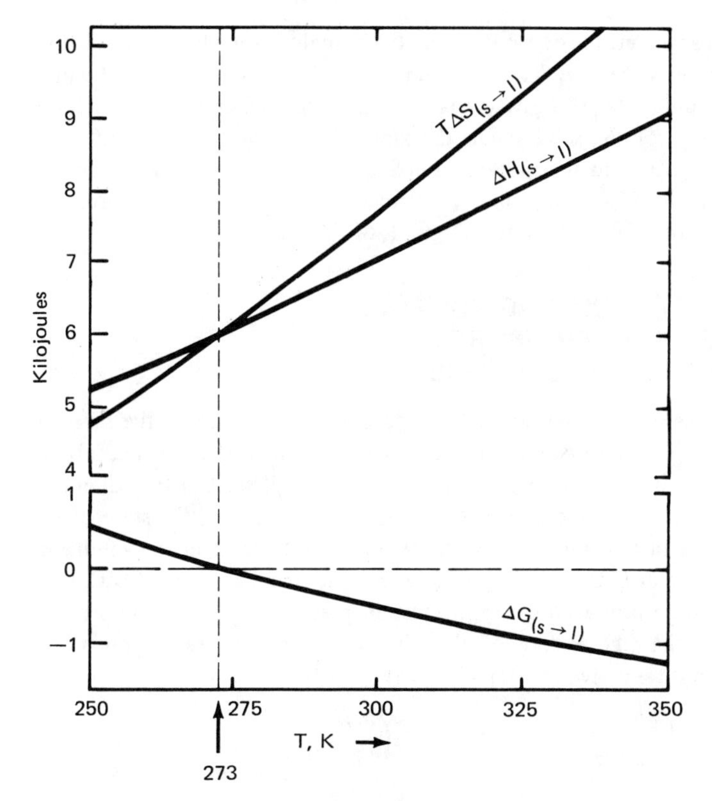

Fig. 7.5. The variations with temperature of the molar free energy of fusion, heat of fusion, and $T \times$ the entropy of fusion of water at 1 atm pressure.

$T = T_m = 273$ K, which is thus the temperature at which solid and liquid water coexist at equilibrium at 1 atm pressure.

Equilibrium between the two phases thus occurs as the result of a compromise between enthalpy considerations and entropy considerations. For equilibrium it is required that G' for the system be a minimum at the fixed values of T and P, and from Eq. (5.2) it is seen that minimization of G' requires that H' be small (or a large negative number) and S' be large. Figure 7.3*a* shows that at all temperatures $H_{(l)} > H_{(s)}$, and from the point of view of the enthalpy contribution to the free energy, it is seen that, in the absence of any other considerations, the solid would always be stable with respect to the liquid. However, Fig. 7.4 shows that, at all temperatures, $S_{(l)} > S_{(s)}$; and from the point of view of the entropy contribution to the free energy it is seen that, again in the absence of any other consideration, the liquid phase would always be stable with respect to the solid phase. However, as the entropy contribution to G, being TS, is temperature-dependent, a unique temperature T_m occurs above which the entropy contribution outweighs the enthalpy contribution and below which the reverse is the case. The temperature T_m is that temperature at which $H_{(l)} - T_m S_{(l)}$ equals $H_{(s)} - T_m S_{(s)}$; that is, T_m is the temperature at which the molar free energy of the solid equals the molar free energy of the liquid. This discussion is analogous to that presented in Sec. 5.3 where, at constant T and V, equilibrium between a solid and its vapor was examined in terms of minimization of the Helmholtz free energy, A, of the system.

7.3 THE VARIATION OF GIBBS FREE ENERGY WITH PRESSURE AT CONSTANT TEMPERATURE

Consider an ice + water system at 0°C and one atm pressure. If the pressure acting on the ice + water system is increased above 1 atm at 0°C, then according to Le Chatelier's principle, the state of the system will shift in that direction which tends to nullify the effect of the pressure increase; i.e., the state of the system will move in that direction which results in a decrease in the volume of the system. As ice at 0°C has a larger molar volume than has water at 0°C, then the change of state produced by an increase in pressure is the melting process. The effect (on the free energy of the phases) of an increase in pressure at constant temperature is given by Eq. (5.25) as

$$\left(\frac{\partial G_{(l)}}{\partial P}\right)_T = V_{(l)} \quad \text{and} \quad \left(\frac{\partial G_{(s)}}{\partial P}\right)_T = V_{(s)}$$

i.e., the rate of increase of G with increase in pressure at the constant temperature T equals the molar volume of the phase at the temperature T and the pressure P. For the reaction solid → liquid

$$\left(\frac{\partial \Delta G_{(s \to l)}}{\partial P}\right)_T = \Delta V_{(s \to l)}$$

and as $\Delta V_{(s \to l)}$ is a negative quantity for H_2O at 0°C, then the ice melts as the result of the pressure increase. Thus, corresponding to Fig. 7.1, which showed the variation of $G_{(s)}$ and $G_{(l)}$ with T at constant P, Fig. 7.6 shows the variation of $G_{(s)}$ and $G_{(l)}$ with P at constant T. It is to be noted that the volume behavior of H_2O at its melting point is anomalous; i.e. normally melting occurs with an increase in molar volume.

7.4 FREE ENERGY AS A FUNCTION OF TEMPERATURE AND PRESSURE

Consideration of Figs. 7.1 and 7.6 shows that it is possible to maintain equilibrium between the solid and liquid phases by simultaneously varying the

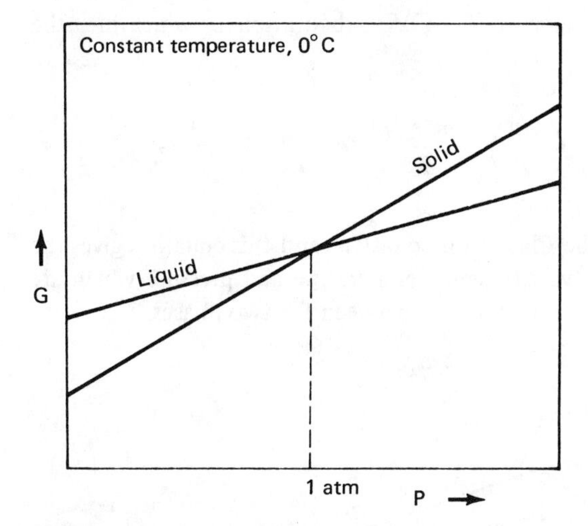

Fig. 7.6. Schematic representation of the molar Gibbs free energies of liquid and solid water as a function of pressure at constant temperature.

pressure and the temperature in such a manner that $\Delta G_{(s \to l)}$ remains zero. For equilibrium to be maintained,

$$G_{(l)} = G_{(s)}$$

or, for any infinitesimal change in T and P,

$$dG_{(l)} = dG_{(s)}$$

From Eq. (5.12)

$$dG_{(l)} = -S_{(l)}dT + V_{(l)}dP$$

and

$$dG_{(s)} = -S_{(s)}dT + V_{(s)}dP$$

Thus, for equilibrium to be maintained between the two phases,

$$-S_{(l)}dT + V_{(l)}dP = -S_{(s)}dT + V_{(s)}dP$$

or

$$\left(\frac{dP}{dT}\right)_{\text{eq}} = \frac{S_{(s)} - S_{(l)}}{V_{(s)} - V_{(l)}} = \frac{\Delta S_{(l \to s)}}{\Delta V_{(l \to s)}}$$

At equilibrium $\Delta G = 0$, and hence $\Delta H = T\Delta S$, substitution of which into the above equation gives,

$$\left(\frac{dP}{dT}\right)_{\text{eq}} = \frac{\Delta H}{T\Delta V} \tag{7.5}$$

Equation (7.5) is known as the Clapeyron equation, and this equation gives the required relationship between variations of temperature and pressure which are necessary for the maintenance of equilibrium between the two phases.

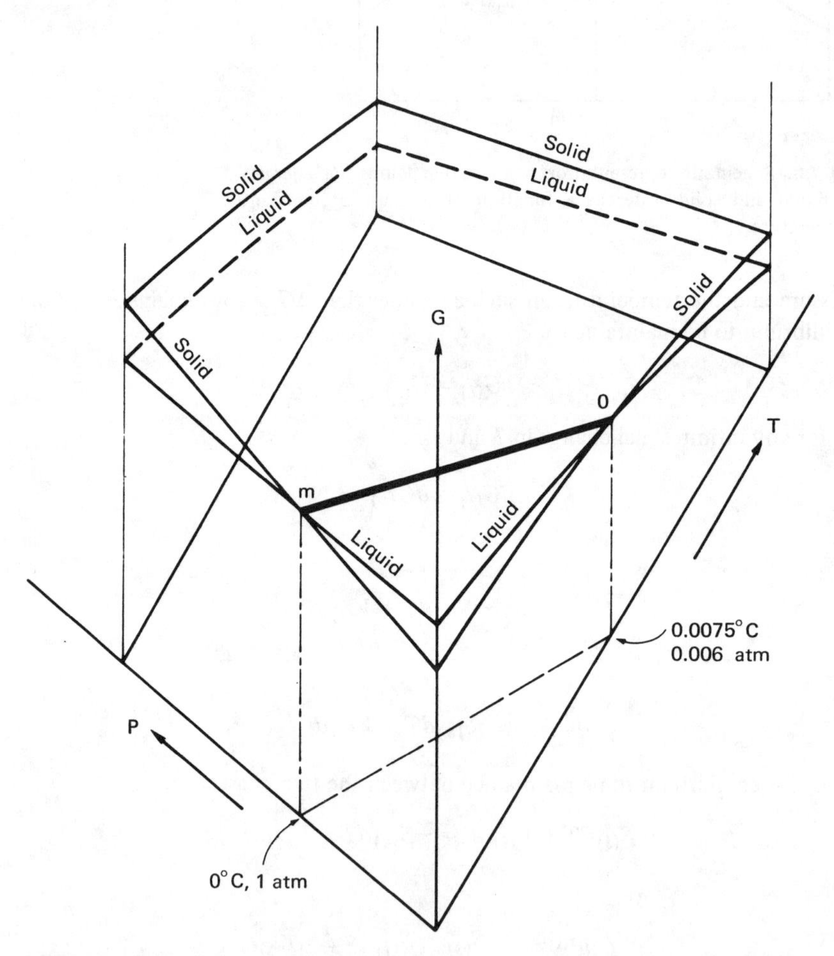

Fig. 7.7. Schematic representation of the equilibrium surfaces of the liquid and solid phases of water in *G-T-P* space.

In the case of the ice-water system, $\Delta V_{(s \to l)}$ is negative, and in any system, $\Delta H_{(s \to l)}$ is positive. Thus, $(dP/dT)_{eq}$ is a negative quantity; i.e., an increase in pressure decreases the equilibrium melting temperature. It is for this reason that ice-skating is possible. The pressure of the skate on the solid ice decreases the melting temperature of the ice; if the melting temperature is decreased below the actual temperature of the ice, then the ice melts and hence provides liquid water as a lubricant for the skate. With normal systems $(dP/dT)_{eq}$ is a positive quantity on account of the fact that $\Delta V_{(s \to l)}$ is positive, and in such cases an increase in pressure increases the equilibrium melting temperature.

The states of the solid and liquid phases can be represented on a three-dimensional diagram with G, T, and P as the coordinates. Such a diagram, drawn schematically for the H_2O system, is shown in Fig. 7.7. In this figure, each phase occurs as a surface in G-T-P space, and the line along which the phase surfaces intersect determines the P-T relationship required for the maintenance of equilibrium between the two phases. For any chosen state, which is determined by fixing the values of T and P, the equilibrium phase is that which has the lower value of G. It is to be noted that Fig. 7.1, if drawn for $P = 0.006$ atm, corresponds to the right front face of Fig. 7.7, and Fig. 7.6, drawn for $T = 0°C$, corresponds to the left front face of Fig. 7.7.

7.5 EQUILIBRIUM BETWEEN THE VAPOR PHASE AND A CONDENSED PHASE

If Eq. (7.5) is applied to vapor-condensed phase equilibria, then ΔV is the molar volume change accompanying the evaporation or sublimation, and ΔH is the molar latent heat of evaporation or sublimation, depending on whether the condensed phase is, respectively, the liquid or the solid. In either case

$$\Delta V = V_{\text{vapor}} - V_{\text{condensed phase}}$$

and as $V_{\text{vapor}} >> V_{\text{condensed phase}}$, then with the introduction of an insignificant error,

$$\Delta V = V_{\text{vapor}}$$

Thus for condensed phase–vapor equilibria, Eq. (7.5) can be written as

$$\left(\frac{dP}{dT}\right)_{eq} = \frac{\Delta H}{TV_{(v)}}$$

If it is further assumed that the vapor in equilibrium with the condensed phase behaves ideally, that is, $PV = RT$, then

$$\left(\frac{dP}{dT}\right)_{\text{eq}} = \frac{P\Delta H}{RT^2}$$

rearrangement of which gives

$$\frac{dP}{P} = \frac{\Delta H}{RT^2}\,dT$$

or,

$$d\ln P = \frac{\Delta H}{RT^2}\,dT \tag{7.6}$$

Equation (7.6) is known as the Clausius–Clapeyron equation.

If ΔH is independent of temperature, i.e., if $c_p(\text{vapor}) = c_p(\text{condensed phase})$, then integration of Eq. (7.6) gives

$$\ln P = -\frac{\Delta H}{RT} + \text{constant} \tag{7.7}$$

As equilibrium is maintained between the vapor phase and the condensed phase, then the value of P at any temperature T in Eq. (7.7) is the saturated vapor pressure exerted by the condensed phase at the temperature T. Equation (7.7) thus shows that the saturated vapor pressure exerted by a condensed phase increases exponentially with increasing temperature, as was noted in Sec. 5.3. If Δc_p for the evaporation or sublimation is not zero, but is independent of temperature, then from Eq. (6.9), ΔH_T in Eq. (7.6) is given as

$$\begin{aligned}\Delta H_T &= \Delta H_{298} + \Delta c_p(T - 298)\\ &= [\Delta H_{298} - 298\Delta c_p] + \Delta c_p T\end{aligned}$$

in which case integration of Eq. (7.6) gives

$$\ln P = \left[\frac{298\Delta c_p - \Delta H_{298}}{R}\right]\frac{1}{T} + \frac{\Delta c_p}{R}\ln T + \text{constant}$$

which is normally expressed in the form

$$\ln P = \frac{A}{T} + B\ln T + C \tag{7.8}$$

In Eq. (7.8),

$$\Delta H_T = -AR + BRT$$

7.6 GRAPHICAL REPRESENTATION OF PHASE EQUILIBRIA IN A ONE-COMPONENT SYSTEM

In the case of liquid-vapor equilibrium, the normal boiling point is defined as that temperature at which the saturation vapor pressure exerted by the liquid equals 1 atm. Knowledge of the heat capacities of the liquid and vapor phases, the heat of evaporation at any one temperature, $\Delta H_{evap,T}$, and the boiling temperature allows the saturated (liquid) vapor pressure-temperature relationship to be determined for the particular substance. For example, in the case of water,

$$c_{p,\,H_2O(v)} = 30 + 10.7 \times 10^{-3}T + 0.33 \times 10^5 T^{-2} \text{ joules/degree}$$

in the temperature range 298–2500 K

$$c_{p,\,H_2O(l)} = 75.44 \text{ joules/degree in the temperature range 273-373 K}$$

and thus

$$\Delta c_{p(l\rightarrow v)} = -45.44 + 10.7 \times 10^{-3}T + 0.33 \times 10^5 T^{-2} \text{ joules/degree}$$

$$\Delta H_{evap} = 41090 \text{ joules/mole at the normal boiling temperature of 373 K}$$

Thus,

$$\Delta H_{evap\ T} = \Delta H_{evap\ 373} + \int_{373}^{T} \Delta c_{p\,(l\rightarrow v)}\,dT$$

$$= 41090 - 45.44(T - 373) + 5.35 \times 10^{-3}(T^2 - 373^2)$$
$$-0.33 \times 10^5\left(\frac{1}{T} - \frac{1}{373}\right)$$

$$= 58872 - 45.44T + 5.35 \times 10^{-3}T^2 - 0.33 \times \frac{10^5}{T} \text{ joules}$$

Now $d \ln p = \dfrac{\Delta H_{evap}}{RT^2}\,dT$

and so, with $R = 8.3144$ joules/degree·mole

$$\ln p = -\frac{58872}{RT} - \frac{45.44 \ln T}{R} + \frac{5.35 \times 10^{-3}T}{R} + \frac{0.33 \times 10^5}{2RT^2}$$
$$+ \text{constant}$$

At the boiling point of 373 K, $p = 1$ atm, and hence the integration constant is evaluated as 51.092.

In terms of logarithms to the base 10, this gives

$$\log p(\text{atm}) = -\frac{58872}{2.303RT} - \frac{45.44}{R}\log T + \frac{5.35 \times 10^{-3}T}{2.303R} + \frac{0.33 \times 10^5}{2 \times 2.303RT^2} + \frac{51.092}{2.303}$$

$$= -\frac{3075}{T} - 5.465 \log T + 0.279 \times 10^{-3}T + \frac{862}{T^2} + 22.185 \qquad (7.9)$$

which is thus the saturated vapor pressure-temperature relationship for H_2O in the temperature range 273 to 373 K. Curve fitting of experimentally measured vapor pressures of liquid water to an equation of the form

$$\log p(\text{atm}) = \frac{A}{T} + B \log T + C$$

gives

$$\log p(\text{atm}) = -\frac{2900}{T} - 4.65 \log T + 19.732 \qquad (7.10)$$

Equations (7.9) and (7.10), as plots of log p(atm) against $1/T$, are illustrated in Fig. 7.8, which shows them to be indistinguishable. In Fig. 7.8 the slope of the line at any temperature T equals $-\Delta H_{\text{evap},T}/4.575$. The vapor pressures of several of the more common elements are presented in Fig. 7.9, where again, log p is plotted against the reciprocal of the absolute temperature.

Figure 7.10 is a one-component phase diagram which uses T and P as the coordinate axes. Line AOA' is a graphical representation of Eq. (7.5), which expresses the P-T relationship for solid-liquid equilibrium. The normal melting point is the melting temperature at 1 atm pressure (which is represented by the point m in Fig. 7.7), and the slope of the line dP/dT, is given as $\Delta H_{\text{melt}}/T\Delta V_{\text{melt}}$. The line BOB' is the liquid-vapor equilibrium line, given by Eqs. (7.7) or (7.8), where ΔH_T is $\Delta H_{\text{evap},T}$. The line BOB' passes through the normal boiling point (which is represented by the point b in the figure) and intersects with the line AOA' at the *triple point*, O. The triple point is the state represented by the invariant values of P and T at which the solid, liquid, and vapor phases are in equilibrium with each other. Knowledge of the triple point,

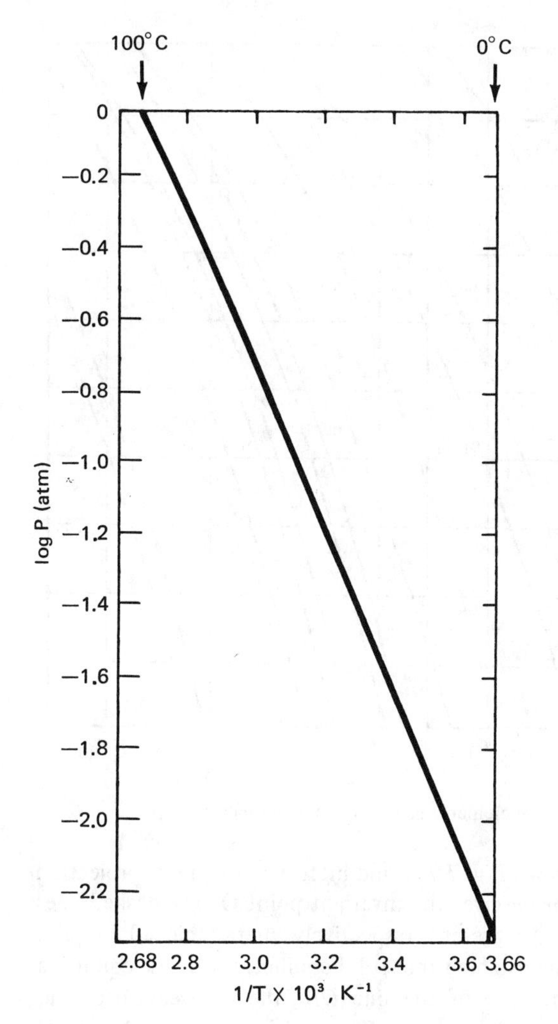

Fig. 7.8. The vapor pressure of liquid water as a function of temperature.

together with the value of $\Delta H_{\text{sublim},T}$, then allows the saturated (solid) vapor pressure-temperature relationship to be determined. This equilibrium line is drawn as *COC'* in Fig. 7.10.

If the *G-T-P* surface for the states of existence of the vapor phase were included in Fig. 7.7, it would intersect with the solid-state surface along a line and would intersect with the liquid state surface along a line. Projection of these lines, together with the line of intersection of the solid- and liquid-state surfaces, onto the two-dimensional *P-T* basal plane of Fig. 7.7 would produce Fig. 7.10. All

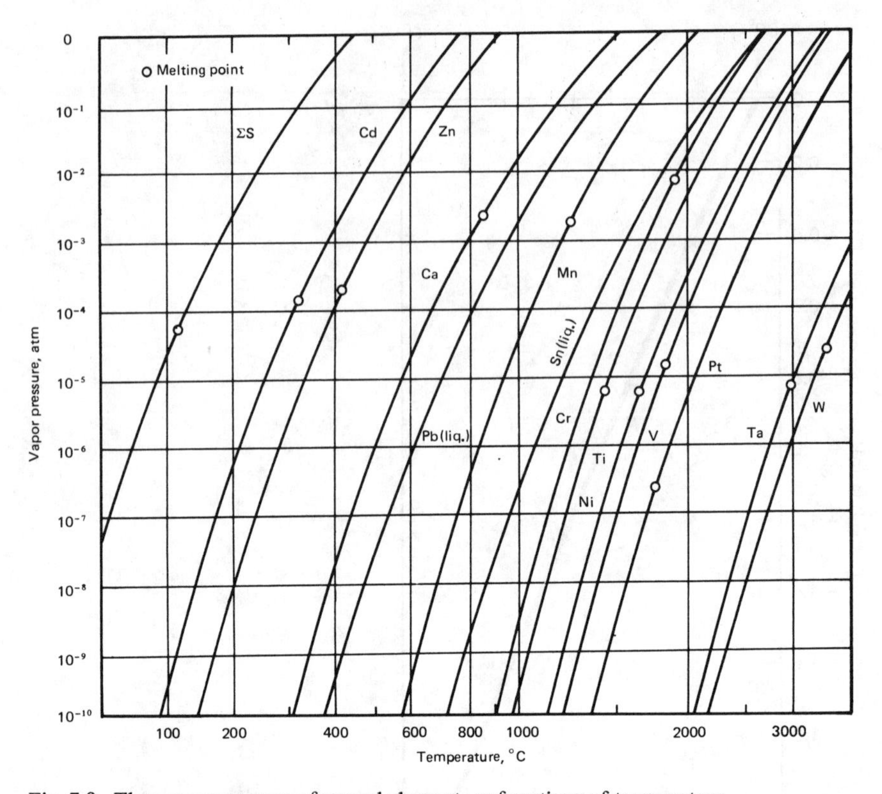

Fig. 7.9. The vapor pressures of several elements as functions of temperature.

three state surfaces in the redrawn Fig. 7.7 would intersect at a point, projection of which onto the *P-T* basal plane gives the invariant point O. The dashed lines OA', OB', and OC' in Fig. 7.10 represent, respectively, metastable solid-liquid, metastable vapor-liquid, and metastable vapor-solid equilibrium. These equilibria are metastable because, in the case of the line OB', the intersection of the liquid- and vapor-state surfaces in the redrawn Fig. 7.7 lies at higher values of G than does the solid-state surface for the same values of P and T. Similarly the solid-liquid equilibrium OA' is metastable with respect to the vapor phase, and the solid-vapor equilibrium OC' is metastable with respect to the liquid phase.

Figure 7.11*a* shows three isobaric sections of the redrawn Fig. 7.7 at $P_1 > P_{\text{triple point}}$, $P_2 = P_{\text{triple point}}$, and $P_3 < P_{\text{triple point}}$; and Fig. 7.11*b* shows three isothermal sections of the redrawn Fig. 7.7 at $T_1 < T_{\text{triple point}}$, $T_2 = T_{\text{triple point}}$, and $T_3 > T_{\text{triple point}}$. In Fig. 7.11*a*, the slopes of the three lines in any isobaric section increase negatively in the order solid, liquid, vapor, in accordance with the fact that $S_{(s)} < S_{(l)} < S_{(v)}$. Similarly in Fig. 7.11*b* the slopes of the lines in any isothermal section increase in the order liquid, solid, vapor, in accordance with the fact that, for H_2O, $V_{(l)} < V_{(s)} < V_{(v)}$.

The lines *OA*, *OB*, and *OC* divide Fig. 7.10 into three distinct areas within each of which only one phase is stable. Within these areas the pressure exerted on the phase and the temperature of the phase can be independently varied without upsetting the one-phase equilibrium. The system is thus said to have two *degrees of freedom*, where the number of degrees of freedom that an equilibrium has is the maximum number of variables which may be independently varied without upsetting the equilibrium. Along the lines *OA*, *OB*, and *OC*, two phases coexist in equilibrium, and for continued maintenance of any of these equilibria only one variable (either *P* or *T*) can be independently varied. Two-phase equilibrium

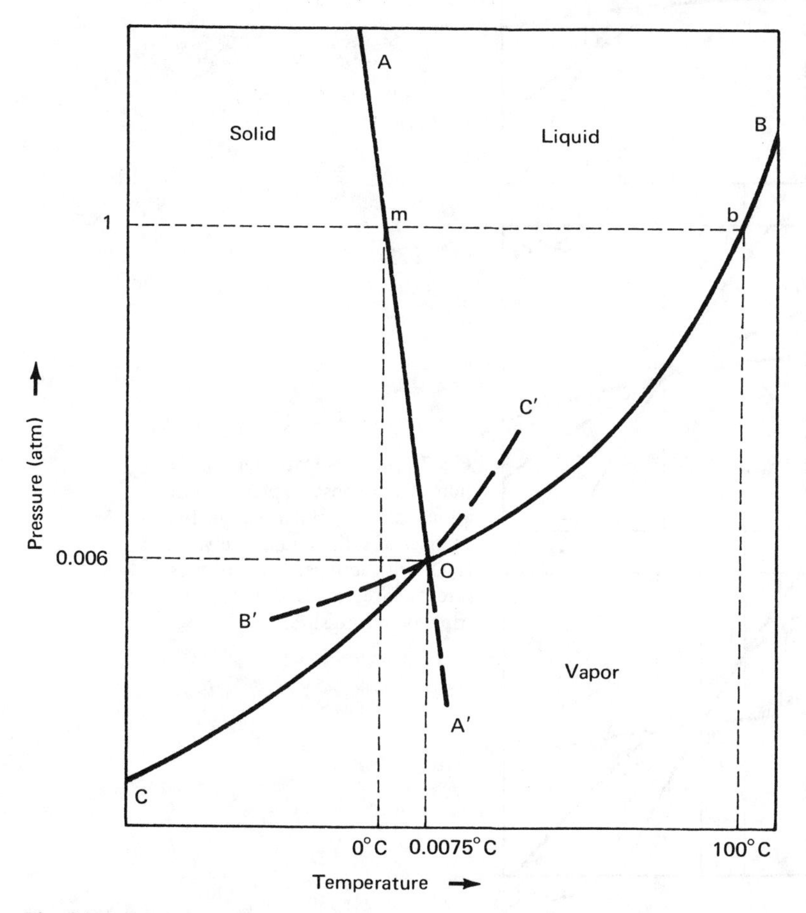

Fig. 7.10. Schematic representation of part of the H_2O phase diagram.

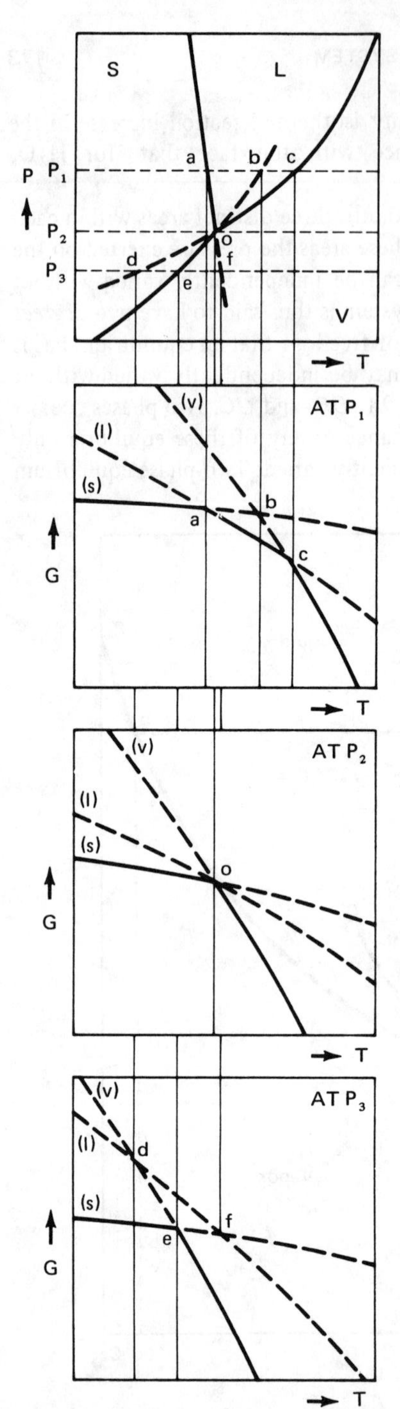

Fig. 7.11*a*. Schematic representation of the constant-pressure variations of the molar Gibbs free energies of solid, liquid, and vapor H_2O with temperature at pressures above, at, and below the triple-point pressure.

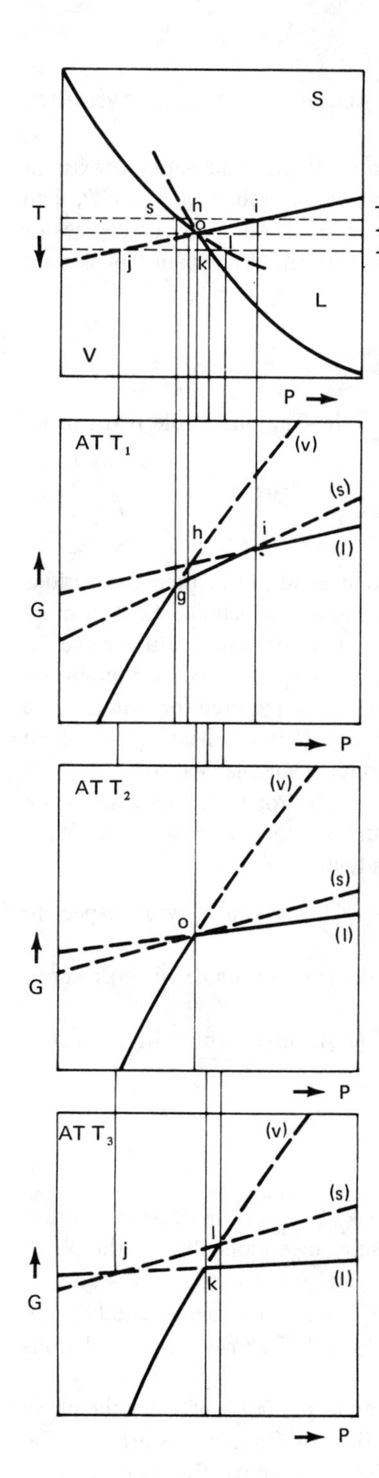

Fig. 7.11*b*. Schematic representation of the constant-temperature variations of the molar Gibbs free energies of solid, liquid, and vapor H_2O with pressure at temperatures above, at, and below the triple-point temperature.

thus has only one degree of freedom. As solid, liquid, and vapor coexist in equilibrium only at the invariant point, i.e., at unique values of T and P, then three-phase equilibrium has no degrees of freedom. The number of degrees of freedom, F, that a system containing C components can have, when P phases are in equilibrium, is given as

$$F = C - P + 2$$

This expression is known as the Gibbs phase rule. The phase rule is discussed fully in Chap. 13.

7.7 SOLID–SOLID EQUILIBRIA

If the substance can exist in more than one solid form, then the various solid-solid equilibria must be considered. An element which can exist in more than one solid form is said to exhibit allotropy, and a compound which can exist in more than one solid form is said to exhibit polymorphism. As the equilibrium is between two condensed phases, the P-T relationship required for maintenance of the equilibrium is given by Eq. (7.5), where ΔH is the heat of the phase change solid I → solid II, and ΔV is the corresponding volume change.

As was mentioned in Chap. 6, sulfur has two allotropes, a low-temperature rhombohedral form and a higher temperature monoclinic form. The phase diagram for sulfur is shown in Fig. 7.12. In this figure,

ABC – liquid-vapor equilibrium (*BC* stable, *AB* metastable with respect to monoclinic)
ADE – rhombic-vapor equilibrium (*ED* stable, *DA* metastable with respect to monoclinic)
AFG – rhombic-liquid equilibrium (*FG* stable, *AF* metastable with respect to monoclinic)
FD – stable rhombic-monoclinic equilibrium
FB – stable monoclinic-liquid equilibrium
DB – stable monoclinic-vapor equilibrium

The phase diagram shows that $V_{\text{monoclinic}} > V_{\text{rhombic}}$ (an increase in pressure at constant temperature brings about the phase change monoclinic → rhombic). Four triple points, three stable at D, B, and F, and one metastable at A, occur; and it is to be noted that four phases cannot coexist together in equilibrium, which is in accordance with the phase rule. Figure 7.13 shows the G-T relations of the four phases at P'.

The polymorphism of solid water is illustrated in Fig. 7.14, which is the phase diagram drawn to include pressures up to 12,000 atm. On this pressure scale the phase area of water vapor is negligibly small. The figure shows five modifications

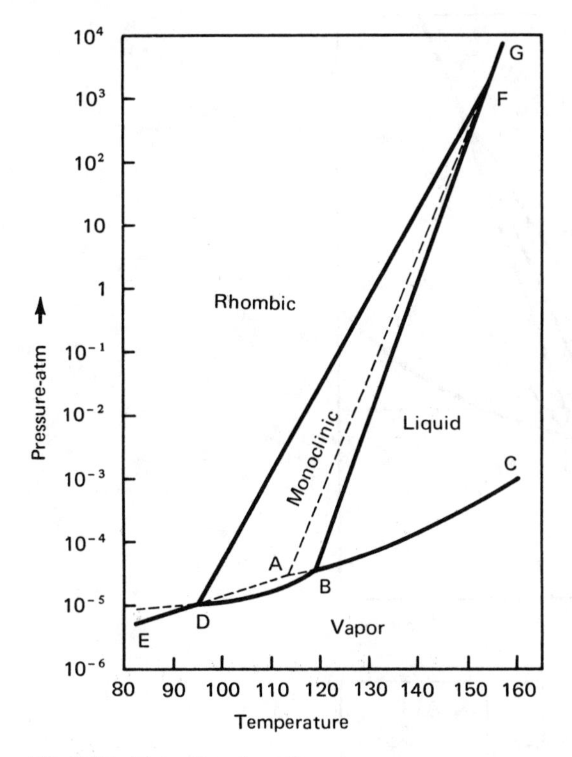

Fig. 7.12. The sulfur phase diagram.

of solid water. If the ice I phase is isothermally compressed at −20°C, eventually the liquid phase forms when that pressure is reached at which −20°C is the melting temperature, and this phase change occurs with a decrease in volume. Further increase in the pressure takes the system, in turn, through the ice III, ice V, and ice VI modifications, with each phase transition occurring with a decrease in specific volume. The figure also shows that −22°C is the minimum temperature at which liquid water can exist. Once the melting temperature of ice I has been decreased to this value, as a result of an increase in pressure, the ice I–liquid equilibrium becomes metastable with respect to the ice III phase.

The phase diagram for silica is shown in Fig. 7.15. In the crystal structures of quartz, tridymite, cristobalite, and coesite, each silicon is coordinated by four oxygens located at the corners of a tetrahedron and each oxygen forms bonds with two silicons. The crystal structures are thus obtained by arranging SiO_4 tetrahedra such that they share corners with one another. Increasing the pressure on tridymite at 1400°C from 1 atm causes the transition tridymite → cristobalite at about 12 atm, the transition cristobalite → high quartz at about 2300 atm, the transition high quartz → coesite at about 23,000 atm, and the transition

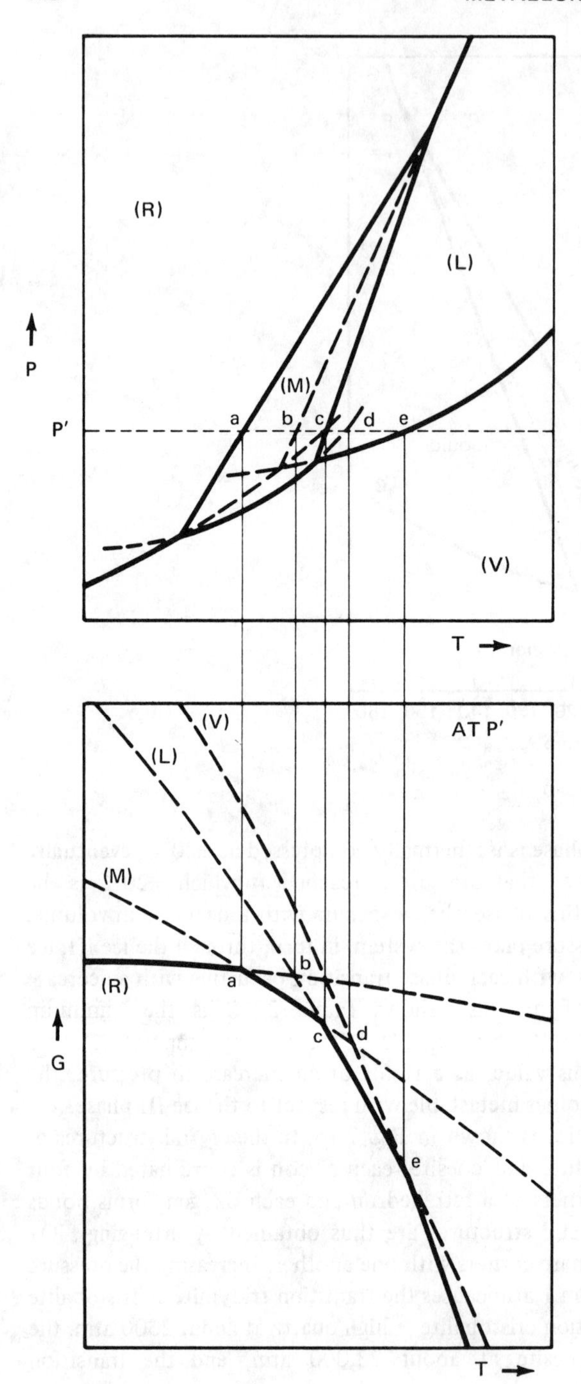

Fig. 7.13. Schematic representation of a constant-pressure variation of the molar Gibbs free energies of rhombic, monoclinic, liquid, and vapor sulfur with temperature.

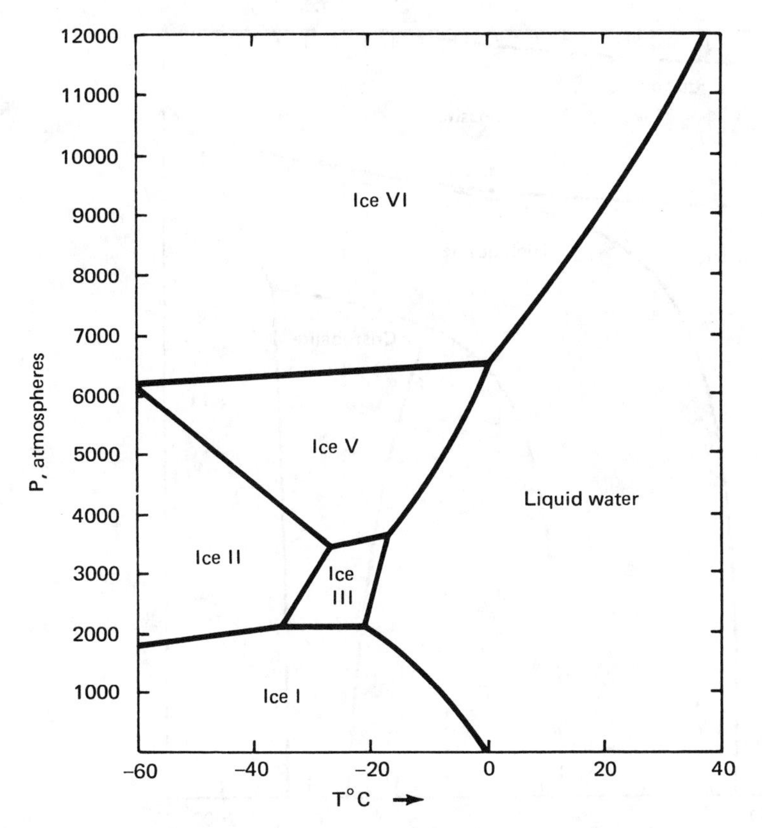

Fig. 7.14. The H_2O phase diagram.

coesite → stishovite at about 180,000 atm. This indicates that the efficiency with which the SiO_4 tetrahedra fill space in the crystal structures increases in the order tridymite, cristobalite, high quartz, coesite, i.e., that the molar volume of SiO_2 decreases in this order. The high-pressure phase, stishovite, which has the lowest molar volume, has the rutile (TiO_2) structure, in which each silicon is coordinated by six oxygens located at the corners of an octahedron and each oxygen forms bonds with three silicons. Stishovite, which was first synthesized in a laboratory in 1961, has subsequently been discovered as a metastable product of shock metamorphism of sandstone by meteorite impact.

Carbon, the phase diagram for which is shown in Fig. 7.16, has three allotropic forms: graphite, diamond, and a high pressure phase "solid III." The tetrahedral diamond structure is significantly more compact than the layered hexagonal structure of graphite and hence the diamond structure only becomes thermodynamically stable at high pressures (see worked example 2). Diamonds are thus

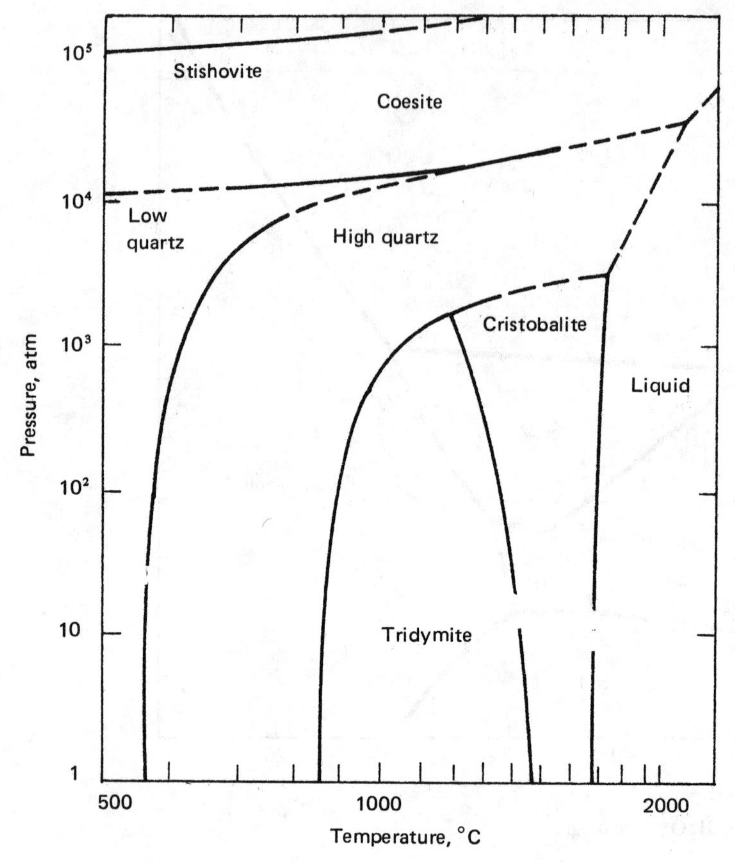

Fig. 7.15. The phase diagram for silica.

metastable under normal atmospheric conditions. The melting curve of graphite is of interest; along *ab* the slope of the curve is positive, indicating that $V_{(l)} > V_{(\text{graphite})}$, and along *bc* the slope is negative, indicating that $V_{(\text{graphite})} > V_{(l)}$. At the state *b*, dP/dT is infinite which indicates that, in this state, $V_{(\text{graphite})} = V_{(l)}$. The melting curve of diamond also has a negative slope which indicates that, on melting, dense diamond forms an even denser liquid.

7.8 SUMMARY

From knowledge of the temperature and pressure dependencies of the enthalpy and entropy differences between the phases of existence of a substance, it is possible to calculate the corresponding free energy changes. As a closed one-component system has only two independent variables, the state dependency of G can be examined most simply by choosing T and P as the independent variables (these are the natural independent variables when G is the dependent variable). The phases of the substance can thus be represented in a three-dimensional diagram using G, T, and P as the coordinates, and in such a diagram the stable states of the phases occur as surfaces. In any state (any choice of P and T), the phase which has the lowest free energy is the stable phase. The surfaces intersect with one another along lines in the G-T-P diagram; these lines represent the P-T relationships for which the two phases, the surfaces of which intersect, coexist in equilibrium with one another. In the case of the intersection of the solid and liquid surfaces, the line of intersection represents the variation of the melting temperature with pressure; similarly the line of intersection of the liquid and vapor lines represents the variation of the boiling temperature of the liquid with pressure. At $P = 1$ atm on these intersection lines, the two temperatures are, respectively, the normal melting temperature and normal boiling temperature of the solid and liquid phases. Three surfaces intersect at a point, i.e., at a unique combination of P and T, and such a state, the so-called invariant point, is that unique state in which three phases can coexist in mutual equilibrium. In a one-component system no more than three phases can coexist in equilibrium.

The three-dimensional G-T-P diagram gives a good illustration of the differences between stable, metastable, and unstable states, and hence illustrates

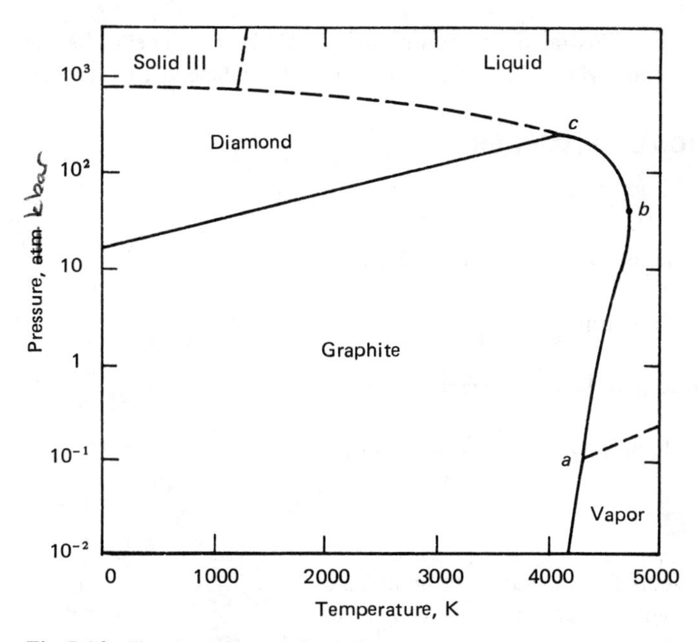

Fig. 7.16. The phase diagram for carbon.

the difference between reversible and irreversible process paths. At any value of P and T, the phase with the lowest free energy is the stable phase, and phases which exist on surfaces which lie higher on the G axis than this lowest surface are metastable with respect to the lowest surface. However, phases occurring with a value of G (at given T and P) which do not lie on any surface are unstable. A reversible process path (involving the variation of P and/or T) lies on a phase surface, and the state of any phase is reversibly changed if, during the process, the state at no time leaves the phase surface. If the process path does not at all times lie on the phase surface, then the change of state, which necessarily passes through nonequilibrium states, is irreversible.

As perspective representation, in two-dimensions, of a three-dimensional diagram–e.g., Fig. 7.7–is difficult, it is normal practice to draw the phase diagram for a one-component system as the basal plane of the G-T-P diagram–i.e., a two-dimensional P-T diagram–on to which are projected the lines of two-surface intersection (equilibrium between two phases) and the points of three-surface intersection (equilibrium among three phases). Such a diagram contains areas in which a single phase is stable separated by lines along which two phases exist and points at the intersection of the lines where three phases exist. The vapor-condensed phase equilibrium lines in such a diagram are called vapor pressure lines, and they are exponential in form. A phase diagram can often be presented in more useful form as a plot of $\log p$ versus $1/T$ rather than P versus T.

This discussion has illustrated the utility of the Gibbs free energy as the criterion of equilibrium when T and P are chosen as the independent variables.

7.9 NUMERICAL EXAMPLES

Example 1

The vapor pressure of solid zinc varies with temperature as

$$\ln p(\text{atm}) = -\frac{15775}{T} - 0.755 \ln T + 19.25$$

and the vapor pressure of liquid zinc varies with temperature as

$$\ln p(\text{atm}) = -\frac{15246}{T} - 1.255 \ln T + 21.79$$

Calculate:

(*a*) The normal boiling temperature of liquid zinc
(*b*) The triple-point temperature
(*c*) The heat of evaporation of zinc at the normal boiling temperature

(*d*) The heat of fusion of zinc at the triple-point temperature
(*e*) The difference between the heat capacities of solid and liquid zinc

The phase diagram is drawn schematically in Fig. 7.17.

(*a*) The normal boiling temperature T_b is defined as that temperature at whicl the vapor pressure of the liquid equals 1 atm. Therefore, from the liquid-vapor equilibrium equation,

$$\ln p(\text{atm}) = -\frac{15246}{T} - 1.255 \ln T + 21.79$$

$$= 0 \text{ at } T = T_b$$

Thus

$$-\frac{15246}{T} + 21.79 = 1.255 \ln T$$

and plots of each side of this equation, against T, intersect at $T = 1181\text{ K} = T_b$.

(*b*) The liquid and solid vapor-pressure curves intersect at the triple point, at which temperature T_{tp}, solid, liquid, and vapor at p_{tp} are in equilibrium. Therefore at $T = T_{tp}$

$$-\frac{15775}{T} - 0.755 \ln T + 19.25 = -\frac{15246}{T} - 1.255 \ln T + 21.79$$

or

$$\frac{529}{T} + 2.54 = 0.5 \ln T$$

Plots of each side of the above equation, against T, intersect at $T = 708\text{ K} = T_{tp}$.

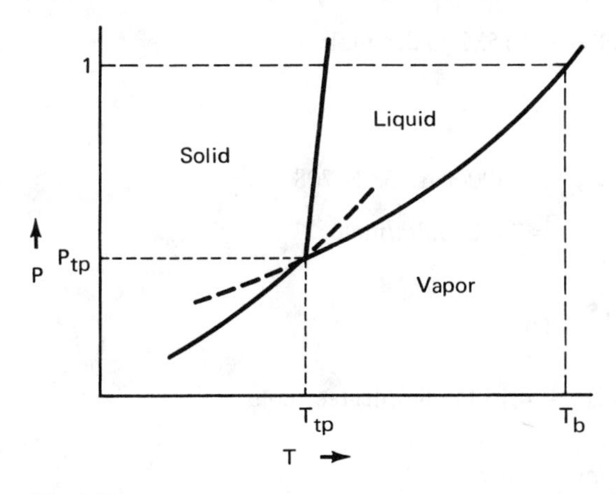

Fig. 7.17.

(*c*) For vapor in equilibrium with liquid,

$$\ln p(\text{atm}) = -\frac{15246}{T} - 1.255 \ln T + 21.79$$

$$\frac{d \ln p}{dT} = \frac{\Delta H}{RT^2} = \frac{15246}{T^2} - \frac{1.255}{T}$$

and so $$\Delta H_{(l\to v)} = (15246 \times 8.3144) - (1.255 \times 8.3144)T$$
$$= 126760 - 10.43T$$

Thus at the normal boiling temperature of 1181 K,

$$\Delta H_{(l\to v)} = 126760 - 10.43 \times 1181 = 114{,}440 \text{ joules/mole}$$

(*d*) For vapor in equilibrium with solid,

$$\ln p(\text{atm}) = -\frac{15775}{T} - 0.755 \ln T + 19.25$$

Thus $$\Delta H_{(s\to v)} = (8.3144 \times 15775) - (0.755 \times 8.3144)T$$
$$= 131160 - 6.277T \text{ joules/mole}$$

At any temperature

$$\Delta H_{(s\to l)} + \Delta H_{(l\to v)} = \Delta H_{(s\to v)}$$

and thus $$\Delta H_{(s\to l)} = 131160 - 6.277T - 126760 + 10.43T$$
$$= 4400 + 4.153T \text{ joules/mole}$$

At triple point

$$\Delta H_{(s\to l)} = 4400 + 4.153 \times 708$$
$$= 7340 \text{ joules/mole}$$

(*e*) $\Delta H_{(s\to l)} = 4400 + 4.153T$

$$\frac{d\Delta H}{dT} = \Delta c_p = 4.153 \text{ joules/degree·mole}$$
$$= c_{p(l)} - c_{p(s)}$$

Example 2

Carbon has two allotropes, graphite and diamond. At 25°C and 1 atm pressure, graphite is the stable form. Calculate the pressure which must be applied to graphite at 25°C in order to bring about its transformation to diamond. Given:

$H_{298\,(\text{graphite})} - H_{298(\text{diamond})} = -1900$ joules/mole
$S_{298\,(\text{graphite})} = 5.73$ joules/degree·mole
$S_{298\,(\text{diamond})} = 2.43$ joules/degree·mole
the density of graphite at 25°C is 2.22 gram/cm^3
the density of diamond at 25°C is 3.515 gram/cm^3

For the phase transformation graphite → diamond at 298 K

$$\Delta G = \Delta H - T\Delta S$$

$$= 1900 - 298(2.43 - 5.73) = 2883 \text{ joules/mole}$$

For the same phase change at the temperature T

$$\left(\frac{\partial \Delta G}{\partial P}\right)_T = \Delta V \quad \text{where} \quad \Delta V = V_{\text{diamond}} - V_{\text{graphite}}$$

$$V_{\text{graphite}} = \frac{12}{2.22} = 5.405 \frac{\text{cm}^3}{\text{mole}}$$

and

$$V_{\text{diamond}} = \frac{12}{3.515} = 3.415 \frac{\text{cm}^3}{\text{mole}}$$

Thus

$$\Delta V = -1.99 \frac{\text{cm}^3}{\text{mole}}$$

The problem involves calculating that pressure at which $\Delta G_{(\text{graphite} \to \text{diamond})}$ at 298 K is zero. As

$$\left(\frac{\partial \Delta G}{\partial P}\right)_T = \Delta V$$

then

$$\Delta G(P, T = 298) = \Delta G(P = 1, T = 298) + \int_1^P \Delta V dP$$

If the isothermal compressibilities of the two phases are negligible, that is, if ΔV is not a function of pressure, then

$$\Delta G(P, T = 298) = 2883 + \frac{(-1.99)(P-1)}{9.87} = 0 (9.87 \text{ cm}^3/\text{atm} = 1 \text{ joule})$$

Thus $$(P - 1) = P = 2883 \times \frac{9.87}{1.99} = 14{,}300 \text{ atm}$$

Thus in order to transform graphite to diamond at 25°C, the applied pressure must exceed 14,300 atm.

PROBLEMS

7.1 From the vapor pressure-temperature relationship for liquid silver, calculate the heat of evaporation of liquid silver at the normal boiling temperature of 2147 K and the heat capacity difference between liquid and gaseous silver.

7.2 Calculate the approximate pressure required to distill mercury at 100°C.

7.3 The molar heat capacity of liquid iron exceeds that of iron vapor by 10.55 joules/degree, the molar heat of evaporation of liquid iron at 1600°C is 358 kilojoules and the vapor pressure of liquid iron at 1600°C is 5.13×10^{-5} atm. Calculate the vapor pressure-temperature relationship of liquid iron.

7.4 The densities of solid and liquid lead at the normal melting temperature of 327°C are 10.94 and 10.65 grams/cm^3 respectively. Calculate the pressure which must be applied to lead in order to increase its melting temperature by 20 centigrade degrees. The atomic weight of Pb is 207.

7.5 One liter of iodine vapor at 0.04 atm pressure and 150°C is isothermally compressed. Determine the nature of the first phase change which eventually occurs in the system as a result of this process, and determine the pressure at which this phase change occurs. If the system had been cooled at constant volume from its initial state, determine the nature of the first phase change which would have occurred and the temperature and pressure at which this phase change would have started. Assume that iodine vapor behaves as an ideal gas in the vapor states of interest.

7.6 Below the triple point (−56.2°C) the vapor pressure of solid CO_2 is given as

$$\ln p(\text{atm}) = -\frac{3116}{T} + 16.01$$

The molar heat of melting of CO_2 is 8330 joules. Calculate the vapor pressure exerted by liquid CO_2 at 25°C, and illustrate why solid CO_2, sitting on the laboratory bench, evaporates rather than melts.

7.7 The variation of melting temperature and volume change from solid to liquid with pressure for potassium is tabulated below. From this data calculate the heat of fusion of K at 5000 kg/cm² pressure. The atomic weight of K is 39.

P kg/cm²	1	1000	2000	3000	4000	5000	6000
T°C	62.5	78.7	92.4	104.7	115.8	126.0	135.4
ΔV cm³/gram	.0268	.02368	.02105	.01877	.01676	.01504	.01347

P kg/cm²	7000	8000	9000	10,000
T°C	144.1	152.5	160.1	167.0
ΔV cm³/gram	.01205	.01073	.0095	.00838

CHAPTER EIGHT

THE BEHAVIOR OF GASES

8.1 INTRODUCTION

The application of thermodynamics to the determination of the equilibrium state of a chemical reaction system requires a knowledge of the thermodynamic properties of the reactants in, and the products of, the reaction. The thermodynamic properties of these individual reactants and products are most conveniently expressed by means of their equations of state which relate the thermodynamic properties of interest (e.g., free energy, enthalpy, entropy, etc.), to the operational independent variables, pressure, temperature, and composition. As a prelude to the examination of the thermodynamics of reaction systems, which is started in Chap. 9, this chapter is concerned with the thermodynamic behavior of the simplest states of existence, namely, gases. Thus far frequent mention has been made of the so-called "ideal gas" as a system for use in the discussion and illustration of thermodynamic arguments. In this chapter real gases are compared with the ideal gas, and the origins of the frequently occurring differences between the two behaviors are sought in the atomic or molecular properties of the real gases. Thus, although for the purposes of obtaining a purely thermodynamic description of a gas, knowledge of the physical properties of the specific gas is not necessary, comparison between the two aids in providing a better understanding of the thermodynamic behavior.

8.2 THE P-V-T RELATIONSHIPS OF GASES

For all gases it is experimentally found that

$$\frac{PV}{RT} \rightarrow 1 \quad \text{limit } P \rightarrow 0 \tag{8.1}$$

where P = the pressure of the gas
V = the molar volume of the gas
R = the Gas Constant
T = the temperature of the gas

Thus isotherms, plotted on a P-V diagram, become hyperbolic, being given by the equation

$$PV = RT \tag{8.2}$$

This equation, in giving the relationship among the state variables of the system, is the equation of state of the gas and is called the ideal gas law. A gas which obeys this law over a range of states is said to behave ideally in this range of states, and a gas which obeys this law in all possible states is called a perfect gas. The perfect gas is a convenient model with which the behavior of actual gases can be compared.

The variation of V with P at several temperatures for a typical real gas is shown in Fig. 8.1. This figure shows that, as the temperature is decreased, the character

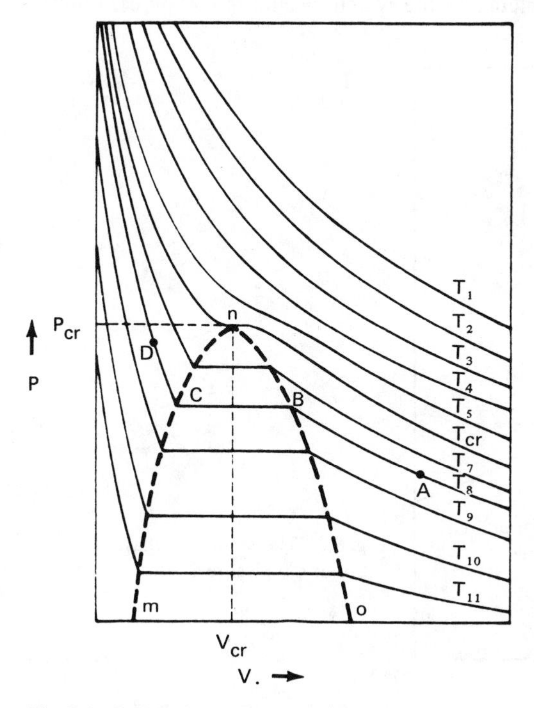

Fig. 8.1. P-V isotherms for a typical real gas.

of the P-V isotherms changes and eventually a value of $T = T_{\text{critical}}$ is reached, at which, at some fixed values of $P = P_{\text{critical}}$ and $V = V_{\text{critical}}$, a horizontal inflexion occurs in the isotherm; i.e.,

$$\left(\frac{\partial P}{\partial V}\right)_{T_{\text{cr}}} = 0 \quad \text{and} \quad \left(\frac{\partial^2 P}{\partial V^2}\right)_{T_{\text{cr}}} = 0$$

At temperatures below T_{cr} two phases can exist. For example, if 1 mole of vapor, initially in the state A (in Fig. 8.1), is isothermally compressed at T_8, the state of the vapor moves along the isotherm until the state B is reached. At the state B the pressure of the vapor is the saturated vapor pressure at the temperature T_8, and further decrease in the volume of the system causes condensation of the vapor and consequent appearance of the liquid phase. The liquid phase, in equilibrium with the vapor phase, appears at the state C (V_C = molar volume of the liquid at P_8 and T_8). Vapor and liquid thus coexist in equilibrium, and further decrease in the volume of the system causes further condensation, during which the states of the liquid and vapor phases remain fixed at C and B respectively; the aggregate volume of the system, which is determined by the proportions of liquid and vapor present, moves from B to C. Eventually condensation is complete and the system occurs as 100 percent liquid

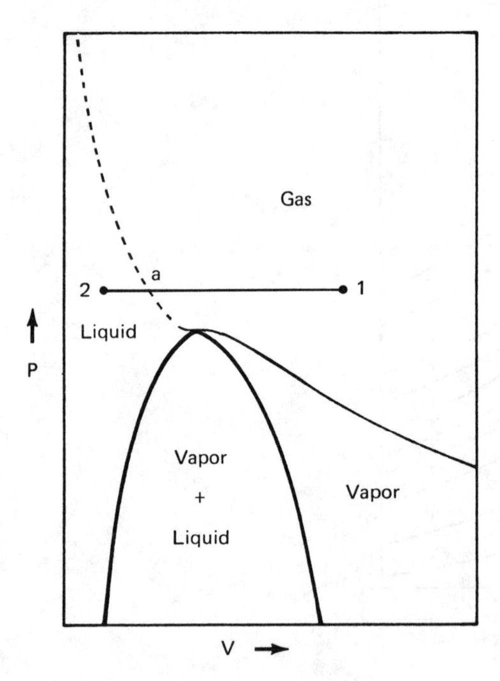

Fig. 8.2. The states of phase stability of a typical real gas.

at the state C. Further increase of pressure moves the state of the system along the isotherm towards D. The high values of $-(\partial P/\partial V)_T$ in the range of liquid states and the low values of $-(\partial P/\partial V)_T$ in the range of vapor states reflect the low compressibility of the liquid phase and the high compressibility of the vapor phase.

Figure 8.1 also shows that, as the temperature is increased up to T_{cr}, the molar volume of the liquid in equilibrium with the vapor (corresponding to the point C) progressively increases and the molar volume of the vapor in equilibrium with the liquid (corresponding to the point B) progressively decreases. The consequence is that the molar volumes of the coexistent vapor and liquid phases progressively approach one another. Thus, as the temperature is increased up to T_{cr}, the vapor in equilibrium with the liquid becomes more dense and the liquid in equilibrium with the vapor becomes less dense. Eventually, when T_{cr} is reached, the molar volumes of the coexistent phases coincide at the state P_{cr}, T_{cr}. The critical point is thus the meeting point of the locus of the point C with temperature (given by the line mn) and the locus of the point B with temperature (given by the line on), and the complete locus line mno defines the field of liquid-vapor equilibrium.

Above T_{cr}, distinct two-phase equilibrium (in which two coexistent phases are separated by a definite boundary surface across which the properties of the system abruptly change) does not occur; and above T_{cr}, the gaseous state cannot be liquefied by isothermal compression alone. As the vapor state below T_{cr} can be liquefied by isothermal pressure increase alone, T_{cr} also serves to distinguish between the gaseous state and the vapor state and hence defines the gaseous state phase field. The phase regions are shown in Fig. 8.2.

Liquefaction of a gas requires that the gas be cooled. Consider the process path $1 \rightarrow 2$ in Fig. 8.2. According to this path, which represents the cooling of the gas at constant pressure, the phase change gas $\rightarrow$ liquid occurs at the point a, where the temperature falls below T_{cr}. In fact, at pressures greater than P_{cr} the critical-temperature isotherm has no physical significance. In passing from the state 1 to the state 2, the molar volume of the system progressively decreases, and hence the density of the system progressively increases. No phase separation occurs between the states 1 and 2, and, in physical reality, the system at the point 2 can equivalently be regarded as being either a liquid of normal density or a gas of high density. Similarly at the point 1, the system can equivalently be regarded as being a gas of normal density or a liquid of low density. Physically, then, no distinction can be made between the liquid and gaseous states at temperatures in excess of T_{cr} and pressures in excess of P_{cr}, and the distinction made between the two on the basis of the critical isotherm is entirely arbitrary. Thus, in the P-T phase diagram of the system (e.g., Fig. 7.10), the liquid-vapor equilibrium line (OB in Fig. 7.10) terminates at the state P_{cr}, T_{cr}.

8.3 DEVIATION FROM IDEALITY AND EQUATIONS OF STATE OF REAL GASES

The deviation of an actual gas from ideal behavior can be measured as the deviation of the compressibility factor Z from unity. The compressibility factor Z is defined as

$$Z = \frac{PV}{RT} \tag{8.3}$$

which, it is seen, is unity for a perfect gas in all states of existence. Z itself is a function of the state of the system and hence is dependent on any two chosen dependent variables, e.g., $Z = Z(P,T)$. Figure 8.3 shows the variation of Z with P at constant T for several gases.

For all of the gases in Fig. 8.3 the variation of Z with P is linear up to about 10 atm and hence can be expressed as

$$Z = mP + 1$$

or

$$\frac{PV}{RT} = mP + 1$$

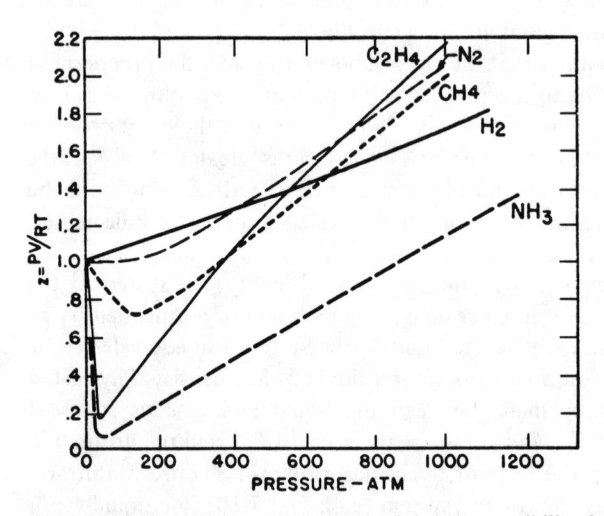

Fig. 8.3. The variations with pressure of the compressibility factors of several gases at 0°C. *(From W. J. Moore, "Physical Chemistry," Longmans, Green and Co., London, 1950.)*

which can be written as

$$P(V - mRT) = RT$$

or

$$P(V - b') = RT \tag{8.4}$$

where $b' = mRT$, and has the dimensions of volume.

Equation (8.4) is an expression of the equation of state for the gases up to pressures at which deviation from linear Z-P behavior begins. Comparison with Eq. (8.3) shows that the deviations from ideal behavior, in the stated pressure range, can be dealt with by making a correction to the volume term in the ideal gas equation of state. The need for such a correction is reasonable in view of the fact that an ideal gas comprises a system of noninteracting, volumeless particles, whereas the particles of a real gas have a small, but nevertheless finite, volume. Thus, in the case of a real gas, the volume available to the movement of an Avogadro's number of particles is actually less than the molar volume of the gas by an amount equal to the volume excluded by the particles themselves; hence the ideal gas equation must be corrected for this effect. At first sight it might appear that the constant b' in Eq. (8.4) is the volume excluded by the particles; but inspection of Fig. 8.3 shows that, with the exception of hydrogen, b' is a negative quantity, and hence the interpretation of b', using the above reasoning, is incorrect. Equation (8.4) is thus to be regarded as being a purely empirical equation which, by suitable choice of the constant b', can be made to describe the behavior of real gases over a narrow range of low pressures in the vicinity of 0°C.

If Fig. 8.3 is replotted as Z versus the reduced pressure P_R (where $P_R = P/P_{cr}$) for fixed values of the reduced temperature $T_R (= T/T_{cr})$, then it is found that all gases lie on a single curve. Figure 8.4 shows a series of such plots. The superposition, in Fig. 8.4, gives rise to the law of corresponding states, which states that all gases obey the same equation of state when expressed in terms of the reduced variables P_R, T_R, and V_R instead of P,T, and V. If two gases have identical values of two reduced variables, then they have approximately equal values of the third, and the two gases are then said to be in corresponding states. Figure 8.4 shows that the compressibility factor Z is the same function of the reduced variables for all gases (see Prob. 8.1).

8.4 THE VAN DER WAALS GAS

An ideal gas is characterized by its compliance with the ideal gas law and the fact that its internal energy U is a function only of temperature. Physically an ideal gas is thus an assemblage of volumeless noninteracting particles, the energy

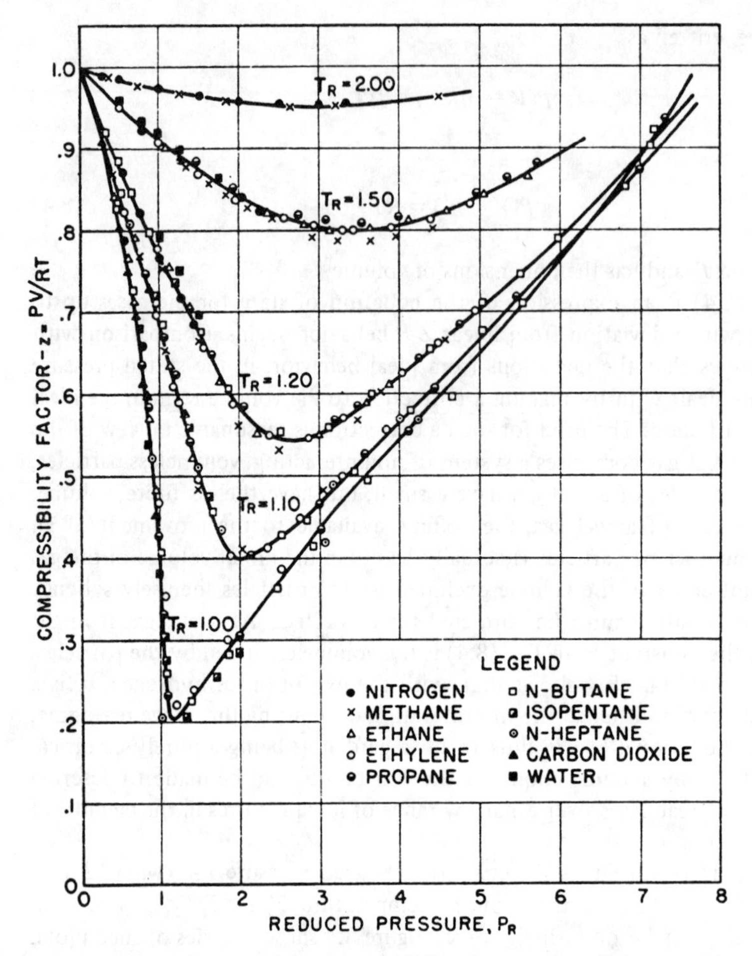

Fig. 8.4. The variations of the compressibility factors of several gases with reduced pressure at several reduced temperatures. *(From G.-J. Su, "Modified Law of Corresponding States for Real Gases," Ind. Eng. Chem. 39:803, 1946.)*

of which is entirely the translational energy of motion of the constituent particles. Attempts to calculate equations of state of real gases have generally attempted to modify the ideal gas equation by taking into consideration

1. The fact that the particles of a real gas occupy a finite volume
2. The fact that interactions occur among the particles of a real gas

The quantitative importance of these two considerations depends on the state of the gas. For example, if the molar volume of the gas is large, then the volume fraction occupied by the gas particles themselves is small, and the magnitude of this effect on the behavior of the gas will be correspondingly small. Similarly, as the molar volume increases, then the average particle-particle separation in the gas increases, and hence the effect of particle-particle interactions on the gas behavior decreases. For a fixed number of moles of gas, an increase in the volume corresponds to a decrease in the density (n/V'), and such states of existence occur at low pressure and at high temperature, as can be seen from the ideal gas equation; i.e.,

$$\frac{n}{V'} = \frac{P}{RT}$$

Thus approach towards ideal behavior is to be expected as the pressure is decreased and as the temperature is increased.

The most celebrated equation of state for nonideal gases, which was derived from considerations 1 and 2 above, is the van der Waals equation. This equation, for 1 mole of gas, is written as

$$\left(P + \frac{a}{V^2}\right)(V - b) = RT$$

where P is the measured pressure of the gas, a/V^2 is a correction term for particle-particle interactions, V is the measured volume of the gas, and b is a correction term for the finite volume of the particles.* The term b is determined by considering collisions between two particles. When two spherical particles, each of radius r, collide, they exclude a volume of

$$\tfrac{4}{3}\,\pi(2r)^3$$

to all other particles, as is seen in Fig. 8.5*a*. The excluded volume per particle is thus

$$\tfrac{1}{2} \times \tfrac{4}{3}\,\pi(2r)^3 = 4 \times \tfrac{4}{3}\,\pi r^3$$
$$= 4 \times \text{the volume of one particle}$$

The excluded volume is thus four times the volume of all the particles present, and this equals the constant b. Hence the volume $(V - b)$ is the volume available to the movement of the gas particles and is the molar volume which the gas

*For n moles of van der Waals gas, the equation of state is

$$\left(P + \frac{n^2 a}{V'^2}\right)(V' - nb) = nRT \qquad \text{where } V' = nV$$

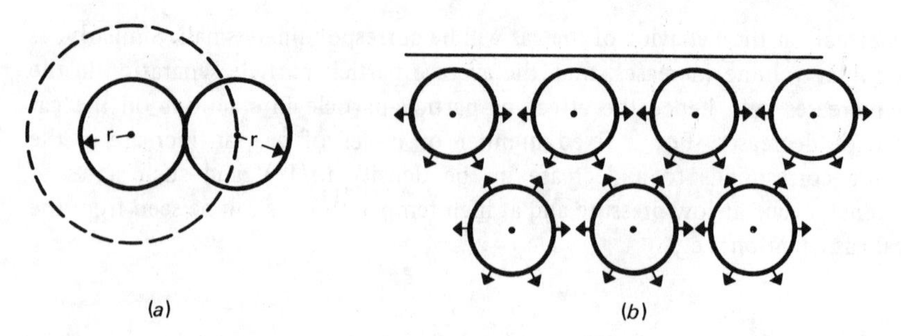

Fig. 8.5. Illustration of the origins of van der Waals gas behavior.

would have were the gas ideal, i.e., if the particles were volumeless. The attractive forces operating between the gas particles decrease the pressure of the gas below that which would be exerted in the absence of the attractive forces. The necessary correction is arrived at in the following way. The particles in the "layer" adjacent to the wall of the vessel containing the gas experience a net inward pull due to the influence of the particles in the next adjacent "layer." These attractive forces give rise to the phenomena of "internal pressure," and the magnitude of this net inward pull (i.e., the decrease in the pressure exerted by the gas on the wall of the containing vessel) is proportional to the number of particles in the "surface layer" and the number of particles in the "next-to-the-surface layer." Both of these quantities are proportional to the density of the gas, n/V', and hence the net inward pull is proportional to the square of the gas density, or is equal to a/V^2 where a is a constant. If P is the measured pressure of the gas, i.e., the pressure which it actually exerts on the walls of its containing vessel, then $P + a/V^2$ is the pressure which the gas would exert were it ideal, i.e., in the absence of the particle-particle interactions. The effect is illustrated in Fig. 8.5*b*.

The van der Waals equation can be written as

$$PV^3 - (Pb + RT)V^2 + aV - ab = 0$$

which, being cubic in V, has three roots. Plotting V against P for different values of T gives the series of isotherms shown in Fig. 8.6. As the temperature is increased the minimum and maximum approach one another until, at T_{cr}, they coincide and produce an inflexion on the P-V curve. At this, the critical, point $T = T_{cr}$, $P = P_{cr}$, and $V = V_{cr}$; and hence the van der Waals equation gives,

$$P_{cr} = \frac{RT_{cr}}{(V_{cr} - b)} - \frac{a}{V_{cr}^2}$$

$$\left(\frac{\partial P}{\partial V}\right)_{T_{cr}} = \frac{-RT_{cr}}{(V_{cr}-b)^2} - \frac{2a}{V_{cr}^3} = 0$$

$$\left(\frac{\partial^2 P}{\partial V^2}\right)_{T_{cr}} = \frac{2RT_{cr}}{(V_{cr}-b)^3} - \frac{6a}{V_{cr}^4} = 0$$

Solving these equations gives

$$T_{cr} = \frac{8a}{27bR}\ , \qquad V_{cr} = 3b\ , \qquad P_{cr} = \frac{a}{27b^2} \tag{8.5}$$

and hence the constants a and b for any gas can be evaluated from a knowledge of the values of T_{cr} and P_{cr}. The critical states, van der Waals constants, and the values of Z at the critical point for several gases are listed in Table 8.1.

Consider the isothermal P-V line, from the van der Waals equation, shown in Fig. 8.7. When the pressure exerted on a system is increased, the volume of the system decreases, that is, $(\partial P/\partial V)_T < 0$. This is a condition of intrinsic stability of the system, and it can be seen that, in Fig. 8.7, this condition is violated over the portion JHF. This portion of the curve thus has no physical significance. The effect of pressure on the equilibrium state of the system can be obtained from a consideration of the variation of the free energy with P along the isotherm. Equa-

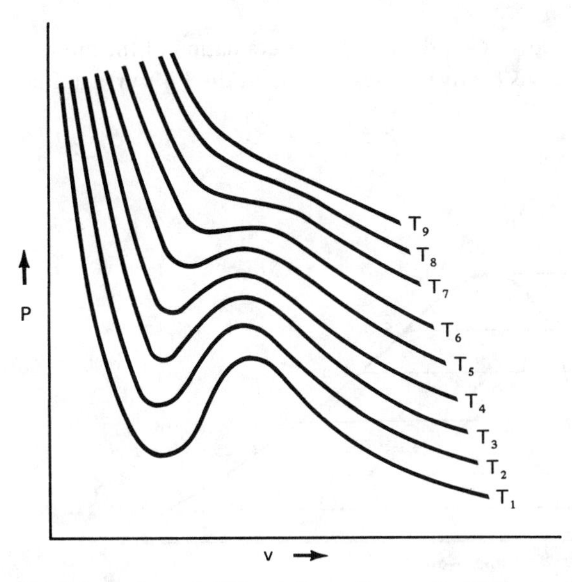

Fig. 8.6. Isothermal P-V variations of a van der Waals gas at several temperatures.

Table 8.1

Gas	T_{cr}, K	P_{cr}, atm	V_{cr}, cm³/mole	a, (liter²/atm)	b, cm³/mole	Z_{cr}
He	5.3	2.26	57.6	.0341	23.7	.299
H_2	33.3	12.8	65.0	.2461	26.7	.304
N_2	126.1	33.5	90.0	1.39	39.1	.292
CO	134.0	35.0	90.0	1.49	39.9	.295
O_2	153.4	49.7	74.4	1.36	31.8	.293
CO_2	304.2	73.0	95.7	3.59	42.7	.280
NH_3	405.6	111.5	72.4	4.17	37.1	.243
H_2O	647.2	217.7	45.0	5.46	30.5	.184

tion (5.12) gives $dG = V\,dP$ for an incremental change of state at constant temperature, and integration of this equation between the states (P, T) and (P_A, T) gives,

$$G(P, T) - G(P_A, T) = \int_{P_A}^{P} V dP$$

or

$$G = G_A + \int_{P_A}^{P} V dP \tag{8.6}$$

If an arbitrary value is assigned to G_A, then graphical evaluation of the integral from Fig. 8.7 allows the G versus P curve, corresponding to the P-V curve in Fig.

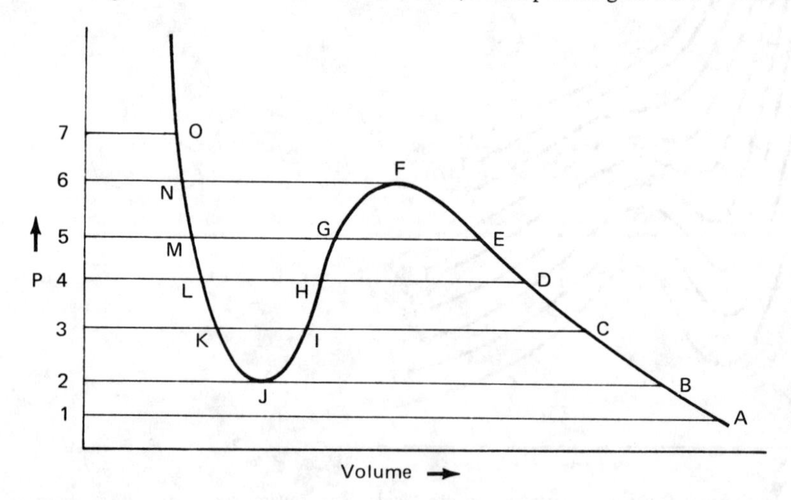

Fig. 8.7. The isothermal P-V variation of a van der Waals gas at a temperature below the critical temperature.

8.7, to be drawn. The values of the integrals are listed in Table 8.2, and the *G-P* curves are shown in Fig. 8.8.

Figure 8.8 shows that, as the pressure is increased from P_1, the value of G increases. At pressures in excess of P_2 three states of existence become available to the system; e.g., at P_3 the three states are given by the points I, K, and C. The stable, or equilibrium, state is that of lowest free energy, and hence, over the pressure range P_2 to P_4, the stable states lie on the line BCD. As the pressure is increased above P_4 the state of lowest free energy no longer lies on the original line, i.e., on the continuation of the line BCD, but lies on the line LMN. The change of stability at P_4 corresponds to a phase change at this point; i.e., at pressure less than P_4 one phase is stable, and at pressures greater than P_4 another phase is stable. In reality the stable phase below P_4 is the vapor phase, and the stable phase above P_4 is the liquid phase. At P_4, G_D (that is, G_{vapor}) = G_L (that is, G_{liquid}), and hence at the state P_4, T vapor and liquid coexist in equilibrium. In Fig. 8.7 a tie-line connects the points D and L across a two-phase region. In Fig. 8.8 the lines DF and LJ represent, respectively, the metastable vapor and metastable liquid states; i.e., in the absence of nucleation of the liquid phase from the vapor phase at the state D, supersaturated vapor would exist along the line DEF, and in the absence of nucleation of the vapor phase from the liquid phase at the state L, supersaturated liquid would exist along the line LKJ. In view of the violation of the intrinsic stability criterion over the states path JHF, the states represented by this line in both Fig. 8.7 and Fig. 8.8 have no physical significance.

It is thus seen that the van der Waals equation predicts the phase change which occurs in the system at temperatures less than T_{cr}. At any temperature below

Table 8.2. Graphical Integration of Figure 8.7

$G_B = G_A + \int_{P_A}^{P_B} V\,dP$	$= G_A$ + area 1AB2			
G_C	$= G_A$ + area 1AC3			
G_D	$= G_A$ + area 1AD4			
G_E	$= G_A$ + area 1AE5			
G_F	$= G_A$ + area 1AF6			
G_G	$= G_A$ + area 1AE5	+ area EFG		
G_H	$= G_A$ + area 1AD4	+ area DFH		
G_I	$= G_A$ + area 1AC3	+ area CFI		
G_J	$= G_A$ + area 1AB2	+ area BFJ		
G_K	$= G_A$ + area 1AC3	+ area CFI	− area IJK	
G_L	$= G_A$ + area 1AD4	+ area DFH	− area HJL	
G_M	$= G_A$ + area 1AEF	+ area EFG	− area GJM	
G_N	$= G_A$ + area 1AF6		− area FJN	
G_O	$= G_A$ + area 1AF6		− area FJN	+ area 6NO7

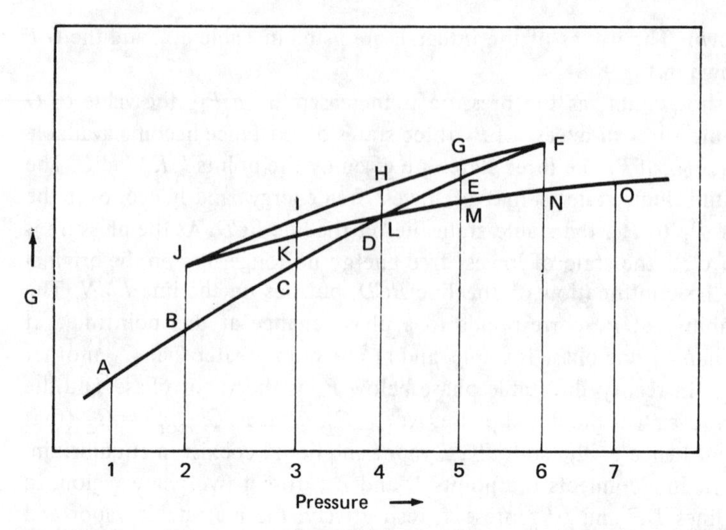

Fig. 8.8. Schematic representation of the isothermal *G-P* variation of a van der Waals gas at a temperature lower than the critical temperature.

T_{cr}, the value of P for liquid-vapor equilibrium, e.g., P_4 in Figs. 8.7 and 8.8, is determined as that pressure which gives equal areas *HFD* and *LJH* in Fig. 8.7.

The experimentally determined values of T_{cr} and P_{cr} for carbon dioxide gas are, respectively, 31°C and 72.9 atm. Thus, from Eq. (8.5),

$$b = \frac{RT_{cr}}{8P_{cr}} = 0.0427 \text{ liters}$$

and

$$a = 27b^2\, P_{cr} = 3.59 \text{ liter}^2/\text{atm}$$

in which case the van der Waals equation for CO_2 is given as

$$\left(P + \frac{3.59}{V^2}\right)(V - 0.0427) = RT$$

The variation of P with V, given by this equation, is shown, at several temperatures, in Fig. 8.9, in which it is seen that the 304 K isotherm exhibits an inflexion at the critical point. At temperatures lower than the critical temperature of 304 K, the isotherms show the expected maxima and minima. The variation of the vapor pressure of liquid CO_2 with temperature could be determined by finding that tie-line on each isotherm which gives equal areas *DFH* and *LJH* as explained with reference to Fig. 8.7. Alternatively the G versus

P relationships for each isotherm could be determined by graphical integration of the P-V relationships according to Eq. (8.6). These relationships, for several temperatures, are shown in Fig. 8.10, which gives a clear indication of the variation of the vapor pressure of liquid CO_2 (the points p) with temperature. Figure 8.10 further indicates that, as the temperature increases towards the critical point, the range of nonphysical states (J to F in Fig. 8.8) diminishes and finally disappears at T_{cr}. Above T_{cr} the continuous-line relationship indicates that, in the entire pressure range, only one phase occurs. It must be pointed out that as G_A in Eq. (8.6) is a function of temperature, the positions of the isothermal G-P lines with respect to one another on the G axis in Fig. 8.10 are arbitrary; i.e., only the P axis is quantitatively significant.

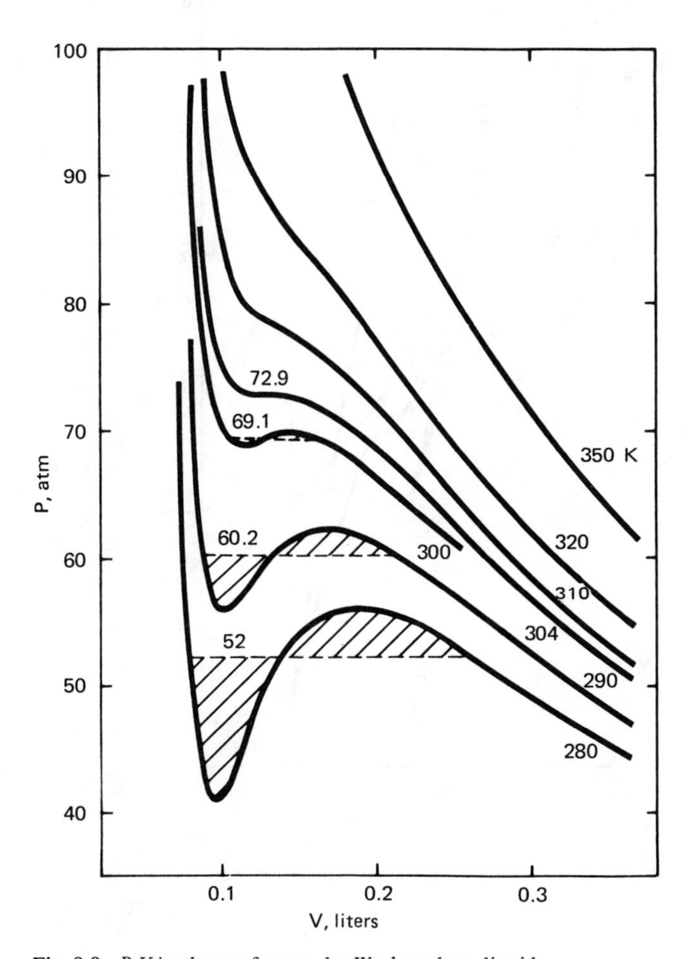

Fig. 8.9. *P-V* isotherms for van der Waals carbon dioxide.

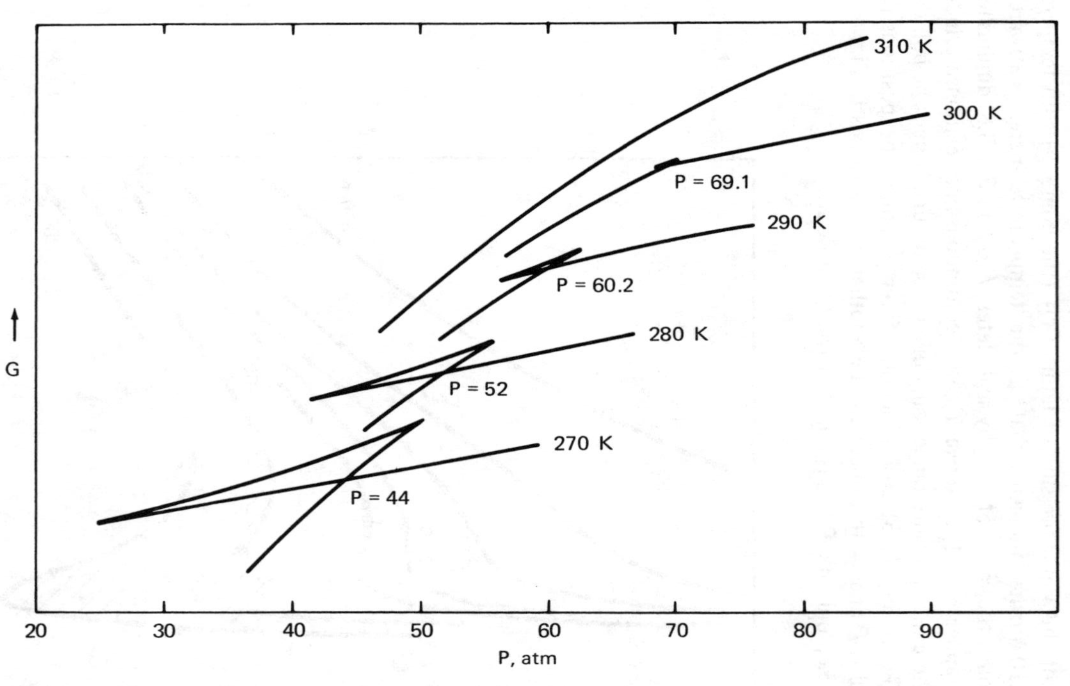

Fig. 8.10. *G-P* variations for van der Waals carbon dioxide at several temperatures.

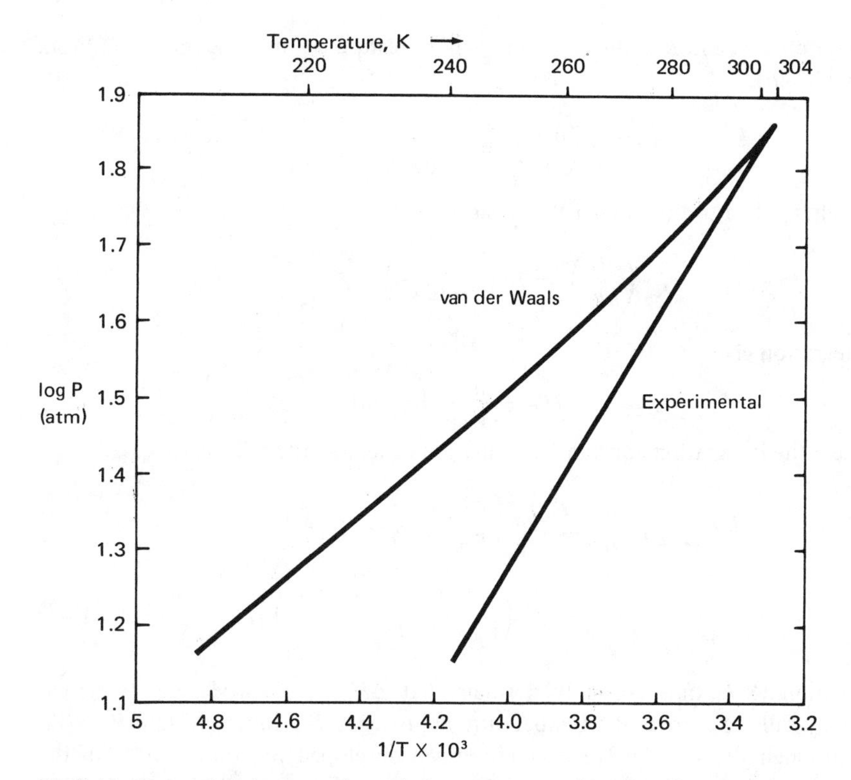

Fig. 8.11. Comparison between the variation with temperature of the vapor pressure of van der Waals liquid carbon dioxide and the actual vapor pressure of liquid carbon dioxide.

The variation of the saturated vapor pressure of liquid CO_2 with temperature, obtained from the van der Waals equation and plotted as $\log p$ versus $1/T$, is shown in Fig. 8.11. Also drawn in this figure is the experimentally determined variation of the saturated vapor pressure of liquid CO_2 with temperature. Comparison between the two lines shows that the van der Waals equation overestimates the vapor pressure, with the extent of this overestimation increasing as the temperature decreases. Consequently the van der Waals equation underestimates the latent heat of evaporation which, from Eq. (7.7), is obtained as $-R \times$ (the slope of the line). The latent heat of evaporation of the liquefied van der Waals gas can be calculated as follows:

$$\Delta H_{\text{evap}} = H_v - H_l = U_v - U_l + P(V_v - V_l)$$

where V_v and V_l are, respectively, the molar volumes of the coexisting vapor and liquid phases, and P is the saturated vapor pressure, at the temperature T. From Eq. (5.34*b*),

$$\left(\frac{\partial U}{\partial V}\right)_T = T\left(\frac{\partial P}{\partial T}\right)_V - P$$

which, applied to the van der Waals gas, gives

$$\left(\frac{\partial U}{\partial V}\right)_T = T\left(\frac{R}{(V-b)}\right) - P = \frac{a}{V^2}$$

Integration gives

$$U = -\frac{a}{V} + \text{constant}$$

where the integration constant is a function of temperature. Thus

$$\Delta H_{\text{evap}} = -\frac{a}{V_v} + \frac{a}{V_l} + P(V_v - V_l)$$

$$= -a\left(\frac{1}{V_v} - \frac{1}{V_l}\right) + P(V_v - V_l) \tag{8.7}$$

Equation (8.7) thus correctly predicts that ΔH_{evap} for a van der Waals gas rapidly falls to zero as the temperature approaches T_{cr}, in which state $V_v = V_l$.

Although the van der Waals equation was developed via consideration of the physical factors effecting nonideal behavior, the requirement that the critical properties be known for the calculation of the constant a means that the equation is empirical. This, however, does not detract from the extreme usefulness of the equation in describing the behavior of a gas which exhibits a relatively small departure from ideality.

8.5 OTHER EQUATIONS OF STATE FOR NONIDEAL GASES

Other examples of derived equations of state for nonideal gases are

$$P(V - b')e^{a'/RTV} = RT \qquad \text{the Dieterici equation}$$

$$\left(P + \frac{A}{TV^2}\right)(V - B) = RT \qquad \text{the Berthelot equation}$$

Neither of these equations has a real fundamental basis, and general empirical equations are normally used. Examples of such equations are the Beattie-Bridgeman equation, which has five constants in addition to R and fits the P-V-T

relationships over a wide range of pressures and temperatures, and the Kamerlingh Onnes, or virial, equation of state. In this latter equation it is assumed that PV/RT is a power series of P or $1/V$, that is

$$\frac{PV}{RT} = 1 + BP + CP^2 + \cdots$$

or

$$\frac{PV}{RT} = 1 + \frac{B'}{V} + \frac{C'}{V^2} + \cdots$$

PV is termed the virial, B or B' is termed the first virial coefficient, C or C' is termed the second virial coefficient, etc., and the virial coefficients are functions of temperature. In both equations, as $P \rightarrow 0$, and hence as $V \rightarrow \infty$, then $PV/RT \rightarrow 1$. The virial equation converges in the gas phase, and thus the equation of state can adequately be represented by the virial expansion over the entire range of densities and pressures. In practice, however, the virial equation is only used when the first few terms need be retained. At low pressures or densities,

$$\frac{PV}{RT} = 1 + BP$$

or

$$\frac{PV}{RT} = 1 + \frac{B'}{V}$$

both of which are expressions of Eq. (8.4).

8.6 THE THERMODYNAMIC PROPERTIES OF IDEAL GASES AND MIXTURES OF IDEAL GASES

The Isothermal Free Energy–Pressure Relationship of an Ideal Gas

For an infinitesimal change of state of a closed system of fixed composition at constant temperature, the fundamental equation, Eq. (5.25), gives

$$dG = VdP$$

For 1 mole of an ideal gas, this can be written as

$$dG = \frac{RT}{P}\, dP = RTd \ln P \tag{8.8}$$

and hence for an isothermal change of state from P_1 to P_2 at the temperature T,

$$G(P_2, T) - G(P_1, T) = RT \ln \frac{P_2}{P_1} \tag{8.9}$$

As Classical Thermodynamics can say nothing about the absolute values of $G(P_2, T)$ or $G(P_1, T)$, it is convenient to choose an arbitrary reference state from which free energy changes can be measured. This reference state is called the *standard state* and is chosen as being the state of 1 mole of pure gas at 1 atm pressure and the temperature of interest. The free energy of 1 mole of gas in the standard state, $G(P = 1, T)$ is designated $G^0(T)$ and hence, from Eq. (8.9), the free energy of 1 mole of gas at any other pressure P and the temperature T is given as

$$G(P, T) = G^0(T) + RT \ln P$$

or simply

$$G = G^0 + RT \ln P \tag{8.10}$$

In Eq. (8.10) it is to be realized that the logarithm of a dimensionless ratio, $P/1$, occurs in the right-hand term.

Mixtures of Perfect Gases

Before discussing the thermodynamic properties of perfect gas mixtures, it is necessary to introduce the concepts of *partial pressure*, *mole fraction*, and *partial molar quantities*.

Mole Fraction

When more than one component is present in a system, i.e., when the composition of the system is variable, it is necessary to invent a means of expressing the composition. Several composition variables are in common use, of which only one–the mole fraction–has any theoretical significance. The mole fraction, X_i, of the component species i is defined as the ratio of the number of moles of i in the system to the total number of moles of all component species in the system. For example, if the system contains n_A moles of A, n_B moles of B, and n_C moles of C, then,

$$X_A = \frac{n_A}{n_A + n_B + n_C}$$

$$X_B = \frac{n_B}{n_A + n_B + n_C}$$

$$X_C = \frac{n_C}{n_A + n_B + n_C}$$

The particular convenience of the use of the mole fraction as a composition variable lies in the fact that the sum of the mole fractions of all component species in a system equals unity; e.g., in the above system, $X_A + X_B + X_C = 1$.

Dalton's Law of Partial Pressures

The pressure P exerted by a mixture of perfect gases is equal to the sum of the pressures exerted by each of the individual component gases; the contribution made to the total pressure P by each individual gas is called the *partial pressure* of that gas. The partial pressure exerted by a component gas, p_i, is thus the pressure which it would exert if it alone were present. In a mixture of perfect gases A, B, and C,

$$P = p_A + p_B + p_C$$

Consider a fixed volume V', at the temperature T, which contains n_A moles of a perfect gas A. The pressure exerted is thus,

$$P = \frac{n_A RT}{V'} \tag{8.11}$$

If to this volume containing n_A moles of gas A, n_B moles of perfect gas B are added, then the pressure increases to

$$P = p_A + p_B = (n_A + n_B)\frac{RT}{V'} \tag{8.12}$$

Division of Eq. (8.11) by Eq. (8.12) gives

$$\frac{p_A}{p_A + p_B} = \frac{n_A}{n_A + n_B}$$

which, for the gas A in the mixture, can be written as

$$\frac{p_A}{P} = X_A$$

or

$$p_A = X_A P \tag{8.13}$$

Thus in a mixture of perfect gases the partial pressure of a component gas is the product of the mole fraction of the component gas and the total pressure of the gas mixture.

Partial Molar Properties

The value of any extensive state property of a component species in a mixture, per mole of that component species, is called the *partial molar value of the property*. This value is not necessarily equal to the value of the property per mole of the pure component species. The partial molar value of an extensive property, Q, of the component i in a mixture of components $i, j, k, \ldots$ is formally defined as

$$\bar{Q}_i = \left(\frac{\partial Q'}{\partial n_i}\right)_{T, P, n_j, n_k, \ldots} \tag{8.14}$$

where Q' is the value of the extensive property for the arbitrary quantity of the mixture.

$\bar{Q}_i$ is thus defined as the change in the value of Q' of the mixture for an infinitesimal addition of the component i, at constant temperature, pressure, and mole numbers of all other components, per mole of i added. The statement that the addition of i be infinitesimal is necessitated by the fact that as $\bar{Q}_i$, being a state property, is dependent on composition, the composition of the system must remain virtually constant during the addition of i. This restriction can be removed if it is considered that the addition is made to a quantity of mixture which is sufficiently large that the addition of a further mole of i to the mixture makes virtually no change in the overall composition of the mixture. In such a case the change in Q' for the mixture, resulting from the addition of 1 mole of i at constant T and P, equals $\bar{Q}_i$. In the case of the extensive property being the free energy, then

$$\bar{G}_i = \left(\frac{\partial G'}{\partial n_i}\right)_{T, P, n_j, n_k, \ldots}$$

and from Eq. (5.16) it is seen that

$$\bar{G}_i = \mu_i$$

i.e., the partial molar free energy of a component in a mixture equals the chemical potential of the component in the mixture.

The relationships between the various state functions developed in the preceding chapters are applicable to the partial molar properties of the components of a system. For example, the fundamental equation, Eq. (5.25), at constant T and composition, gives

$$\left(\frac{\partial G'}{\partial P}\right)_{T,\text{comp}} = V'$$

where G' is the free energy of the system and V' is the volume of the system. For a variation in n_i, the number of moles of component i in the system, at constant T, P, and n_j,

$$\left[\frac{\partial}{\partial n_i}\left(\frac{\partial G'}{\partial P}\right)_{T,\text{comp}}\right]_{T,P,n_j} = \left(\frac{\partial V'}{\partial n_i}\right)_{T,P,n_j}$$

But, by definition,

$$\left(\frac{\partial V'}{\partial n_i}\right)_{T,P,n_j} = \bar{V}_i$$

and, as G is a state function,

$$\left[\frac{\partial}{\partial n_i}\left(\frac{\partial G'}{\partial P}\right)_{T,\text{comp}}\right]_{T,P,n_j} = \left[\frac{\partial}{\partial P}\left(\frac{\partial G'}{\partial n_i}\right)_{T,P,n_j}\right]_{T,\text{comp}}$$

$$= \left(\frac{\partial \bar{G}_i}{\partial P}\right)_{T,\text{comp}}$$

Hence

$$\left(\frac{\partial \bar{G}_i}{\partial P}\right)_{T,\text{comp}} = \bar{V}_i$$

which is simply the application of Eq. (5.25) to the component i in the system. Thus for the perfect gas A in a mixture of perfect gases,

$$d\bar{G}_A = \bar{V}_A \, dP$$

The partial molar volume, $\bar{V}_A$, in a gas mixture is

$$\bar{V}_A = \frac{V'}{\Sigma n_i} = \frac{X_A RT}{p_A}$$

Differentiation of Eq. (8.13) at constant T and composition gives $dp_A = X_A\, dP$, and hence,

$$d\bar{G}_A = \bar{V}_A\, dP = \frac{X_A RT}{p_A}\frac{dp_A}{X_A} = RT\, d\,(\ln p_A)$$

Integration between the limits $p_A = p_A$ and $p_A = 1$ gives

$$\begin{aligned} \bar{G}_A &= G_A^\circ + RT \ln p_A \\ &= G_A^\circ + RT \ln X_A + RT \ln P \end{aligned} \tag{8.15}$$

Equation (8.15) could also have been obtained by integrating Eq. (8.8) from the standard state $p_A = P_A = 1$, $X_A = 1$, T to the state $p_A = p_A$, $X_A = X_A$, T.

The Heat of Mixing of Perfect Gases

For each component gas in a perfect gas mixture,

$$\bar{G}_i = G_i^0 + RT \ln X_i + RT \ln P$$

where P is the pressure of the gas mixture at the temperature T. Dividing by T and differentiating with respect to T at constant pressure and composition, gives

$$\frac{\partial(\bar{G}_i/T)}{\partial T} = \frac{\partial(G_i^0/T)}{\partial T} \tag{8.16}$$

But, from Eq. (5.37),

$$\left[\frac{\partial(G_i^0/T)}{\partial T}\right]_{P,\,\mathrm{comp}} = -\frac{H_i^0}{T^2} \quad \text{and} \quad \left[\frac{\partial(\bar{G}_i/T)}{\partial T}\right]_{P,\,\mathrm{comp}} = -\frac{\bar{H}_i}{T^2} \tag{8.17}$$

and thus,

$$\bar{H}_i = H_i^0 \tag{8.18}$$

i.e., the molar enthalpy of pure i equals the partial molar enthalpy of i in the gas mixture, and thus the enthalpy of the gas mixture equals the enthalpy of the unmixed component gases; i.e.,

$$\Delta H'^{\text{mix}} = \sum_i n_i \bar{H}_i - \sum_i n_i H_i^0 = 0 \tag{8.19}$$

where $\Delta H'^{\text{mix}}$ is the change in the enthalpy as a result of the mixing process.

As G_i^0 is, by definition, a function only of temperature, then from Eqs. (8.16) and (8.17) it is seen that $\bar{H}_i$ is a function only of temperature. Thus, in addition to being independent of composition, $\bar{H}_i$ is independent of pressure. This zero heat of mixing of ideal gases is a consequence of the fact that ideal gases are assemblages of noninteracting particles.

The Free Energy of Mixing of Perfect Gases

For each component gas i in a perfect gas mixture,

$$\bar{G}_i = G_i^0 + RT \ln p_i$$

and for each component gas before mixing,

$$G_i = G_i^0 + RT \ln P_i$$

where p_i is the partial pressure of i in the gas mixture, and P_i is the pressure of pure i before mixing.

The mixing process, being a change of state, can be written as

$$\text{Unmixed components (state 1)} \longrightarrow \text{mixed components (state 2)}$$

and

$$\begin{aligned} \Delta G(1 \rightarrow 2) &= G'(\text{mixture}) - G'(\text{unmixed components}) \\ &= \Delta G'^{\text{mix}} \\ &= \sum_i n_i \bar{G}_i - \sum_i n_i G_i \\ &= \sum_i n_i RT \ln\left(\frac{p_i}{P_i}\right) \end{aligned} \tag{8.20}$$

The value of $\Delta G'^{\text{mix}}$ depends, thus, on the value of p_i and P_i for each gas. If, before mixing, the gases are all at the same pressure, i.e., if $P_i = P_j = P_k = \ldots$ and mixing is carried out at constant total volume such that the total pressure of the mixture, P_{mix}, equals the initial pressures of the individual unmixed gases, then, as $p_i/P_i = X_i$,

$$\Delta G'^{\text{mix}} = \sum_i n_i RT \ln X_i \tag{8.21}$$

As the values of X_i are less than unity, $\Delta G'^{\text{mix}}$ is a negative quantity, which corresponds with the fact that the mixing of gases is a spontaneous process.

The Entropy of Mixing of Perfect Gases

As $\Delta H'^{\text{mix}} = 0$

and

$$\Delta G'^{\text{mix}} = \Delta H'^{\text{mix}} - T\Delta S'^{\text{mix}}$$

then

$$\Delta S'^{\text{mix}} = -\sum_i n_i R \ln\left(\frac{p_i}{P_i}\right) \tag{8.22}$$

or, if $P_i = P_j = P_k = \ldots = P$, then

$$\Delta S'^{\text{mix}} = -\sum_i n_i R \ln X_i \tag{8.23}$$

which is seen to be a positive quantity, in accordance with the fact that gas mixing is a spontaneous process.

8.7 THE THERMODYNAMIC TREATMENT OF IMPERFECT GASES

Equation (8.10) showed that, at any temperature, the molar free energy of an ideal gas is a linear function of the logarithm of the pressure of the gas. This property is a direct result of the ideal gas law, which was used in the derivation of Eq. (8.10); hence if a gas deviates from ideality, then the relation between its free energy and the logarithm of its pressure is no longer linear. However, the simple form of Eq. (8.10) is such as to warrant the invention of a function of the state of the gas, which when used in place of the pressure in Eq. (8.10) ensures linearity between G and the logarithm of this function in any state for any gas. This function is called the *fugacity*, f, and is partially defined by the equation

$$dG = RTd \ln f$$

The integration constant is chosen such that the fugacity approaches the pressure as the pressure approaches zero; i.e.,

$$\frac{f}{P} \to 1 \text{ as } P \to 0$$

in which case

$$G = G^0 + RT \ln f \tag{8.24}$$

where G^0 is the molar free energy of the gas in its standard state, which now is defined as that state in which $f = 1$ at the temperature T. (The standard state for an ideal gas was defined as being $P = 1, T$.)

Consider a gas which obeys the equation of state,

$$V = \frac{RT}{P} - \alpha$$

where α is a function only of temperature and is a measure of the deviation of the gas from ideality. Equation (5.12) gives $dG = V\,dP$ at constant T and, Eq. (8.24) gives $dG = RT\,d \ln f$ at constant T. Thus, at constant T,

$$VdP = RTd \ln f$$

and hence

$$d \ln\left(\frac{f}{P}\right) = -\frac{\alpha}{RT}\,dP \tag{8.25}$$

Integration between the states $P = P$ and $P = 0$, at constant T gives

$$\ln\left(\frac{f}{P}\right)_{P=P} - \ln\left(\frac{f}{P}\right)_{P=0} = -\frac{\alpha P}{RT} \tag{8.26}$$

As $f/P = 1$ when $P = 0$, then $\ln(f/P) = 0$ when $P = 0$, and hence

$$\ln\left(\frac{f}{P}\right) = -\frac{\alpha P}{RT} \quad \text{or} \quad \frac{f}{P} = e^{-\alpha P/RT}$$

In order that α can be regarded as being independent of pressure, the deviation from ideality must be small, in which case α is a small number. Thus,

$$e^{-\alpha P/RT} \doteqdot 1 - \frac{\alpha P}{RT}$$

and hence
$$\frac{f}{P} = 1 - \frac{\alpha P}{RT} = 1 - \left(\frac{RT}{P} - V\right)\frac{P}{RT} = \frac{PV}{RT}$$

If the gas behaved ideally, then the ideal pressure, P_{id}, would be given as RT/V.

Thus

$$\frac{f}{P} = \frac{P}{P_{id}} \tag{8.27}$$

which indicates that the actual pressure of the gas is the geometric mean of the fugacity and the ideal pressure. Also it is seen that the percentage error involved in assuming the fugacity to be equal to the pressure is the same as the percentage departure from the ideal gas law.

Alternatively, the fugacity can be dealt with in terms of the compressibility factor Z. From Eq. (8.25)

$$d\ln\left(\frac{f}{P}\right) = -\frac{\alpha}{RT}\,dP = \left(\frac{V}{RT} - \frac{1}{P}\right)dP$$

But $Z = PV/RT$, and hence

$$d\ln\left(\frac{f}{P}\right) = \frac{Z-1}{P}\,dP$$

and

$$\ln\left(\frac{f}{P}\right)\Bigg|_{P=P} = \int_{P=0}^{P=P} \frac{Z-1}{P}\,dP \tag{8.28}$$

This may be evaluated either by graphical integration of a plot of $(Z-1)/P$ versus P at constant T or by direct integration if Z is known as a function of P, i.e., if the virial equation of state of the gas is known.

For example, it has been determined that the variation of PV (cm^3-atm) with P in the range 0 to 200 atm for nitrogen gas at 0°C can be expressed by the equation

$$PV = 22{,}414.6 - 10.281P + 0.065189P^2 + 5.1955 \times 10^{-7}P^4 - 1.3156 \times 10^{-11}P^6 + 1.009 \times 10^{-16}P^8$$

Thus, dividing by $RT = 22{,}414.6$ at 0°C gives

$$\frac{PV}{RT} = Z = 1 - 4.5867 \times 10^{-4}P + 2.9083 \times 10^{-6}P^2 + 2.3179 \times 10^{-11}P^4 - 5.8694 \times 10^{-15}P^6 + 4.5015 \times 10^{-21}P^8$$

This variation of Z with P is shown graphically in Fig. 8.3. From integration of Eq. (8.28), ln (f/P) is obtained as

$$\ln\left(\frac{f}{P}\right) = -4.5867 \times 10^{-4}P + 1.4542 \times 10^{-6}P^2 + 5.7948 \times 10^{-12}P^4 - 0.9782 \times 10^{-16}P^6 + 5.627 \times 10^{-22}P^8$$

This variation, as the ratio f/P versus P, is illustrated in Fig. 8.12.

The change in the free energy of a nonideal gas resulting from an isothermal pressure change can be calculated from either

$$dG = VdP$$

or

$$dG = RT\, d \ln f$$

The correspondence between these two approaches can be illustrated as follows.

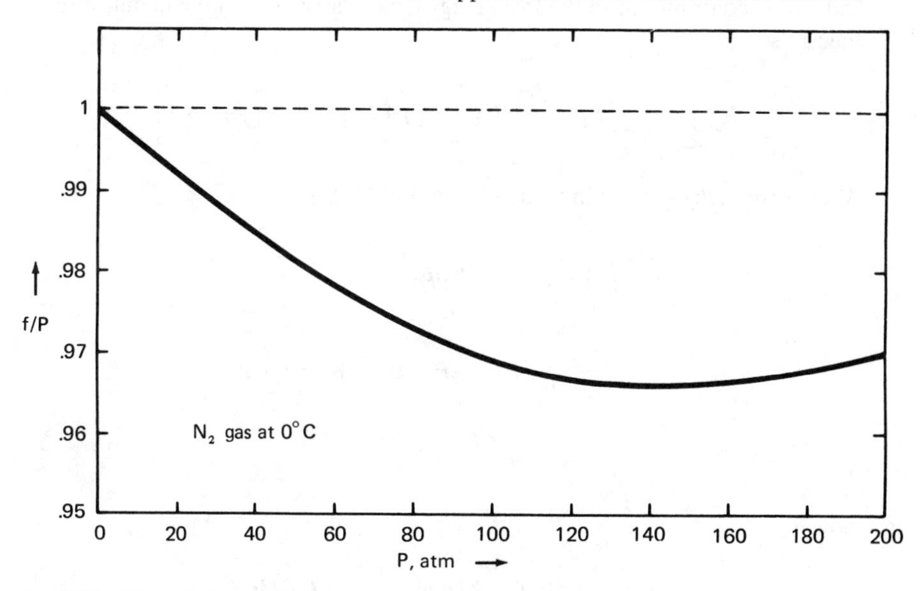

Fig. 8.12. The variation of f/P with pressure for nitrogen gas at 0°C.

Suppose the virial equation of state of the gas is

$$\frac{PV}{RT} = Z = 1 + BP + CP^2 + DP^3 + \cdots$$

Then

$$V = RT\left(\frac{1}{P} + B + CP + DP^2 + \cdots\right)$$

and so for the change of state of 1 mole of gas from (P_1, T) to (P_2, T),

$$\Delta G = \int_{P_1}^{P_2} V dP = RT\int_{P_1}^{P_2}\left(\frac{1}{P} + B + CP + DP^2 + \cdots\right)dP$$

$$= RT\left[\ln\frac{P_2}{P_1} + B(P_2 - P_1) + \frac{C}{2}(P_2^2 - P_1^2) + \frac{D}{3}(P_2^3 - P_1^3) + \cdots\right]$$

If the gas had been ideal, then

$$\Delta G = RT\ln\left(\frac{P_2}{P_1}\right)$$

and so the contribution to the free energy change arising from the nonideality of the gas is

$$RT\left[B(P_2 - P_1) + \frac{C}{2}(P_2^2 - P_1^2) + \frac{D}{3}(P_2^3 - P_1^3) + \cdots\right]$$

Alternatively, $dG = RT\,d\ln f$ where, from Eq. (8.29),

$$\ln\left(\frac{f}{P}\right) = \int_0^P \frac{Z-1}{P}\,dP$$

$$= \int_0^P (B + CP + DP^2 + \cdots)\,dP$$

$$= BP + \frac{CP^2}{2} + \frac{DP^3}{3}$$

Now

$$dG = RTd\ln f = RTd\ln\left(\frac{f}{P}\right) + RTd\ln P$$

and so

$$\Delta G = RT\left[B(P_2 - P_1) + \frac{C}{2}(P_2^2 - P_1^2) + \frac{D}{3}(P_2^3 - P_1^3) + \cdots\right] + RT\ln\left(\frac{P_2}{P_1}\right)$$

in agreement with the above.

Thus for 1 mole of N_2 at 0°C, the free energy difference between the state $P = 150$ atm and $P = 1$ atm is

$$\Delta G = RT\left[\ln\left(\frac{f}{P}\right)_{150} - \ln\left(\frac{f}{P}\right)_1\right] + RT\ln 150$$

$$= 8.3144 \times 273\,(-0.034117 + 0.000457) + 8.3144 \times 273 \times 5.011 \text{ joules}$$

$$= -76 + 11373$$

$$= 11{,}297 \text{ joules}$$

The contribution due to the nonideality of nitrogen is thus seen to be only 76 joules in almost 11,300 joules.

The number of terms which must be retained in the virial equation depends on the extent of the pressure range over which it must be applied; e.g., with the virial equation for N_2 at 0°C, only the first term need be used up to 6 atm, and only the first two terms need be used up to 20 atm. When only the first term need be retained, i.e.,

$$\frac{PV}{RT} = 1 + BP$$

then

$$V = \frac{RT}{P} + BRT$$

and hence $-BRT = \alpha$ in Eq. (8.25).

Consider a nonideal gas that obeys the equation of state $PV = RT(1 + BP)$. The work done by this nonideal gas in a reversible, isothermal expansion from P_1 to P_2 is the same as that done when an ideal gas is reversibly and isothermally expanded from P_1 to P_2, but the work done by the nonideal gas in a reversible expansion from V_1 to V_2 is greater than that done when an ideal gas is reversibly and isothermally expanded from V_1 to V_2. Let us consider why this is so.

For the ideal gas $V = RT/P$, and for the nonideal gas $V = RT/P + BRT$. Thus, on a P-V diagram any isotherm for the nonideal gas is displaced from the isotherm for the ideal gas by the constant increment BRT, as shown in Fig. 8.13.

Because of this constant displacement the area under the isotherm for the ideal gas between P_1 and P_2 (the area *abcd*) equals the area under the isotherm for the nonideal gas between the same pressures (the area *efgh*). Thus the same work is done by both gases in expanding isothermally from P_1 to P_2.

For the ideal gas,

$$w_{\text{ideal gas}} = \int_{V_1}^{V_2} P\,dV = RT \ln\left(\frac{V_2}{V_1}\right) = RT \ln\left(\frac{P_1}{P_2}\right)$$

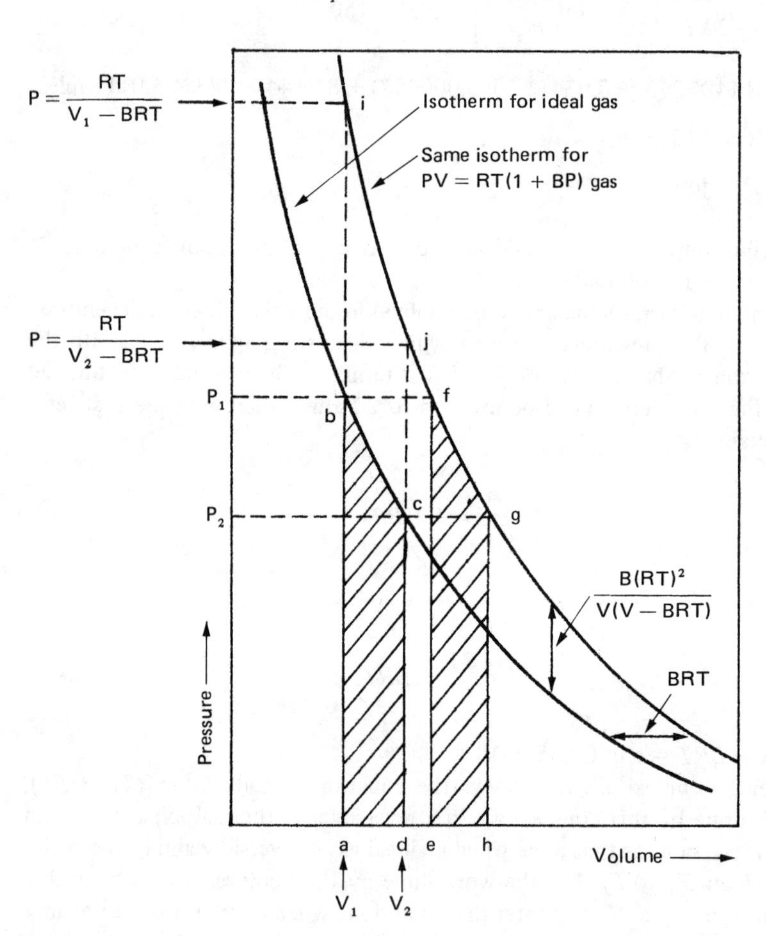

Fig. 8.13.

and for the nonideal gas

$$w_{\text{nonideal gas}} = \int_{V_1}^{V_2} P\,dV$$

but as $V = RT/P + BRT$ and, hence, at constant T, $dV = -RT(dP/P^2)$, then

$$w_{\text{nonideal gas}} = -\int_{P_1}^{P_2} RT\left(\frac{dP}{P}\right) = RT\ln\left(\frac{P_1}{P_2}\right) = w_{\text{ideal gas}}$$

However, as any isotherm for the nonideal gas also lies above the isotherm for the ideal gas (for a positive value of B), the work done by the nonideal gas in expanding isothermally from V_1 to V_2 (the area *aijd*) is greater than that done by the ideal gas in isothermally expanding between V_1 and V_2 (the area *abcd*). The vertical separation between the two isotherms is

$$P_{\text{nonideal gas}} - P_{\text{ideal gas}} = \frac{RT}{V - BRT} - \frac{RT}{V} = \frac{B(RT)^2}{V(V - BRT)}$$

For the ideal gas, $w_{\text{ideal gas}} = RT\ln(V_2/V_1)$, and for the nonideal gas,

$$w = \int_{V_1}^{V_2} P\,dV$$

where

$$P = \frac{RT}{V - BRT}$$

such that

$$w_{\text{nonideal gas}} = RT\ln\left(\frac{V_2 - BRT}{V_1 - BRT}\right) > w_{\text{ideal gas}}$$

Let us compare hydrogen, for which $PV = RT(1 + 0.0064P)$, with an ideal gas in reversible isothermal expansions of one mole between $P_1 = 100$ atm and $P_2 = 50$ atm at 298 K.

$$V_{1(\text{ideal}, P_1 = 100, T = 298\text{ K})} = \frac{RT}{P_1} = \frac{0.08206 \times 298}{100} = 0.2445 \text{ liters}$$

$$V_{1(\text{H}_2, P_1 = 100, T = 298\text{ K})} = \frac{RT}{P_1} + RTB = 0.2445 + 0.08206 \times 298 \times 6.4 \times 10^{-4}$$

$$= 0.2445 + 0.0157 = 0.2602 \text{ liters}$$

$$V_{2(\text{ideal}, P_2=50,\, T=298\text{ K})} = \frac{0.08206 \times 298}{50} = 0.4890 \text{ liters}$$

$$V_{2(\text{H}_2, P=50,\, T=298\text{ K})} = 0.4890 + 0.0157 = 0.5047 \text{ liters}$$

Thus, for the change of state

$$V_1 = 0.2445, T = 298\text{ K} \rightarrow V_2 = 0.4890, T = 298\text{ K}$$

$$w_{\text{ideal gas}} = RT \ln \frac{V_2}{V_1} = 8.3144 \times 298 \times \ln\left(\frac{0.4890}{0.2445}\right)$$

$$= 1717 \text{ joules}$$

and

$$w_{\text{H}_2} = RT \ln\left(\frac{V_2 - RTB}{V_1 - RTB}\right) = 8.3144 \times 298 \times \ln\left(\frac{0.4890 - 0.0157}{0.2445 - 0.0157}\right)$$

$$= 1801 \text{ joules}$$

At $V = 0.2445$, $T = 298$ K, $P_{\text{ideal gas}} = 100$ atm, and

$$P_{\text{H}_2} = \frac{RT}{V - RTB} = \frac{0.08206 \times 298}{0.2445 - 0.0157}$$

$$= 106.4 \text{ atm}$$

and at $V = 0.489$, $T = 298$ K, $P_{\text{ideal gas}} = 50$ atm, and

$$P_{\text{H}_2} = \frac{0.08206 \times 298}{0.4890 - 0.0157}$$

$$= 51.7 \text{ atm}$$

8.8 SUMMARY

1. An ideal gas is an assemblage of volumeless noninteracting particles. Such an assemblage obeys the equation of state $PV = RT$, which is consequently termed the ideal gas law. The internal energy of an ideal gas, being solely that of the translational motion of the constituent particles, is a function only of temperature. Similarly the enthalpy of an ideal gas is a function only of temperature.

2. As a direct result of the ideal gas law, the free energy of an ideal gas is a

linear function of the logarithm of its pressure. Because Classical Thermodynamics is not concerned with the absolute values of free energy, it is convenient to consider free energy changes from some arbitrary reference state. Such a state is termed the standard state, and for an ideal gas it is chosen as being 1 atm pressure at the temperature of interest. Thus the difference between the molar free energy of an ideal gas at the pressure P and the molar free energy in the standard state, at the temperature T, is $\Delta G = RT \ln P$.

3. The deviation of real gases from ideality arises from the fact that real gases comprise atoms or molecules of finite volume among which occur finite interactions. Various attempts have been made to correct the ideal gas law for these effects, and, for gases that show only slight deviation from ideality, the best known of these equations is the van der Waals equation of state. This equation predicts the condensation of the vapor resulting from pressure increase at temperatures less than T_{cr}, but does not give the correct temperature dependency of the saturated vapor pressure of the liquid phase.

4. Generally the experimentally determined state behavior of a nonideal gas is expressed as a power series (in P or $1/V$) of the function PV. Such equations, which are completely empirical, are termed virial equations.

5. The compressibility factor $Z = PV/RT$ of all real gases at constant reduced temperature $T_R = T/T_{cr}$ is the same function of the reduced pressure $P_R = P/P_{cr}$. This gives rise to the law of corresponding states, which states that if two gases have identical values of two reduced variables, then they have almost identical values of the third reduced variable.

6. For nonideal gases a function f, the fugacity, is introduced. This is defined by the equation $dG = RT\, d \ln f$ and the condition that $f/P \rightarrow 1$ as $P \rightarrow 0$. Thus the standard state is that in which the fugacity of the gas is unity at the temperature T. For small deviations from ideality, the actual pressure exerted by the gas is the geometric mean of the fugacity f and P_{id}, the pressure which the gas would exert if it were ideal.

7. The composition of a gas mixture is most conveniently expressed in terms of the mole fractions of its constituents; if the mixture is ideal, then the partial pressures exerted by the component gases (being equal to the pressure which each gas would exert if it alone were present) are related to the total pressure P and the mole fraction X_i by $p_i = X_i P$. This equation is known as Dalton's law of partial pressures, and it is seen that the sum of the partial pressures is equal to the total pressure.

8. In a perfect gas mixture the partial molar free energy of a constituent gas is a linear function of the logarithm of its partial pressure, and in a nonideal gas mixture it is a linear function of the logarithm of its fugacity.

9. No heat effects result from the mixing of ideal gases, and this is to be expected in view of the fact that the physical source of such effects, namely forces of interaction between atoms, are absent in an ideal gas. The entropy of

mixing of ideal gases arises solely from complete randomization of the different types of particles in the available volume, and hence, as $\Delta H^{\text{mix}} = 0$, $\Delta G^{\text{mix}} = -T\,\Delta S^{\text{mix}}$.

PROBLEMS

8.1 Demonstrate the law of corresponding states by writing the van der Waals equation in terms of the reduced variables. Calculate the compressibility factor for a van der Waals gas at its critical point, and compare the result with the values obtained for real gases at their critical points listed in Table 8.1.
Obtain the value of $(\partial U/\partial V)_T$ for a van der Waals gas.

8.2 n moles of an ideal gas A and (1-n) moles of an ideal gas B, each at 1 atm pressure, are mixed at constant total pressure. What ratio of A to B in the mixture maximizes the free energy decrease of the system? If this free energy decrease is ΔG^M, to what value must the pressure be increased in order to increase the free energy of the gas mixture by $\frac{1}{2}\Delta G^M$?

8.3 For sulfur dioxide gas, $T_{cr} = 430.7$ K and $P_{cr} = 77.8$ atm. Calculate:

a. The van der Waals constants for the gas.
b. The critical volume of van der Waals SO_2.
c. The pressure exerted by 1 mole of SO_2 occupying a volume of 500 cm^3 at 500 K. Compare this with the pressure which would be exerted by an ideal gas occupying the same molar volume at the same temperature.

8.4 One hundred moles of hydrogen gas at 298 K are isothermally compressed from 30 liters to 10 liters. The van der Waals constants for hydrogen are $a = 0.2461$ liter^2/atm and $b = 0.02668$ liters, and in the pressure range 0 to 1500 atm the virial equation for hydrogen is given as $PV = RT(1 + 6.4 \times 10^{-4}\,P)$.
Calculate the work that must be done on the system to effect the required volume change; compare this with the values that would be calculated assuming (*a*) hydrogen to behave as a van der Waals gas and (*b*) hydrogen to behave as an ideal gas.

8.5 Using the virial equation for hydrogen gas at 298 K given in Prob. 8.4, calculate:

a. The fugacity of hydrogen at 500 atm and 298 K
b. The pressure at which the fugacity is twice the pressure
c. The free energy change resulting from the compression of 1 mole of hydrogen at 298 K from 1 atm to 500 atm.

What is the magnitude of the contribution to (*c*) arising from the nonideality of hydrogen?

CHAPTER NINE

REACTIONS INVOLVING GASES

9.1 INTRODUCTION

In Chap. 8 it was seen that, as a result of there being no interactive forces among the particles of perfect gases, no heat effects occur when two (or more) such gases are mixed together. This situation represents one extreme of a range of possible situations. Towards the other extreme of this range is the situation where gaseous species which exhibit marked chemical affinity for one another are mixed. For example, when hydrogen and oxygen gases are mixed in the presence of a suitable catalyst, the heat released is of considerable magnitude. The thermodynamics of such a system can be treated in either of two ways. Either the mixture can be regarded as being a highly nonideal gas mixture of H_2 and O_2, the thermodynamic equilibrium state of which, at given temperature and total pressure, can be described in terms of the fugacities of the two components H_2 and O_2; or it can be considered that the H_2 and O_2 have reacted with one another to some extent to give rise to the physical appearance of the product species H_2O. In this case, if the pressure of the system is low enough, the equilibrium state at the given temperature, can be described in terms of the partial pressures exerted by the three species H_2, O_2, and H_2O, in the system. Although both treatments are thermodynamically equivalent, the latter, by virtue of its correspondence with physical reality, is by far the more convenient and practical.

As with any constant-pressure, constant-temperature system, the equilibrium state is that in which the free energy of the system is a minimum. If the gases initially present in the system react to form distinct product species, the total free energy change in the system comprises a contribution arising from the free energy change due to chemical reaction and a contribution arising from the mixing of the appearing product gases with the remaining reactant gases. Knowledge of this total free energy change with composition (which ranges from the pure unmixed reactant gases to the pure unmixed product gases) allows determination of the equilibrium state in any system of reactive gases. This determination is facilitated by means of the introduction of the equilibrium constant for the reaction, and it will be seen that the relation between this constant and the standard free energy change of the reaction is one of the more important relationships in reaction equilibrium thermodynamics.

9.2 REACTION EQUILIBRIUM IN A GAS MIXTURE AND THE EQUILIBRIUM CONSTANT

Consider the reaction,

$$A_{(g)} + B_{(g)} = 2C_{(g)}$$

occurring at constant temperature and constant pressure P. At any moment during the reaction, the free energy of the system is

$$G' = n_A\bar{G}_A + n_B\bar{G}_B + n_C\bar{G}_C \tag{9.1}$$

where n_A, n_B, and n_C are respectively the numbers of moles of A, B, and C present at that moment; and $\bar{G}_A$, $\bar{G}_B$, and $\bar{G}_C$ are respectively the partial molar free energies of A, B, and C in the gas mixture which occurs at that moment. The problem is to determine the values of n_A, n_B, and n_C which minimize the value of G' in Eq. (9.1), as this state of minimum free energy is the equilibrium state of the system at the given temperature and pressure. That is, once chemical reaction between A and B has proceeded to the extent that the free energy of the system has been minimized, then the reaction ceases.

The stoichiometric chemical reaction equation allows the numbers of moles of all of the species present at any instant to be expressed in terms of the number of moles of any one of the species. Starting with 1 mole of A and 1 mole of B (i.e., 2 moles of gas), as 1 atom of A reacts with 1 atom of B to produce 2 molecules of C, then at any time,

$$n_A = n_B$$

and

$$n_C = 2 - n_A - n_B = 2(1 - n_A)$$

Equation (9.1) can thus be written as,

$$G' = n_A \bar{G}_A + n_A \bar{G}_B + 2(1 - n_A)\bar{G}_C$$

From Eq. (8.15), $\bar{G}_i = G_i^0 + RT \ln P + RT \ln X_i$

and

$$X_A = \frac{n_A}{2} \qquad X_B = \frac{n_B}{2} \qquad X_C = \frac{2(1 - n_A)}{2} = (1 - n_A)$$

substitution of which gives

$$G' = n_A(G_A^0 + G_B^0 - 2G_C^0) + 2G_C^0 + 2RT \ln P + 2RT\left[n_A \ln\left(\frac{n_A}{2}\right) + (1 - n_A) \ln (1 - n_A)\right]$$

or,

$$G' - 2G_C^0 = n_A[-\Delta G^0] + 2RT \ln P + 2RT\left[n_A \ln\left(\frac{n_A}{2}\right) + (1 - n_A) \ln (1 - n_A)\right] \tag{9.2}$$

where $\Delta G^0 = 2G_C^0 - G_A^0 - G_B^0$ is the *standard free energy change* for the chemical reaction at the temperature T. The standard free energy change for any reaction is the difference between the sum of the free energies of the reaction products in their standard states and the sum of the free energies of the reactants in their standard states. In the present case ΔG^0 is the difference in free energy between 2 moles of C at 1 atm pressure and the temperature T, and 1 mole of A and 1 mole of B each at 1 atm pressure and the temperature T. If the total pressure of the system is 1 atm, then Eq. (9.2) simplifies to

$$G' - 2G_C^0 = n_A[-\Delta G^0] + 2RT\left[n_A \ln\left(\frac{n_A}{2}\right) + (1 - n_A) \ln (1 - n_A)\right] \tag{9.3}$$

The left-hand side of Eq. (9.3) is the difference between the free energy of the 2-mole system when $n_A = n_A$ and the free energy of the system when it comprises 2 moles of C. This free energy difference is determined by two factors:

1. The free energy change due to chemical reaction, i.e., due to the disappearance of the reactants and the appearance of the products, given by the first term on the right-hand side of Eq. (9.3)
2. The free energy decrease due to mixing of the gases, given by the second term on the right-hand side of Eq. (9.3)

Figure 9.1 is drawn for the reaction

$$A_{(g)} + B_{(g)} = 2C_{(g)}$$

at 500 K and 1 atm pressure. ΔG^0_{500} for the reaction is taken as being −5000 joules. If the reference free energy is arbitrarily chosen as being $(G^0_A + G^0_B) = 0$, then $2G^0_C = -5000$ joules. In Fig. 9.1 the ordinate $\Delta G'$ is plotted as the difference between the free energy of the system when $n_A = n_A$ and the free energy of the system comprising 1 mole of A and 1 mole of B before mixing of A and B occurs. Hence the point L, ($n_A = 1, n_B = 1$ before mixing) is located at $\Delta G' = 0$, and the point Q, ($n_C = 2$) is located at $\Delta G' = -5000$ joules. The point M represents the free energy decrease due to mixing of 1 mole of A and 1 mole of B before any chemical reaction between the two occurs; i.e., from Eq. (8.20),

$$\Delta G(L \rightarrow M) = \sum_i n_i RT \ln\left(\frac{p_i}{P_i}\right)$$

$$= RT\left[n_A \ln\left(\frac{p_A}{P_A}\right) + n_B \ln\left(\frac{p_B}{P_B}\right)\right]$$

But $n_A = n_B = 1$, and

$$P_A = P_B = P_{mix} = 1$$

in which case

$$p_A = p_B = ½$$

Thus

$$\Delta G(L \rightarrow M) = RT(\ln 0.5 + \ln 0.5)$$

$$= 8.3144 \times 500 \times 2 \times \ln 0.5$$

$$= -5763 \text{ joules}$$

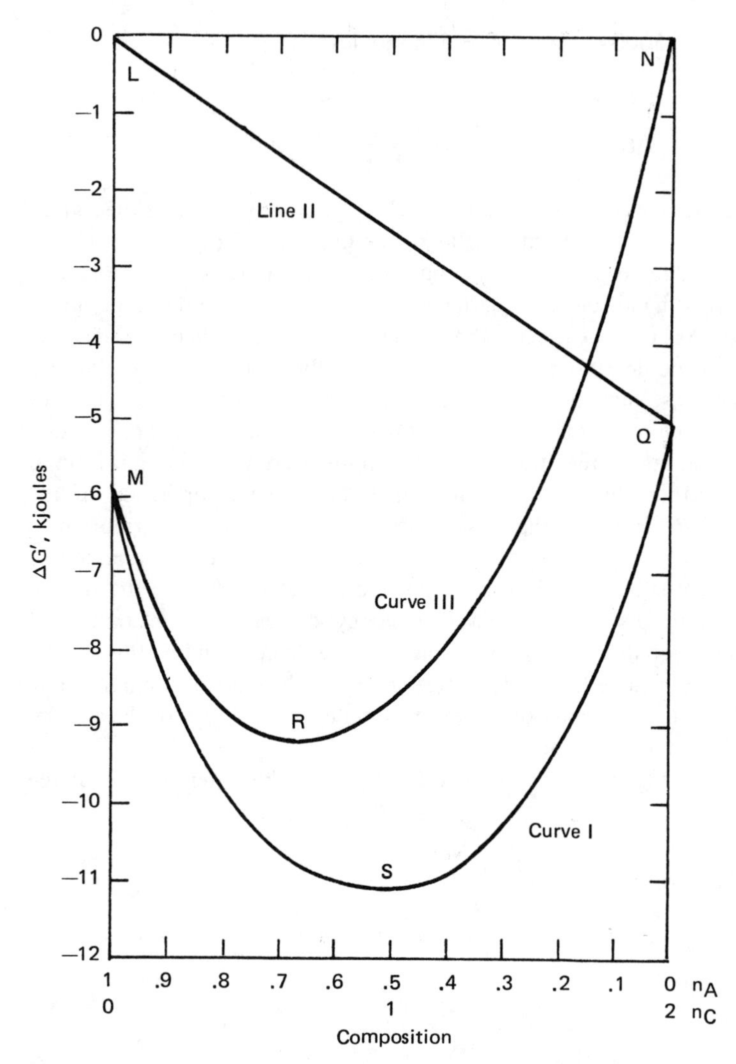

Fig. 9.1. The variations of the contribution to the free energy decrease due to chemical reaction (line II), the contribution to the free energy decrease due to gas mixing (curve III), and the sum of these two contributions (curve I) with the extent of the reaction $A_{(g)} + B_{(g)} = 2C_{(g)}$, for which $\Delta G^0 = -5000$ joules at 500 K.

which equals the value of the second term on the right-hand side of Eq. (9.3) when $n_A = 1$. Thus, from Eq. (9.3),

$$G' - 2G_C^0 = n_A[-\Delta G^0] + 2RT\left[n_A \ln\left(\frac{n_A}{2}\right) + (1 - n_A)\ln(1 - n_A)\right]$$

or $$G' + 5000 = 1 \times 5000 - 5763 \text{ joules}$$

and hence $$G' = -5763 \text{ joules}$$

and so $$\Delta G' = G' - (G_A^0 + G_B^0) = -5763 \text{ joules}$$

Curve I, which represents the variation of G' with n_A, is obtained as the sum of line II [given by the first term on the right-hand side of Eq. (9.3)–the free energy decrease due to chemical reaction] and curve III [given by the second term on the right-hand side of equation (9.3)–the free energy decrease due to gas mixing]. As can be seen, the magnitude of the chemical reaction contribution to the decrease in the free energy of the system increases linearly with increasing n_C, but the magnitude of the contribution to the total free energy decrease due to gas mixing passes through a maximum at the composition R. R is that state, the composition of which permits maximum randomization of the system. Further chemical reaction, which takes the composition of the system beyond R, decreases the magnitude of the gas-mixing contribution as further increase in n_C, at the expense of n_A and n_B, decreases the randomness of the system. Eventually the composition S is reached at which state the sum of the two contributions to the total free energy decrease is a maximum. If chemical reaction continued beyond S, then, as the decrease in line II is smaller than the increase in curve III, the total free energy of the system would increase. Composition S thus represents the state of minimum free energy of the system, and hence is the equilibrium state.

The position of the minimum in curve I is fixed by the criterion that, at the minimum,

$$\left(\frac{\partial G'}{\partial n_A}\right)_{T,P} = 0$$

and as

$$G' = n_A \bar{G}_A + n_A \bar{G}_B + 2(1 - n_A)\bar{G}_C$$

then*

$$\left(\frac{\partial G'}{\partial n_A}\right) = \bar{G}_A + \bar{G}_B - 2\bar{G}_C = 0$$

i.e., the criterion of reaction equilibrium is that

$$\bar{G}_A + \bar{G}_B = 2\bar{G}_C \tag{9.4}$$

*From the Gibbs-Duhem equation, Eq. (11.19), $n_A d\bar{G}_A + n_B d\bar{G}_B + n_C d\bar{G}_C = 0$.

Equation (9.4) can be written as

$$G_A^0 + RT \ln p_A + G_B^0 + RT \ln p_B = 2G_C^0 + 2RT \ln p_C \tag{9.5}$$

where p_A, p_B, and p_C are, respectively, the partial pressures of A, B, and C *which occur at reaction equilibrium*. Rearrangement of Eq. (9.5) gives

$$2G_C^0 - G_A^0 - G_B^0 = -RT \ln \frac{p_C^2}{p_A p_B}$$

or

$$\Delta G^0 = -RT \ln\left(\frac{p_C^2}{p_A p_B}\right) \tag{9.6}$$

The quotient of the equilibrium partial pressures of the reactants and products occurring as the logarithmic term in Eq. (9.6) is termed the *equilibrium constant*, K_p; i.e.,

$$\left(\frac{p_C^2}{p_A p_B}\right)^{eq} = K_p \tag{9.7}$$

and hence

$$\Delta G^0 = -RT \ln K_p \tag{9.8}$$

As ΔG^0 is a function only of temperature, it follows from Eq. (9.8) that K_p is a function only of temperature. For the example used in Fig. 9.1,

$$\ln K_p = -\frac{\Delta G^0}{RT} = -\frac{\Delta G^0}{8.3144T} = \frac{5000}{8.3144 \times 500}$$

$$= 1.203$$

Therefore $\qquad K_p = 3.329$

Now

$$K_p = \frac{p_C^2}{p_A p_B} = \frac{X_C^2 P^2}{X_A P X_B P} = \frac{X_C^2}{X_A X_B}$$

$$= \frac{(1 - n_A)^2}{n_A^2/4}$$

and hence

$$n_A = 0.523 \qquad \text{(the other solution, } n_A = 11.4\text{, is nonphysical)}$$

i.e., reaction equilibrium is established when $n_A = 0.523$, $n_B = 0.523$ and $n_C = 0.954$, or when the chemical reaction A + B = 2C has proceeded to 47.7% completion. Inspection of Fig. 9.1 shows that the minimum in curve I occurs at $n_A = 0.523$.

If the temperature T was such that ΔG^0 for the reaction was zero, then there would be no chemical reaction contribution to $\Delta G'$, and the variation of $\Delta G'$ with n_A would be given by curve III; i.e., the criterion of reaction equilibrium would be maximization of the randomness of the system, which occurs at the composition R. From Eq. (9.8), if $\Delta G^0_{500} = 0$, then $K_p = 1$, and hence,

$$1 = \frac{4(1 - n_A)^2}{n_A^2} \quad \text{and} \quad n_A = 0.67$$

Inspection of Fig. 9.1 shows that the minimum in curve III occurs at $n_A = n_B = n_C = 2/3$. It is thus seen that maximum randomization of the system occurs when all three species are present in equal amount.

It is to be realized that the minimum in curve I in Fig. 9.1, in representing the equilibrium state of the system, is the only point on curve I which has any significance within the scope of Classical Thermodynamics. The point S is the only point on the curve I which lies on the equilibrium surface in P-T-composition space for the fixed values of $P = 1$ and $T = 500$. By altering P and/or T, the equilibrium composition of the system moves over the equilibrium surface in P-T-composition space. It is of interest to determine the effects of pressure and temperature on the equilibrium composition of the system.

9.3 THE EFFECT OF TEMPERATURE ON THE EQUILIBRIUM CONSTANT

The position of the minimum in curve I in Fig. 9.1 is determined by the difference between LM and NQ. The length LM equals the decrease in free energy due to mixing of the reactant gases before reaction commences, and the length NQ equals the standard free energy change, ΔG^0, for the reaction. The lengths of both these lines are dependent on temperature, that is, $LM = 2\,RT \ln 0.5$ and $NQ = \Delta G^0$. The effect of temperature on the position of the minimum in curve I (and hence on the value of K_p) thus depends on the relative effects of temperature on the lengths of LM and NQ. For given reactants, the length of LM increases linearly with increasing temperature, and the variation in the length NQ with temperature is determined by the sign and magnitude of the standard entropy change of the reaction according to

$$\left(\frac{\partial \Delta G^0}{\partial T}\right)_P = -\Delta S^0$$

In the case of ΔS^0 being negative, an increase in temperature increases the length *LM* and decreases the length *NQ*. Thus the minimum in the curve shifts to the left, indicating that K_p decreases with increasing temperature.

The exact variation of K_p with temperature is obtained from a consideration of the Gibbs-Helmholtz equation, [Eq. (5.37a)],

$$\left[\frac{\partial(\Delta G^0/T)}{\partial T}\right]_P = -\frac{\Delta H^0}{T^2}$$

As $\Delta G^0 = -RT \ln K_p$, then

$$\frac{\partial \ln K_p}{\partial T} = \frac{\Delta H^0}{RT^2} \tag{9.9}$$

or

$$\frac{\partial \ln K_p}{\partial\left(\frac{1}{T}\right)} = -\frac{\Delta H^0}{R} \tag{9.10}$$

Equation (9.9) is known as the *van't Hoff equation* and indicates that the effect of temperature on K_p is determined by the sign and magnitude of ΔH^0 for the reaction. If ΔH^0 is positive, i.e., if the reaction is endothermic, then K_p increases with increasing temperature; and, conversely, if ΔH^0 is negative, i.e., if the reaction is exothermic, then K_p decreases with increasing temperature. Integration of Eq. (9.9) requires a knowledge of the temperature dependence of ΔH^0, which, as was seen in Chap. 6, depends on the value of Δc_p, for the reaction.

The direction of the variation of K_p with temperature can be deduced from consideration of Le Chatelier's principle, i.e., if heat is added to a system at reaction equilibrium, then the equilibrium is displaced in that direction which involves the absorption of heat. Consider the simple gaseous reaction,

$$Cl_2 = 2Cl$$

This reaction is endothermic and hence has a positive ΔH^0. Thus the equilibrium constant, $K_p = p_{Cl}^2/p_{Cl_2}$, increases with increasing temperature; i.e., the equilibrium shifts in that direction which involves the absorption of heat. Conversely, if the reaction was written as

$$2Cl = Cl_2$$

then, as the reaction has a negative ΔH^0, $K_p = p_{Cl_2}/p_{Cl}^2$ decreases with increasing temperature; i.e., equilibrium shifts in that direction which involves the absorption of heat. Thus in either case an increase in temperature increases p_{Cl} and decreases p_{Cl_2}.

Equation (9.10) shows that, if ΔH^0 is independent of temperature, then $\ln K_p$ varies linearly with $1/T$.

9.4 THE EFFECT OF PRESSURE ON THE EQUILIBRIUM CONSTANT

K_p, as defined by Eq. (9.7), is independent of pressure. This is a consequence of the fact that ΔG^0, being the free energy difference between the pure products, each at unit pressure, and the pure reactants, each at unit pressure, is by definition independent of pressure. However, reaction equilibrium expressed in terms of the numbers of moles of species present rather than in terms of the partial pressures of the species present is dependent on the total pressure if the chemical reaction under consideration involves a change in the total number of moles present.

Consider again the reaction,

$$Cl_2 = 2Cl$$

Completion of this reaction involves a doubling of the number of moles present. The effect of a pressure change on the equilibrium of such a system can again be qualitatively deduced from a consideration of Le Chatelier's principle. If the pressure exerted on a system at reaction equilibrium is increased, then the equilibrium will shift in that direction which tends to decrease the pressure exerted by the system (i.e., will shift in that direction which decreases the number of moles present). Thus if the pressure exerted on the Cl-Cl_2 system is increased, the equilibrium will shift towards the Cl_2 side, as thereby the total number of moles present will be decreased to accommodate the increased pressure. Specifically, the effect of pressure on the reaction equilibrium expressed in terms of the numbers of moles present (or in terms of the mole fractions) can be seen as follows.

$$K_p = p_{Cl}^2/p_{Cl_2} \quad \text{which is independent of pressure}$$

$$= \frac{X_{Cl}^2 P^2}{X_{Cl_2} P} = \frac{X_{Cl}^2 P}{X_{Cl_2}} = K_x P$$

where K_x is the equilibrium constant expressed in terms of the mole fractions. Thus if the pressure is increased, then K_x decreases in order to maintain K_p

constant, and the decrease in K_x is achieved by the reaction equilibrium shifting towards the Cl_2 side such that X_{Cl} decreases and X_{Cl_2} increases.

In the case of the reaction $A + B = 2C$, the system at all times contains 2 moles of gas, and hence the reaction equilibrium, expressed in terms of the mole fractions, is independent of pressure; i.e.,

$$K_p = \frac{p_C^2}{p_A p_B} = \frac{X_C^2 P^2}{X_A P X_B P} = \frac{X_C^2}{X_A X_B} = K_x$$

This can also be seen from Eq. (9.2), as, if $P \neq 1$, then the effect of the nonzero term $2RT \ln P$ is the raising or lowering of curve I in Fig. 9.1, without affecting the position of the minimum with respect to the composition axis.

The magnitude of the effect, on the value of K_x, of a pressure change, depends on the magnitude of the change in the number of moles present in the system, occurring as a result of the chemical reaction. For the general reaction

$$aA + bB = cC + dD$$

$$K_p = \frac{p_C^c p_D^d}{p_A^a p_B^b} = \frac{X_C^c X_D^d}{X_A^a X_B^b} \frac{P^c P^d}{P^a P^b} = K_x P^{c+d-a-b}$$

and only if $c+d-a-b = 0$ will K_x be independent of the total pressure of the system. K_p is normally used when gas phase reactions are being considered, and K_x is used when reactions involving condensed phases are being considered, the equilibria of which, as will be seen, are relatively insensitive to pressure variation.

9.5 REACTION EQUILIBRIUM AS A COMPROMISE BETWEEN ENTHALPY AND ENTROPY FACTORS

The Gibbs free energy of a system is defined as

$$G = H - TS$$

and hence low values of G are obtained with low values of H and high values of S. It was seen, in the discussion of one-component systems, that equilibrium occurs as the result of a compromise between enthalpy and entropy considerations. Similar discussions can be made concerning chemical reaction equilibria.

Consider again the reaction $Cl_2 = 2Cl$. This reaction has a positive ΔH^0 (ΔH^0 is the energy required to break Avogadro's number of Cl–Cl bonds), and has a positive ΔS^0 (2 moles of chlorine atoms are produced from 1 mole of chlorine molecules). Thus the system occurring exclusively as Cl_2 molecules has a low H and a low S, and the system occurring exclusively as Cl atoms has a high H and a high S. The minimum value of G thus occurs somewhere between the two extreme states. This compromise between the enthalpy and entropy is analogous to the compromise between the chemical reaction contribution to the free energy change and the gas-mixing contribution to the free energy decrease illustrated in Fig. 9.1.

For the reaction A + B = 2C, Eq. (9.3) can be written as

$$G' - 2G_C^0 = n_A[-\Delta H^0] + n_A[T\Delta S^0] + 2RT\left[n_A \ln\left(\frac{n_A}{2}\right) + (1-n_A)\ln(1-n_A)\right]$$

$$= \left\{n_A(-\Delta H^0)\right\} + T\left\{n_A\Delta S^0 + 2R\left[n_A \ln\left(\frac{n_A}{2}\right) + (1-n_A)\ln(1-n_A)\right]\right\} \quad (9.11)$$

The term in the first brackets is the enthalpy contribution to the free energy change, and the second term on the right-hand side is the entropy contribution, $n_A\Delta S^0$ being the entropy change due to chemical reaction, and $2R[n_A \ln(n_A/2) + (1-n_A)\ln(1-n_A)]$ being the entropy change due to gas mixing. In the previous example, ΔG^0 equaled -5000 joules at 500 K. Let it be that $\Delta H^0 = -2500$ joules and $\Delta S^0 = 5$ joules/K, in which case Fig. 9.2 can be drawn from Eq. (9.11). In Fig. 9.2 the $\Delta H'$ line is the first term on the right-hand side of Eq. (9.11), and the $-T\Delta S'$ line is the second term. The sum of these two gives $G'-2G_C^0$, the scale of which is given on the left-hand edge of Fig. 9.2. The scale marked on the right-hand edge of the figure is $\Delta G'$ where, as before, the reference zero of free energy is chosen as $G_A^0 + G_B^0 = 0$, such that $\Delta G' = G'$ (i.e., the scale is displaced by $2G_C^0 = -5000$). On this scale the $\Delta G'$ curve in Fig. 9.2 is identical with curve I in Fig. 9.1. As can be seen, the minimum in the $\Delta G'$ curve is determined as the compromise between the minimum value of H' at $n_A = 0$ and the maximum value of $T\Delta S'$ at $n_A = 0.597$ (the point M in Fig. 9.2). If the temperature is increased, then the $T\Delta S'$ term becomes relatively more important, and hence the equilibrium value of n_A increases (the minimum in the $\Delta G'$ curve shifts to the left). Thus K_p decreases with increasing temperature in accordance with Eq. (9.9) (ΔH^0 is negative).

The effect of temperature on the $\Delta G'$ curve is illustrated in Fig. 9.3, in which curves are drawn for the reaction A + B = 2C at 500 K, 1000 K, and 1500 K. It

It is assumed that Δc_p for the reaction is zero, in which case ΔH^0 and ΔS^0 are independent of temperature. As $\Delta G^0 = \Delta H^0 - T\Delta S^0$, then

$$\Delta G^0_{500} = -2500 - (500 \times 5) = -5000 \text{ joules} \qquad K_{p,500} = 3.329$$

$$\Delta G^0_{1000} = -2500 - (1000 \times 5) = -7500 \text{ joules} \qquad K_{p,1000} = 2.465$$

$$\Delta G^0_{1500} = -2500 - (1500 \times 5) = -10{,}000 \text{ joules} \qquad K_{p,1500} = 2.229$$

The equilibrium values of n_A at 500 K, 1000 K, and 1500 K are thus, respectively, 0.523, 0.560, and 0.572.

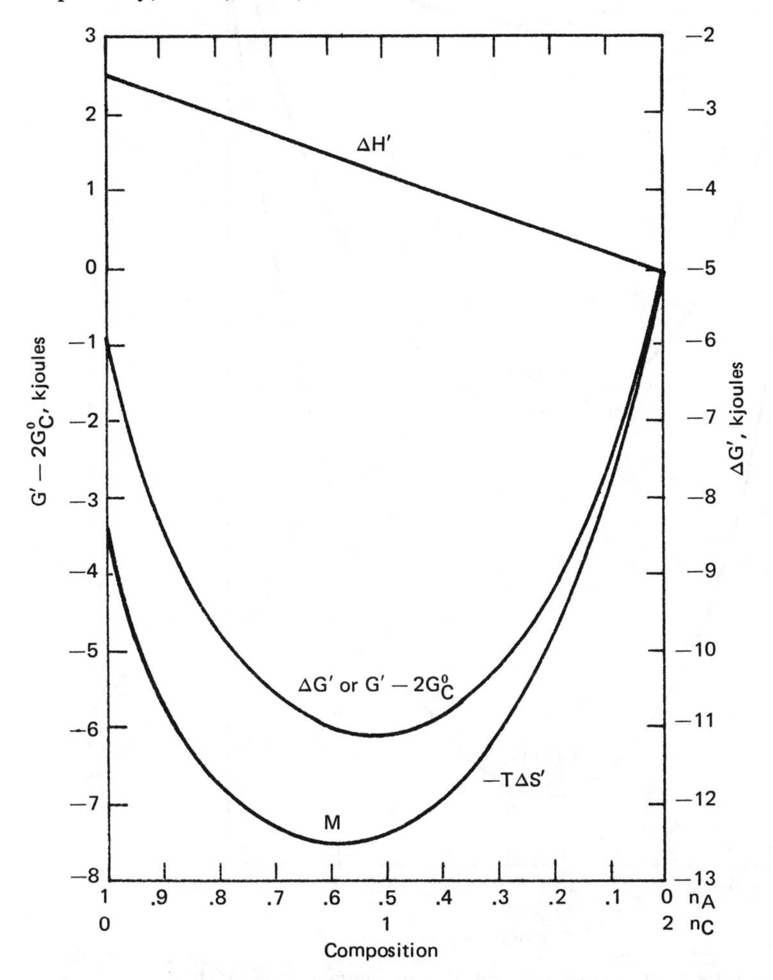

Fig. 9.2. The variations at 500 K of $\Delta H'$, $T\,\Delta S'$, and $\Delta G'$ with the extent of reaction $A_{(g)} + B_{(g)} = 2C_{(g)}$ for which $\Delta G^0 = -2500 - 5T$ joules.

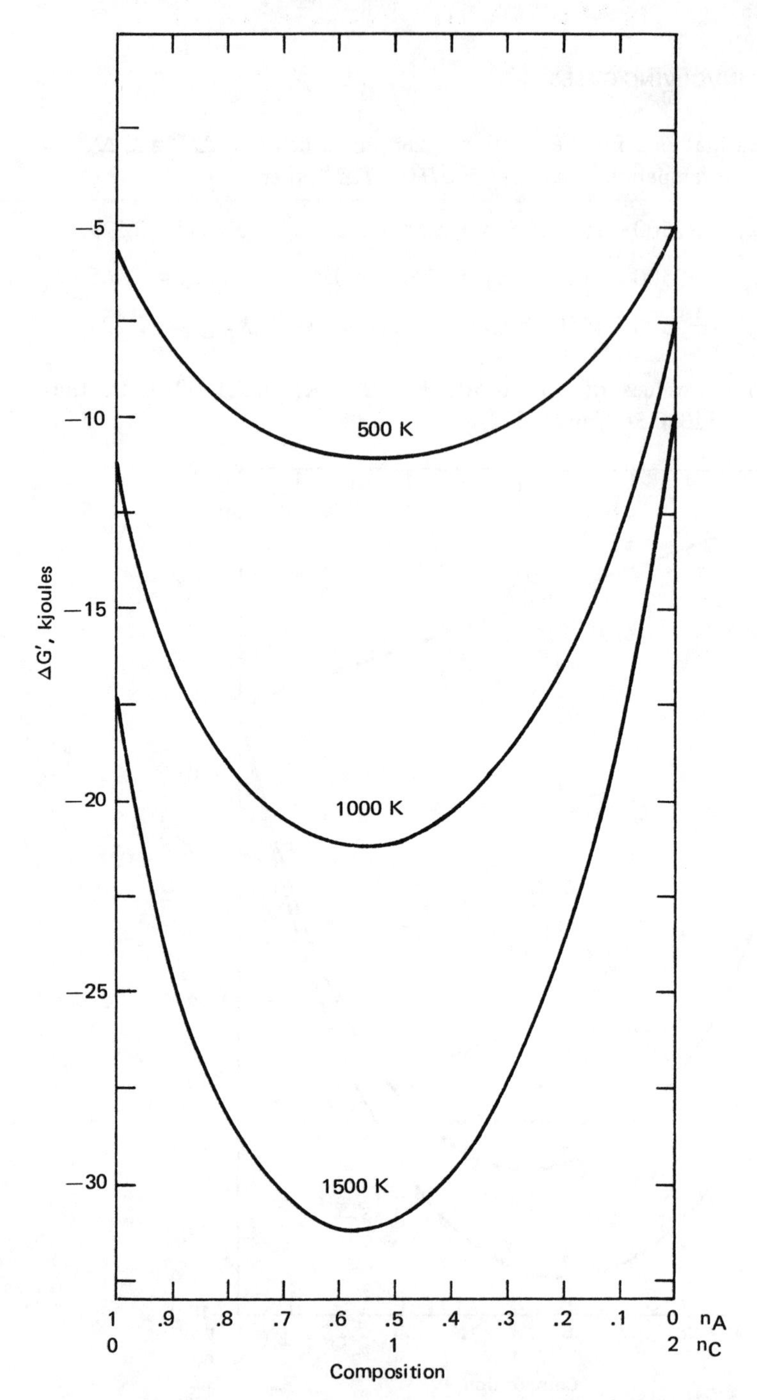

Fig. 9.3. The effect of temperature on the equilibrium state of the reaction $A_{(g)} + B_{(g)} = 2C_{(g)}$ for which $\Delta G^0 = -2500 - 5T$ joules.

Although K_p is constant at constant temperature, it is to be realized that an infinite set of partial pressures of reactants and products correspond to the fixed value of K_p. If the reaction involves three species, then arbitrary choice of the partial pressures of two of the species uniquely fixes the equilibrium partial pressure of the third.

9.6 REACTION EQUILIBRIUM IN THE SYSTEM $SO_{2\,(g)}$-$SO_{3\,(g)}$-$O_{2\,(g)}$

Consider the equilibrium

$$SO_{2\,(g)} + \tfrac{1}{2}O_{2\,(g)} = SO_{3\,(g)} \tag{9.12}$$

For this reaction the standard free energy change is given as

$$\Delta G^0 = -94{,}600 + 89.37T \text{ joules}$$

Thus, at 1000 K, $\Delta G^0_{1000} = -5230$ joules

$$\ln K_p = +\frac{5230}{8.3144 \times 1000}$$

$$= +0.629$$

and $$K_p = 1.876 = \frac{p_{SO_3}}{p_{SO_2} p_{O_2}^{1/2}}$$

Consider the reaction between 1 mole of SO_2 gas at 1 atm pressure and $\frac{1}{2}$ mole of O_2 gas at 1 atm pressure to form an equilibrium mixture of SO_2, SO_3, and O_2 at 1 atm pressure and 1000 K. From the stoichiometry of chemical reaction in Eq. (9.12), x moles of SO_3 are formed from reaction of x moles of SO_2 with $\frac{1}{2}x$ moles of O_2 and any reacting mixture contains x moles of SO_3, $(1-x)$ moles of SO_2, and $\frac{1}{2}(1-x)$ moles of O_2; i.e.,

$$\begin{array}{ccccc} SO_2 & + & \tfrac{1}{2}O_2 & = & SO_3 \\ (1-x) & & \left(\tfrac{1}{2}-\tfrac{1}{2}x\right) & & x \end{array}$$

The total number of moles in the system, n_T, is

$$n_T = 1 - x + \tfrac{1}{2} - \tfrac{1}{2}x + x = \tfrac{1}{2}(3-x)$$

and, as $$p_i = \frac{n_i}{n_T} P$$

then

$$p_{SO_2} = \frac{2(1-x)P}{(3-x)} \qquad p_{O_2} = \frac{(1-x)P}{(3-x)} \qquad \text{and} \qquad p_{SO_3} = \frac{2xP}{(3-x)}$$

Thus

$$K_p^2 = \frac{p_{SO_3}^2}{p_{SO_2}^2 p_{O_2}} = \frac{(3-x)x^2}{(1-x)^3 P}$$

or

$$(1 - PK_p^2)x^3 + (3PK_p^2 - 3)x^2 - 3PK_p^2 x + PK_p^2 = 0 \tag{9.13}$$

Although Eq. (9.13) is cubic in x the required solution must lie in the range $0 < x < 1$. With $P = 1$ and $K_p = 1.876$, plotting the left-hand side of Eq. (9.13) against x gives the solution as $x = 0.463$. Thus, at equilibrium, there are 0.537 moles of SO_2, 0.269 moles of O_2, and 0.463 moles of SO_3, such that

$$p_{SO_2} = \frac{2(1-0.463)}{(3-0.463)} = 0.423 \text{ atm}$$

$$p_{O_2} = \frac{(1-0.463)}{(3-0.463)} = 0.212 \text{ atm}$$

and

$$p_{SO_3} = \frac{2 \times 0.463}{(3-0.463)} = 0.365 \text{ atm}$$

As a check,

$$K_p = \frac{0.365}{0.423 \times (0.212)^{1/2}} = 1.874$$

Of the moles of gas present in the equilibrium mixture, 42.3% are SO_2, 21.2% are O_2, and 36.5% are SO_3.

The Effect of Temperature

As ΔH^0 for the reaction in Eq. (9.12) is negative (−94,600 joules) and Le Chatelier's principle indicates that a decrease in temperature at constant pressure shifts the equilibrium in that direction which involves an evolution of heat, i.e., a decrease in temperature will shift the equilibrium toward the SO_3 side.

At 900 K,

$$\Delta G^0_{900} = -14{,}167 \text{ joules}$$

$$\ln K_p = +\frac{14{,}167}{8.3144 \times 900} = +1.893$$

and thus

$$K_p = 6.64$$

Substituting $K_p = 6.64$ and $P = 1$ into Eq. (9.13) gives, as the solution, $x = 0.7$, whence,

$$p_{SO_2} = 0.261, p_{O_2} = 0.13 \quad \text{and} \quad p_{SO_3} = 0.609$$

or, of the total number of moles present,

26.1% are SO_2 (which is a decrease from 42.3% at 1000 K)
13% are O_2 (which is a decrease from 21.2% at 1000 K)
60.9% are SO_3 (which is an increase from 36.5% at 1000 K)

Thus it is seen that a decrease in temperature has shifted the equilibrium towards the SO_3 side.

The Effect of Pressure

Although K_p is independent of pressure, Le Chatelier's principle indicates that an increase in total pressure at constant temperature will shift the equilibrium in that direction which involves a decrease in the number of moles in the system, i.e., toward the SO_3 side.

Determine the equilibrium mixture at $P = 10$ atm and 1000 K. Substituting $P = 10$ and $K_p = 1.876$ into Eq. (9.13) gives as the solution $x = 0.686$, in which case $n_{SO_2} = 0.314$, $n_{O_2} = 0.157$, and $n_{SO_3} = 0.686$. Thus

$$p_{SO_2} = \frac{2(1 - 0.686) \times 10}{(3 - 0.686)} = 2.714 \text{ atm}$$

$$p_{O_2} = \frac{(1 - 0.686) \times 10}{(3 - 0.686)} = 1.357 \text{ atm}$$

and

$$p_{SO_3} = \frac{2 \times 0.686 \times 10}{(3 - 0.686)} = 5.929 \text{ atm}$$

As a check, $K_p = 5.929/[2.714 \times (1.357)^{1/2}] = 1.875$ indicating that, indeed, K_p is not a function of pressure. Of the total number of moles present

27.14% are SO_2 (which is a decrease from 42.3% at $P = 1$)
13.57% are O_2 (which is a decrease from 21.2% at $P = 1$)
59.29% are SO_3 (which is an increase from 36.5% at $P = 1$)

Thus it is seen that an increase in total pressure has shifted the equilibrium towards the SO_3 side.

In order to simultaneously vary the temperature and total pressure in such a manner that the numbers of moles of the three gas species present remain constant, Eq. (9.13) indicates that the variation must be such that PK_p^2 remains constant, where

$$K_p = 10^{\frac{94{,}600}{2.303 \times 8.3144T}} \times 10^{\frac{-89.37}{2.303 \times 8.3144T}}$$

It is apparent that, by suitable mixing of SO_3 gas and SO_2 gas, a mixture with a known oxygen partial pressure can be obtained. For example, consider that it was required to have an SO_2-SO_3-O_2 mixture at 1 atm total pressure in which $p_{O_2} = 0.1$ atm. In order to obtain this gas mixture, SO_3 and SO_2 gas, both at 1 atm pressure, would be mixed in the volume (and hence molar) ratio $SO_2/SO_3 = a$ and allowed to equilibrate. If a moles of SO_2 and 1 mole of SO_3 are mixed, then, from the stoichiometry of Eq. (9.12), x moles of SO_3 would decompose to form x moles of SO_2 and $\frac{1}{2}x$ moles of O_2 such that, at equilibrium, the numbers of moles present would be

$$SO_2 + \tfrac{1}{2}O_2 = SO_3$$

$$(a + x) \quad \tfrac{1}{2}x \quad (1 - x)$$

with

$$n_T = a + x + \tfrac{1}{2}x + 1 - x = \tfrac{1}{2}(2a + 2 + x)$$

In this gas mixture

$$p_{O_2} = \frac{n_{O_2}}{n_T} P = \frac{x}{2a + 2 + x} P$$

which, for $P = 1$ and $p_{O_2} = 0.1$, gives $a = 4.5x - 1$

Also

$$K_p^2 = \frac{p_{SO_3}^2}{p_{SO_2}^2 p_{O_2}} = \frac{(1 - x)^2(2a + 2 + x)}{(a + x)^2 xP}$$

At 1000 K, $K_p = 1.876$ thus, substituting for a in terms of x and taking

$K_p^2 = 3.519$ gives

$$96.45x^3 - 18.709x^2 - 6.481x = 0$$

or, since $x \neq 0$,

$$96.45x^2 - 18.709x - 6.481 = 0$$

which has the solution $x = 0.374$.

Thus $$a = (4.5 \times 0.374) - 1 = 0.683$$

and

$$p_{O_2} = \frac{xP}{2a + 2 + x} = 0.1 \text{ atm}$$

$$p_{SO_3} = \frac{2(1-x)}{2a + 2 + x} = 0.335 \text{ atm}$$

and

$$p_{SO_2} = \frac{2(a + x)}{2a + 2 + x} = 0.565 \text{ atm}$$

In the equilibrium mixture, $p_{SO_2}/p_{SO_3} = 1.7$, compared with $p_{SO_2}/p_{SO_3} = a = 0.683$ in the initial mixture. As a check,

$$K_p = 0.335/[0.565 \times (0.1)^{1/2}] = 1.875$$

If it had been required to have $p_{O_2} = 0.212$ (the equilibrium p_{O_2} in an equilibrated gas at 1 atm pressure of initial composition 1 mole of SO_2 + $\frac{1}{2}$ mole of O_2), then solution of the stoichiometric condition would have given $a = 0$; this would indicate that pure SO_3 at 1000 K and 1 atm pressure decomposes to the same equilibrium gas mixture as is formed from an initial mixture of 1 mole of SO_2 + $\frac{1}{2}$ mole of O_2 at 1 atm. Thus, 0.212 atm is the maximum value of p_{O_2} which can be produced in an SO_2-SO_3 mixture at 1000 K and 1 atm total pressure.

9.7 EQUILIBRIUM IN H_2O-H_2 AND CO_2-CO MIXTURES

H_2O-H_2 and CO_2-CO gas mixtures are extensively used in experimental chemical metallurgy, where it is often necessary to fix the partial pressure of oxygen in a gas phase at an extremely low value. For example, if it were required to have a gas atmosphere containing a partial pressure of oxygen of 10^{-10} atm, then such an oxygen potential can be obtained with relative ease via establishment of equilibrium

$$H_2 + \tfrac{1}{2}O_2 = H_2O$$

for which

$$\Delta G^0 = -239{,}500 + 8.14T \ln T - 9.25T \text{ joules}$$

From Eq. (9.8),

$$\ln K_p = \frac{239{,}500}{8.3144T} - \frac{8.14 \ln T}{8.314} + \frac{9.25}{8.3144}$$

If it is required to have an atmosphere containing $p_{O_2} = 10^{-10}$ at 2000 K, then, at this temperature,

$$\ln K_p = \frac{239{,}500}{8.3144 \times 2000} - \frac{8.14 \times 7.6}{8.3144} + \frac{9.25}{8.3144}$$

$$= 8.075$$

Therefore

$$K_p = 3.212 \times 10^3 = \frac{p_{H_2O}}{p_{H_2} p_{O_2}^{1/2}}$$

and for $p_{O_2} = 10^{-10}$ atm,

$$\frac{p_{H_2O}}{p_{H_2}} = 3.212 \times 10^3 \times 10^{-5} = 3.212 \times 10^{-2}$$

Thus if, in the H_2-H_2O gas mixture, $p_{H_2} = 1$, then p_{H_2O} must equal 0.03212 atm. As 0.03212 atm is the saturated water vapor pressure at 25.47°C, then the required gas mixture can be produced by bubbling hydrogen gas at 1 atm through pure liquid water at 25.47°C, thereby saturating the hydrogen with water vapor. The establishment of reaction equilibrium at 2000 K will give $p_{O_2} = 10^{-10}$ atm in the gas.

Similarly the partial pressure of oxygen in a gaseous atmosphere could be determined by means of the reaction

$$CO + \frac{1}{2}O_2 = CO_2$$

For the reaction $CO_{(g)} = \frac{1}{2}O_{2(g)} + C_{(s)}$,

$$\Delta G^0 = 111{,}700 + 87.65T \text{ joules}$$

and for the reaction $C_{(g)} + O_{2(g)} = CO_{2(g)}$,

$$\Delta G^0 = -394{,}100 - 0.84T \text{ joules}$$

summation of which gives $\Delta G^0 = -282{,}400 + 86.81T$ for the reaction $CO_{(g)} + \frac{1}{2}O_{2(g)} = CO_{2(g)}$. Thus

$$\ln K_p = \frac{282{,}400}{8.3144T} - \frac{86.81}{8.3144}$$

If it were required to have $p_{O_2} = 10^{-20}$ at 1000 K, then

$$\ln K_p = \frac{282{,}400}{8.3144 \times 1000} - \frac{86.81}{8.3144} = 23.52$$

Therefore

$$K_p = 1.646 \times 10^{10} = \frac{p_{CO_2}}{p_{CO}p_{O_2}^{1/2}} = \frac{p_{CO_2}}{p_{CO} \times 10^{-10}}$$

and so

$$\frac{p_{CO_2}}{p_{CO}} = 1.646$$

If the total pressure $P = 1$ atm, then as $p_{CO_2} + p_{CO} = 1$, $p_{CO_2} = 1.646(1 - p_{CO_2}) = 0.622$, and $p_{CO} = 0.378$. The required mixture can be produced by mixing CO_2 and CO in the volume ratio 1.646/1, that is, 62.2 volume percent CO_2 and 37.8 volume percent CO.

In the cases of both of the above equilibria, the oxygen pressure in the equilibrated gas is so small that the p_{H_2}/p_{H_2O} and p_{CO_2}/p_{CO} ratios in the equilibrated gas mixtures are negligibly different from the corresponding ratios in the initial mixtures.

The equality of the ratio of the volume percentages with the ratio of the partial pressures in a gas mixture can be demonstrated as follows. Consider *a*

cm^3 of gas A at 1 atm pressure and b cm^3 of gas B at 1 atm being mixed at constant pressure (and hence at constant total volume, $a + b$). The number of moles of $A = n_A = (1 \times a)/RT = a/RT$, and the number of moles of $B = n_B = (1 \times b)/RT = b/RT$. Therefore, in the mixture,

$$p_A = \frac{n_A RT}{V} = \frac{n_A RT}{(a+b)}$$

and

$$p_B = \frac{n_B RT}{(a+b)}$$

and so

$$\frac{p_A}{p_B} = \frac{n_A}{n_B} = \frac{a}{b} = \frac{\text{volume percentage of A}}{\text{volume percentage of B}}$$

9.8 GASEOUS REACTION EQUILIBRIA AND FUGACITY

In the introductory paragraph of this chapter it was stated that, at equilibrium, a system of gases which undergo chemical reaction with one another could be considered to be either a nonideal mixture of the reactant gases or an ideal mixture of the product gases and the residual reactant gases. In the former case, the equilibrium state, at given temperature and pressure, is thus described in terms of the fugacities of the reactant gases; and in the latter case, the equilibrium state is described in terms of the partial pressures of the product gases and the residual reactant gases. As the latter case has been the subject of the present chapter, it is desirable, for the sake of completeness, to at least introduce the former case.

Consider $\frac{1}{2}$ mole of $A_{(g)}$ and $\frac{1}{2}$ mole of $B_{(g)}$, each at 1 atm pressure, which, when mixed at constant temperature and pressure, react according to

$$\tfrac{1}{2}A + \tfrac{1}{2}B = C$$

for which the standard free energy change is ΔG^0. When x moles of A have reacted, the free energy decrease in the system is

$$\Delta G = 2x\Delta G^0 + RT[(0.5-x)\ln(0.5-x) + (0.5-x)\ln(0.5-x) + 2x\ln 2x] \tag{9.14}$$

and reaction equilibrium is established when

$$\left(\frac{\partial \Delta G}{\partial x}\right)_{T,P} = 2\Delta G^{0} + 2RT \ln \frac{2x}{0.5 - x} = 0$$

i.e., when

$$x = \frac{e^{-\Delta G^{0}/RT}}{4 + 2e^{-\Delta G^{0}/RT}} \tag{9.15}$$

Substituting Eq. (9.15) into Eq. (9.14) gives the free energy decrease to reach the equilibrium state as

$$\Delta G = RT \ln \frac{1}{2 + e^{-\Delta G^{0}/RT}} \tag{9.16}$$

or

$$\Delta G = RT\left(0.5 \ln \frac{1}{2 + e^{-\Delta G^{0}/RT}} + 0.5 \ln \frac{1}{2 + e^{-\Delta G^{0}/RT}}\right)$$

But this is also the free energy of mixing of ½ mole of A and ½ mole of B given as

$$\Delta G^{M} = RT(0.5 \ln f_{A} + 0.5 \ln f_{B}) \tag{9.17}$$

and so it is seen that

$$f_{A} = \frac{1}{2 + e^{-\Delta G^{0}/RT}}$$

and

$$f_{B} = \frac{1}{2 + e^{-\Delta G^{0}/RT}}*$$

As $P = 1$ atm, the equilibrium partial pressure of A, p_{A}, is

*If A and B react to form no compound other than C, then $f_{A} = f_{B}$ only when, in the original mixture, $n_{A} = n_{B}$. It can be shown that $f_{A} < f_{B}$ when $n_{A} < n_{B}$ and vice versa.

$$p_{\mathrm{A}} = (0.5 - x)$$
$$= 0.5 - \frac{e^{-\Delta G^0/RT}}{4 + 2e^{-\Delta G^0/RT}}$$
$$= 0.5 - f_{\mathrm{A}} \frac{e^{-\Delta G^0/RT}}{2}$$

or

$$f_{\mathrm{A}} = \frac{2(0.5 - p_A)}{e^{-\Delta G^0/RT}}$$

Thus if the temperature was such that $\Delta G^0 = 0$, then

$$x = \tfrac{1}{6}$$
$$f_{\mathrm{A}} = f_{\mathrm{B}} = 1 - 2p_{\mathrm{A}}$$

and

$$p_{\mathrm{A}} = p_{\mathrm{B}} = p_{\mathrm{C}} = \tfrac{1}{3}$$

Thus from Eq. (9.14)

$$\Delta G^M = RT\left(\tfrac{1}{3} \ln \tfrac{1}{3} + \tfrac{1}{3} \ln \tfrac{1}{3} + \tfrac{1}{3} \ln \tfrac{1}{3}\right)$$
$$= RT \ln \tfrac{1}{3}$$

or from Eq. (9.17)

$$\Delta G^M = RT\left(\tfrac{1}{2} \ln \tfrac{1}{3} + \tfrac{1}{2} \ln \tfrac{1}{3}\right)$$
$$= RT \ln \tfrac{1}{3}$$

It is apparent that, provided the temperature of the system is high enough and the total pressure is low enough that the system behaves as an ideal mixture of the product and residual reactant gases, there is no advantage in using the concept of fugacity. The vast majority of gaseous systems of metallurgical interest fall into this category.

9.9 NUMERICAL EXAMPLES

Example 1

Consider the equilibrium

$$P_{4(g)} = 2P_{2(g)}$$

and calculate:

(*a*) The temperature at which $X_{P_4} = X_{P_2} = 0.5$ at a total pressure of 1 atm.
(*b*) The total pressure at which $X_{P_4} = X_{P_2} = 0.5$ at 2000 K.
(*a*) For the reaction $P_{4(g)} = 2P_{2(g)}$,

$$\Delta G^0 = 225{,}400 + 7.90T \ln T - 209.4T \text{ joules}$$

and hence

$$\ln K_p = \frac{-27{,}109}{T} - 0.95 \ln T + 25.18 \tag{i}$$

Decomposition of 1 mole of P_4 produces $1 - x$ moles of P_4 and $2x$ moles of P_2.

Thus, for $X_{P_4} = (1 - x)/(1 + x) = 0.5$, $x = \frac{1}{3}$, and hence $p_{P_4} = 0.5P$ and $p_{P_2} = 0.5P$.

Thus,
$$K_p = \frac{p_{P_2}^2}{p_{P_4}} = 0.5P \tag{ii}$$

With $P = 1$ atm, Eq. (i) gives

$$\ln K_p = \ln (0.5) = \frac{-27{,}109}{T} - 0.95 \ln T + 25.18$$

which has the solution $T = 1429$ K.

Thus at $P = 1$ atm and $T = 1429$ K, $X_{P_2} = X_{P_4} = 0.5$ in a P_4–P_2 mixture.

(*b*) From Eqs. (i) and (ii),

$$K_{p,2000\,K} = 81.83 = 0.5P$$

which gives $P = 163.6$ atm.

Thus at $P = 163.6$ atm and $T = 2000$ K, $X_{P_2} = X_{P_4}$ in the mixture.

Example 2

Consider the cracking of gaseous ammonia according to the reaction

$$2NH_{3(g)} = N_{2(g)} + 3H_{2(g)} \quad \text{(i)}$$

under conditions of

(*a*) Constant total pressure and
(*b*) Constant volume at 400°C.
(*a*) For reaction (i), $\Delta G^0 = 87{,}030 - 25.8T \ln T - 31.7T$ joules. Therefore,

$$\Delta G^0_{673\,\text{K}} = -47{,}370 \text{ joules} \quad \text{and} \quad K_{p,\,673\,\text{K}} = 4748$$

From the stoichiometry of the reaction, decomposition of 1 mole of NH_3 produces $3x$ moles of H_2, x moles of N_2, and $(1 - 2x)$ moles of NH_3. Thus for

$$2NH_3 = N_2 + 3H_2$$

reaction gives $\quad (1 - 2x) \quad x \quad 3x$

and $n_T = 1 - 2x + x + 3x = 1 + 2x$ moles. Thus,

$$p_{H_2} = \frac{3x}{1+2x}P \quad p_{N_2} = \frac{x}{1+2x}P \quad \text{and} \quad p_{NH_3} = \frac{1-2x}{1+2x}P$$

such that

$$K_{p,\,673\,\text{K}} = 4748 = \frac{p^3_{H_2} \cdot p_{N_2}}{p^2_{NH_3}} = \frac{27x^4P^2}{(1+2x)^2(1-2x)^2} \quad \text{(ii)}$$

Using the identity $(1 - y)(1 + y) = 1 - y^2$, Eq. (ii) can be written as

$$K_p = \frac{27x^4P^2}{[1-(2x)^2]^2}$$

or

$$K_p^{1/2} = \frac{5.196x^2P}{(1-4x)^2}$$

Thus, for a constant total pressure of 1 atm,

$$(4748)^{1/2}(1 - 4x^2) = 5.196x^2$$

which gives $x = 0.4954$.

Thus, at equilibrium,

$$p_{H_2} = \frac{3x}{1+2x} = 0.7465 \text{ atm}$$

$$p_{N_2} = \frac{x}{1+2x} = 0.2488 \text{ atm}$$

and
$$p_{NH_3} = \frac{1-2x}{1+2x} = 0.0047 \text{ atm}$$

in which state the NH_3 is 99.08% decomposed.

An alternative approach to the problem is as follows. From the stoichiometry of the reaction, at all times,

$$p_{H_2} = 3p_{N_2} \qquad \text{(iii)}$$

and
$$P = p_{NH_3} + p_{N_2} + p_{H_2} \qquad \text{(iv)}$$

Eliminating p_{NH_3} and p_{H_2} from Eqs. (iii) and (iv) and substituting into Eq. (ii) gives

$$K_p = \frac{27p_{N_2}^4}{(P-4p_{N_2})^2}$$

or
$$K_p^{1/2} = \frac{(27)^{1/2}p_{N_2}^2}{(P-4p_{N_2})}$$

With $P = 1$ atm,

$$(4784)^{1/2}(1-4p_{N_2}) = (27)^{1/2}p_{N_2}^2$$

which gives
$$p_{N_2} = 0.2488 \text{ atm}$$

$$p_{H_2} = 3p_{N_2} = 0.7464 \text{ atm}$$

and
$$p_{NH_3} = 1 - p_{H_2} - p_{N_2} = 0.0048 \text{ atm}$$

(*b*) Now consider that the decomposition occurs at constant volume. As the decomposition reaction results in an increase in the number of moles of gas from 1 to $1 + 2x$, the reaction at constant P increases the volume of the gas by the factor $(1 + 2x)$. Thus a decrease in this volume to the original value raises the total pressure by the factor $(1 + 2x)$. From Le Chatelier's Principle, an increase in pressure shifts the equilibrium in that direction which decreases the number of moles of gas present, i.e., in the direction $3H_2 + N_2 \rightarrow 2NH_3$. Thus the equilibrium decomposition of ammonia under conditions of constant volume will produce a smaller extent of decomposition and will cause an increase in pressure.

As before,

$$p_{H_2} = \frac{3x}{1+2x} P'$$

$$p_{N_2} = \frac{x}{1+2x} P'$$

and

$$p_{NH_3} = \frac{1-2x}{1+2x} P' \qquad \text{(v)}$$

where P' is the pressure of the reacting mixture.

Before decomposition begins, the one mole of NH_3 obeys the relation $PV = RT$. The decomposition reaction, at constant V and T, increases the number of moles of gas to $1 + 2x$ and hence increases the total pressure of the gas to P', where

$$P'V = (1 + 2x)RT$$

Therefore,

$$V = \text{constant} = \frac{RT}{P} = \frac{(1+2x)RT}{P'}$$

such that, in Eq. (v),

$$\frac{P'}{1+2x} = \text{the original pressure of } NH_3 \text{ at volume } V \text{ before decomposition started}$$

Thus, for an original pressure of $P = 1$ atm, at equilibrium,

$$p_{H_2} = 3x$$

$$p_{N_2} = x$$

and

$$p_{NH_3} = 1 - 2x$$

Hence,

$$K_p = \frac{27x^4}{(1-2x)^2}$$

or
$$K_p^{1/2} = (4784)^{1/2} = \frac{(27)^{1/2}x^2}{(1-2x)}$$

which gives $x = 0.4909$. Thus

$$p_{H_2} = 3x = 1.4727 \text{ atm}$$
$$p_{N_2} = x = 0.4909 \text{ atm}$$
$$p_{NH_3} = 1 - 2x = 0.0182 \text{ atm}$$
$$P = \sum_i p_i = 1.9819 = 1 + 2x$$

in which state the NH_3 is 98.18% decomposed.

It can be shown that, in forming NH_3 by reaction of N_2 with H_2, the yield of NH_3 is a maximum when the reactants H_2 and N_2 are mixed in the molar ratio 3:1.

In the mixture, $P = p_{H_2} + p_{N_2} + p_{NH_3}$, and, at reaction equilibrium, let it be that

$$p_{H_2} = ap_{N_2}$$

Thus
$$p_{NH_3} = P - (a+1)p_{N_2}$$

or
$$p_{N_2} = \frac{P - p_{NH_3}}{a+1} \quad \text{and} \quad p_{H_2} = \frac{a(P - p_{NH_3})}{(a+1)}$$

and hence

$$K_p = \frac{p_{NH_3}^2}{p_{H_2}^3 \cdot p_{N_2}} = \frac{p_{NH_3}^2}{[a(P-p_{NH_3})/(a+1)]^3\,[(P-p_{NH_3})/(a+1)]}$$
$$= \frac{p_{NH_3}^2(a+1)^4}{a^3(P-p_{NH_3})^4} \quad \text{(vi)}$$

It is necessary to show that p_{NH_3} has its maximum value when $a = 3$, i.e., that

$$\frac{dp_{NH_3}}{da} = 0 \quad \text{when} \quad a = 3$$

The derivative is most easily obtained by taking the logarithms of Eq. (vi),

$$\ln K_p + 3 \ln a + 4 \ln (P - p_{NH_3}) = 2 \ln p_{NH_3} + 4 \ln (a + 1)$$

and differentiating to obtain

$$\left[\frac{3}{a} - \frac{4}{(1+a)}\right] da = \left[\frac{2}{p_{NH_3}} + \frac{4}{(P - p_{NH_3})}\right] dp_{NH_3}$$

Thus, for $dp_{NH_3}/da = 0$, $3/a = 4/(a+1)$, or $a = 3$.

The stoichiometry of the reaction shows that, for the ratio p_{H_2}/p_{N_2} to have the value of 3 in the equilibrium mixture, H_2 and N_2 must be mixed, in the reaction mixture, in the ratio 3:1.

Example 3

Determine the state of equilibrium when CH_4 and CO_2 are mixed in the molar ratio 1:1 and are allowed to react at 1000 K and 1 atm pressure to form H_2 and CO according to

$$CH_4 + CO_2 = 2H_2 + 2CO$$

For $CH_{4(g)} = C_{(gr)} + 2H_{2(g)}$,

$$\Delta G^0 = 69{,}120 - 22.25T \ln T + 65.35T \text{ joules}$$

for $2C_{(gr)} + O_{2(g)} = 2CO_{(g)}$,

$$\Delta G^0 = -223{,}400 - 175.3T \text{ joules}$$

and for $CO_{2(g)} = C_{(gr)} + O_{2(g)}$,

$$\Delta G^0 = 394{,}100 + 0.8T \text{ joules}$$

Summation of these three free energy changes gives

$$\Delta G^0 = 239{,}820 - 22.25T \ln T - 109.15T \text{ joules}$$

for the reaction

$$CH_{4(g)} + CO_{2(g)} = 2H_{2(g)} + 2CO_{(g)} \qquad \text{(i)}$$

Thus $\Delta G^0_{1000\,K} = -23{,}027$ joules and $K_{p(i)} = 15.95$.

For

$$CH_4 + CO_2 = 2H_2 + 2CO$$

initially, 1 1 0 0

and, on reaction, $1-x$ $1-x$ $2x$ $2x$

with $n_T = 2(1+x)$ moles.

Thus, $p_{CH_4} = p_{CO_2} = \dfrac{(1-x)P}{2(1+x)}$ and $p_{H_2} = p_{CO} = \dfrac{xP}{1+x}$

and, hence, at equilibrium,

$$K_{p,(i)} = 15.95 = \frac{p_{H_2}^2 p_{CO}^2}{p_{CH_4} p_{CO_2}}$$

$$= \frac{(2x)^4 P^2}{[2(1+x)]^2 (1-x)^2} \quad \text{(ii)}$$

or

$$K_{p,(i)}^{1/2} = 3.99 = \frac{(2x)^2 P}{2(1+x)(1-x)} = \frac{2x^2 P}{(1-x^2)}$$

which, with $P = 1$ atm, gives $x = 0.8163$.

Thus the equilibrium state is

$$p_{H_2} = p_{CO} = \frac{0.8163}{1 + 0.8163} = 0.4494 \text{ atm}$$

$$p_{CH_4} = p_{CO_2} = \frac{1 - 0.8163}{2(1 + 0.8163)} = 0.0506 \text{ atm}$$

As a check,

$$K_{p,(i)} = \frac{(0.4494)^4}{(0.0506)^2} = 15.93$$

Consider, now, that the product H_2 reacts with the reactant CO_2 to establish the separate equilibrium

$$H_2 + CO_2 = H_2O + CO$$

For $H_{2(g)} + \frac{1}{2}O_{2(g)} = H_2O_{(g)}$,

$$\Delta G^0 = -246{,}400 + 54.8T \text{ joules}$$

for $C_{(gr)} + \frac{1}{2}O_{2(g)} = CO_{(g)}$,

$$\Delta G^0 = -111{,}700 - 87.65T \text{ joules}$$

and for $CO_{2(g)} = O_{2(g)} + C_{(gr)}$

$$\Delta G^0 = 394{,}100 + 0.8T \text{ joules}$$

Summation of these free energy changes gives

$$\Delta G^0 = 36{,}000 - 32.05T \text{ joules}$$

for the reaction

$$H_{2(g)} + CO_{2(g)} = H_2O_{(g)} + CO_{(g)} \qquad \text{(iii)}$$

Hence

$$\Delta G^0_{1000\,K} = 3950 \text{ joules}$$

and

$$K_{p,\text{(iii)}} = 0.62 = \frac{p_{H_2O}p_{CO}}{p_{CO_2}p_{H_2}} \qquad \text{(iv)}$$

Thus complete reaction equilibrium within the system at 1000 K and $P = 1$ atm requires that the partial pressures of the species CH_4, CO_2, CO, H_2, and H_2O be such that Eqs. (ii) and (iv) be simultaneously satisfied.

Following the original approach, equilibrium in reaction (i) is established when the system contains $1 - x$ moles of each of CH_4 and CO_2 and $2x$ moles of each of CO and H_2, i.e.,

$$\begin{array}{cccc} CH_4 & + CO_2 & = 2CO & + 2H_2 \\ 1-x & 1-x & 2x & 2x \end{array}$$

Thus, initially, with respect to reaction (iii) we have

$$\begin{array}{lcccc} & CO_2 & + H_2 & = CO & + H_2O \\ & 1-x & 2x & 2x & 0 \\ \text{and, on reaction,} & 1-x-y & 2x-y & 2x+y & y \end{array}$$

Thus, at complete equilibrium,

$$n_{CH_4} = 1 - x \qquad n_{CO_2} = 1 - x - y \qquad n_{H_2} = 2x - y$$

$$n_{CO} = 2x + y \qquad n_{H_2O} = y \qquad \text{and} \qquad n_T = 2(1 + x)$$

Thus

$$p_{CH_4} = \frac{1 - x}{2(1 + x)} P \text{ atm}$$

$$p_{CO_2} = \frac{1 - x - y}{2(1 + x)} P \text{ atm}$$

$$p_{H_2} = \frac{2x - y}{2(1 + x)} P \text{ atm}$$

$$p_{CO} = \frac{2x + y}{2(1 + x)} P \text{ atm}$$

$$p_{H_2O} = \frac{y}{2(1 + x)} P \text{ atm}$$

Thus, in Eq. (ii),

$$K_{p,(\text{i})} = 15.95 = \frac{(2x - y)^2 (2x + y)^2 P^2}{(1 - x)(1 - x - y)(2 + 2x)^2} \qquad \text{(v)}$$

and in Eq. (iv),

$$K_{p,(\text{iii})} = 0.62 = \frac{(2x + y)y}{(2x - y)(1 - x - y)} \qquad \text{(vi)}$$

Simultaneous solution of Eqs. (v) and (vi) gives the equilibrium state. It may be noted that the required simultaneous solution is not particularly simple. An easier approach to the problem can be made via an examination of the mole balance in the system. The system contains the species CH_4, CO_2, CO, H_2, and H_2O and thus

$$n_C\text{, the number of moles of C, } = n_{CH_4} + n_{CO_2} + n_{CO} \qquad \text{(vii)}$$

$$n_O\text{, the number of moles of O, } = 2n_{CO_2} + n_{CO} + n_{H_2O} \qquad \text{(viii)}$$

and

$$n_H\text{, the number of moles of H, } = 2n_{H_2} + 2n_{H_2O} + 4n_{CH_4} \qquad \text{(ix)}$$

Initially, 1 mole of CH_4 was mixed with 1 mole of CO_2, and thus, in the closed system,

$$\frac{n_C}{n_O} = 1 \tag{x}$$

and

$$\frac{n_H}{n_O} = 2 \tag{xi}$$

From (xi), (viii), and (ix),

$$n_{H_2} + n_{H_2O} + 2n_{CH_4} = 2n_{CO_2} + n_{CO} + n_{H_2O}$$

or, as $n_i \propto p_i$ at constant P,

$$p_{H_2} + 2p_{CH_4} = 2p_{CO_2} + p_{CO} \tag{xii}$$

From (x), (vii), and (viii),

$$n_{CH_4} + n_{CO_2} + n_{CO} = 2n_{CO_2} + n_{CO} + n_{H_2O}$$

or

$$p_{CH_4} = p_{CO_2} + p_{H_2O} \tag{xiii}$$

Also,

$$P = p_{CH_4} + p_{CO_2} + p_{CO} + p_{H_2} + p_{H_2O} \tag{xiv}$$

Eliminating p_{CH_4} from (xiii) and (xii) gives

$$p_{H_2O} = \tfrac{1}{2}p_{CO} - \tfrac{1}{2}p_{H_2} \tag{xv}$$

and eliminating p_{CH_4} from (xiii) and (xiv) gives

$$P = 2p_{CO_2} + 2p_{H_2O} + p_{CO} + p_{H_2} \tag{xvi}$$

From (xv) and (xvi)

$$p_{CO_2} = \tfrac{1}{2}P - p_{CO} \tag{xvii}$$

and from (xiii), (xv), and (xvii),

$$p_{CH_4} = \tfrac{1}{2}P - \tfrac{1}{2}p_{CO} - \tfrac{1}{2}p_{H_2} \tag{xviii}$$

Thus all five partial pressures are expressed in terms of the total pressure P, and p_{H_2} and p_{CO}, i.e.,

$$p_{CO_2} = \tfrac{1}{2}(P - 2p_{CO}) \tag{xvii}$$

$$p_{H_2O} = \tfrac{1}{2}(p_{CO} - p_{H_2}) \tag{xv}$$

$$p_{CH_4} = \tfrac{1}{2}(P - p_{CO} - p_{H_2}) \tag{xviii}$$

Thus Eqs. (ii) and (iv) can be written as

$$K_{p,(i)} = \frac{p_{CO}^2 p_{H_2}^2}{p_{CH_4} p_{CO_2}} = \frac{4p_{CO}^2 p_{H_2}^2}{(P - p_{CO} - p_{H_2})(P - 2p_{CO})} \tag{xix}$$

and

$$K_{p,(iii)} = \frac{p_{CO} p_{H_2O}}{p_{CO_2} p_{H_2}} = \frac{p_{CO}(p_{CO} - p_{H_2})}{p_{H_2}(P - 2p_{CO})} \tag{xx}$$

and the equilibrium state is determined by simultaneous solution of Eqs. (xix) and (xx). As a result of the choice of P, p_{CO}, and p_{H_2} as the independent variables, Eqs. (xix) and (xx) are less frightening than Eqs. (v) and (vi), and the search for the solutions is best started by setting $p_{CO} = p_{H_2} = 0.4494$ (the original solution) and noting that the establishment of equilibrium (iii) increases p_{CO} above 0.4494 and decreases p_{H_2} below 0.4494. Hunting and pecking with $P = 1$ atm gives

$$p_{H_2} = 0.4165 \text{ atm} \quad \text{and} \quad p_{CO} = 0.4606 \text{ atm}$$

Thus the equilibrium state is

$$p_{H_2} = 0.4165 \text{ atm}$$

$$p_{CO} = 0.4606 \text{ atm}$$

$$p_{CO_2} = \tfrac{1}{2}(1 - 0.9212) = 0.0394 \text{ atm}$$

$$p_{H_2O} = \tfrac{1}{2}(0.4606 - 0.4165) = 0.02205 \text{ atm}$$

$$p_{CH_4} = \tfrac{1}{2}(1 - 0.4165 - 0.4606) = 0.06145 \text{ atm}$$

As a check,

$$K_{p,(i)} = \frac{(0.4606)^2(0.4165)^2}{0.0394 \times 0.06145} = 15.2$$

and

$$K_{p,(iii)} = \frac{0.4606 \times 0.02205}{0.0394 \times 0.4165} = 0.62$$

The partial pressure of oxygen in the gas can be calculated from consideration of either the CO/CO_2 ratio or the H_2/H_2O ratio in the equilibrium state. For $CO + \frac{1}{2}O_2 = CO_2$,

$$\Delta G^0 = -282{,}400 + 86.85T \text{ joules}$$

Thus, $\Delta G^0_{1000\text{ K}} = -195{,}550$ joules and

$$K_{p,1000\text{ K}} = 1.64 \times 10^{10} = \frac{p_{CO_2}}{p_{CO}p_{O_2}^{1/2}} = \frac{0.0394}{0.4606 \times p_{O_2}^{1/2}}$$

Therefore $p_{O_2} = 2.7 \times 10^{-23}$ atm.

Alternatively, for $H_2 + \frac{1}{2}O_2 = H_2O$,

$$\Delta G^0 = -246{,}400 + 54.8T \text{ joules}$$

Thus $\Delta G^0_{1000\text{ K}} = -191{,}600$ joules and

$$K_{p,1000\text{ K}} = 1.02 \times 10^{10} = \frac{p_{H_2O}}{p_{H_2}p_{O_2}^{1/2}} = \frac{0.02205}{0.4165 \times p_{O_2}^{1/2}}$$

Therefore $p_{O_2} = 2.7 \times 10^{-23}$ atm.

The occurrence of reaction (iii) is indicated if an attempt is made to calculate the value of p_{O_2} in the originally considered CH_4-CO-CO_2-H_2 mixture; i.e., the existence of the equilibrium

$$CO + \tfrac{1}{2}O_2 = CO_2$$

requires the production of H_2O to establish the equilibrium

$$H_2 + \tfrac{1}{2}O_2 = H_2O$$

9.10 SUMMARY

1. The condition for equilibrium in the reaction $aA + bB = cC + dD$ is that $a\bar{G}_A + b\bar{G}_B = c\bar{G}_C + d\bar{G}_D$. The equilibrium state of a gaseous chemical reaction system is thus determined by the value of ΔG^0 for the reaction and is quantified by the equilibrium constant K_p, where $K_p = (p_C^c\, p_D^d / p_A^a\, p_B^b)_{eq}$ when the standard state for each of the reactant and product gases is the pure gas at 1 atm pressure. ΔG^0 and K_p are related as $\Delta G^0 = -RT \ln K_p$. This equation is

one of the more powerful equations of chemical thermodynamics and will be used extensively in the subsequent chapters. For increasingly negative values of ΔG^0, K_p becomes increasingly greater than unity; and conversely, for increasingly positive values of ΔG^0, K_p becomes increasingly less than unity.

2. As ΔG^0 is a function only of temperature, then K_p is a function only of temperature, and the temperature dependence of K_p is determined by the value of ΔH^0 for the reaction; i.e.,

$$\Delta G^0 = \Delta H^0 - T\Delta S^0 = -RT \ln K_p$$

and hence

$$\ln K_p = -\frac{\Delta H^0}{RT} + \frac{\Delta S^0}{R}$$

or

$$\frac{\partial \ln K_p}{\partial T} = \frac{\Delta H^0}{RT^2}$$

or

$$\frac{\partial \ln K_p}{\partial (1/T)} = -\frac{\Delta H^0}{R}$$

Thus for an exothermic reaction, K_p decreases with increasing temperature, and for an endothermic reaction, K_p increases with increasing temperature.

3. As $p_i = X_i P$ in an ideal gas mixture, the equilibrium constant can be written in terms of the mole fractions rather than in terms of the equilibrium partial pressures; i.e.,

$$K_p = \frac{p_C^c p_D^d}{p_A^a p_B^b} = \frac{X_C^c X_D^d}{X_A^a X_B^b} P^{(c+d-a-b)} = K_x P^{(c+d-a-b)}$$

Although K_p, by definition, is independent of pressure, K_x is only independent of pressure if $c+d-a-b = 0$, that is, if the gas mixture, at all states along the reaction coordinate, contains the same number of moles. If forward progression of the reaction decreases the number of moles, i.e., if $(c+d-a-b) < 0$, then an increase in pressure increases the value of K_x; and conversely, if $(c+d-a-b) > 0$, an increase in pressure decreases the value of K_x.

Both the pressure dependence of K_x and the temperature dependence of K_p are manifestations of Le Chatelier's principle.

PROBLEMS

9.1 Calculate ΔG^0_{1000} for the reaction

$$CH_{4(g)} + H_2O_{(g)} = CO_{(g)} + 3H_{2(g)}$$

At what temperature does $K_p = 1$? In which direction does the equilibrium shift when:

a. The temperature of an equilibrated CH_4-H_2O-CO-H_2 gas mixture is increased?

b. The total pressure is decreased?

9.2 A gas mixture of 50% CO, 25% CO_2 and 25% H_2 (by volume) is fed into a furnace at 900°C. Find the composition of the equilibrium CO-CO_2-H_2O-H_2 gas if the total pressure in the furnace is 1 atm.

9.3 How much heat is evolved when 1 mole of SO_2 and $\frac{1}{2}$ mole of O_2, each at 1 atm pressure, react to form the equilibrium SO_3-SO_2-O_2 mixture at 1000 K and 1 atm pressure.

9.4 A CO_2-CO-H_2-H_2O gas mixture at a total pressure of 1 atm exerts an oxygen partial pressure of 10^{-7} atm at 1600°C. In what ratio was CO_2 and H_2 mixed to produce this value of p_{O_2}? What oxygen partial pressure is exerted by the equilibrium gas mixture produced by mixing CO_2 and H_2 in the ratio 3:1?

9.5 By establishing the equilibrium

$$PCl_5 = PCl_3 + Cl_2$$

in a mixture of PCl_5 and PCl_3, it is required to obtain a partial pressure of Cl_2 of 0.1 atm at 500 K when the total pressure is 1 atm. In what ratio must PCl_5 and PCl_3 be mixed?

9.6 Calculate the partial pressure of monatomic hydrogen in hydrogen gas at 2000 K and 1 atm pressure. For

$$\tfrac{1}{2}H_{2(g)} = H_{(g)}$$

$$\Delta H^0_{298} = 217{,}990 \text{ joules} \quad \text{and} \quad \Delta S^0_{298} = 49.35 \text{ joules/degree}$$

Assume the heat capacity of monatomic hydrogen to be that of an ideal gas.

CHAPTER TEN

REACTIONS INVOLVING PURE CONDENSED PHASES AND A GASEOUS PHASE

10.1 INTRODUCTION

In the previous chapter the criterion for equilibrium in a gaseous reaction system was discussed. The question now arises: How is the situation altered if one or more of the reactants or products of the reaction is present in a condensed phase? As an introduction to the problem, it will first be considered that the condensed phases are those of pure species, i.e., are of fixed composition. Many practical systems occur in this category, e.g., the reaction of pure metals with gaseous elements or compounds to form pure metal oxides, sulfides, halides, etc. Questions of interest include what is the maximum oxygen pressure which can be tolerated in a gaseous atmosphere without oxidation of a given metal occurring at a given temperature, or, to what temperature must a given carbonate be heated in a gaseous atmosphere of given partial pressure of carbon dioxide before decomposition of the carbonate occurs? The first question is of interest in the bright annealing of copper, and the second is of interest in the production of burnt lime from limestone.

In such systems, complete equilibrium entails the establishment of (1) phase equilibrium between the individual condensed phases and the gas phase, and (2) reaction equilibrium among the various species present in the gas phase. As phase equilibrium is established when the pure condensed species exert their saturated vapor pressures (these saturated vapor pressures are uniquely fixed when the temperature of the system is fixed), then only the pressures of those species which exist solely in the gas phase are amenable to temperature-independent variation. The unique temperature variation of the saturated vapor pressures of pure condensed phase species together with the relative insensitivity to pressure variation of the free energy of species occurring in condensed states, considerably simplifies the thermodynamic treatment of reaction equilibria in systems containing both gaseous and pure condensed phases.

10.2 REACTION EQUILIBRIUM IN A SYSTEM CONTAINING CONDENSED PHASES AND A GASEOUS PHASE

Consider the reaction equilibrium between a pure solid metal M, its pure oxide MO and oxygen gas at the temperature T and the pressure P,

$$M_{(s)} + \tfrac{1}{2}O_{2(g)} = MO_{(s)}$$

It is considered that oxygen is insoluble in the solid metal. Both the metal M and the oxide MO exist as vapor species in the gas phase, as is required by the criteria for phase equilibria; i.e.,

$$\bar{G}_M \text{ (in the gas phase)} = G_M \text{ (in the solid metal phase)}$$

and

$$\bar{G}_{MO} \text{ (in the gas phase)} = G_{MO} \text{ (in the solid oxide phase)}$$

and hence reaction equilibrium is established in the gas phase. The equilibrium of interest is thus

$$M_{(g)} + \tfrac{1}{2}O_{2(g)} = MO_{(g)}$$

for which Eq. (9.6) is written as

$$G^0_{MO(g)} - \tfrac{1}{2}G^0_{O_2(g)} - G^0_{M(g)} = -RT\ln\frac{p_{MO}}{p_M p^{\frac{1}{2}}_{O_2}} \tag{10.1}$$

or

$$\Delta G^0 = -RT\ln\frac{p_{MO}}{p_M p^{\frac{1}{2}}_{O_2}}$$

where ΔG^0 is the difference between the free energy of 1 mole of *gaseous MO at 1 atm pressure*, and the sum of the free energies of $\frac{1}{2}$ mole of oxygen gas at 1 atm pressure and 1 mole of *gaseous M at 1 atm pressure*, all at the temperature T. As M and MO are present in the system as pure solid phases, phase equilibrium requires that p_{MO} in Eq. (10.1) be the equilibrium vapor pressure of solid MO at the temperature T, and that p_M be the equilibrium vapor pressure of solid M at the temperature T. Thus the values of p_{MO} and p_M in the gas phase are uniquely fixed by the temperature T, and so the value of p_{O_2} in Eq. (10.1) is uniquely fixed at the temperature T. As has been stated, phase equilibrium in the system requires that

$$\bar{G}_M(\text{in the gas}) = G_M(\text{in the solid metal phase}) \tag{10.2}$$

and

$$\bar{G}_{MO}\ (\text{in the gas}) = G_{MO}\ (\text{in the solid oxide phase}) \tag{10.3}$$

Equation (10.2) can be written as

$$G^0_{M(g)} + RT \ln p_{M(g)} = G^0_{M(s)} + \int_{P=1}^{P=p_{M(g)}} V_{M(s)} dP \tag{10.4}$$

and Eq. (10.3) can be written as

$$G^0_{MO(g)} + RT \ln p_{MO(g)} = G^0_{MO(s)} + \int_{P=1}^{P=p_{MO(g)}} V_{MO(s)} dP \tag{10.5}$$

Consider the implications of Eq. (10.4). $G^0_{M(s)}$ is the molar free energy of solid M under a pressure of 1 atm at the temperature T. The integral $\int_1^P V_{M(s)} dP$ (where $V_{M(s)}$ is the molar volume of the solid metal at the pressure P and temperature T) is the effect on the value of the molar free energy of solid M at the temperature T, of a change in pressure from $P = 1$ to $P = P$. Consider iron as a typical metal at a temperature of 1000°C. The vapor pressure of solid iron at 1000°C is 6×10^{-10} atm, and hence the term $RT \ln p_{M(g)}$ has the value $8.3144 \times 1273 \times \ln 6 \times 10^{-10} = -224{,}750$ joules. The molar volume of solid iron at 1000°C is 7.34 cm^3, which, in the range 0 to 1 atm is independent of pressure. The value of the integral for $P = 6 \times 10^{-10}$ atm is -7.34×1 cm$^3 \cdot$ atm $= -0.74$ joules. It is thus seen that $G^0_{Fe(g)}$ at 1000°C is considerably larger in value than is $G^0_{Fe(s)}$ at 1000°C, which is to be expected in view of the high metastability, with respect to the solid, of iron vapor at 1 atm pressure and a temperature of 1000°C. Secondly, it is to be noted that the value of the integral $\int_1^P V_{M(s)} dP$ is small enough to be considered negligible, in which case Eq. (10.4) can be written as

$$G^0_{M(g)} + RT \ln p_{M(g)} = G^0_{M(s)}$$

As a result of the negligible effect of pressure on the free energy of a condensed phase, the standard state of a species occurring as a condensed phase can be defined as the *pure species at the temperature* T; i.e., the specification that the pressure be unity is no longer required, and $G^0_{M(s)}$ is now simply the molar free energy of pure solid M at the temperature T. Similarly, Eq. (10.5) can be written as

$$G^0_{MO(g)} + RT \ln p_{MO(g)} = G^0_{MO(s)}$$

and hence Eq. (10.1) can be written as

$$G^0_{MO(s)} - \tfrac{1}{2}G^0_{O_2(g)} - G^0_{M(s)} = -RT \ln \left(\frac{1}{p^{\frac{1}{2}}_{O_2}} \right)$$

or

$$\Delta G^0 = -RT \ln K \qquad (10.6)$$

where $K = 1/p^{1/2}_{O_2}$, and ΔG^0 is the *standard free energy of the reaction* $M_{(s)} + ½O_{2(g)} = MO_{(s)}$.

Thus in the case of reaction equilibria involving pure condensed phases and a gas phase, the equilibrium constant K can be written solely in terms of those species which occur only in the gas phase. Again, as ΔG^0 is a function only of temperature, then K is a function only of temperature, and hence at any fixed temperature the establishment of reaction equilibrium occurs at a unique value of $p_{O_2} = p_{O_2(\text{eq},T)}$. The equilibrium thus has one degree of freedom, as can be seen from application of the phase rule. $P = 3$ (two pure solids and a gas phase), $C = 2$ [metal M + oxygen or $C =$ the number of species (3) minus the number of independent reaction equilibria (1) = 2], and thus $F = C + 2 - P = 2 + 2 - 3 = 1$.

If, at any temperature T, the actual oxygen partial pressure in a closed metal-metal oxide-oxygen system is greater than $p_{O_2(\text{eq},T)}$, then spontaneous oxidation of the metal will occur, thus consuming oxygen and decreasing the oxygen pressure in the gas phase. When the actual oxygen pressure has thus been lowered to $p_{O_2(\text{eq},T)}$, then, provided that both solid phases are still present, the oxidation reaction ceases and equilibrium prevails. Similarly, if the oxygen partial pressure in the closed system was originally less than $p_{O_2(\text{eq},T)}$, then spontaneous reduction of the oxide would occur until $p_{O_2(\text{eq},T)}$ was reached.

Extractive metallurgical processes involving the reduction of oxide ores depend on the achievement and maintenance of an oxygen pressure less than $p_{O_2(eq,T)}$ in the reaction vessel.

For example, the standard free energy change for the reaction

$$4Cu_{(s)} + O_{2(g)} = 2Cu_2O_{(s)}$$

is

$$\Delta G^0 = -339{,}000 - 14.2T \ln T + 247T \text{ joules}$$

in the temperature range 298 to 1356 K ($T_{m,Cu}$). Thus,

$$-\ln K = \ln p_{O_2(eq,T)} = \frac{\Delta G^0}{RT}$$

or

$$\log p_{O_2(eq,T)} = -\frac{339{,}000}{2.303 \times 8.3144T} - \frac{14.2 \log T}{8.3144} + \frac{247}{2.303 \times 8.3144}$$

This variation of $\log p_{O_2(eq,T)}$ with $1/T$ is drawn as the line *ab* in Fig. 10.1*a*; all points on this line represent the unique oxygen pressure $p_{O_2(eq,T)}$ required for equilibrium between $Cu_{(s)}$, $Cu_2O_{(s)}$, and $O_{2(g)}$ at the particular temperature T. Thus *ab* divides the diagram into two regions: above *ab* (where $p_{O_2} > p_{O_2(eq,T)}$) the metal phase is unstable, and hence the system exists as $Cu_2O_{(s)} + O_{2(g)}$; and below *ab* (where $p_{O_2} < p_{O_2(eq,T)}$) the oxide is unstable, and hence the system exists as $Cu_{(s)} + O_{2(g)}$.

Other equilibria among two condensed pure phases and a gas phase include the formation of carbonates and hydroxides. For example, at the temperature T the equilibrium

$$MO_{(s)} + H_2O_{(g)} = M(OH)_{2(s)}$$

occurs when

$$G^0_{MO(s)} + G^0_{H_2O(g)} + RT \ln p_{H_2O} = G^0_{M(OH)_2(s)}$$

i.e., when
$$\Delta G^0 = -RT \ln K = RT \ln p_{H_2O(eq,T)}$$

and similarly the equilibrium

$$MO_{(s)} + CO_{2(g)} = MCO_{3(s)}$$

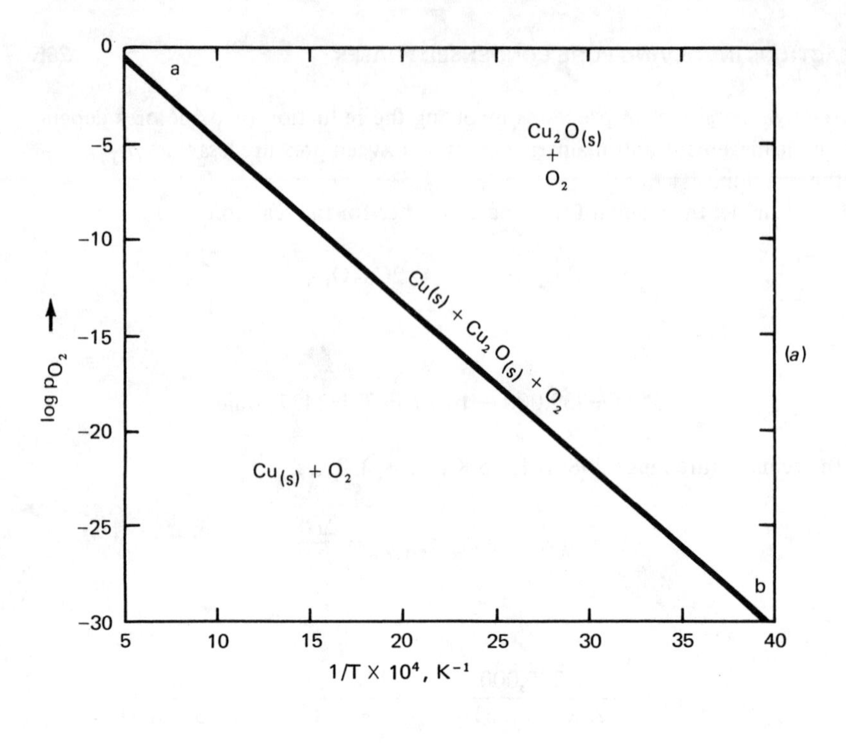

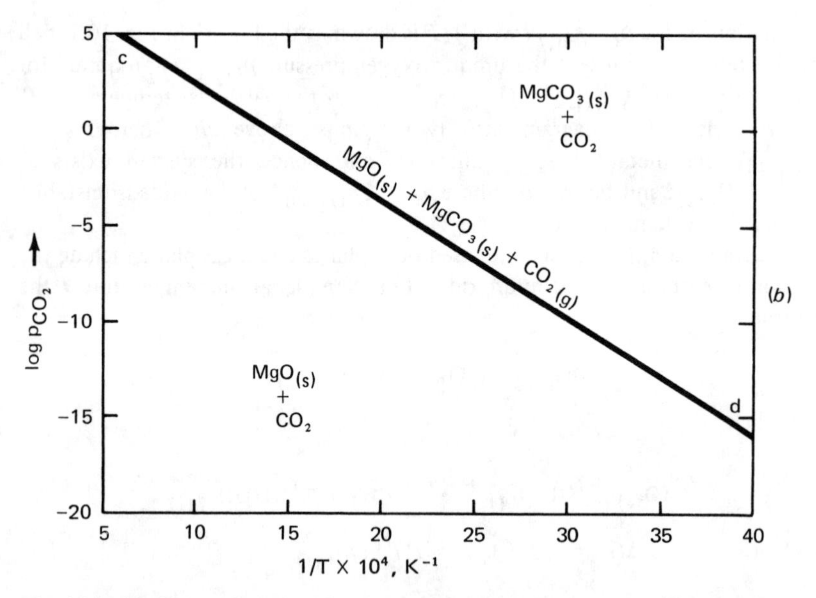

Fig. 10.1. (*a*) The variation with temperature of the oxygen partial pressure required for maintenance of the equilibrium $4Cu_{(s)} + O_2(g) = 2Cu_2O_{(s)}$. (*b*) The variation with temperature of the carbon dioxide pressure required for maintenance of the equilibrium $MgO_{(s)} + CO_2(g) = MgCO_3(s)$.

occurs when

$$G^0_{MO_{(s)}} + G^0_{CO_2\,(g)} + RT \ln p_{CO_2} = G^0_{MCO_3\,(s)}$$

i.e., when $\Delta G^0 = -RT \ln K = RT \ln p_{CO_2\,(eq,T)}$

For the reaction

$$MgO_{(s)} + CO_{2\,(g)} = MgCO_{3\,(s)}$$

$$\Delta G^0 = -117{,}600 + 170T \text{ joules}$$

in the temperature range 298 to 1000 K, and thus

$$\log p_{CO_2\,(eq,T)} = -\frac{117{,}600}{2.303 \times 8.3144T} + \frac{170}{2.303 \times 8.3144}$$

This variation is shown in the Fig. 10.1*b* as the line *cd*, which again divides the diagram into two regions: one in which $MgO_{(s)} + CO_{2\,(g)}$ are stable, and one in which $MgCO_{3\,(s)} + CO_{2\,(g)}$ are stable.

10.3 THE VARIATION OF STANDARD FREE ENERGY WITH TEMPERATURE

For any chemical reaction, combination of Eqs. (6.11) and (6.12) gives ΔG^0 for the reaction as a function of temperature.

$$\Delta G^0_T = \Delta H^0_T - T\Delta S^0_T$$

$$= \Delta H^0_{298} + \int_{298}^{T} \Delta c_p dT - T\Delta S^0_{298} - T\int_{298}^{T} \frac{\Delta c_p}{T} dT \qquad (10.7)$$

from which it is seen that the deviation from linearity between ΔG^0 and T depends on the sign and magnitude of Δc_p for the reaction. Generally, however, the variation of ΔG^0 with T is considered as follows. For each of the individual reactants and products of the reaction, the molar heat capacity c_p is expressed, over a stated range of temperature, in the form

$$c_p = a + bT + cT^{-2}$$

Hence for the reaction, again within a stated range of temperature,

$$\Delta c_p = \Delta a + \Delta bT + \Delta cT^{-2}$$

Kirchhoff's equation is

$$\left(\frac{\partial \Delta H^0}{\partial T}\right)_P = \Delta c_p = \Delta a + \Delta b T + \Delta c T^{-2}$$

where ΔH^0 is the standard enthalpy change for the reaction. Integration gives

$$\Delta H_T^0 = \Delta H_0 + \Delta a T + \frac{\Delta b T^2}{2} - \frac{\Delta c}{T} \tag{10.8}$$

where ΔH_0 is an integration constant which would only equal the standard enthalpy of reaction at 0 K if the analytical expression for Δc_p as a function of T was valid down to 0 K. ΔH_0 is normally evaluated by substituting a known value of ΔH_T^0 into Eq. (10.8).

For the reaction, the Gibbs-Helmholtz equation is

$$\frac{\partial\left(\frac{\Delta G^0}{T}\right)}{\partial T} = -\frac{\Delta H^0}{T^2}$$

$$= -\frac{\Delta H_0}{T^2} - \frac{\Delta a}{T} - \frac{\Delta b}{2} + \frac{\Delta c}{T^3}$$

integration of which gives

$$\frac{\Delta G^0}{T} = I + \frac{\Delta H_0}{T} - \Delta a \ln T - \frac{\Delta b T}{2} - \frac{\Delta c}{2T^2}$$

or

$$\Delta G^0 = IT + \Delta H_0 - \Delta a T \ln T - \frac{\Delta b T^2}{2} - \frac{\Delta c}{2T} \tag{10.9}$$

where I is an integration constant.

As $\Delta G^0 = -RT \ln K$, then Eq. (10.9) gives

$$\ln K = -\frac{\Delta H_0}{8.3144T} - \frac{I}{8.3144} + \frac{\Delta a \ln T}{8.3144}$$

$$+ \frac{\Delta b T}{(2 \times 8.3144)} + \frac{\Delta c}{(2 \times 8.3144 T^2)} \tag{10.10}$$

The value of I can be determined if K, at any temperature T, is known. For the reaction

$$4Cu_{(s)} + O_{2(g)} = 2Cu_2O_{(s)}$$

$$\Delta H^0_{298} = -335{,}000 \text{ joules}$$

$$\Delta S^0_{298} = -152.2 \text{ joules/degree}$$

and thus

$$\Delta G^0_{298} = -335{,}000 + (298 \times 152.2) = -289{,}600 \text{ joules}$$

$$c_{p,Cu_{(s)}} = 22.6 + 6.3 \times 10^{-3} T \text{ joules/degree·mole in the range 298 to 1356 K}$$

$$c_{p,Cu_2O_{(s)}} = 62.34 + 24 \times 10^{-3} T \text{ joules/degree·mole in the range 298 to 1200 K}$$

$$c_{p,O_{2(g)}} = 30 \times 4.2 \times 10^{-3} T - 1.7 \times 10^5 T^{-2} \text{ joules/degree·mole in the range 298 to 3000 K}$$

such that, in the temperature range 298 to 1200 K,

$$\Delta c_p = 2c_{p,Cu_2O_{(s)}} - 4c_{p,Cu_{(s)}} - c_{p,O_{2(g)}}$$
$$= 4.28 + 18.6 \times 10^{-3} T + 1.7 \times 10^5 T^{-2} \text{ joules/degree}$$

Thus,

$$\Delta H^0_T = \Delta H_0 + 4.28T + 9.3 \times 10^{-3} T^2 - 1.7 \times 10^5 T^{-1} \text{ joules}$$

As $\Delta H^0_{298} = -335{,}000$ joules, substitution gives $\Delta H_0 = -336{,}500$ joules

such that

$$\Delta H^0_T = -336{,}500 + 4.28T + 9.3 \times 10^{-3} T^2 - 1.7 \times 10^5 T^{-1} \text{ joules}$$

Dividing by $-T^2$, integrating with respect to T, and multiplying through by T gives

$$\Delta G^0_T = -336{,}500 - 4.28T \ln T - 9.3 \times 10^{-3} T^2 - 0.85 \times 10^5 T^{-1} + IT \text{ joules}$$

As $\Delta G^0_{298} = -289{,}600$ joules, substitution gives $I = 185.5$, and hence

$$\Delta G^0_T = -336{,}500 - 4.28T \ln T - 9.3 \times 10^{-3} T^2 - 0.85 \times 10^5 T^{-1} + 185.5T \text{ joules} \quad (a)$$

and
$$-\ln K = \ln p_{O_2(eq,T)}$$
$$= -\frac{336{,}500}{8.3144T} - \frac{4.28 \ln T}{8.3144} - \frac{9.3 \times 10^{-3} T}{8.3144}$$
$$-\frac{0.85 \times 10^5}{8.3144 \times T^3} + \frac{185.5}{8.3144}$$

The variation of ΔG_T^0, as calculated from the experimentally measured variation of $\log p_{O_2(eq,T)}$ with temperature, can be fitted to an equation of the form

$$\Delta G^0 = A + BT \ln T + CT$$

For the oxidation of $4Cu_{(s)}$ to $2Cu_2O_{(s)}$, this gives

$$\Delta G^0 = -338{,}000 - 14.2T \ln T + 246T \text{ joules} \qquad (b)$$

This can be approximated, in linear form, by

$$\Delta G^0 = -333{,}000 + 126T \text{ joules} \qquad (c)$$

in the temperature range 298 to 1356 K.

The values of ΔG^0 given by expressions (*a*), (*b*), and (*c*), at various temperatures, are listed in Table 10.1. The table shows that $\Delta G_{(a)}^0$ (which is calculated purely from thermochemical data) differs at most by 630 joules from $\Delta G_{(b)}^0$ (which is obtained from experimental equilibrium data). On the other hand, $\Delta G_{(c)}^0$ varies by as much as 18 kilojoules from $\Delta G_{(b)}^0$.

It can be noticed that Eq. (10.10) is similar to the vapor pressure equation, Eq. (7.8). The connection between the two can be seen as follows. Consider the evaporation of A,

$$A_{(l)} = A_{(v)}$$

At the temperature T, equilibrium occurs when

Table 10.1

T, K	$\Delta G_{(a)}^0$	$\Delta G_{(b)}^0$	$\Delta G_{(c)}^0$	Δab	Δbc
400	−274,260	−273,630	−282,600	630	8,970
600	−245,120	−244,900	−257,400	220	12,500
800	−217,050	−217,140	−232,200	90	15,060
1000	−189,950	−190,090	−207,000	140	16,910
1200	−163,780	−163,610	−181,800	170	18,190

$$G^0_{A(l)} = G^0_{A(v)} + RT \ln p_A$$

i.e., when

$$\Delta G^0 = -RT \ln p_A = -RT \ln K$$

If the heat capacities of the liquid and vapor are equal, then

$$\ln p_A = -\frac{\Delta G^0}{RT} = -\frac{\Delta H^0}{RT} + \frac{\Delta S^0}{R}$$

This is to be compared with Eq. (7.7), which gave

$$\ln p_A = -\frac{\Delta H_{evap}}{RT} + \text{constant}$$

If the vapor behaves ideally, then at constant temperature, $H_{(v)}$ is independent of pressure, and hence

$$\Delta H^0 = H^0_{(v)} - H^0_{(l)} = H_{(v)} - H_{(l)} = \Delta H_{evap}$$

However, from Eq. (6.15), for the vapor

$$S(T, p_A) = S^0_T - R \ln p_A$$

and thus

$$\begin{aligned}\Delta S^0_{(l \to v)} &= S^0_{(v)} - S^0_{(l)} \\ &= S_{(v)} - S^0_{(l)} + R \ln p_A \\ &= \Delta S_{evap} + R \ln p_A\end{aligned}$$

The constant in Eq. (7.7) thus has the value $\Delta S_{evap}/R + \ln p_A$ such that Eq. (7.7) becomes

$$\ln p_A = -\frac{\Delta H_{evap}}{RT} + \frac{\Delta S_{evap}}{R} + \ln p_A$$

indicating, thus, that when equilibrium prevails between the liquid and the vapor at the temperature T,

$$\Delta S_{evap} = \frac{\Delta H_{evap}}{T}$$

or

$$\Delta G_{evap} = 0$$

10.4 ELLINGHAM DIAGRAMS

Ellingham* plotted the experimentally determined ΔG^0-T relationships for the oxidation and sulfidation of a series of metals. He found that, in spite of the terms involving $\ln T$, T^2, and T^{-1} in Eq. (10.9), the general forms of the relationships approximated to straight lines over temperature ranges in which no change in physical state occurred. The relations could thus be expressed by means of the simple equation

$$\Delta G^0 = A + BT \tag{10.11}$$

where the constant A is identified with the temperature-independent standard enthalpy change of the reaction, ΔH^0, and the constant B is identified with the negative of the temperature-independent standard entropy change of the reaction, $-\Delta S^0$.

In Fig. 10.2 the variation of ΔG^0 with T at constant total pressure is plotted

*H. J. T. Ellingham, Reducibility of Oxides and Sulfides in Metallurgical Processes, *J. Soc. Chem. Ind.*, **63**:125 (1944).

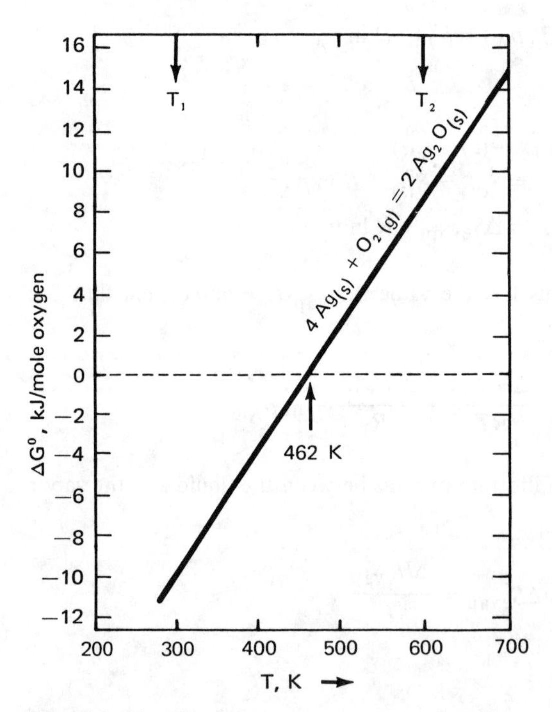

Fig. 10.2. The Ellingham line for the oxidation of silver.

for the oxidation reaction

$$4Ag_{(s)} + O_{2(g)} = 2Ag_2O_{(s)}$$

and Fig. 10.2 is known as an Ellingham diagram. From Eq. (10.11), ΔH^0 is the intercept of the line with the $T = 0$ K axis, and ΔS^0 is the negative of the slope of the line. As ΔS^0 is a negative quantity (the reaction involves the disappearance of a mole of gas), the line has a positive slope. At the temperature 462 K, ΔG^0 for the reaction is zero; i.e., at this temperature pure solid silver and oxygen gas at 1 atm pressure are in equilibrium with pure solid silver oxide. From Eq. (10.6), $\Delta G^0 = -RT \ln K = RT \ln p_{O_2(eq,T)} = 0$ at 462 K, and therefore $p_{O_2(eq,462)} = 1$. If the temperature of the system (pure $Ag_{(s)}$, pure $Ag_2O_{(s)}$, and O_2 at 1 atm) is decreased to T_1, then ΔG^0 for the oxidation becomes negative; i.e., Ag_2O is more stable than are Ag and O_2 at 1 atm, and hence the Ag spontaneously oxidizes. The value of $p_{O_2(eq,T_1)}$ is calculated from $\Delta G^0_{T_1} = RT_1 \ln p_{O_2(eq,T_1)}$, and as $\Delta G^0_{T_1}$ is a negative quantity, then $p_{O_2(eq,T_1)}$ is less than unity. Similarly, if the temperature of the system is increased from 462 K to T_2, then ΔG^0 for the oxidation becomes positive; i.e., Ag_2O is less stable than Ag and O_2 at 1 atm, and Ag_2O decomposes to Ag and O_2. As $\Delta G^0_{T_2}$ is a positive quantity, then $p_{O_2(eq,T_2)}$ is greater than unity.

The value of ΔG^0 for an oxidation reaction is thus a measure of the chemical affinity of the metal for oxygen, and the more negative the value of ΔG^0 at any temperature, then the more stable the oxide.

For the oxidation reaction $A_{(s)} + O_{2(g)} = AO_{2(s)}$,

$$\Delta S^0 = S^0_{AO_2(s)} - S^0_{O_2(g)} - S^0_{A(s)}$$

and as, generally, in the temperature range where A and AO_2 are solid, $S^0_{O_2}$ is considerably greater than either S^0_A or $S^0_{AO_2}$, (see, for example Fig. 6.14), then

$$\Delta S^0 \doteqdot -S^0_{O_2}$$

Hence the standard entropy changes of oxidation reactions involving solid phases are almost the same, corresponding, essentially, to the entropy decrease resulting from the disappearance of 1 mole of oxygen gas initially at 1 atm pressure. As the slopes of the lines in an Ellingham diagram are equal to $-\Delta S^0$, then the lines are more or less parallel to one another, as is seen in Fig. 10.13.

ΔG^0 at any temperature is the sum of the enthalpy contribution ΔH^0 (which, if $\Delta c_p = 0$, is independent of temperature) and the entropy contribution $-T\Delta S^0$ (which, if $\Delta c_p = 0$, is linearly dependent on temperature). The two contributions are illustrated in Fig. 10.3 for the oxidation reactions,

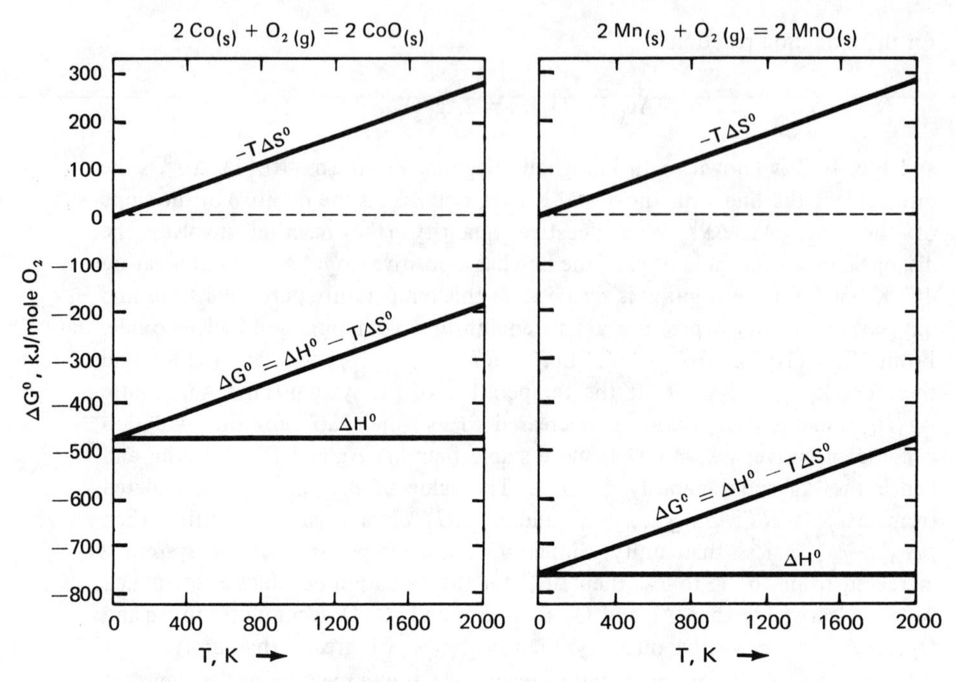

Fig. 10.3. Illustration of the effect of the magnitude of ΔH^0 on the ΔG^0-T relationships for reactions of the type $2M_{(s)} + O_{(g)} = 2MO_{(s)}$.

$$2Co_{(s)} + O_{2\,(g)} = 2CoO_{(s)}$$

for which

$$\Delta G^0 = -467{,}800 + 143.7T \text{ joules}$$

in the temperature range 298 to 1763 K, and

$$2Mn_{(s)} + O_{2(g)} = 2MnO_{(s)}$$

for which

$$\Delta G^0 = -769{,}400 + 145.6T \text{ joules}$$

in the temperature range 298 to 1500 K.

As the values of ΔS^0 for these reactions are virtually equal to one another, Fig. 10.3 shows that the relative stabilities of the oxides CoO and MnO parallel the values of ΔH^0 for the respective oxidation reactions; i.e., the more negative the value of ΔH^0, then the more negative the value of ΔG_T^0, and hence the more stable the oxide.

As $$\ln K = -\frac{\Delta H^0}{RT} + \frac{\Delta S^0}{R} = -\ln p_{O_2(eq,T)}$$

then $p_{O_2(eq,T)} = \exp(\Delta H^0/RT)\exp(-\Delta S^0/R) = \text{constant} \times \exp(\Delta H^0/RT)$, i.e., as ΔH^0 is a negative quantity, $p_{O_2(eq,T)}$ increases exponentially with increasing temperature and, at any temperature, decreases as ΔH^0 becomes more negative.

Consider two oxidation reactions, the Ellingham lines of which intersect one another, e.g.,

$$2X + O_2 = 2XO \qquad \text{(i)}$$

and

$$Y + O_2 = YO_2 \qquad \text{(ii)}$$

which are shown in Fig. 10.4. From Fig. 10.4 it can be seen that $\Delta H^0_{(ii)}$ is more negative than $\Delta H^0_{(i)}$ and that $\Delta S^0_{(ii)}$ is more negative than $\Delta S^0_{(i)}$. Subtraction of reaction (i) from reaction (ii) gives

$$Y + 2XO = 2X + YO_2 \qquad \text{(iii)}$$

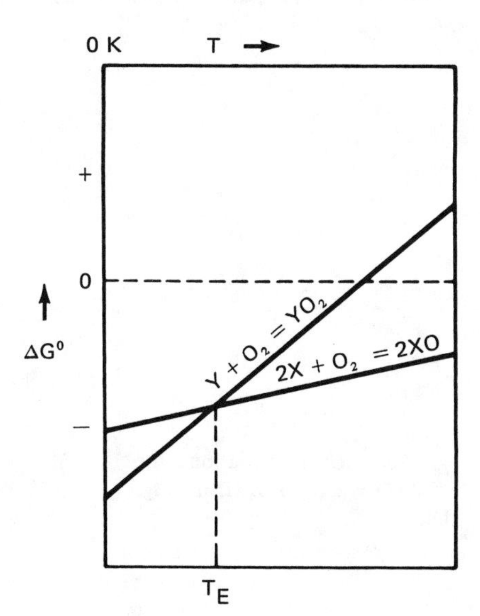

Fig. 10.4. Intersecting Ellingham lines for two hypothetical oxidation reactions.

for which the variation of ΔG^0 with T is as shown in Fig. 10.5. Below T_E, X and YO_2 are stable with respect to Y and XO, and above T_E the reverse is the case. At T_E, X, Y, XO, and YO_2, occurring in their standard states, are in equilibrium with one another. The equilibrium at T_E (as with any equilibrium) results from a compromise between enthalpy and entropy considerations. As $\Delta H^0_{(iii)}$ is negative (being equal to $\Delta H^0_{(ii)} - \Delta H^0_{(i)}$) and $\Delta S^0_{(iii)}$ is negative (being equal to $\Delta S^0_{(ii)} - \Delta S^0_{(i)}$), the system $X + Y + O_2$ has minimum enthalpy when it occurs as $X + YO_2$ and has maximum entropy when it occurs as $Y + XO$. At T_E, $\Delta H^0_{(iii)}$ equals $T\Delta S^0_{(iii)}$, and hence $\Delta G^0_{(iii)}$ equals zero. Below T_E the enthalpy contribution to $\Delta G^0_{(iii)}$ outweighs the entropy contribution, and hence $\Delta G^0_{(iii)}$ is negative such that $X + YO_2$ is the stable state. Above T_E the reverse is the case, $\Delta G^0_{(iii)}$ is positive and $Y + XO$ is the stable state. The Ellingham diagram, Fig. 10.4, thus shows that if pure X were to be used as a reducing agent to reduce the pure oxide YO_2 to form pure Y and pure XO, then the required reduction could only be effected at temperatures in excess of T_E. The foregoing discussion also illustrates that, in order for comparison to be made between the stabilities of different oxides, the Ellingham diagrams must be drawn for oxidation reactions involving the consumption of 1 mole of oxygen. The units of ΔG^0 for the oxidation reaction thus must be energy per mole of oxygen, e.g., calories per mole of oxygen or joules per mole of oxygen.

In order to avoid the necessity of calculating the value of $p_{O_2(eq,T)}$ for any oxidation reaction, Richardson* added a nomographic scale to the Ellingham

*F. D. Richardson and J. H. E. Jeffes, The Thermodynamics of Substances of Interest in Iron and Steel Making from 0°C to 2400°C; I–Oxides, *J. Iron and Steel Inst.* **160**:261 (1948).

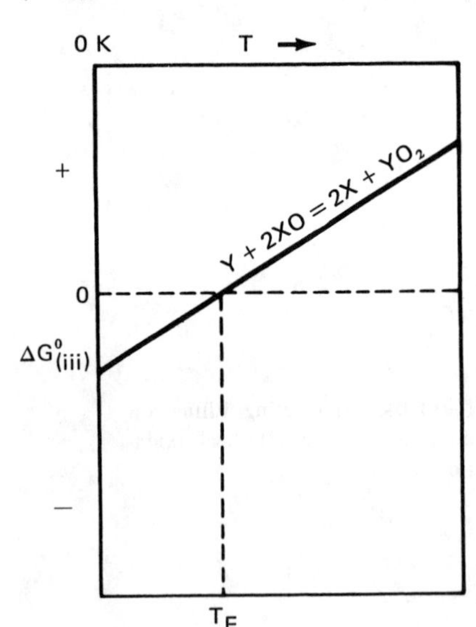

Fig. 10.5. ΔG^0-T variation for $Y + 2XO = 2X + YO_2$ from Fig. 10.4.

diagram. This scale is constructed as follows. At any temperature T, the standard free energy for an oxidation reaction, ΔG_T^0, is given by Eq. (10.6), as $RT \ln p_{O_2(eq,T)}$. But, from Eq. (8.10), $G = G^0 + RT \ln P$, ΔG_T^0 is seen to be numerically equal to the decrease in the free energy of one mole of oxygen gas when its pressure is decreased from 1 atm to $p_{O_2(eq,T)}$ atm at the temperature T. Consider the variation of ΔG with T in Eq. (8.10). For a given pressure decrease of 1 mole of an ideal gas, ΔG is a linear function of temperature; i.e., ΔG becomes increasingly negative with increasing temperature, and the slope of the ΔG-T line is given as $R \ln P$, (which, with $P < 1$, is a negative number). Similarly, for a given increase in pressure, ΔG becomes increasingly positive with increasing temperature. Thus a series of lines can be drawn for given pressure changes (always from $P = 1$ atm to $P = P$ atm) as a function of temperature. These lines radiate from $\Delta G = 0$, $T = 0$ as shown in Fig. 10.6. Superimposition of Fig. 10.6 with a typical Ellingham diagram is shown in Fig. 10.7. In Fig. 10.7:

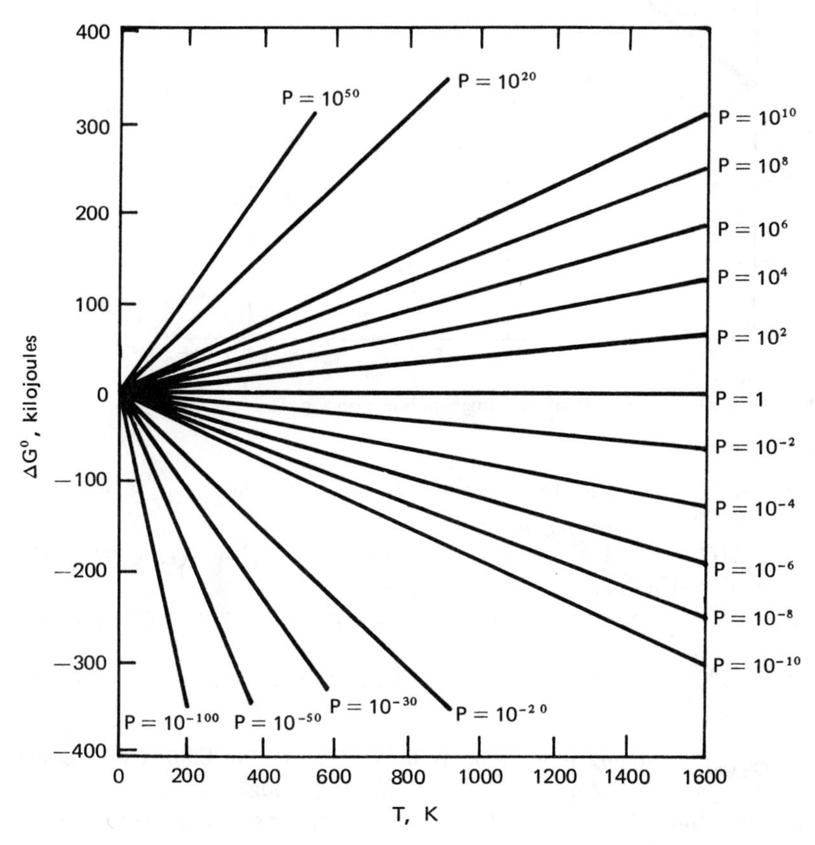

Fig. 10.6. The variation with temperature of the Gibbs free energy difference between 1 mole of ideal gas at the state ($P = P, T$) and 1 mole of ideal gas at the state ($P = 1, T$).

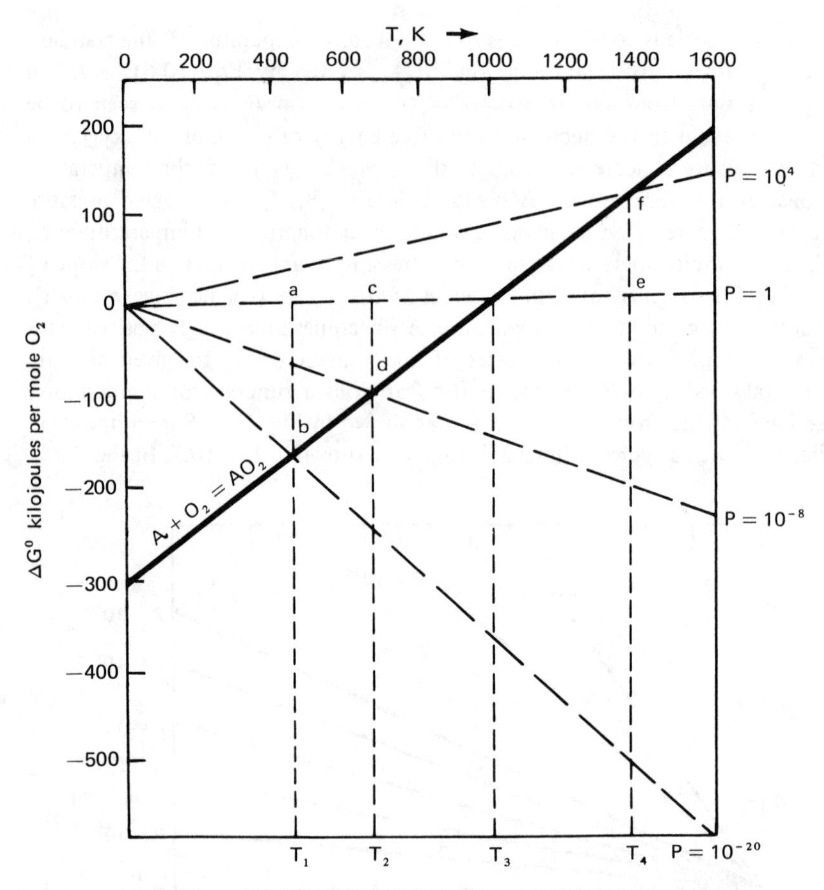

Fig. 10.7. The superimposition of an Ellingham line with Fig. 10.6.

At T_1, $\Delta G^0 = ab =$ the free energy decrease at T_1 when p_{O_2} is decreased from 1 to 10^{-20}. Thus $p_{O_2(\text{eq},T_1)} = 10^{-20}$.

At T_2, $\Delta G^0 = cd =$ the free energy decrease at T_2 when p_{O_2} is decreased from 1 to 10^{-8}. Thus $p_{O_2(\text{eq},T_2)} = 10^{-8}$.

At T_3, $\Delta G^0 =$ zero, which corresponds to no change from unity of p_{O_2}. Thus $p_{O_2(\text{eq},T_3)} = 1$.

At T_4, $\Delta G^0 = ef =$ the free energy increase at T_4 when p_{O_2} is increased from 1 to 10^4.

The $p_{O_2(\text{eq},T)}$ nomographic scale is thus added to the Ellingham diagram along the right-hand edge and along the bottom edge. For any oxidation reaction, the value of $p_{O_2(\text{eq},\,T)}$ is read off the graph as that value on the scale which is collinear with the points $\Delta G^0 = 0$, $T = 0$ and ΔG_T^0, $T = T$. That the lines in Fig. 10.6 are drawn for 1 mole of oxygen gas further illustrates the requirement that the Ellingham line be drawn for the reaction involving the consumption of 1 mole of oxygen.

The reactions (i) and (ii) shown in Fig. 10.4 can be reexamined in light of the $p_{O_2(\text{eq}, T)}$ nomographic scale. Figure 10.4 is reproduced with a nomographic oxygen pressure scale in Fig. 10.8.

At any temperature less than T_E (say T_1), it is seen that

$$p_{O_2}[\text{eq for reaction (ii) at } T_1] < p_{O_2}[\text{eq for reaction (i) at } T_1]$$

Thus if metal X and metal Y are placed in a closed system in an atmosphere of

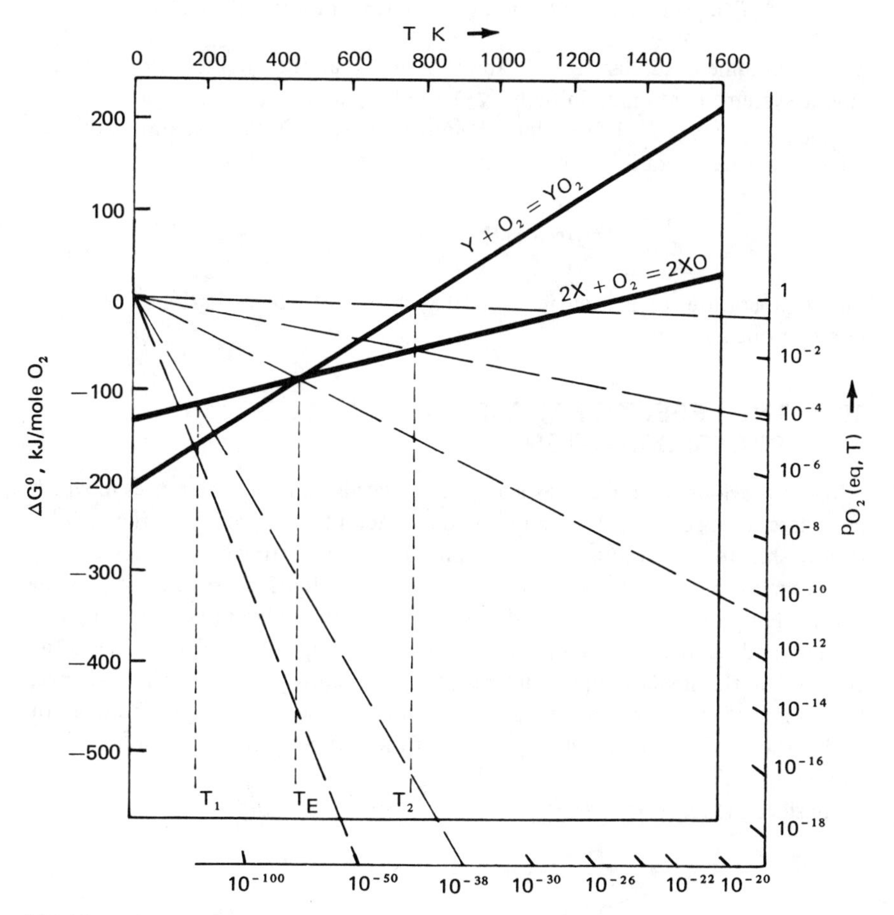

Fig. 10.8. Illustration of the addition of the Richardson oxygen pressure nomographic scale to an Ellingham diagram.

oxygen at unit pressure at T_1, both metals spontaneously oxidize, and, as a result of the consumption of oxygen, the oxygen pressure in the system decreases. Oxidation of both metals occurs until p_{O_2} is decreased to the value p_{O_2} [eq for reaction (i) at T_1], at which point oxidation of X ceases. However, as $Y + O_2$ (at p_{O_2} [eq for reaction (i) at T_1]) is still unstable with respect to YO_2, the oxidation of Y continues until p_{O_2} [eq for reaction (ii)] occurs in the gas phase. As p_{O_2} falls below p_{O_2} [eq for reaction (i)], then XO becomes unstable with respect to X and O_2 at the prevailing pressure, and hence XO decomposes. When complete equilibrium is achieved, the state of the system, at T_1, is $X + YO_2 + O_2$ at p_{O_2} [eq for reaction (ii) at T_1]. At any temperature higher than T_E, say T_2,

$$p_{O_2}\,[\text{eq for reaction (i) at } T_2] < p_{O_2}\,[\text{eq for reaction (ii) at } T_2]$$

and an argument similar to that above shows that the equilibrium state of a closed system, containing initially X, Y, and O_2 at 1 atm, is $Y + XO + O_2$ (at p_{O_2}[eq reaction (i) at T_2]). It is thus obvious that X, Y, XO, YO_2, and an oxygen atmosphere are in equilibrium only at that temperature T at which

$$p_{O_2}\,[\text{eq (i)}, T] = p_{O_2}\,[\text{eq (ii)}, T]$$

This temperature, as is seen in Fig. 10.8, is that at which the Ellingham lines intersect one another.

10.5 THE EFFECT OF PHASE TRANSFORMATION

In the previous section it was stated that, within temperature ranges in which no phase change occurs in any of the reactants or products, the ΔG^0-T relationship for the reaction can be approximated by a straight line. However, as the enthalpy of a high-temperature phase, (e.g., the liquid phase) exceeds that of a lower-temperature phase (e.g., the solid phase) by the latent heat of the phase change, and similarly the entropy of the higher-temperature phase exceeds the entropy of the lower-temperature phase by the entropy of the phase change, then, at the temperature of a phase change (be it melting, boiling, or sublimation, of a reactant in or product of the reaction), the Ellingham line will exhibit an "elbow."

Consider the reaction

$$X_{(s)} + O_{2(g)} = XO_{2(s)}$$

for which the standard enthalpy change is ΔH^0 and the standard entropy change is ΔS^0. At $T_{m,X}$, the melting point of X, the reaction

$$X_{(s)} = X_{(l)}$$

occurs for which the standard enthalpy change (the latent heat of melting) is $\Delta H^0_{m,X}$ and the corresponding entropy change is $\Delta S^0_{m,X} = \Delta H^0_{m,X}/T_{m,X}$. Thus for the reaction

$$X_{(l)} + O_{2\,(g)} = XO_{2\,(s)}$$

the standard enthalpy change is $\Delta H^0 - \Delta H^0_{m,X}$, and the standard entropy change is $\Delta S^0 - \Delta S^0_{m,X}$. As $\Delta H^0_{m,X}$ and $\Delta S^0_{m,X}$ are always positive quantities (melting is an endothermic process), then $\Delta H^0 - \Delta H^0_{m,X}$ is a larger negative quantity than is ΔH^0, and similarly $\Delta S^0 - \Delta S^0_{m,X}$ is a larger negative quantity than is ΔS^0. Hence the Ellingham line for the oxidation of liquid X to form solid XO_2 has a greater slope than the corresponding line for the oxidation of solid X to form solid XO_2, and the line shows an "elbow upwards" at $T_{m,X}$. This is shown in Fig. 10.9*a*. The line shows no discontinuity as, at $T_{m,X}$, $G^0_{X(s)} = G^0_{X(l)}$. If the melting point of the oxide, T_{m,XO_2}, is lower than the melting point of the metal, then at T_{m,XO_2} the reaction

$$XO_{2\,(s)} = XO_{2\,(l)}$$

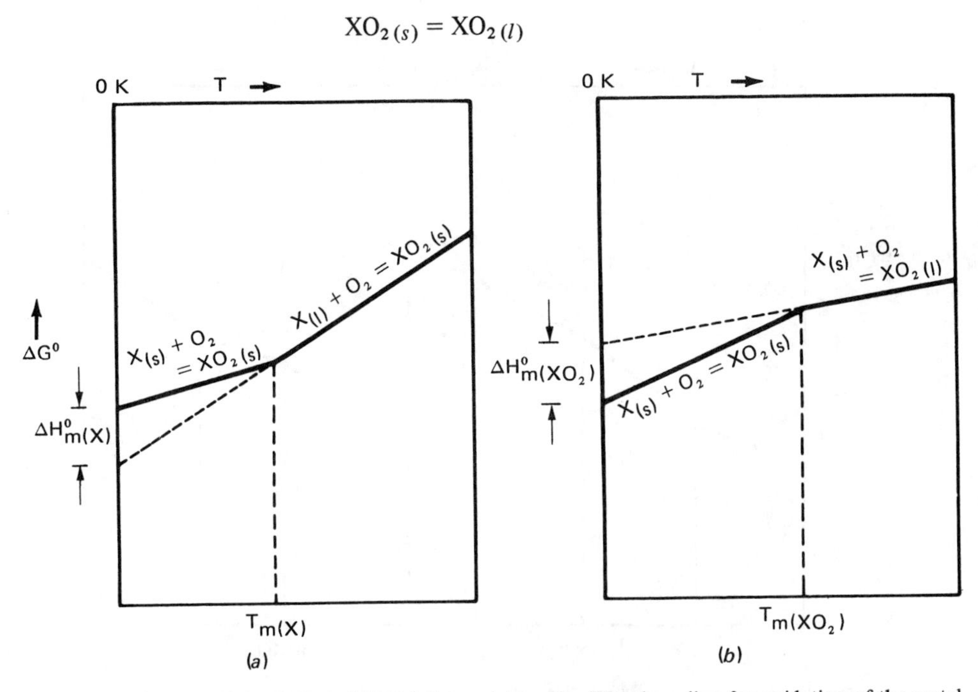

Fig. 10.9. (*a*) The effect of the melting of the metal on the Ellingham line for oxidation of the metal. (*b*) The effect of melting of the metal oxide on the Ellingham line for the metal oxidation.

occurs for which the standard enthalpy and entropy changes are, respectively, $\Delta H^0_{m,XO_2}$ and $\Delta S^0_{m,XO_2}$. Thus for the reaction

$$X_{(s)} + O_{2\,(g)} = XO_{2\,(l)}$$

the standard enthalpy change is $\Delta H^0 + \Delta H^0_{m,XO_2}$, and the standard entropy change is $\Delta S^0 + \Delta S^0_{m,XO_2}$, both of which are less negative than the corresponding quantities ΔH^0 and ΔS^0. In this case the Ellingham line for oxidation of the solid metal to produce the liquid oxide has a lesser slope than that for oxidation of the solid metal to produce the solid oxide; hence at T_{m,XO_2} the line has an elbow downwards, as is shown in Fig. 10.9*b*. Thus if $T_{m,X} < T_{m,XO_2}$, the Ellingham line is as shown in Fig. 10.10*a*, and if $T_{m,X} > T_{m,XO_2}$, then the line is as shown in Fig. 10.10*b*.

Copper is a metal which melts at a lower temperature than its lowest oxide Cu_2O. Measurements of the oxygen pressure in equilibrium with solid Cu and solid Cu_2O in the temperature range of stability of solid Cu, and in equilibrium with liquid Cu and solid Cu_2O in the temperature range of stability of liquid Cu, give

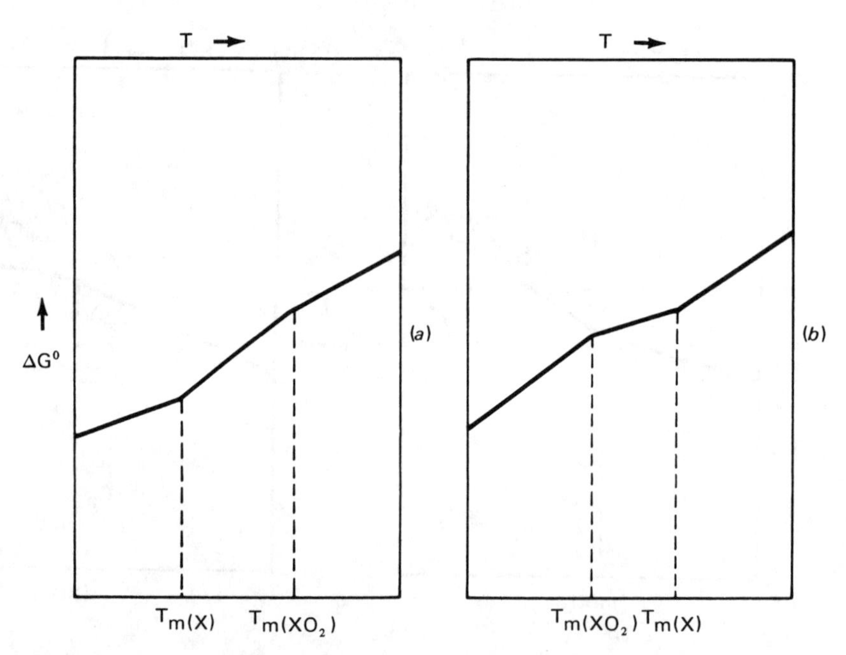

Fig. 10.10. Illustration of the effects of phase changes of the reactants and products of a reaction on the Ellingham line for the reaction.

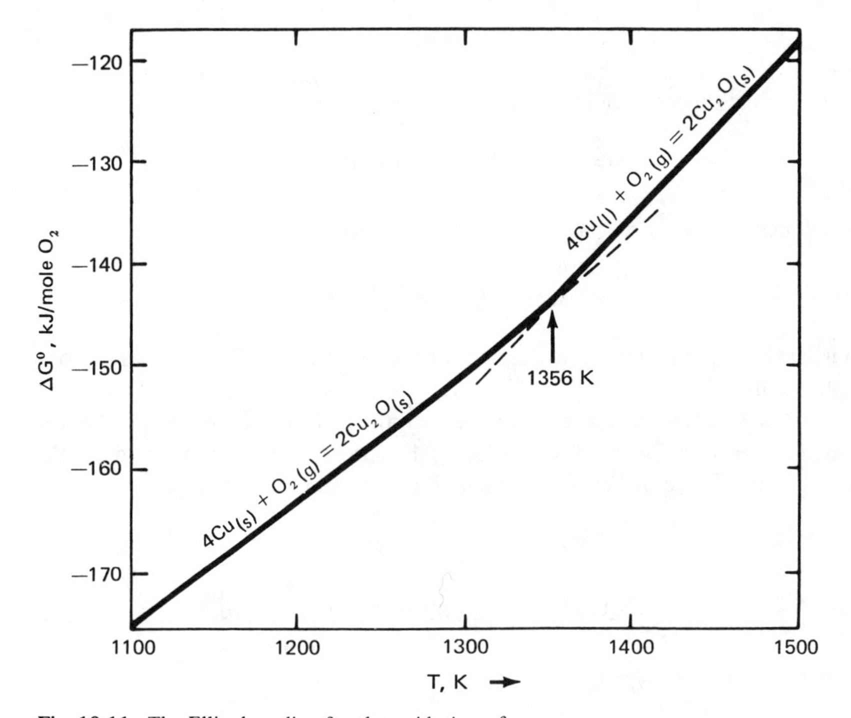

Fig. 10.11. The Ellingham line for the oxidation of copper.

$$\Delta G^0 = -338{,}900 - 14.2T \ln T + 247T \text{ joules} \quad \text{(i)}$$

for

$$4Cu_{(s)} + O_{2\,(g)} = 2Cu_2O_{(s)}$$

in the range 298 to $T_{m,Cu}$, and

$$\Delta G^0 = -390{,}800 - 14.2T \ln T + 285.3T \text{ joules} \quad \text{(ii)}$$

for

$$4Cu_{(l)} + O_{2\,(g)} = 2Cu_2O_{(s)}$$

in the range $T_{m,Cu} - 1503$ K.

These two lines, which are drawn in Fig. 10.11, intersect at 1356 K which is thus the melting temperature of copper. $\Delta G^0_{(i)} - \Delta G^0_{(ii)}$ gives

$$\Delta G = 51{,}900 - 38.3T \text{ joules}$$

for the phase change

$$4\text{Cu}_{(s)} = 4\text{Cu}_{(l)}$$

or, for the melting of 1 mole of Cu

$$\Delta G_{m,\text{Cu}} = 12{,}970 - 9.58T \text{ joules}$$

from which

$$\Delta H_{m,\text{Cu}} = 12{,}970 \text{ joules}$$

and

$$\Delta S_{m,\text{Cu}} = 9.58 \text{ joules/degree}$$

Thus, at $T_{m,\text{Cu}}$ the Ellingham diagram for the oxidation of Cu increases in slope by 9.58 joules/degree.

As $FeCl_2$ boils at a lower temperature than the melting temperature of Fe, the Ellingham diagram for the chlorination of Fe shows "elbows downward" at the melting temperature of $FeCl_2$ and the boiling temperature of $FeCl_2$.

For

$$\text{Fe}_{(s)} + \text{Cl}_{2(g)} = \text{FeCl}_{2(s)}$$

$$\Delta G^0 = -346{,}300 - 12.68T \ln T + 212.9T \text{ joules} \tag{iii}$$

in the range 298 to T_{m,FeCl_2}

For

$$\text{Fe}_{(s)} + \text{Cl}_{2(g)} = \text{FeCl}_{2(l)}$$

$$\Delta G^0 = -286{,}400 + 63.68T \text{ joules} \tag{iv}$$

in the range T_{m,FeCl_2} to T_{b,FeCl_2}

and for

$$\text{Fe}_{(s)} + \text{Cl}_{2(g)} = \text{FeCl}_{2(g)}$$

$$\Delta G^0 = -105{,}600 + 41.8T \ln T - 375.1T \text{ joules} \tag{v}$$

in the range T_{b,FeCl_2} to $T_{m,\text{Fe}}$.

Lines (iii), (iv), and (v) are shown in Fig. 10.12, which shows that

$$T_{m,\text{FeCl}_2} = 969 \text{ K} \quad \text{and} \quad T_{b,\text{FeCl}_2} = 1298 \text{ K}$$

For $\text{FeCl}_{2(s)} = \text{FeCl}_{2(l)}$

$$\Delta G^0_{(\text{iv})} - \Delta G^0_{(\text{iii})}$$

gives

$$\Delta G_{m,\text{FeCl}_2} = 59{,}900 + 12.68T \ln T - 149.0T \text{ joules}$$

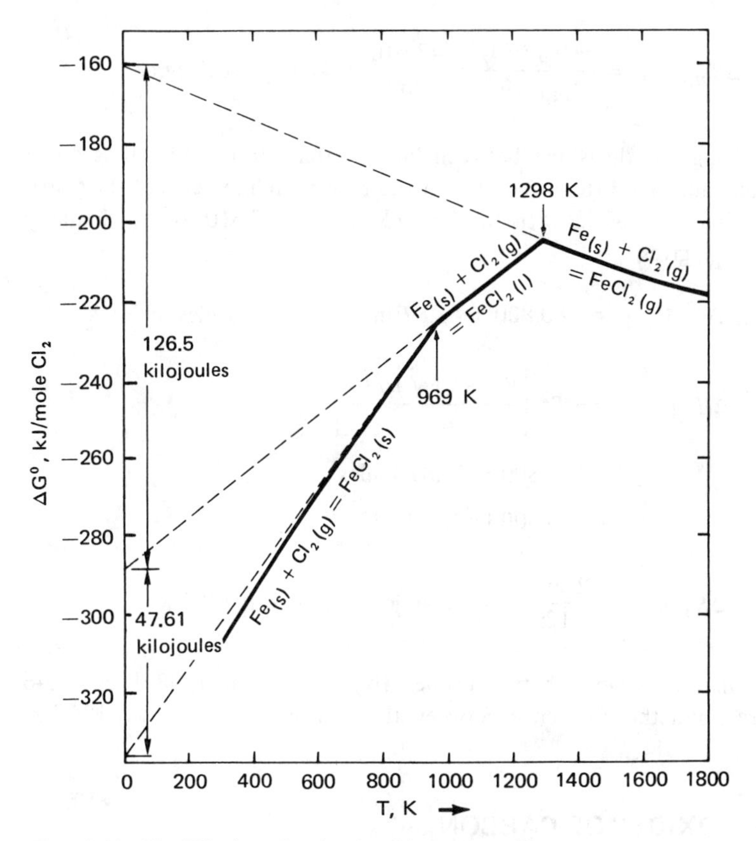

Fig. 10.12. The Ellingham line for the chlorination of iron.

Thus

$$\Delta H_{m,\mathrm{FeCl_2}} = -T^2 \left[\frac{\partial(\Delta G_{m,\mathrm{FeCl_2}}/T)}{\partial T}\right] = 59{,}900 - 12.68T \text{ joules}$$

which at 969 K gives

$$\Delta H_{m,\mathrm{FeCl_2}} = 47{,}610 \text{ joules}$$

$$\Delta S_{m,\mathrm{FeCl_2}} = -\frac{\partial \Delta G_{m,\mathrm{FeCl_2}}}{\partial T}$$

$$= -12.68 \ln T - 12.68 + 149.0 \text{ joules/degree}$$

$$= 49.13 \text{ joules/degree at 969 K}$$

or $$\Delta S_{m,\mathrm{FeCl}_2} = \frac{\Delta H_{m,\mathrm{FeCl}_2}}{T_{m,\mathrm{FeCl}_2}} = \frac{47{,}610}{969} = 49.13 \text{ joules/degree}$$

Thus the change in the slope between lines (iii) and (iv) at 969 K is 49.13 joules/degree, and the difference between the tangential intercepts of the slopes of the two lines at 969 K, with the $T = 0$ K axis, is 47,610 joules. Similarly $\Delta G^0_{(v)} - \Delta G^0_{(iv)}$ gives

$$\Delta G_{b,\mathrm{FeCl}_2} = 180{,}800 + 41.8T \ln T - 438.8T \text{ joules}$$

Thus $$\Delta H_{b,\mathrm{FeCl}_2} = -T^2 \left[\frac{\partial(\Delta G_{b,\mathrm{FeCl}_2}/T)}{\partial T}\right]$$

$$= 180{,}800 - 41.8T \text{ joules}$$

$$= 126{,}500 \text{ joules at } 1298 \text{ K}$$

and $$\Delta S_{b,\mathrm{FeCl}_2} = \frac{126{,}500}{1298} = 97.46 \text{ joules/degree at } 1298 \text{ K}$$

Thus the change in slope between lines (iv) and (v) at 1298 K is 97.46 joules/degree, and the difference between the tangential intercepts is 126,500 joules.

10.6 THE OXIDES OF CARBON

Carbon forms two gaseous oxides, CO and CO_2, according to

$$C_{(gr)} + O_{2(g)} = CO_{2(g)} \tag{i}$$

for which $\Delta G^0_{(i)} = -394{,}100 - 0.84T$ joules and

$$2C_{(gr)} + O_{2(g)} = 2CO_{(g)} \tag{ii}$$

for which $\Delta G^0_{(ii)} = -223{,}400 - 175.3T$ joules.

Combination of reactions (i) and (ii) gives

$$2CO_{(g)} + O_{2(g)} = 2CO_{2(g)} \tag{iii}$$

for which $\Delta G^0_{(iii)} = 2\Delta G^0_{(i)} - \Delta G^0_{(ii)} = -564{,}800 + 173.62T$ joules.

The Ellingham lines for reactions (i), (ii), and (iii) are included in Fig. 10.13, where it is seen that:

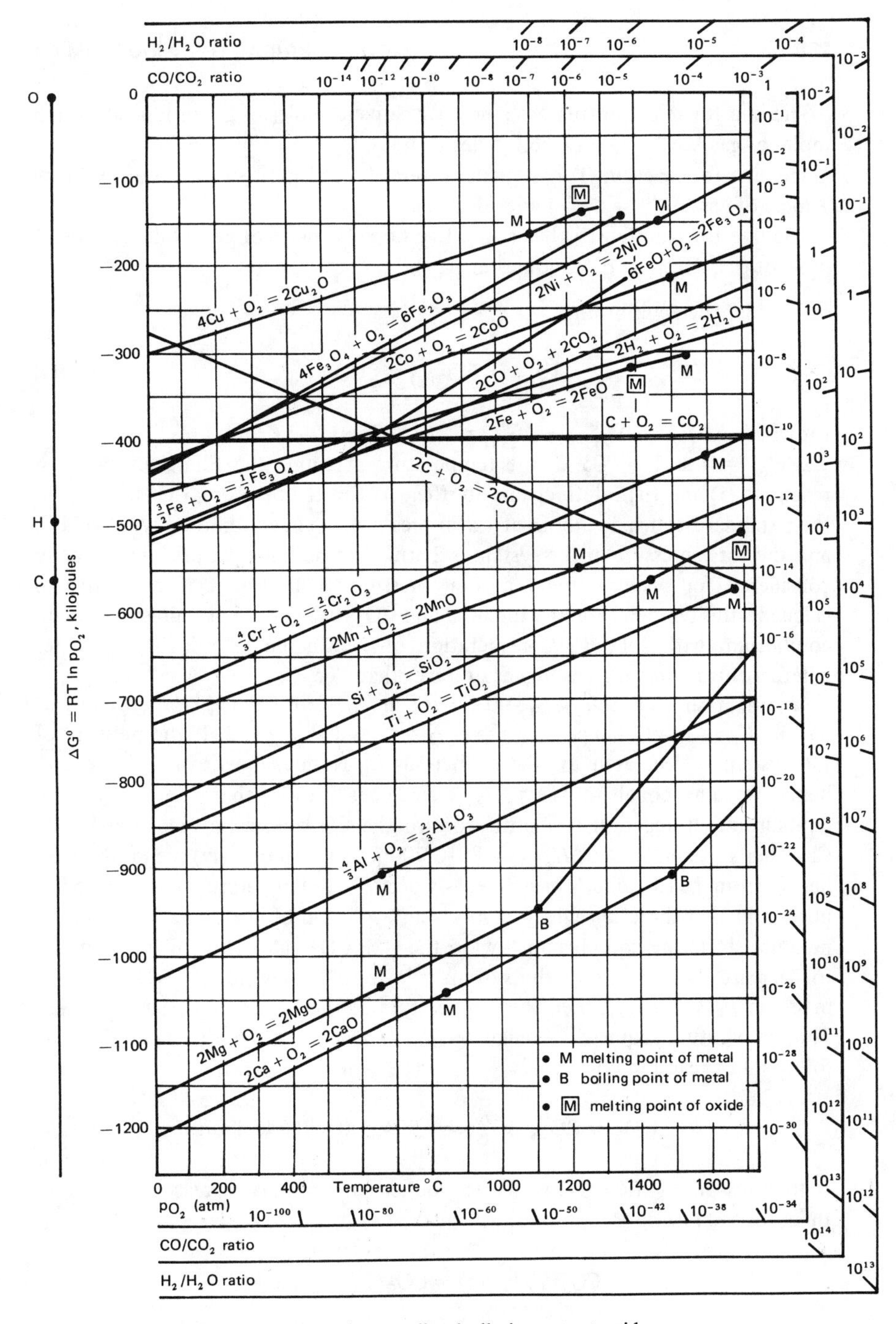

Fig. 10.13. The Ellingham diagram for metallurgically important oxides.

The line for reaction (iii) has a positive slope (2 moles of gas produced from 3 moles of gas, $\Delta S^0_{(iii)} = -173.62$ joules/degree).

The line for reaction (i) has virtually zero slope (1 mole of gas produced from 1 mole of gas, $\Delta S^0_{(i)} = 0.84$ joules/degree).

The line for reaction (ii) has a negative slope (2 moles of gas produced from 1 mole of gas, $\Delta S^0_{(ii)} = 175.3$ joules/degree).

Consider the equilibrium

$$C_{(gr)} + CO_{2(g)} = 2CO_{(g)} \tag{iv}$$

for which $\Delta G^0_{(iv)} = \Delta G^0_{(ii)} - \Delta G^0_{(i)} = 170{,}700 - 174.5T$ joules.

$\Delta G^0_{(iv)} = 0$ at $T = 705°C$, the temperature at which the Ellingham lines for reactions (i) and (ii) intersect one another. At this temperature, CO and CO_2, in their standard states, i.e., both at 1 atm pressure, are in equilibrium with solid C, and the total pressure of the system is 2 atm. As reaction equilibria are normally considered for systems under a total pressure of 1 atm, it is instructive to calculate the temperature at which CO and CO_2, each at 0.5 atm pressure, are in equilibrium with solid C. Consideration of Le Chatelier's principle indicates whether this temperature is higher or lower than 705°C.

For reaction (iv) $\Delta G^0_{(975\text{ K})} = 0 = -RT \ln K_p = -RT \ln (p^2_{CO}/p_{CO_2})$ i.e., at 978 K, $K_p = 1$, and hence $p_{CO_2} = p_{CO} = 1$, and $P_{total} = 2$. If the pressure of the system is decreased to 1 atm, then, as K_p is independent of pressure, and hence remains equal to unity, p_{CO} becomes greater than p_{CO_2}; i.e., the equilibrium of reaction (iv) shifts toward the CO side, as is predicted by Le Chatelier's principle. As $\Delta H^0_{(iv)} = +170{,}700$ joules, reaction (iv) is endothermic, hence, from Le Chatelier's principle, as a decrease in temperature shifts the equilibrium in that direction which involves an evolution of heat, a decrease in temperature shifts the equilibrium toward the C + CO_2 side. Thus, if it is required to decrease the pressure of the system from 2 to 1 atm and, at the same time, maintain $p_{CO} = p_{CO_2}$, then the temperature of the system must be decreased. The temperature required to make $p_{CO(eq)} = p_{CO_2(eq)} = 0.5$ atm is calculated as follows. For reaction (i)

$$C + O_2 = CO_2, \ \Delta G^0_{(i)} = -394{,}100 - 0.84T \text{ joules}$$

If the pressure of the CO_2, which is produced at 1 atm, is decreased to 0.5 atm, then, for the reaction

$$CO_2(T, P = 1) = CO_2(T, P = 0.5) \tag{v}$$

the free energy decrease is $\Delta G_{(v)} = RT \ln 0.5$, and hence for the reaction,

$$C_{(gr)} + O_{2(g,P=1)} = CO_{2(g,P=0.5)}$$

$$\Delta G_{(vi)} = \Delta G^0_{(i)} + \Delta G_{(v)}$$
$$= -394{,}100 - 0.84T + RT \ln 0.5 \text{ joules} \qquad \text{(vi)}$$

This line, on the Ellingham diagram, is obtained by rotating the line for reaction (i) clockwise about its point of intersection with the $T = 0$ axis until, at the temperature T, the vertical separation between line (i) and line (vi) is $RT \ln 0.5$. This is illustrated in Fig. 10.14. Similarly the Ellingham line for the reaction

$$2C_{(gr)} + O_{2(g,P=1)} = 2CO_{(g,P=0.5)} \qquad \text{(vii)}$$

is obtained as the sum of $\Delta G^0_{(ii)}$ and ΔG for the change

$$2CO(T, P = 1) = 2CO(T, P = 0.5)$$

i.e., $$\Delta G_{(vii)} = -223{,}400 - 175.3T + 2\,RT \ln 0.5 \text{ joules}$$

This line is obtained by rotating the line for reaction (ii) clockwise about its point of intersection with the $T = 0$ axis until, at any temperature T, the vertical separation between line (ii) and line (vii) is $2\,RT \ln 0.5$.

Combination of reactions (vi) and (vii) gives

$$C_{(gr)} + CO_{2(g,0.5\text{ atm})} = 2CO_{(g,0.5\text{ atm})} \qquad \text{(viii)}$$

for which $$\Delta G_{(viii)} = \Delta G^0_{(iv)} + RT \ln 0.5$$

Hence CO_2 and CO, each at 0.5 atm pressure, are in equilibrium with solid C at that temperature, which makes $\Delta G_{(viii)} = 0$, that is, at the temperature of intersection of the lines (vi) and (vii) in Fig. 10.14 (the point c).

The temperature at which CO (at 0.25 atm) and CO_2 (at 0.75 atm) are in equilibrium with solid C is obtained in a similar fashion as the intersection of line (i) rotated clockwise until, at T, it has been displaced a vertical distance $RT \ln 0.75$, and line (ii) rotated clockwise until, at T, it has been displaced a vertical distance $2\,RT \ln 0.25$. In Fig. 10.14 this is the point b. Similarly the point d in Fig. 10.14 is the temperature at which CO at 0.75 atm pressure and CO_2 at 0.25 atm pressure are in equilibrium with solid C. For a gas mixture of CO and CO_2 at 1 atm total pressure in equilibrium with solid C, the variation of p_{CO}/p_{CO_2} with T is given in Table 10.2; the variation of percent CO by volume in the gas with temperature is shown in Fig. 10.15. Figure 10.15 includes the points a, b, c, d, and e drawn in Fig. 10.14.

Figure 10.15 indicates that below 400°C the equilibrium gas is virtually CO_2

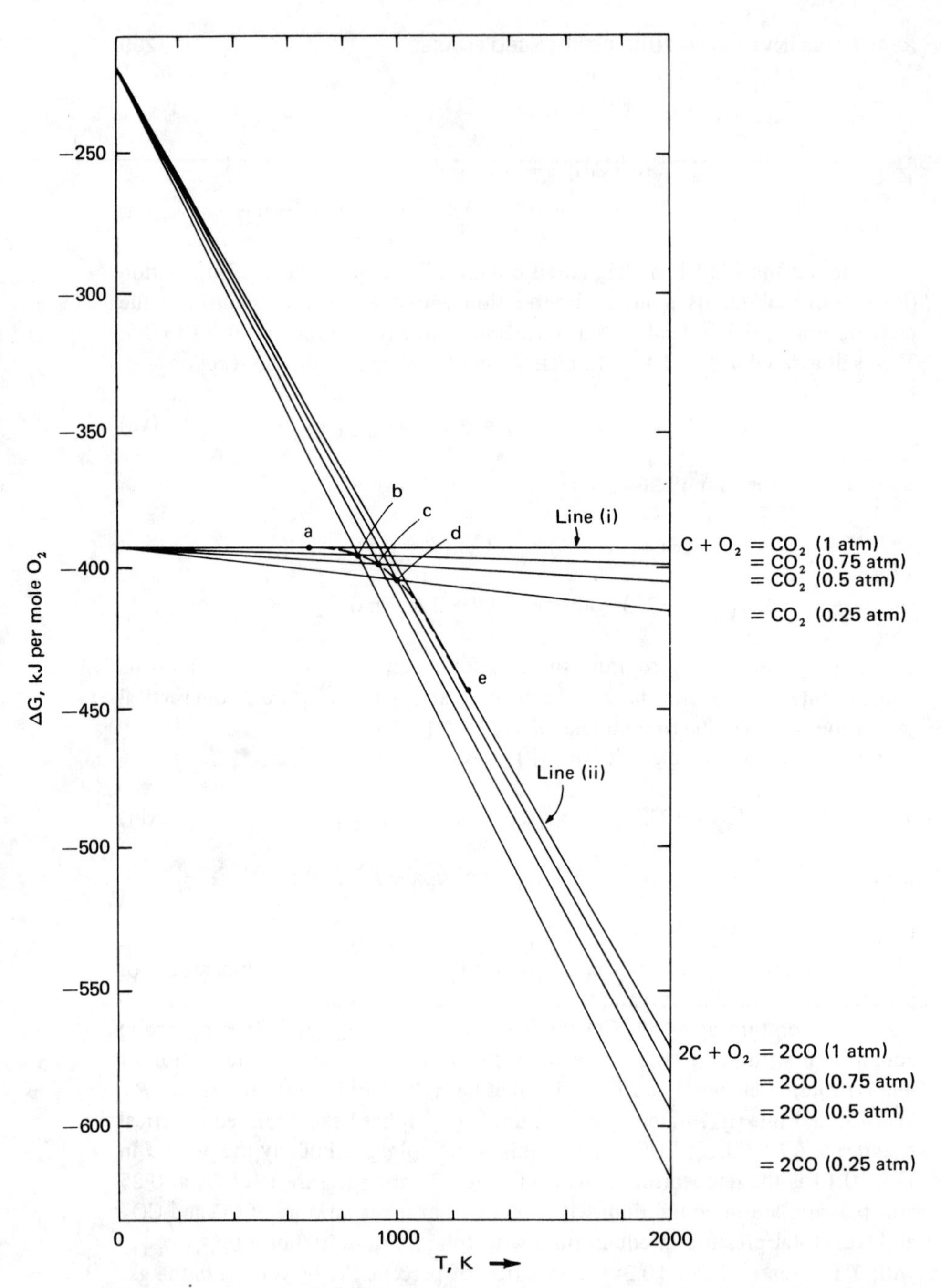

Fig. 10.14. The effect of varying the pressures of the product gases of the reactions $C_{(gr)} + O_2{}_{(g,P=1)} = CO_2{}_{(g)}$ and $2C_{(gr)} + O_2{}_{(g,P=1)} = 2CO_{(g)}$ on the ΔG-T variations for the two reactions.

Table 10.2

Temp °C	1893	1470	1185	979	820	674	526	406	317
p_{CO}/p_{CO_2}	$10^5/1$	$10^4/1$	$10^3/1$	$10^2/1$	10/1	1/1	1/10	$1/10^2$	$1/10^3$

at 1 atm pressure, and above 1000°C the equilibrium gas is virtually CO at 1 atm pressure. These points are respectively the points *a* and *e* in Fig. 10.14; hence the variation with temperature of the free energy of oxidation of solid C to produce a CO–CO_2 mixture at 1 atm pressure which is in equilibrium with solid C, is given by line (i) up to the point *a*, then by line *abcde*, and then by line (ii) beyond the point *e*.

It is to be realized that, at any temperature T, the CO/CO_2 mixture in equilibrium with C exerts an equilibrium oxygen pressure via the equilibrium

$$2CO + O_2 = 2CO_2$$

for which $$\Delta G^0_{(iii)} = -564{,}800 + 173.62T \text{ joules} = -RT \ln\left(\frac{p^2_{CO_2}}{p^2_{CO}p_{O_2}}\right)$$

$$= 2RT \ln\left(\frac{p_{CO}}{p_{CO_2}}\right)_{\text{eq with C}} + RT \ln p_{O_2(eq)}$$

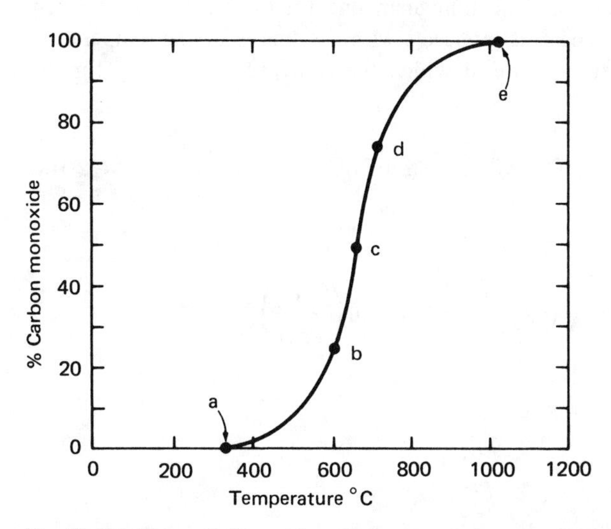

Fig. 10.15. The variation with temperature of the composition of the CO–CO_2 gas mixture in equilibrium with solid graphite at $P_{total} = 1$ atm.

Thus $$\ln p_{O_2(\text{eq},T)} = -\frac{564{,}800}{8.3144T} + \frac{173.62}{8.3144} + 2\ln\left(\frac{p_{CO_2}}{p_{CO}}\right)_{\text{eq with C}} \tag{10.12}$$

If it is required that solid carbon be used as a reducing agent to reduce the metal oxide MO_2 at the temperature T, then $p_{O_2(\text{eq},T)}$ in Eq. (10.12) must be lower than $p_{O_2(\text{eq},T)}$ for the equilibrium $M + O_2 = MO_2$ (see Sec. 10.7).

The Equilibrium $2CO + O_2 = 2CO_2$

The Ellingham line for the above reaction is shown in Fig. 10.16 as the line *cs*. Being the variation with T of the standard free energy of formation, ΔG^0, this line is for the reaction which produces CO_2 at 1 atm pressure from CO at 1 atm and O_2 at 1 atm. The effect of producing the CO_2 at any pressure P, other than 1 atm, (from CO and O_2 each at 1 atm) is the rotation of the Ellingham line *cs* about the point *c*, clockwise if $P < 1$ and anticlockwise if $P > 1$. For the given value of P this rotation is such that, as before, at the temperature T, the vertical displacement of *cs* is $2\,RT\ln P$. A series of lines, radiating from the point *c*, can thus be drawn for different pressures of CO_2 produced from CO and O_2, each at 1 atm. Figure 10.16 shows four of these lines, *cq* for CO_2 produced at 10^2 atm, *cr* for CO_2 produced at 10 atm, *cu* for CO_2 produced at 0.1 atm, and *cv* for CO_2 produced at 10^{-2} atm. The significance of this series of lines, with respect to the possibility of using CO-CO_2 gas mixtures as reducing agents for the metal oxide MO_2, is illustrated as follows. The Ellingham line for the reaction $M + O_2 = MO_2$ is drawn in Fig. 10.16. This intersects with the line *cs* at the temperature T_s, which is thus the temperature at which the standard free energy for the reaction

$$MO_2 + 2CO = M + 2CO_2 \tag{ix}$$

is zero, i.e.,

$$\Delta G^0_{(\text{ix})} \text{ at } T_s = 0 = -RT\ln\left(\frac{p_{CO_2}}{p_{CO}}\right)^2$$

and thus

$$\frac{p_{CO_2}}{p_{CO}} = 1$$

Above the temperature T_s a CO-CO_2 mixture of $p_{CO}/p_{CO_2} = 1$ is reducing with respect to MO_2, and below T_s it is oxidizing with respect to the metal M. If

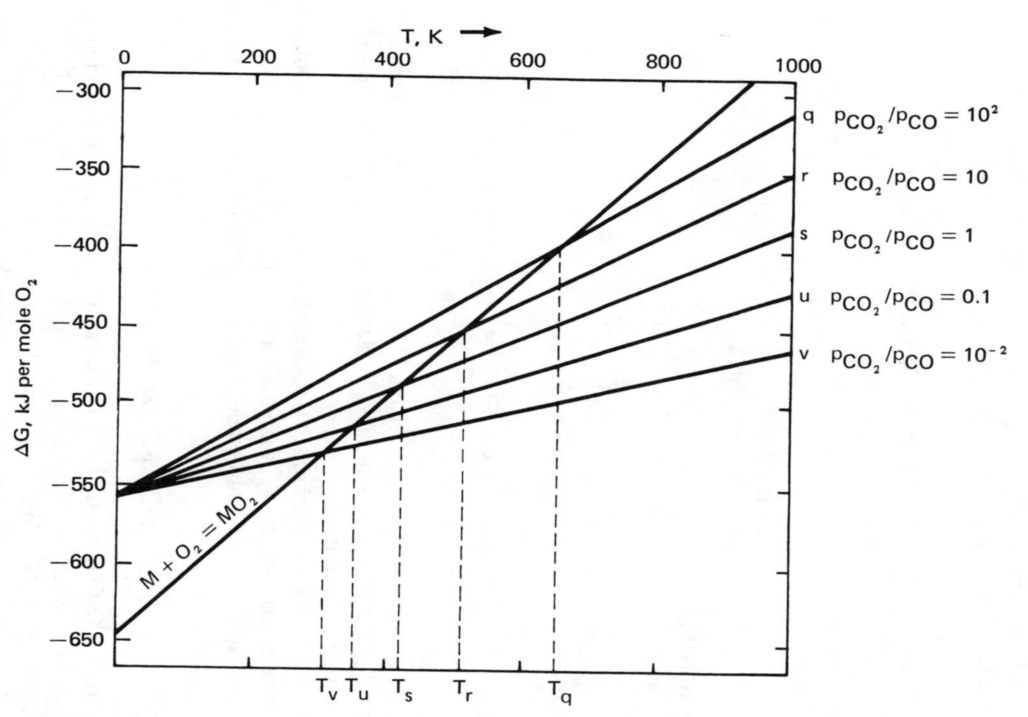

Fig. 10.16. Illustration of the effect of the p_{CO_2}/p_{CO} ratio in a CO–CO_2 gas mixture on the temperature at which the reaction equilibrium $M + CO_2 = MO + CO$ is established.

it is required that a CO-CO_2 mixture be made reducing with respect to MO_2 at temperatures lower than T_s, then it is obvious that the p_{CO_2}/p_{CO} ratio must be decreased below unity. The Ellingham line for the reaction $M + O_2 = MO_2$ intersects with the line *cu* at the temperature T_u, and T_u is thus the temperature at which the reaction

$$MO_2 + 2CO(1 \text{ atm}) = M + 2CO_2(0.1 \text{ atm}) \qquad \text{(x)}$$

is at equilibrium, i.e.,

$$\Delta G_{(x)} \text{ at } T_u = \Delta G^0_{(ix)} + 2\,RT \ln 0.1$$

But by definition,

$$\Delta G^0_{(ix)} = -RT \ln \left(\frac{p_{CO_2}}{p_{CO}}\right)^2_{eq}$$

and hence $$\Delta G_{(x)} = 0 = -2\,RT \ln \left(\frac{p_{CO_2}}{p_{CO}}\right)_{eq} + 2\,RT \ln 0.1$$

Thus $(p_{CO_2}/p_{CO})_{eq}$ at T_u equals 0.1, and hence by decreasing the temperature from T_s to T_u, the CO_2/CO ratio must be decreased from 1 to 0.1 in order to maintain reaction equilibrium.

Similarly at T_v the equilibrium CO_2/CO ratio is 0.01, at T_r the equilibrium CO_2/CO ratio is 10, and at T_q the equilibrium CO_2/CO ratio is 100. Thus a CO/CO_2 nomographic scale can be added to the Ellingham diagram and for any reaction

$$MO_2 + 2CO = M + 2CO_2$$

the equilibrium CO/CO_2 ratio at any temperature T is read off the nomographic scale as that value which is collinear with the point c and the point ΔG^0_T, $T = T$, for the reaction $M + O_2 = MO_2$. This scale is drawn in Fig. 10.13.

Figure 10.15 is generated by reading off the equilibrium CO/CO_2 ratios for the reaction $C + O_2 = CO_2$ up to the point *a* in Fig. 10.14, then the CO/CO_2 ratios along the line *abcde*, and finally the CO/CO_2 ratios for the reaction $2C + O_2 = 2CO$ beyond the point *e*.

In an exactly similar fashion the H_2/H_2O nomographic scale is added to Fig. 10.13 by considering the effect of variation of the H_2O pressure on the reaction equilibrium

$$2H_2 + O_2 = 2H_2O$$

The equilibrium H_2/H_2O ratio at the temperature T for the reaction

$$MO_2 + 2H_2 = M + 2H_2O$$

is read off the H_2/H_2O scale as that value which is collinear with the points H and $\Delta G^0_T, T = T$, for the reaction $M + O_2 = MO_2$.

10.7 GRAPHICAL REPRESENTATION OF EQUILIBRIA IN THE SYSTEM METAL-OXYGEN-CARBON

The main criteria for graphical representation of equilibria within a system are (1) amount of information provided and (2) clarity. Both of these considerations are largely dependent on the particular coordinates chosen for use in the graphical representation. As the nomographic scale for the ratio CO/CO_2 in Fig. 10.13 shows that the range of interest of p_{CO_2}/p_{CO} values is 10^{-14} to 10^8, it is obviously desirable to express this ratio on a logarithmic scale. Figure 10.17, which uses the coordinates $\log(p_{CO_2}/p_{CO})$ and T, represents a convenient method of clearly presenting equilibrium data in the carbon-oxygen and carbon-oxygen-metal systems. From Eq. (iii), for

$$2CO_{(g)} + O_{2\,(g)} = 2CO_{2\,(g)}$$

$$\Delta G^0_{(iii)} = -564{,}800 + 173.62T \text{ joules}$$

$$= -RT \ln\left(\frac{p^2_{CO_2}}{p^2_{CO}p_{O_2}}\right)$$

and thus

$$\log\left(\frac{p_{CO_2}}{p_{CO}}\right) = \frac{1}{2}\log p_{O_2} + \frac{564{,}800}{2 \times 2.303 \times 8.3144T} - \frac{173.62}{2 \times 2.303 \times 8.3144} \qquad \text{(xi)}$$

and for any given value of p_{O_2} this expresses the variation of this given oxygen isobar with $\log(p_{CO_2}/p_{CO})$ and temperature. The oxygen isobars in the range 10^{-29} to 10^{-4} are drawn as a function of $\log(p_{CO_2}/p_{CO})$ and T in Fig. 10.17.

The equilibrium

$$C_{(s)} + CO_{2\,(g)} = 2CO_{(g)} \qquad \text{(iv)}$$

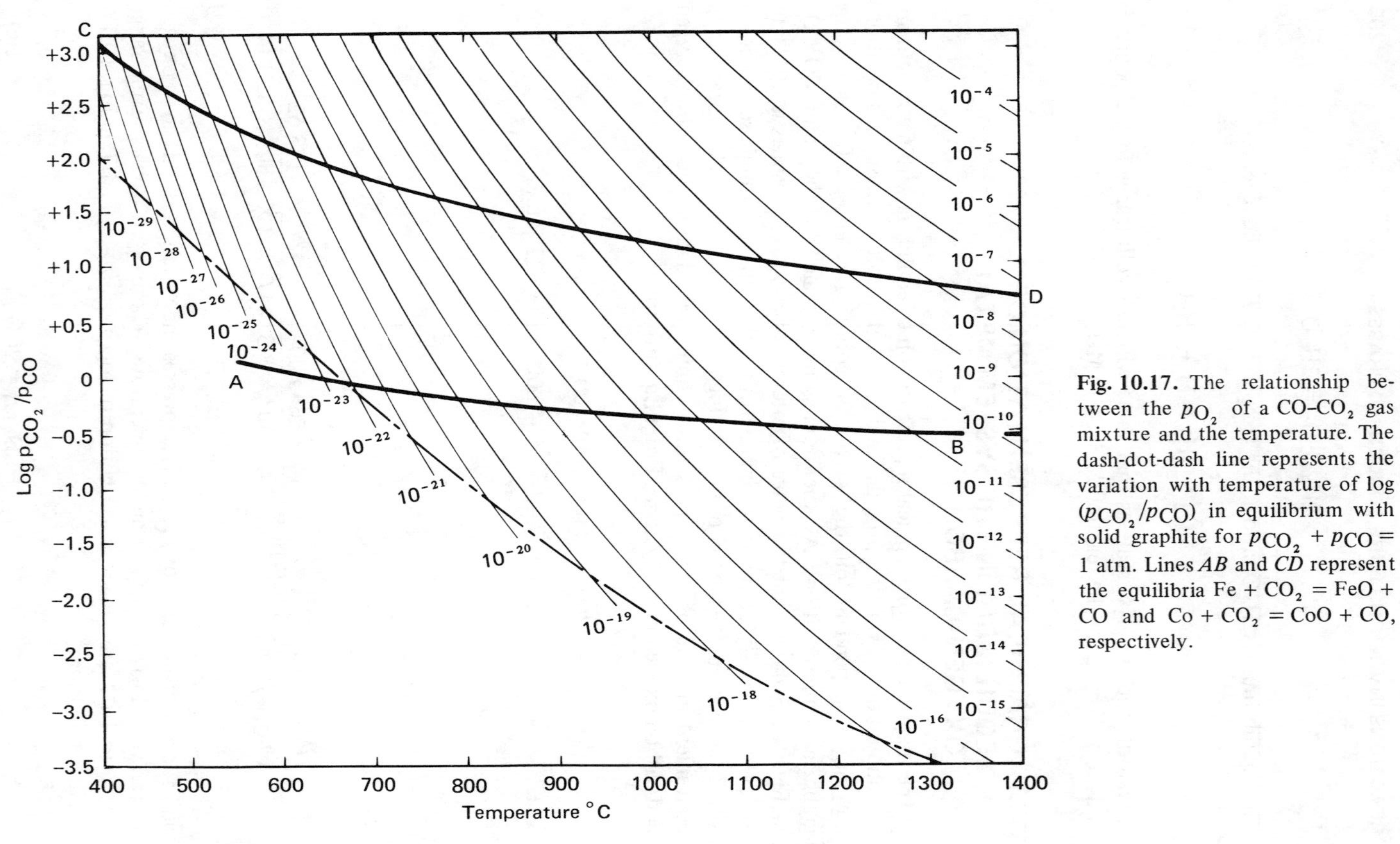

Fig. 10.17. The relationship between the p_{O_2} of a CO–CO_2 gas mixture and the temperature. The dash-dot-dash line represents the variation with temperature of log (p_{CO_2}/p_{CO}) in equilibrium with solid graphite for $p_{CO_2} + p_{CO} =$ 1 atm. Lines *AB* and *CD* represent the equilibria Fe + CO_2 = FeO + CO and Co + CO_2 = CoO + CO, respectively.

sets a lower limit on the CO_2/CO ratio which can be obtained at any temperature.

$$\Delta G^0_{(iv)} = 170{,}700 - 174.5T \text{ joules} = -RT \ln\left(\frac{p^2_{CO}}{p_{CO_2}}\right)$$

and hence, at 1 atm total pressure, the minimum obtainable ratio of p_{CO_2}/p_{CO} at the temperature T is

$$\frac{2 + x - \sqrt{(2+x)^2 - 4}}{\sqrt{(2+x)^2 - 4} - x} \qquad \text{(xii)}$$

where $x = \exp(-170{,}700/8.3144T)\exp(174.5/8.3144)$.

If an attempt is made to mix CO and CO_2 (at $P_{total} = 1$) in a ratio lower than (xii), carbon will precipitate until, thereby, the ratio has been increased to its unique value for equilibrium with carbon at the temperature T. The variation of $\log(p_{CO_2}/p_{CO})_{eq,C/CO/CO_2}$ with temperature is shown as the dash-dot line in Fig. 10.17. This ratio, via equilibrium (iii), necessarily sets a lower limit to the oxygen pressure which can be obtained in a CO–CO_2 gas mixture at 1 atm total pressure. The variation of this minimum value of p_{O_2} with temperature is illustrated in Fig. 10.17 by the intersections of the oxygen isobars with the carbon deposition line.

Equilibria such as

$$MO + CO = M + CO_2$$

can readily be represented on such plots as Fig. 10.17. For example, for

$$FeO_{(s)} + CO_{(g)} = Fe_{(s)} + CO_{2\,(g)}$$

$\Delta G^0 = -22{,}800 + 24.26T$ joules and hence the variation of the equilibrium CO_2/CO ratio is given as

$$\log\left(\frac{p_{CO_2}}{p_{CO}}\right)_{eq,FeO/Fe} = \frac{22{,}800}{2.303 \times 8.3144T} - \frac{24.26}{2.303 \times 8.3144}$$

This variation is drawn in Fig. 10.17 as the line AB, and thus any gas, the state of which lies above the line AB, is oxidizing with respect to Fe, and states below AB are reducing with respect to FeO. The variation of $\log p_{O_2(eq,T,Fe/FeO)}$

with temperature is given by the intersections of the oxygen isobars with *AB*. The temperature at which *AB* intersects with the carbon deposition line is the minimum temperature at which solid FeO can be reduced to solid Fe by solid C. Strictly, this is the temperature at which $Fe_{(s)}$, $FeO_{(s)}$, $C_{(s)}$, and the CO-CO_2 atmosphere at unit pressure coexist in equilibrium, i.e., is the temperature at which

$$p_{O_2(\text{eq,C/CO/CO}_2)} = p_{O_2(\text{eq,Fe/FeO})}$$

The line *AB* stops at 560°C, below which temperature FeO (wustite) dissociates into Fe and Fe_3O_4 (magnetite) (see Chap. 13 and Prob. 10.12).

In Fig. 10.17, line *CD* represents the variation of $\log(p_{CO_2}/p_{CO})$ with *T* for the equilibrium

$$CoO_{(s)} + CO_{(g)} = Co_{(s)} + CO_{2\,(g)}$$

for which $$\Delta G^0 = -48{,}500 + 14.9T \text{ joules}$$

and hence

$$\log\left(\frac{p_{CO_2}}{p_{CO}}\right)_{\text{eq,CoO/Co}} = \frac{48{,}500}{2.303 \times 8.3144T} - \frac{14.9}{2.303 \times 8.3144}$$

As, for the equilibrium

$$MO + CO = M + CO_2$$

the equilibrium constant *K* is given by $p_{CO_2}/p_{CO(\text{eq},T,\text{M/MO})}$, a plot of $\log(p_{CO_2}/p_{CO})$ versus $1/T$ is necessarily a plot of log *K* versus $1/T$. Figure 10.18 shows the information given in Fig. 10.17 as such a plot. With respect to the amount of information which can conveniently be obtained from a graphical representation of equilibria within a system, Fig. 10.18 is a better representation than is Fig. 10.17. As

$$\left[\frac{\partial \ln K}{\partial\left(\frac{1}{T}\right)}\right] = -\frac{\Delta H^0}{R}$$

the tangential slope of an equilibrium line at the temperature *T* gives the value of $-\Delta H^0/R$. In the case that ΔH^0 for the reaction is not a function of temperature (that is, $\Delta c_p = 0$) linear variations of log *K* with $1/T$ occur. Thus the slope of the

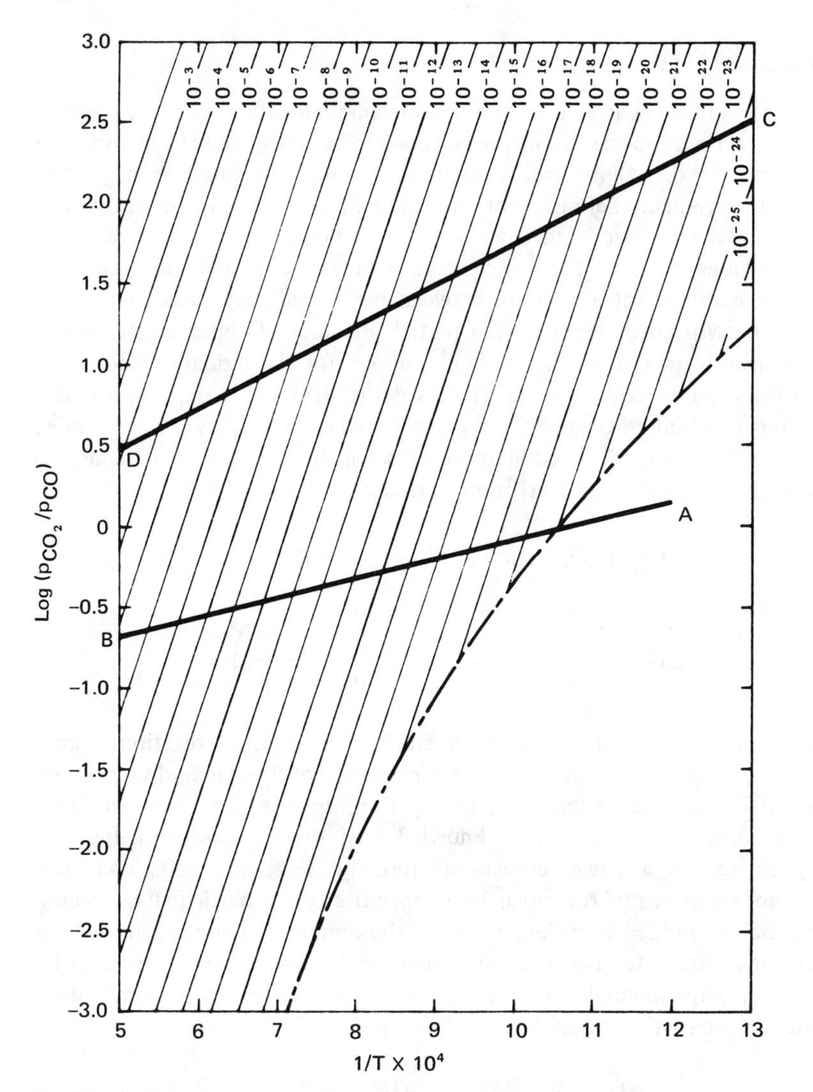

Fig. 10.18. Figure 10.17 reproduced as $\log(p_{CO_2}/p_{CO})$ versus $1/T$.

line AB in Fig. 10.18 equals $-\Delta H^0/R$ for the reaction $FeO + CO = Fe + CO_2$, and similarly the slope of the line CD equals $-\Delta H^0/R$ for the reaction $CoO + CO = Co + CO_2$. Also the intercepts of the tangential slopes, or the lines themselves if they are linear, with the $1/T = 0$ axis give the respective values of $\Delta S^0/R$ for the particular reactions. From equation (xi) the slope of any oxygen isobar equals $-\Delta H^0/2R$ for the reaction $2CO + O_2 = 2CO_2$, and hence the oxygen isobars in Fig. 10.18 are parallel lines.

10.8 SUMMARY

1. The facts that (*a*) at a given temperature a unique saturated vapor pressure is exerted by a pure species occurring in a condensed phase, and (*b*) the pressure dependencey of the Gibbs free energy of a condensed phase is negligible, facilitate a convenient definition of the standard state of pure species which occur as condensed phases. This standard state is simply the pure species in its stable condensed state at the temperature T. Using this standard state, the equilibrium constant for a reaction between pure condensed phases and a gas phase can be written in terms of the partial pressures of those species which occur exclusively in the gas phase. For example, for the oxidation of a pure metal to its pure oxide (where the oxide is of fixed composition), the equilibrium constant is given by $1/p_{O_2(\text{eq},T)}$, where $p_{O_2(\text{eq},T)}$ is the unique oxygen pressure required for equilibrium between the metal, its oxide, and the gas phase at the temperature T. This pressure is such that

$$G^0_{\text{M}} + G^0_{O_2} + RT \ln p_{O_2(\text{eq},T)} = G^0_{\text{MO}_2}$$

or

$$\Delta G^0_T = -RT \ln K = -RT \ln\left(\frac{1}{p_{O_2(\text{eq},T)}}\right)$$

2. Determination of the equilibrium state of a chemical reaction system requires a knowledge of the temperature dependency of the standard free energy change, ΔG^0_T, of the reaction. This relationship can be obtained from thermochemical data, i.e., from a knowledge of the standard enthalpy and entropy changes at a single temperature (usually ΔH^0_{298} and ΔS^0_{298}) and the temperature variations of the molar heat capacities of the participating species; or it can be determined from knowledge of the temperature dependency of the equilibrium constant for the reaction. For the oxidation of a pure metal to its pure oxide, experimental measurement of $1/p_{O_2(\text{eq},T)}$ with temperature facilitates calculation of the ΔG^0_T-T relationship via

$$\ln\left(\frac{1}{p_{O_2(\text{eq},T)}}\right) = \ln K = -\frac{\Delta H^0}{RT} + \frac{\Delta S^0}{R} = -\frac{\Delta G^0}{RT}$$

If ΔH^0 and ΔS^0 are independent of T (i.e., if $\Delta c_p = 0$), then a linear variation of $\ln K$ with $1/T$ occurs.

3. The fitting of the variation of ΔG^0_T with T to a straight line produces the so-called Ellingham line for the particular reaction, and a diagram of ΔG^0_T versus T for a reaction is known as an Ellingham diagram. Ellingham lines plotted on a single diagram for a series of similar reactions, e.g., for the formation of oxides,

sulfides, etc., provide a convenient representation of the relative stabilities of the compounds. The addition of p_{O_2}, p_{CO}/p_{CO_2}, and p_{H_2O}/p_{H_2} nomographic scales to these diagrams facilitates geometric determination of $p_{O_2(eq,T)}$, $p_{CO}/p_{CO_2(eq,T)}$, and $p_{H_2O}/p_{H_2(eq,T)}$ in the equilibria $2M + O_2 = 2MO$, $M + CO_2 = MO + CO$, and $M + H_2O = MO + H_2$ respectively.

10.9 NUMERICAL EXAMPLES

Example 1

Iron has face-centered-cubic (the γ phase) and body-centered-cubic (the α and δ phases) allotropes. At 1400°C the γ phase transforms to the δ phase with a heat of transformation of 880 joules/mole, and at 1536°C the δ phase melts with a heat of fusion of 13,800 joules/mole. The heat capacities of the three phases are

$c_{p(\mathrm{Fe},\gamma)} = 7.70 + 19.5 \times 10^{-3}T$ joules/degree·mole in the temperature range 1183 to 1673 K

$c_{p(\mathrm{Fe},\delta)} = 43.9$ joules/degree·mole in the temperature range 1673 to 1809 K

$c_{p(\mathrm{Fe},L)} = 41.8$ joules/degree·mole in the temperature range 1809 to 1873 K

Calculate the hypothetical melting temperature of the γ phase and the heat and entropy of fusion of the γ phase.

The hypothetical melting temperature of the γ phase is that temperature at which $G_\gamma = G_L$, or $\Delta G_{(\gamma \to L)} = 0$. As $S_\gamma < S_\delta < S_L$, the G-T relationships for the three phases are as illustrated schematically in Fig. 10.19 (see Fig. 7.11*a*).

The lines for the δ and liquid phases intersect at 1809 K, at which temperature $G_L = G_\delta$; the lines for the γ and δ phases intersect at 1673 K, at which temperature $G_\gamma = G_\delta$; and the lines for the γ and liquid phases intersect at some intermediate temperature at which $G_\gamma = G_L$.

As the relative free energy of the γ and liquid phases, in equilibrium with one another, is higher than the relative free energy of the δ phase at the same temperature, the γ and liquid phases are metastable at this temperature, and so the phase equilibrium is hypothetical.

At the hypothetical melting temperature of the γ phase

$$\Delta G_{(\gamma\to L)} = \Delta G_{(\gamma\to\delta)} + \Delta G_{(\delta\to L)} = 0$$

$\Delta G_{(\gamma\to\delta)}$ and $\Delta G_{(\delta\to L)}$ are calculated as follows. For the phase change $\gamma \to \delta$,

$$\Delta c_p = c_{p(\delta)} - c_{p(\gamma)} = 36.2 - 19.5 \times 10^{-3}T \text{ joules/degree·mole} = \left(\frac{d\Delta H}{dT}\right)_P$$

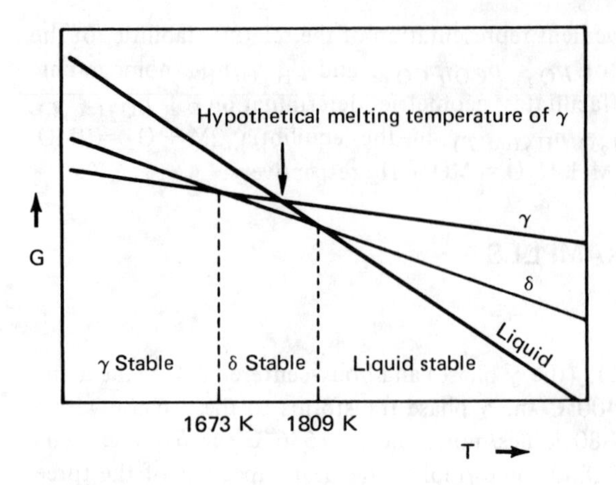

Fig. 10.19.

where ΔH is the latent heat of the phase change. Integrating gives

$$\Delta H_T = \Delta H_0 + 36.2T - 9.75 \times 10^{-3} T^2 \text{ joules}$$

where ΔH_0 is the integration constant.

At 1400°C (1673 K) the latent heat of the phase change is 880 joules.

Thus $$\Delta H_0 = 880 - (36.2 \times 1673) + (9.75 \times 10^{-3} \times 1673^2)$$
$$= -32{,}393 \text{ joules}$$

and hence $$\Delta H_{(\gamma \to \delta)} = -32{,}393 + 36.2T - 9.75 \times 10^{-3} T^2 \text{ joules}$$

The Gibbs-Helmholtz equation is

$$\left[\frac{d(\Delta G/T)}{dT}\right]_P = -\frac{\Delta H}{T^2}$$

Therefore

$$\left[\frac{d(\Delta G/T)}{dT}\right]_P = -\frac{\Delta H}{T^2} = \frac{32{,}393}{T^2} - \frac{36.2}{T} + 9.75 \times 10^{-3}$$

Integrating and multiplying through by T gives

$$\Delta G_{(\gamma \to \delta)} = IT - 32{,}393 - 36.2T \ln T + 9.75 \times 10^{-3} T^2 \text{ joules}$$

and as the γ and δ phases are in equilibrium at 1673 K, that is, $\Delta G_{1673} = 0$, then

$$I = +\frac{32{,}393}{1673} + (36.2 \times \ln 1673) - (9.75 \times 10^{-3} \times 1673)$$

$$= +271.7$$

such that

$$\Delta G_{(\gamma \to \delta)} = 271.7T - 32{,}393 - 36.2T \ln T + 9.75 \times 10^{-3} T^2 \text{ joules}$$

For the phase change $Fe_{(\delta)} \to Fe_{(L)}$,

$$\Delta c_p = -2.1 \text{ joules/degree·mole}$$

Thus $$\Delta H_{(\delta \to L)} = \Delta H_0 - 2.1T \text{ joules}$$

and as the heat of fusion of the δ phase at the melting temperature of 1536°C (1809 K) is 13,800 joules/mole

$$\Delta H_0 = 13{,}800 + (2.1 \times 1809) = 17{,}599 \text{ joules}$$

such that $$\Delta H_{(\delta \to L)} = 17{,}599 - 2.1T \text{ joules}$$

Again, from the Gibbs-Helmholtz equation,

$$\left[\frac{d(\Delta G/T)}{dT}\right]_P = -\frac{\Delta H}{T^2} = -\frac{17{,}599}{T^2} + \frac{2.1}{T}$$

integration of which gives

$$\frac{\Delta G}{T} = I + \frac{17{,}599}{T} + 2.1 \ln T$$

or $$\Delta G_{(\delta \to L)} = IT + 17{,}599 + 2.1T \ln T \text{ joules}$$

As δ and liquid iron are in equilibrium at 1809 K, then

$$I = -\frac{17{,}599}{1809} - 2.1 \ln 1809$$

$$= -25.48$$

such that $\Delta G_{(\delta \to L)} = -25.48T + 17{,}599 + 2.1T \ln T$ joules

Now, at the hypothetical melting temperature of the γ phase,

$$\Delta G_{(\gamma \to \delta)} + \Delta G_{(\delta \to L)} = 0$$

i.e., $$246.22T - 14{,}794 - 34.1T \ln T + 9.75 \times 10^{-3} T^2 = 0$$

Plots of $(14{,}797 - 246.22T)$ and $(9.75 \times 10^{-3}T^2 - 34.1T \ln T)$ against T intersect at $T = 1790$ K (1517°C) which is thus the hypothetical melting temperature. (The fact that this temperature must be between 1673 K and 1809 K decreases the amount of arithmetic necessary in the graphical solution.)

For the phase change $\gamma \to L$,

$$\left(\frac{d\Delta G}{dT}\right)_P = -\Delta S = 246.22 - 34.1 \ln T - 34.1 + 2 \times 9.75 \times 10^{-3} T$$

$$= -8.38 \text{ joules/degree at 1790 K}$$

$$= \text{minus the entropy of fusion of } \gamma \text{ at 1790}$$

and $$\Delta H = T_m \, \Delta S = 1790 \times 8.38 = 15{,}000 \text{ joules/mole}$$

Alternatively,

$$\Delta H_{(\gamma \to L)} = \Delta H_{(\gamma \to \delta)} + \Delta H_{(\delta \to L)}$$

$$= -14{,}794 + 34.1T - 9.75 \times 10^{-3} T^2$$

$$= 15{,}000 \text{ joules/mole at 1790 K}$$

Example 2

If FeO is to be reduced to Fe by solid carbon at 600°C, what is the maximum pressure which can be tolerated in the reaction system? Given:

$$FeO_{(s)} = Fe_{(s)} + \tfrac{1}{2}O_{2(g)} \qquad \Delta G^0 = 259{,}600 - 62.55T \text{ joules}$$

$$C_{(s)} + O_{2(g)} = CO_{2(g)} \qquad \Delta G^0 = -394{,}100 - 0.84T \text{ joules}$$

$$2C_{(s)} + O_{2(g)} = 2CO_{(g)} \qquad \Delta G^0 = -223{,}400 - 175.3T \text{ joules}$$

$$2CO_{(g)} + O_{2(g)} = 2CO_{2(g)} \qquad \Delta G^0 = -564{,}800 + 173.6T \text{ joules}$$

$$C_{(s)} + CO_{2(g)} = 2CO_{(g)} \qquad \Delta G^0 = +170{,}700 - 174.5T \text{ joules}$$

For the reduction of FeO

$$FeO = Fe + \tfrac{1}{2}O_2 \quad \Delta G^0 = 259{,}600 - 62.55T \text{ joules} = -RT \ln K_p$$

Thus $$\Delta G^0_{873} = 204{,}994 \text{ joules} = -8.3144 \times 873 \ln K_p$$

But $$K_p = p^{1/2}_{O_2(eq,T)}$$

and so $$p_{O_2(eq,873)} = 2.9 \times 10^{-25} \text{ atm}$$

This solution, as point A, is illustrated in the Ellingham diagram in Fig. 10.20. Thus in order to reduce FeO at 600°C, the oxygen pressure in the system must be less than 2.9×10^{-25} atm.

If it is considered that C is reducing FeO to form Fe and CO, then the pertinent reaction is

$$FeO + C = Fe + CO$$

for which $$\Delta G^0 = 147{,}900 - 150.2T \text{ joules} = -RT \ln K_p$$

where $$K_p = p_{CO(eq,T)}$$

Thus $$\Delta G^0_{873} = 16{,}775 \text{ joules} = -8.3144 \times 873 \ln p_{CO(eq,873)}$$

and so $$p_{CO(eq,873)} = 0.1 \text{ atm}$$

Thus in order that the reduction of FeO proceed spontaneously, the CO pressure in the system must be less than 0.1 atm. If the CO is in equilibrium with the solid carbon present, then a CO_2 pressure is exerted via the equilibrium

$$C + CO_2 = 2CO$$

for which $$\Delta G^0 = 170{,}700 - 174.5T \text{ joules} = -RT \ln K_p$$

where $$K_p = \left(\frac{p^2_{CO}}{p_{CO_2}}\right)_{eq,T}$$

Thus $$\Delta G^0_{873} = 18{,}362 \text{ joules} = -8.3144 \times 873 \ln K_p$$

Thus $$\left(\frac{p^2_{CO}}{p_{CO_2}}\right)_{eq,873} = 0.08$$

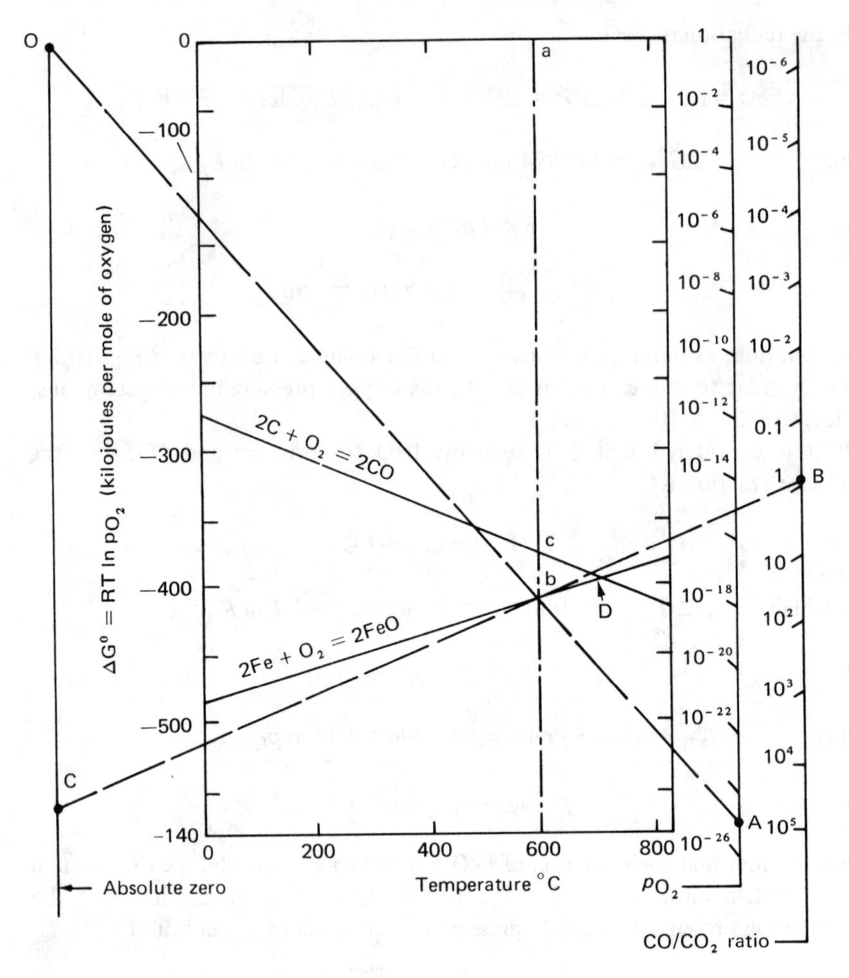

Fig. 10.20.

which, with $p_{CO} = 0.1$ atm, gives

$$p_{CO_2(eq,\,873)} = 0.125 \text{ atm}$$

Thus if the reduction reaction is considered to be

$$FeO + \tfrac{1}{2}C = Fe + \tfrac{1}{2}CO_2$$

then the CO_2 pressure must be lower than 0.125 atm. For this reaction

$$\Delta G^0 = 62{,}550 - 62.97T \text{ joules} = -RT \ln K_p$$

where $$K_p = p^{1/2}_{CO_2(eq,T)}$$

Thus $$\Delta G^0_{873} = 7577 \text{ joules} = -8.3144 \times 873 \ln p^{1/2}_{CO_2(eq,873)}$$

from which $$p_{CO_2(eq,873)} = 0.125 \text{ atm}$$

in agreement with the above.

The CO and CO_2 (which are both in equilibrium with the solid C) give rise to an oxygen pressure via the equilibrium

$$2CO + O_2 = 2CO_2$$

for which $$\Delta G^0 = -564{,}800 + 173.62T \text{ joules} = -RT \ln K_p$$

where $$K_p = \left(\frac{p^2_{CO_2}}{p^2_{CO} p_{O_2}}\right)_{eq,T}$$

$$\Delta G^0_{873} = -413{,}230 \text{ joules}$$

and hence $$\left(\frac{p^2_{CO_2}}{p^2_{CO} p_{O_2}}\right) = 5.305 \times 10^{24}$$

Thus, with $p_{CO_2} = 0.125$ atm and $p_{CO} = 0.1$ atm,

$$p_{O_2(eq,873)} = 2.9 \times 10^{-25} \text{ atm}$$

in agreement with the above.

Thus, at 873 K, for FeO, Fe, and C to be in equilibrium with a gaseous atmosphere, the atmosphere must be of composition

$$p_{CO} = 0.1$$
$$p_{CO_2} = 0.125 \text{ atm}$$
$$p_{O_2} = 2.9 \times 10^{-25} \text{ atm}$$

and if the total pressure of the gas is less than $0.1 + 0.125 = 0.225$ atm, then spontaneous reduction of FeO will occur.

At equilibrium at 873 K, the p_{CO}/p_{CO_2} ratio in the gas is $0.1/0.125 = 0.8$, as is illustrated as the point B in Fig. 10.20.

As the temperature of the system is varied, the equilibrium oxygen pressure for the Fe-FeO-C-CO-CO_2 equilibrium varies as

$$p_{O_2(\text{eq}, T)} = 10^{(2 \times 62.55/8.3144 \times 2.303)} \cdot 10^{-(2 \times 259{,}600/2.303 \times 8.3144T)}$$

the equilibrium CO pressure varies as

$$p_{O_2(\text{eq}, T)} = 10^{(150.2/2.303 \times 8.3144)} \cdot 10^{-(147{,}900/2.303 \times 8.3144T)}$$

and the equilibrium CO_2 pressure varies as

$$p_{CO_2(\text{eq}, T)} = 10^{(2 \times 62.97/2.303 \times 8.3144)} \cdot 10^{-(2 \times 62{,}550/2.303 \times 8.3144T)}$$

Thus as the temperature is decreased, the pressure, above which spontaneous reduction of FeO will not occur, is decreased.

$$\left(\frac{p_{CO}}{p_{CO_2}}\right)_{\text{eq}, T} = 10^{(24.25/2.303 \times 8.3144)} \cdot 10^{-(22{,}800/2.303 \times 8.3144T)}$$

decreases with decreasing temperature as indicated by the variation of the point B with temperature.

Alternatively

The standard free energy lines for the reactions $2Fe + O_2 = 2FeO$ and $2C + O_2 = 2CO$ intersect at that temperature at which the standard free energy change for the reaction $2FeO + 2C = 2Fe + 2CO$ is zero. For this reaction $\Delta G^0 = 295{,}800 - 300.4T$ joules, and ΔG^0 is zero at 985 K (710°C), which is the point D in Fig. 10.20. At this temperature the CO pressure for equilibrium in the system FeO-Fe-C-CO is 1 atm (the standard state). A decrease in the pressure at which the CO is produced rotates the standard free energy line for the reaction $2C + O_2 = 2CO$ in a clockwise direction; as a result the temperature of its intersection with the standard free energy line for the reaction $2Fe + O_2 = 2FeO$ decreases. This rotation is such that, at any temperature T, the vertical displacement from the standard line is $RT \ln p^2_{CO}$. The problem thus requires calculation of that pressure of CO which rotates the standard line to the extent that it intersects with the $2Fe + O_2 = 2FeO$ line at 873 K.

At 873 K, the standard free energy change for $2Fe + O_2 = 2FeO$ is

$$\Delta G^0_{873} = -410{,}000 \text{ joules} = ab$$

$$= \text{the standard free energy change for } 2C + O_2 = 2CO$$

$$+ \text{ the increment } RT \ln p^2_{CO}$$

$$= \Delta G^0_{873} + 8.3144 \times 873 \ln p^2_{CO}$$

$$= (-223{,}400 - 175.3 \times 873) + 8.3144 \times 873 \ln p^2_{CO}$$

$$= ac + cb$$

Thus

$$\ln p_{CO} = \frac{-409{,}988 + 223{,}400 + (175.3 \times 873)}{2 \times 8.3144 \times 873}$$

$$= -2.31$$

and

$$p_{CO(\text{eq}, 873)} = 0.1 \text{ atm.}$$

Example 3

What is the equilibrium state of a CO-CO_2-H_2-H_2O gas mixture produced by mixing CO_2 and H_2 in the molar ratio 1:1 at 1000 K and a total pressure of 1 atm?

The reaction which occurs is

$$CO_2 + H_2 = CO + H_2O \qquad \text{(i)}$$

As the molar ratio of CO_2 to H_2 in the initial mixture is 1:1 and $P = 1$ atm, then, before reaction begins, $p_{CO_2} = p_{H_2} = 0.5$ atm. From the stoichiometry of the reaction, at any time in the reacting mixture, $p_{CO_2} = p_{H_2}$ and $p_{CO} = p_{H_2O}$. At equilibrium,

$$K_{p,(i)} = \frac{p_{CO}p_{H_2O}}{p_{CO_2}p_{H_2}} \qquad \text{(ii)}$$

Now,

$$P = p_{CO_2} + p_{CO} + p_{H_2O} + p_{H_2} = 1$$

but, as

$$p_{CO_2} = p_{H_2} \quad \text{and} \quad p_{CO} = p_{H_2O}$$

then

$$P = 1 \text{ atm} = 2p_{H_2O} + 2p_{H_2}$$

Thus

$$p_{CO} = p_{H_2O} = 0.5 - p_{H_2} \quad \text{and} \quad p_{CO_2} = p_{H_2}$$

substitution of which into Eq. (ii) gives

$$K_{p,(\mathrm{i})} = \frac{(0.5 - p_{H_2})^2}{p_{H_2}^2}$$

For the reaction $H_{2(g)} + \frac{1}{2}O_{2(g)} = H_2O_{(g)}$, $\Delta G^0 = -246{,}000 + 54.8T$ joules, and for the reaction $CO_{2(g)} = \frac{1}{2}O_{2(g)} + CO_{(g)}$, $\Delta G^0 = 282{,}000 - 86.8T$ joules. Summing these reactions gives the standard free energy change for reaction (i) as

$$\Delta G^0_{(\mathrm{i})} = 36{,}000 - 32T \text{ joules}$$

Thus, $\quad \Delta G^0_{(\mathrm{i}),1000} = 4000 \text{ joules} = -8.3144 \times 1000 \ln K_{p,(\mathrm{i}),1000}$

and hence,

$$K_{p,(\mathrm{i}),1000} = 0.618 = \frac{(0.5 - p_{H_2})^2}{p_{H_2}^2}$$

which has the solution $p_{H_2} = 0.28$ atm.

Thus, at reaction equilibrium,

$$p_{H_2} = p_{CO_2} = 0.28 \text{ atm}$$

and

$$p_{H_2O} = p_{CO} = 0.22 \text{ atm}$$

Consider that this equilibrated gas is contained at 1 atm pressure and 1000 K in a rigid vessel of constant volume. What happens if CaO is introduced to the vessel? The reactions

$$CaO_{(s)} + 5CO_{(g)} = CaC_{2(s)} + 3CO_2 \qquad \text{(iii)}$$

$$CaO_{(s)} + H_2O_{(g)} = Ca(OH)_{2(s)} \qquad \text{(iv)}$$

and

$$CaO_{(s)} + CO_{2(g)} = CaCO_{3(s)} \qquad \text{(v)}$$

may occur.

Consider the possible formation of CaC_2 according to (iii). Summing

$\Delta G^0 = -48{,}620 - 36.1T$ joules for $Ca + 2C = CaC_2$

$\Delta G^0 = -1{,}182{,}000 - 2.4T$ joules for $3C + 3O_2 = 3CO_2$

$$\Delta G^0 = 633{,}140 - 99T \text{ joules} \quad \text{for} \quad CaO = Ca + \tfrac{1}{2}O_2$$

$$\Delta G^0 = 560{,}000 + 438.3T \text{ joules} \quad \text{for} \quad 5CO = 5C + \tfrac{5}{2}O_2$$

gives

$$\Delta G^0_{(iii)} = -37{,}480 + 300.7T$$

Thus $\Delta G^0_{(iii),1000} = 263{,}220 \text{ joules} = -8.3144 \times 1000 \ln K_{p,(iii),1000}$

and hence

$$K_{p,(iii),1000} = 1.78 \times 10^{-14} = \left(\frac{p^3_{CO_2}}{p^5_{CO}}\right)_{(eq)}$$

Thus, if the CaO were to react with the CO in the gas mixture, (which exists at $p_{CO} = 0.22$ atm), to form CaC_2 and CO_2, the pressure of CO_2 in the gas mixture would have to be less than

$$[1.76 \times 10^{-14} \times (0.22)^5]^{1/3}$$

i.e., less than 2.09×10^{-6} atm. As the partial pressure of CO_2 in the gas is 0.28 atm, it is seen that reaction (iii) will not occur.

Consider the possible formation of $Ca(OH)_2$ according to reaction (iv):

$$\Delta G^0_{(iv)} = -117{,}600 + 145T \text{ joules}$$

and thus

$$\Delta G^0_{(iv),1000} = 27{,}400 \text{ joules} = -8.3144 \times 1000 \ln K_{p,(iv),1000}$$

or

$$K_{p,(iv),1000} = 0.037 = \frac{1}{p_{H_2O(eq)}}$$

Thus the pressure of water vapor required for equilibrium CaO, $Ca(OH)_2$, and $H_2O_{(g)}$ at 1000 K at 27 atm, or if the CaO were to react with the water vapor to form $Ca(OH)_2$, p_{H_2O} in the gas phase would have to be greater than 27 atm. As the actual water vapor pressure in the gas mixture is 0.22 atm, it is seen that reaction (iv) does not occur.

Consider the possible formation of $CaCO_3$ according to reaction (v).

$$\Delta G^0_{(v)} = -168{,}400 + 144T \text{ joules}$$

Thus $\Delta G^0_{(v),1000} = -24{,}400 \text{ joules} = -8.3144 \times 1000 \ln K_{p,(v),1000}$

and hence
$$K_{p,(v),1000} = 18.82 = \frac{1}{p_{CO_2(eq)}}$$

or, for equilibrium between CaO, $CaCO_3$, and CO_2 at 1000 K, p_{CO_2} must be 0.053 atm. As, in the gas mixture, p_{CO_2} is greater than this value, it is seen that the CaO does react with the CO_2 to form $CaCO_3$. Consider that excess CaO is added to the vessel, i.e., that, by virtue of reaction (v) occurring in the vessel, the partial pressure of CO_2 is decreased to 0.053 atm before the added CaO is completely consumed by the reaction, such that, at the new equilibrium, CaO and $CaCO_3$ exist in equilibrium with the gas mixture in which $p_{CO_2} = 0.053$ atm. Now calculate the state of the newly equilibrated gas.

The removal of CO_2 from the gas has two effects: (a) the pressure exerted by the constant-volume gas mixture is decreased, and (b) the equilibrium of reaction (i) is shifted to the left. However, as all of the hydrogen in the vessel, occurring as H_2 or as H_2O, remains in the constant-volume gas phase, the sum $p_{H_2} + p_{H_2O}$ is not changed by the shift in the equilibrium. Also, from the stoichiometry of reaction (i), during the shift to the left, $p_{CO} = p_{H_2O}$.

Thus, at the new equilibrium,

$$p_{H_2} + p_{H_2O} = 0.5 \text{ atm}$$
$$p_{CO} = p_{H_2O}$$
$$p_{CO_2} = 0.053 \text{ atm}$$

and
$$K_{p,(i),1000} = 0.618 = \frac{p^2_{H_2O}}{0.053 \times (0.5 - p_{H_2O})}$$

This has the solution $p_{H_2O} = 0.113$, and hence the new equilibrium state is

$$p_{H_2O} = p_{CO} = 0.113 \text{ atm}$$
$$p_{H_2} = 0.387 \text{ atm}$$
$$p_{CO_2} = 0.053 \text{ atm}$$
$$P = 0.666 \text{ atm}$$

What happens now if graphite is introduced to the system? If excess graphite is added, the equilibrium

$$C_{(gr)} + CO_{(g)} = 2CO_{(g)} \qquad \text{(vi)}$$

must be established.

$$\Delta G^0_{(vi)} = 170{,}700 - 174.5T \text{ joules}$$

Thus $\quad \Delta G^0_{(vi),1000} = -3800 \text{ joules} = -8.3144 \times 1000 \ln K_{p,(vi)1000}$

and $$K_{p,(vi),1000} = 1.579 = \left(\frac{p^2_{CO}}{p_{CO_2}}\right)_{(eq)}$$

Thus, as $p_{CO_2} = 0.053$ atm is required for the equilibrium CaO-$CaCO_3$-CO_2, p_{CO} in the gas mixture must change from 0.113 atm to the value

$$(1.579 \times 0.053)^{1/2} = 0.289 \text{ atm}$$

to establish the equilibrium C-CO-CO_2. Again, as all the hydrogen remains in the gas phase, $p_{H_2} + p_{H_2O} = 0.5$ and so

$$K_{p,(i),1000} = 0.618 = \frac{0.289 \times (0.5 - p_{H_2})}{0.053 p_{H_2}}$$

which has the solution $p_{H_2} = 0.449$. Thus the newly equilibrated gas mixture, which is now in equilibrium with CaO, $CaCO_3$ and graphite, is

$$p_{H_2} = 0.449 \text{ atm}$$
$$p_{H_2O} = 0.051 \text{ atm}$$
$$p_{CO} = 0.289 \text{ atm}$$
$$p_{CO_2} = 0.053 \text{ atm}$$
$$P = 0.842 \text{ atm}$$

Consider now that the graphite is added *before* the CaO, i.e., graphite is added to the original gas mixture, in which $p_{CO} = p_{H_2O} = 0.22$ atm, and $p_{H_2} = p_{CO_2} = 0.28$ atm, contained in the rigid vessel at 1000 K.

The equilibrium (vi) is established, i.e., p_{CO} and p_{CO_2} in the gas mixture must change to conform with

$$\frac{p^2_{CO}}{p_{CO_2}} = K_{p,(vi),1000} = 1.579$$

In the gas mixture, before any reaction occurs, $p_{CO} = 0.22$ atm, which, for the C-CO-CO_2 equilibrium, would require $p_{CO_2} = (0.22)^2/1.579 = 0.031$ atm (which is lower than the value occurring in the gas mixture), or, for the existing

p_{CO_2} of 0.28, establishment of the C–CO–CO_2 equilibrium would require $p_{CO} = (1.579 \times 0.28)^{1/2} = 0.665$ atm (which is higher than the value occurring in the gas mixture). Thus reaction (vi) must proceed and the gas phase equilibrium (i) must shift such as to increase p_{CO} and decrease p_{CO_2}, until, simultaneously,

$$\frac{p_{CO}^2}{p_{CO_2}} = 1.579 \quad \text{and} \quad \frac{p_{CO}p_{H_2O}}{p_{CO_2}p_{H_2}} = 0.618$$

As before, $p_{H_2} + p_{H_2O} = 0.5$ atm, and the fourth condition, knowledge of which is necessary for determination of the equilibrium state, is obtained from consideration of the oxygen and hydrogen mole balances. In the original mixture, $CO_2/H_2 = 1$, and thus equal numbers of moles of oxygen and hydrogen occur in the gas phase.

The number of moles of oxygen in the gas is

$$2n_{CO_2} + n_{CO} + n_{H_2O}$$

and the number of moles of hydrogen is

$$2n_{H_2O} + 2n_{H_2}$$

Thus, in the gas mixture,

$$2n_{CO_2} + n_{CO} + n_{H_2O} = 2n_{H_2O} + 2n_{H_2}$$

or

$$n_{CO_2} + \tfrac{1}{2}n_{CO} = n_{H_2} + \tfrac{1}{2}n_{H_2O}$$

Under conditions of constant volume and temperature, $p_i \propto n_i$ and thus

$$p_{CO_2} + \tfrac{1}{2}p_{CO} = p_{H_2} + \tfrac{1}{2}p_{H_2O}$$

which, in combination with

$$p_{H_2O} + p_{H_2} = 0.5$$

gives

$$p_{H_2O} = 1 - 2p_{CO_2} - p_{CO}$$

Thus,

$$p_{CO_2} = \frac{p_{CO}^2}{1.579} = 0.633p_{CO}^2$$

$$p_{H_2O} = 1 - 2p_{CO_2} - p_{CO} = 1 - 1.266p_{CO}^2 - p_{CO}$$

$$p_{H_2} = 0.5 - p_{H_2O} = 1.266p_{CO}^2 + p_{CO} - 0.5$$

substitution of which into (ii) gives

$$K_{p,(i),1000} = 0.618 = \frac{(1 - 1.266p_{CO}^2 - p_{CO})p_{CO}}{(1.266p_{CO}^2 + p_{CO} - 0.5) \times 0.633p_{CO}^2}$$

or

$$p_{CO}^3 + 3.346p_{CO}^2 + 1.624p_{CO} - 2.019 = 0$$

which has the solution $p_{CO} = 0.541$ atm. Thus the new equilibrium is

$$p_{CO} = 0.541 \text{ atm}$$
$$p_{CO_2} = 0.185 \text{ atm}$$
$$p_{H_2} = 0.412 \text{ atm}$$
$$p_{H_2O} = 0.088 \text{ atm}$$
$$P = 1.226 \text{ atm}$$

Now add CaO to the system. The partial pressure of CO_2 in the gas ($p_{CO_2} = 0.185$ atm) is greater than the value of 0.053 required for equilibrium between CaO, $CaCO_3$, and CO_2 at 1000 K. Thus the CO_2 reacts with the CaO to form $CaCO_3$ until, thereby, p_{CO_2} is decreased to the equilibrium value of 0.053 atm, and the gas phase equilibrium shifts in order to maintain the C-CO-CO_2 equilibrium. Thus, at the new equilibrium,

$$p_{CO_2} = 0.053 \text{ atm} \quad \text{and} \quad p_{CO} = (1.579 \times 0.053)^{1/2} = 0.289 \text{ atm}$$

Also,

$$K_{p,(i),1000} = 0.618 = \frac{0.289p_{H_2O}}{0.053p_{H_2}}$$

which gives

$$\frac{p_{H_2O}}{p_{H_2}} = 0.133$$

which, with $p_{H_2} + p_{H_2O} = 0.5$, gives $p_{H_2} = 0.449$ and $p_{H_2O} = 0.051$. Thus the new equilibrium gas is,

$$p_{H_2} = 0.449 \text{ atm}$$
$$p_{H_2O} = 0.051 \text{ atm}$$
$$p_{CO} = 0.289 \text{ atm}$$

$$p_{CO_2} = 0.053 \text{ atm}$$

$$P = 0.842$$

which, necessarily, is the same state as that produced by introducing the CaO before the graphite.

PROBLEMS

10.1 To what temperature must $MgCO_3$ be heated in an atmosphere containing a partial pressure of CO_2 of 10^{-2} atm in order that decomposition of the carbonate occurs?

10.2 Using the standard free energy equations for the oxidation of solid and liquid nickel, calculate the melting temperature, the heat of fusion, and the entropy of fusion of nickel.

10.3 A piece of cold-rolled copper sheet has to be annealed. In order to prevent oxidation of the copper, the heat treatment must be performed in a vacuum (which is produced by pumping air out of the gas-tight heat-treatment furnace). If it is required that the heat treatment be carried out at 650°C, calculate the maximum total pressure (i.e., the poorest vacuum) that can be tolerated. Could a vacuum of 10^{-4} atm be used at any temperature?

10.4 Calculate the temperature at which pure Ag_2O decomposes on heating in (*a*) pure oxygen at 1 atm pressure, and (*b*) atmospheric air. Assume the molar heat capacity change for the oxidation reaction to be zero.

10.5 Determine the maximum pressure of water vapor in wet hydrogen at 1 atm pressure in which chromium can be heated without oxidation occurring at 1500 K. Is the oxidation of Cr by water vapor exothermic or endothermic?

10.6 A mixture of argon gas and hydrogen gas at 1 atm total pressure is passed through a reaction chamber containing a mixture of liquid Sn and liquid $SnCl_2$ at 900 K. The composition of the gas leaving the reaction chamber is 50% H_2, 7% HCl, and 43% Ar. Has equilibrium been attained between the gas phase and the liquid phases in the reaction chamber?

10.7 What is the maximum partial pressure of CO_2 which can be tolerated in a CO-CO_2 gas mixture at 1 atm total pressure without the oxidation of Ni occurring at 1500 K?

10.8 The variation, with temperature, of the p_{H_2S}/p_{H_2} ratio in equilibrium with solid Mn and solid MnS is tabulated below. Calculate the standard free energy expression for the reaction

$$MnS_{(s)} + H_{2(g)} = Mn_{(s)} + H_2S_{(g)}$$

T K	p_{H_2S}/p_{H_2}
500	6.07×10^{-18}
600	6.07×10^{-15}
700	8.44×10^{-13}
800	3.42×10^{-11}
900	6.07×10^{-10}
1000	6.08×10^{-9}

10.9 Calculate the vapor pressure of Mg exerted at 1400°C by the system in which the reaction equilibrium

$$4MgO_{(s)} + Si_{(s)} = 2Mg_{(g)} + Mg_2SiO_{4\,(s)}$$

is established.

10.10 One gram of $CaCO_3$ is placed in an evacuated rigid vessel of volume 1 liter at room temperature, and the system is heated. Calculate (*a*) the highest temperature at which the $CaCO_3$ phase is present, (*b*) the pressure inside the vessel at 1000 K, and (*c*) the pressure inside the vessel at 1500 K. The molecular weight of $CaCO_3$ is 100.

10.11 An Ar-H_2O gas mixture of $p_{H_2O} = 0.9$ atm ($P_{total} = 1$ atm) is passed over solid CaF_2, as a result of which CaO forms according to

$$CaF_{2\,(s)} + H_2O_{(g)} = CaO_{(s)} + 2HF_{(g)}$$

This reaction proceeds to equilibrium, and solid CaF_2 and CaO are mutually immiscible. When the gas flow rate (measured at 298 K and 1 atm pressure) over the sample is 1 liter per minute, the measured rates of weight loss of the sample are 2.69×10^{-4} and 8.30×10^{-3} grams per hour at 900 and 1100 K, respectively. From this data calculate the variation of ΔG^0 with temperature for the above reaction, given the atomic weights O = 16, H = 1, F = 19, Ca = 40.08.

10.12 Two of the oxides of iron are FeO and Fe_3O_4. Solid Fe exists in equilibrium with one of these oxides at low temperatures and exists in equilibrium with the other oxide at high temperatures. The molar free energies of formation of these oxides are, for $Fe + \frac{1}{2}O_2 = FeO$,

$$\Delta G^0 = -259{,}600 + 62.55T \text{ joules}$$

and, for $3Fe + 2O_2 = Fe_3O_4$,

$$\Delta G^0 = -1{,}091{,}000 + 312.8T \text{ joules}$$

Determine which of the two oxides is in equilibrium with iron at room temperature and the maximum temperature at which this oxide is in equilibrium with iron.

10.13 In one process for the separation of Zr from Hf, the gaseous tetrachlorides of these metals are allowed to react at 800°C with a gas containing 67% oxygen and 33% chlorine by volume. If the standard free energy changes at this temperature for the oxidation of $HfCl_4$ and $ZrCl_4$ to give HfO_2 and ZrO_2 and Cl_2 gas are respectively +204.4 and −143.5 kilojoules per mole of the chloride, determine the end products of the separation process.

10.14 Three equations for the oxidation of Mg are

$$Mg + \tfrac{1}{2}O_{2(g)} = MgO_{(s)} \qquad \Delta G^0 = -604{,}000 - 5.36T \ln T + 142.0T \text{ joules} \tag{I}$$

$$Mg + \tfrac{1}{2}O_{2(g)} = MgO_{(s)} \qquad \Delta G^0 = -759{,}800 - 13.4T \ln T + 317T \text{ joules} \tag{II}$$

$$Mg + \tfrac{1}{2}O_{2(g)} = MgO_{(s)} \qquad \Delta G^0 = -608{,}100 - 0.44T \ln T + 112.8T \text{ joules} \tag{III}$$

One of these equations is for the oxidation of solid Mg, one is for the oxidation of liquid Mg, and one is for the oxidation of gaseous Mg. Determine which equation is for which oxidation, and calculate the melting and normal boiling temperatures of Mg.

10.15 Solid ZnO and solid ZnS are equilibrated at 2000 K with an H_2S–H_2O–H_2 atmosphere in which $p_{H_2O} = 0.5$ and $p_{H_2} = 0.0421$. Calculate the equilibrium partial pressures of O_2, H_2S, S_2, and Zn in the atmosphere.

10.16 Calculate the vacuum required to cause decomposition of MgO to magnesium vapor and oxygen at 1600°C.

10.17 Hydrogen gas is passed over liquid $MgCl_2$ at 1200°C and the reaction equilibrium

$$MgCl_{2(l)} + H_{2(g)} = Mg_{(g)} + 2HCl_{(g)}$$

is established. Calculate the equilibrium pressures of H_2, Mg, and HCl if the total pressure is maintained constant at 1 atm. Calculate the maximum pressure of water vapor that can be tolerated in the hydrogen without causing oxidation of the magnesium vapor.

10.18 Calculate the pressure of inert gas which must be applied to liquid lead at 1000°C in order to triple the vapor pressure of the lead. The density of liquid lead at 1000°C is 9.79 gm/cm^3.

CHAPTER ELEVEN

THE BEHAVIOR OF SOLUTIONS

11.1 INTRODUCTION

In the previous chapter it was seen that thermodynamic consideration of the equilibria among pure condensed phases and a gas phase was facilitated by the fact that a pure species, occurring in a condensed state, exerts a unique vapor pressure, (its saturated vapor pressure), at the temperature *T*. Thus, for example, at the temperature *T*, by virtue of the unique saturated vapor pressures exerted by a pure metal and its pure oxide, the equilibrium between the pure metal, the pure oxide and oxygen gas occurs at a unique value of the partial pressure of oxygen within the system, $p_{O_2(eq, T)}$, which is easily calculable from a knowledge of the standard free energy of oxidation of the metal. In practice, however, as "pure" species are much less common than "impure" species (indeed the *absolutely* pure state is thermodynamically unattainable–see Chap. 12), the question arises as to how to calculate, say, the oxygen pressure required for equilibrium between an impure metal and its impure oxide. In this context "impure" implies that the specific species occurs in solution, either solid or liquid, with some other species. Consider the case of the oxidation of a solid Fe-Si alloy, (a solid solution of Fe and Si) to form pure SiO_2 at the temperature *T*. This is equivalent to the oxidation of impure Si to form pure SiO_2. As the Si no longer occurs as a pure species, i.e., in its defined standard state, it no longer exerts its unique saturated vapor pressure; hence the relatively simple treatment of reaction equilibria dealt with in the previous chapter is no longer applicable. Calculation of the equilibrium oxygen pressure in this system requires a knowledge of the extent to which the vapor pressure of Si has been decreased as a result of its solution in Fe to form an Fe-Si alloy; i.e., at the temperature *T* the relationship between Si concentration in the alloy and the vapor pressure of silicon exerted by the alloy is required. One step more complex would be the situation where Si in an Fe-Si solution was being oxidized to form SiO_2 dissolved in FeO as an FeO-SiO_2 solution. In this case calculation of the equilibrium oxygen pressure requires a knowledge of the vapor pressure-composition relationship of both Si in Fe and SiO_2 in FeO. As more species are added to the system, e.g., C, Mn, and P, the factors governing the equilibrium become increasingly complex. Such complex equilibria are of considerable importance in metallurgy. In the case of the refining of steel, in order to oxidize Si from the liquid steel (the Fe-C-Si-Mn-P alloy) to form SiO_2 dissolved in the slag (FeO-MnO-CaO-SiO_2-P_2O_5-MgO liquid solution), the oxygen pressure in the system must be greater than $p_{O_2(\text{eq}, T)}$ for the equilibrium

$$\text{Si}_{(\text{dissolved in the steel})} + \text{O}_2 = \text{SiO}_{2(\text{dissolved in the slag})}$$

Similar considerations apply to the removal of the other impurities present in the steel bath.

Such calculations require a knowledge of that body of information which is referred to as "solution thermodynamics." This term encompasses knowledge of the vapor pressure-temperature-composition relationships of the constituents (or components) in a solution, and only when this information is available can an approach be made to calculation of the reaction equilibria in systems containing components in solution. In this chapter the thermodynamic properties and treatment of solutions are examined.

11.2 RAOULT'S LAW AND HENRY'S LAW

If a quantity of pure liquid A is placed in a closed, initially evacuated vessel at the temperature T, the liquid will spontaneously evaporate until the pressure in the vessel is the saturated vapor pressure of liquid A, p_A^0, at the temperature T. At this point a dynamic equilibrium is established between the rate of evaporation of A from the liquid phase and the rate of condensation of A from the vapor phase. The evaporation rate, $r_{e(A)}$, is determined by the magnitude of the attractive interactions operating between the atoms of A at the surface of the liquid. These attractive forces are such that each surface atom is located in the vicinity of the bottom of a potential energy well. In order for an atom to leave the liquid surface and enter the gas phase, it must overcome the attractive forces exerted on it by its neighbors, i.e., acquire a necessary activation energy E^*. The depth of this potential energy well (the magnitude of E^*), at the temperature T, determines the intrinsic evaporation rate, $r_{e(A)}$. On the other hand the condensation rate, $r_{c(A)}$, is proportional to the number of A atoms in the vapor phase which strike the liquid surface in unit time. This, for a fixed temperature, is proportional to the pressure of the vapor. Thus $r_{c(\mathrm{A})} = k\,p^0_{\mathrm{A}}$, and at equilibrium

$$r_{e(\mathrm{A})} = kp^0_{\mathrm{A}} \qquad (11.1)$$

(Equation 11.1 illustrates why vapor pressure increases exponentially with increasing temperature).*

Similarly, for liquid B in an initially evacuated vessel at the temperature T, equilibrium occurs when

$$r_{e(\mathrm{B})} = k'p^0_{\mathrm{B}} \qquad (11.2)$$

Consider the effect of the addition of a small quantity of liquid B to liquid A. If the mole fraction of A in the resultant A-B liquid solution is X_{A} and the atomic diameters of A and B are comparable, then assuming the surface composition of the liquid to be the same as the bulk liquid composition, the fraction of the surface area of the liquid occupied by A atoms is X_{A}. As A can only evaporate from surface sites occupied by A atoms, the evaporation rate of A is decreased by the factor X_{A}; and as, at equilibrium, the rates of evaporation and condensation are equal, the equilibrium vapor pressure exerted by A is necessarily decreased from p^0_{A} to p_{A}. That is,

$$r_{e(\mathrm{A})}X_{\mathrm{A}} = kp_{\mathrm{A}} \qquad (11.3)$$

Similarly, for B containing a small quantity of A,

$$r_{e(\mathrm{B})}X_{\mathrm{B}} = k'p_{\mathrm{B}} \qquad (11.4)$$

Combination of Eq. (11.1) and (11.3) gives

$$p_{\mathrm{A}} = X_{\mathrm{A}}p^0_{\mathrm{A}} \qquad (11.5)$$

*The energies of the surface atoms are quantized, and the distribution of the surface atoms among the available quantized energy levels is given by Eq. (4.13); that is, $n_i = n\exp(-E_i/kT)/P$, where n_i/n is the fraction of atoms in the E_ith energy level, and P, the partition function, is given as $P = \sum_0^\infty \exp(-E_i/kT)$. If the quantized energy levels are closely enough spaced that the summation can be replaced by an integral, then $P = \int_0^\infty \exp(-E_i/kT)\,dE = kT$ (which is thus the average energy per atom); and hence the fraction of surface atoms which have energies greater than the activation energy for evaporation, E^*, is

$$\frac{n_i^*}{n} = \frac{1}{kT}\int_{E^*}^{\infty} \exp\left(\frac{-E_i}{kT}\right) dE = \exp\left(\frac{-E^*}{kT}\right)$$

From Eq. (11.1), as the evaporation rate, $r_{e(\mathrm{A})}$, is proportional to n_i^*/n, it is seen that the vapor pressure increases exponentially with increasing temperature and decreases exponentially with increasing E^*.

and combination of Eq. (11.2) and (11.4) gives

$$p_B = X_B p_B^0 \tag{11.6}$$

Equations (11.5) and (11.6) are expressions of *Raoult's law*, which states that the vapor pressure exerted by a component i in a solution is equal to the product of the mole fraction of i in the solution and the vapor pressure of pure i at the temperature of the solution. The form of the behavior of a solution, the components of which obey Raoult's law, is shown in Fig. 11.1, and the components of such a solution are said to exhibit *Raoultian* behavior.

The derivation of Eqs. (11.3) and (11.4) required that the intrinsic evaporation rates of A and B, $r_{e(A)}$ and $r_{e(B)}$, respectively, be independent of the composition of the solution. This requires that the magnitudes of the A-A, B-B, and A-B interactions be identical, such that the depth of the potential energy well of an atom at the liquid surface is independent of the types of atoms which it has as nearest neighbors.

Consider the case where the A–B attraction is significantly stronger than either the A–A or B–B attractions, and consider a solution of A in B which is sufficiently dilute that every A atom on the surface of the liquid is surrounded only by B atoms. In this case the A atoms at the surface are each located in a deeper potential energy well than are the A atoms at the surface of pure liquid A. Thus, in order to leave the surface and enter the vapor phase, the A atoms have to be lifted from deeper wells, and consequently the intrinsic evaporation

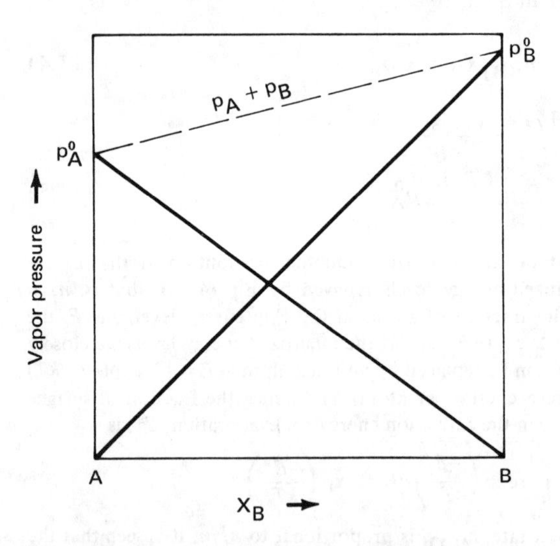

Fig. 11.1. The vapor pressures exerted by the components of a binary Raoultian solution.

rate of A from the solution is decreased from $r_{e(\mathrm{A})}$ to $r'_{e(\mathrm{A})}$. Equilibrium thus occurs when

$$r'_{e(\mathrm{A})}X_{\mathrm{A}} = kp_{\mathrm{A}} \tag{11.7}$$

Combination of Eqs. (11.1) and (11.7) then gives

$$p_{\mathrm{A}} = \frac{r'_{e(\mathrm{A})}}{r_{e(\mathrm{A})}} X_{\mathrm{A}} p_{\mathrm{A}}^{0} \tag{11.8}$$

and as $r'_{e(\mathrm{A})} < r_{e(\mathrm{A})}$, p_{A} in Eq. (11.8) is a smaller quantity than p_{A} in Eq. (11.5). Equation (11.8) can be written as

$$p_{\mathrm{A}} = k'_{\mathrm{A}}X_{\mathrm{A}} \tag{11.9}$$

As X_{A} in the solution of A in B is increased, the probability that all the A atoms on the surface of the liquid are surrounded only by B atoms decreases. The occurrence of a pair of neighboring A atoms on the liquid surface decreases the depth of the potential wells in which they are located and hence increases the value of $r'_{e(\mathrm{A})}$. Beyond a critical value of X_{A}, $r'_{e(\mathrm{A})}$ thus becomes composition-dependent, and hence Eq. (11.9) is no longer obeyed by A in solution. Consequently Eq. (11.9) is obeyed only over an initial concentration range of A in B, the extent of which is dependent on the temperature of the solution and on the relative magnitudes of the A-A, B-B, and A-B interactions. A similar argument concerning dilute solutions of B in A gives

$$p_{\mathrm{B}} = k'_{\mathrm{B}}X_{\mathrm{B}} \tag{11.10}$$

which is obeyed over an initial concentration range. Equations (11.9) and (11.10) are known as *Henry's law*, and in the composition ranges where Henry's law is obeyed the solutes are said to exhibit *Henrian behavior*. If the magnitude of the A-B attraction is greater than those of the A-A and B-B attractions, then, as $r'_{e(\mathrm{A})} < r_{e(\mathrm{A})}$, the Henry's law line lies below the Raoult's law line. Conversely, if the magnitude of the A-B attraction is less than those of the A-A and B-B attractions, the solute atom, surrounded by solvent atoms, is located in a shallower potential energy well than if it were surrounded by atoms of its own kind. In such a case $r'_{e(\mathrm{A})} > r_{e(\mathrm{A})}$, and hence the Henry's law line lies above the Raoult's law line. Henrian solute behavior is illustrated in Fig. 11.2*a* and *b*.

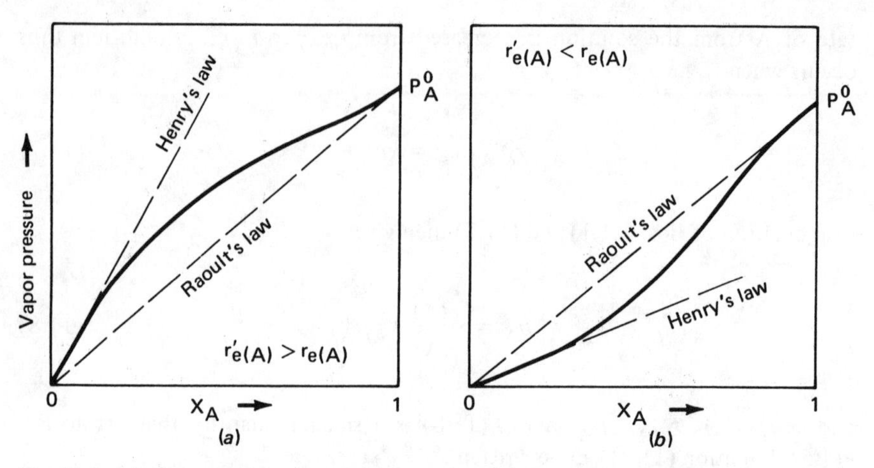

Fig. 11.2. (*a*) The vapor pressure of a component of a binary solution which exhibits positive deviation from Raoultian behavior. (*b*) The vapor pressure of a component of a binary solution which exhibits negative deviation from Raoultian behavior.

11.3 THE THERMODYNAMIC ACTIVITY OF A COMPONENT IN SOLUTION

The thermodynamic activity of a component in any state at the temperature T is formally defined as being the ratio of the fugacity of the substance in that state to its fugacity in its standard state; i.e., for the species or substance i,

$$\text{activity of } i = a_i = \frac{f_i}{f_i^0} \tag{11.11}$$

With respect to a solution, f_i is the fugacity of the component i in the solution at the temperature T, and f_i^0 is the fugacity of pure i (the standard state) at the temperature T. If the vapor above the solution is ideal, then $f_i = p_i$, in which case

$$a_i = \frac{p_i}{p_i^0} \tag{11.12}$$

i.e., the activity of i in a solution, with respect to pure i, is the ratio of the vapor pressure of i exerted by the solution to the vapor pressure of pure i at the same temperature. If the component i behaves ideally in solution, then Eq. (11.5) and Eq. (11.12) give

$$a_i = X_i \tag{11.13}$$

which is thus an alternative expression of Raoult's law. Figure 11.3 shows Raoultian behavior in a binary solution in terms of the activities of the two components. The introduction of activity thus has the effect of normalizing the vapor pressure–composition variations with respect to the vapor pressure exerted in the standard state.

Similarly, over the composition range in which Henry's law is obeyed by the solute i, Eqs. (11.9) and (11.12) give

$$a_i = k_i X_i \tag{11.14}$$

which is an alternative expression of Henry's law. Figure 11.4 illustrates Henrian behavior in terms of the activity of a component of a binary solution.

11.4 THE GIBBS-DUHEM EQUATION

It is frequently found that the extensive thermodynamic properties of only one component in a binary (or multicomponent) solution are amenable to

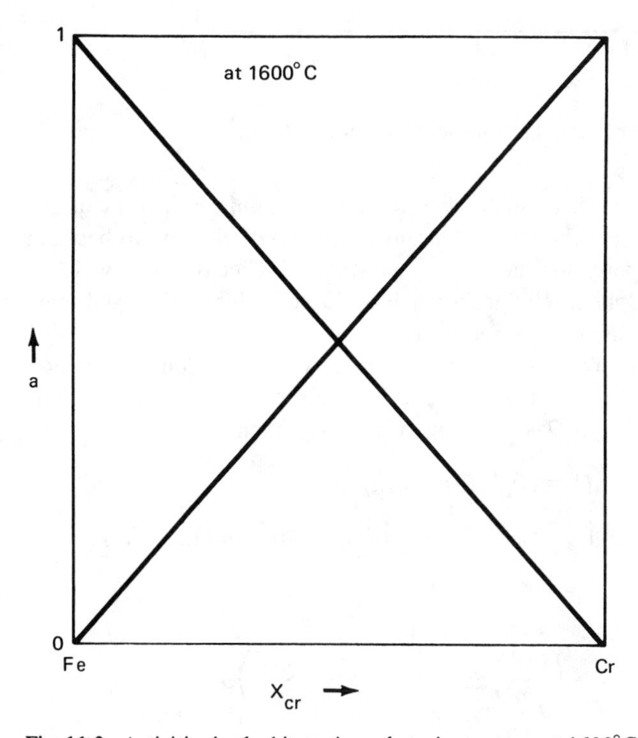

Fig. 11.3. Activities in the binary iron-chromium system at 1600°C.

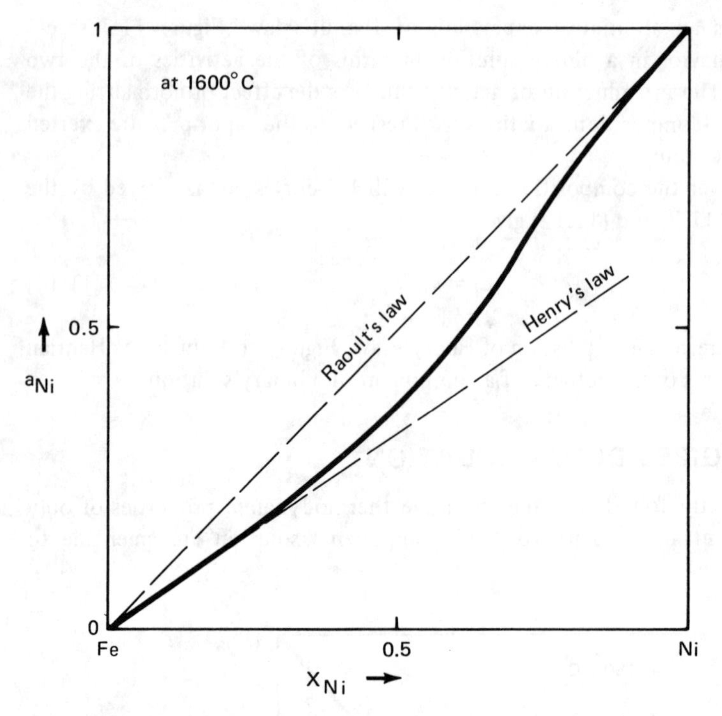

Fig. 11.4. The activity of nickel in the iron-nickel system at 1600°C.

experimental measurement. In such cases the corresponding properties of the other component can be obtained by means of a general relationship between the values of the property for the two components. This relationship, which is known as the Gibbs-Duhem relationship, is introduced in this section, and some of its specific uses are discussed in Sec. 11.8.

The value of an extensive thermodynamic property of a solution is a function of the temperature, the pressure, and the numbers of moles of the various solution components; i.e., if Q is an extensive property, then

$$Q' = Q'(T, P, n_i, n_j, n_k, \ldots)$$

Thus at constant T and P the variation of Q' with variation of solution composition is given as

$$dQ' = \left(\frac{\partial Q'}{\partial n_i}\right)_{T,P,n_j,n_k,\ldots} dn_i + \left(\frac{\partial Q'}{\partial n_j}\right)_{T,P,n_i,n_k,\ldots} dn_j + \left(\frac{\partial Q'}{\partial n_k}\right)_{T,P,n_i,n_j,\ldots} dn_k + \cdots \tag{11.15}$$

In Chap. 8 the partial molar value of an extensive property of a component was defined as

$$\bar{Q}_i = \left(\frac{\partial Q'}{\partial n_i}\right)_{T, P, n_j, n_k, \ldots}$$

in which case Eq. (11.15) can be written as

$$dQ' = \bar{Q}_i dn_i + \bar{Q}_j dn_j + \bar{Q}_k dn_k + \cdots \tag{11.16}$$

Also in Chap. 8 it was seen that $\bar{Q}_i$ is the increase in the value of Q' for the mixture or solution when 1 mole of i is added to a large quantity of the solution at constant T and P. (The stipulation that the quantity of solution be large is necessitated by the requirement that the addition of 1 mole of i to the solution should not cause a measurable change in the composition of the solution.) Thus if $\bar{Q}_i$ is the value of Q per mole of i as it occurs in the solution, then the value of Q' for the solution itself is

$$Q' = n_i \bar{Q}_i + n_j \bar{Q}_j + n_k \bar{Q}_k + \cdots \tag{11.17}$$

differentiation of which gives

$$\begin{aligned} dQ' = {} & n_i d\bar{Q}_i + n_j d\bar{Q}_j + n_k d\bar{Q}_k + \cdots \\ & + \bar{Q}_i dn_i + \bar{Q}_j dn_j + \bar{Q}_k dn_k + \cdots \end{aligned} \tag{11.18}$$

Comparison of Eqs. (11.16) and (11.18) indicates that, at constant T and P,

$$n_i d\bar{Q}_i + n_j d\bar{Q}_j + n_k d\bar{Q}_k + \cdots = 0$$

or, generally,

$$\sum_i n_i d\bar{Q}_i = 0 \tag{11.19}$$

Division of Eq. (11.19) by n, the total number of moles of all solution components, gives

$$\sum_i X_i d\bar{Q}_i = 0 \tag{11.20}$$

Equations (11.19) and (11.20) are equivalent expressions of the *Gibbs-Duhem equation*.

11.5 THE FREE ENERGY OF SOLUTION

The Molar Free Energy of a Solution and the Partial Molar Free Energies of Solution of Components

In terms of the free energy (as an extensive thermodynamic property), Eq. (11.17), for a binary A–B solution at fixed values of T and P, is

$$G' = n_A \bar{G}_A + n_B \bar{G}_B \tag{11.21}$$

where $\bar{G}_A$ and $\bar{G}_B$ are, respectively, the partial molar free energies of A and B in the solution. Written for 1 mole of solution (dividing both sides by $n_A + n_B$), this gives

$$G = X_A \bar{G}_A + X_B \bar{G}_B \tag{11.22}$$

Differentiation gives

$$dG = X_A d\bar{G}_A + X_B d\bar{G}_B + \bar{G}_A dX_A + \bar{G}_B dX_B \tag{11.23}$$

but as the Gibbs-Duhem equation gives $X_A d\bar{G}_A + X_B d\bar{G}_B = 0$, then Eq. (11.23) becomes

$$dG = \bar{G}_A dX_A + \bar{G}_B dX_B \tag{11.24}$$

or

$$\frac{dG}{dX_A} = \bar{G}_A - \bar{G}_B \tag{11.25}$$

($X_A + X_B = 1$, and hence $dX_A = -dX_B$). Multiplying Eq. (11.25) by X_B gives

$$X_B \frac{dG}{dX_A} = X_B \bar{G}_A - X_B \bar{G}_B \tag{11.26}$$

and addition of Eqs. (11.26) and (11.22) gives

$$G + X_B \frac{dG}{dX_A} = \bar{G}_A (X_A + X_B)$$

or

$$\bar{G}_A = G + X_B \frac{dG}{dX_A} \tag{11.27a}$$

Similarly

$$\bar{G}_B = G + X_A \frac{dG}{dX_B} \tag{11.27b}$$

This expression relates the partial molar free energy of a component of a binary solution to the molar free energy of the solution.

The Free Energy Change Due to the Formation of a Solution

The pure component i, occurring in a condensed state at the temperature T, exerts an equilibrium vapor pressure p_i^0. When occurring in a condensed solution at the temperature T, it exerts a lower equilibrium vapor pressure p_i. Consider the three-step process

(*a*) Evaporation of 1 mole of pure condensed i to vapor i at the pressure p_i^0 and the temperature T
(*b*) Decrease in the pressure of 1 mole of vapor i from p_i^0 to p_i at the temperature T
(*c*) Condensation of 1 mole of vapor i from the pressure p_i into the condensed solution at the temperature T

The difference in molar free energy between pure i and i in solution equals $\Delta G_{(a)} + \Delta G_{(b)} + \Delta G_{(c)}$. However, as steps (*a*) and (*c*) are equilibrium processes, $\Delta G_{(a)}$ and $\Delta G_{(c)}$ are both zero. The overall free energy change for the three-step process thus equals $\Delta G_{(b)}$ which, from Eq. (8.9) is given as

$$\Delta G_{(b)} = RT \ln\left(\frac{p_i}{p_i^0}\right)$$

From Eq. (11.12), this can be written as

$$\Delta G_{(b)} = G_i(\text{in solution}) - G_i(\text{pure}) = RT \ln a_i$$

But G_i (in solution) is simply the partial molar free energy of i in the solution, $\bar{G}_i$, and G_i (pure) is the molar free energy of pure i, G_i^0. The difference between the two is the free energy change accompanying the solution of 1 mole of i in the solution. This quantity is designated $\Delta\bar{G}_i^M$, the partial molar free energy of the solution of i. Hence,

$$\Delta\bar{G}_i^M = \bar{G}_i - G_i^0 = RT \ln a_i \tag{11.28}$$

If, at constant T and P, n_A moles of A and n_B moles of B are mixed to form a binary solution,

$$\text{the free energy before mixing} = n_A G_A^0 + n_B G_B^0$$

and

$$\text{the free energy after mixing} = n_A \bar{G}_A + n_B \bar{G}_B$$

The free energy change due to mixing, $\Delta G'^M$, referred to as the integral free energy of mixing, is the difference between these quantities; i.e.,

$$\begin{aligned}\Delta G'^M &= (n_A\bar{G}_A + n_B\bar{G}_B) - (n_A G_A^0 + n_B G_B^0)\\ &= n_A(\bar{G}_A - G_A^0) + n_B(\bar{G}_B - G_B^0)\end{aligned}$$

Substitution from Eq. (11.28) gives

$$\Delta G'^M = n_A \Delta\bar{G}_A^M + n_B \Delta\bar{G}_B^M \tag{11.29}$$

or

$$\Delta G'^M = RT(n_A \ln a_A + n_B \ln a_B) \tag{11.30}$$

In terms of 1 mole of solution, Eqs. (11.29) and (11.30), respectively, become

$$\Delta G^M = X_A \Delta\bar{G}_A^M + X_B \Delta\bar{G}_B^M \tag{11.31}$$

and

$$\Delta G^M = RT(X_A \ln a_A + X_B \ln a_B) \tag{11.32}$$

Equation (11.32) gives the variation of ΔG^M with composition of a binary solution, which, typically, is as shown in Fig. 11.5.

The Method of Tangential Intercepts

In terms of the solution properties, Eqs. (11.27*a*) and (11.27*b*) can be written as

$$\Delta \bar{G}_A^M = \Delta G^M + X_B \frac{d\Delta G^M}{dX_A} \tag{11.33a}$$

and

$$\Delta \bar{G}_B^M = \Delta G^M + X_A \frac{d\Delta G^M}{dX_B} \tag{11.33b}$$

Consider composition X_A (the point p in Fig. 11.5). At X_A,

$$\Delta G^M = pq$$

$$X_B = rq$$

$$\frac{d\Delta G^M}{dX_A} = \text{the slope of the tangent to the } \Delta G^M \text{ curve at } X_A = X_A$$

$$= \frac{rs}{rq}$$

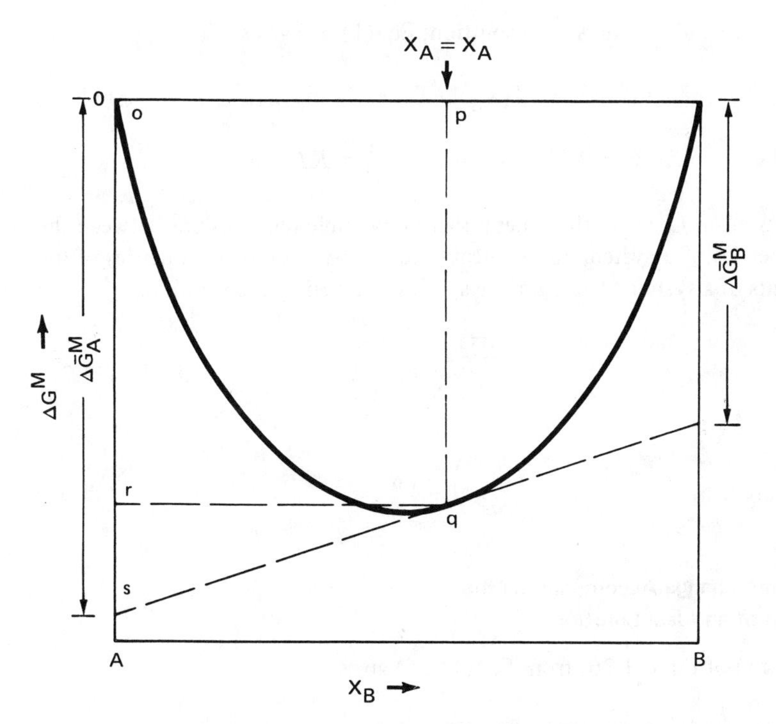

Fig. 11.5. The variation with composition of the free energy of mixing of the components of a binary solution.

From Eq. (11.33*a*),

$$\Delta \bar{G}_A^M = pq + rq\frac{rs}{rq} = pq + rs = or + rs = os$$

$$= \text{the tangential intercept at } X_A = 1$$

Similarly, at $X_A = X_A$, $\Delta \bar{G}_B^M$ = the tangential intercept at $X_B = 1$.

The method of tangential intercepts can be used to obtain the partial molar values of solution of any extensive property from the integral value of the property for the solution.

11.6 THE PROPERTIES OF RAOULTIAN IDEAL SOLUTIONS

The components of a Raoultian ideal solution obey the relation

$$a_i = X_i$$

and hence for an ideal binary A–B solution, Eq. (11.32) gives

$$\Delta G^{M,\mathrm{id}} = RT(X_A \ln X_A + X_B \ln X_B) \tag{11.34}$$

where $\Delta \bar{G}_A^M = RT \ln X_A$ and $\Delta \bar{G}_B^M = RT \ln X_B$

As discussed in Chap. 8, the general thermodynamic relationships between the state properties of a system are applicable to the partial molar properties of the components of a system. Thus, for the species *i* occurring in a solution,

$$\frac{\partial \bar{G}_i}{\partial P} = \bar{V}_i \tag{11.35}$$

and for pure *i*,

$$\left(\frac{\partial G_i^0}{\partial P}\right) = V_i^0 \tag{11.36}$$

The Volume Change Accompanying the Formation of an Ideal Solution

Subtraction of Eq. (11.36) from Eq. (11.35) gives

$$\left(\frac{\partial(\bar{G}_i - G_i^0)}{\partial P}\right)_{T,\,\mathrm{comp}} = (\bar{V}_i - V_i^0)$$

or

$$\left(\frac{\partial \Delta \bar{G}_i^M}{\partial P}\right)_{T,\,\text{comp}} = \Delta \bar{V}_i^M \tag{11.37}$$

For an ideal solution, $\Delta \bar{G}_i^M = RT \ln X_i$; and as X_i is not a function of pressure, then

$$\Delta \bar{V}_i^M = 0$$

The volume change of mixing, $\Delta V'^M$, is obtained as the difference between the volumes of the components in the solution and the volumes of the pure components; i.e., for a binary A-B solution containing n_A moles of A and n_B moles of B,

$$\begin{aligned}\Delta V'^M &= (n_A \bar{V}_A + n_B \bar{V}_B) - (n_A V_A^0 + n_B V_B^0)\\ &= n_A(\bar{V}_A - V_A^0) + n_B(\bar{V}_B - V_B^0)\\ &= n_A \Delta \bar{V}_A^M + n_B \Delta \bar{V}_B^M\end{aligned}$$

But, as for an ideal solution, $\Delta \bar{V}_i^M = 0$, it is seen that the volume change accompanying the formation of an ideal solution is zero; i.e.,

$$\Delta V^{M,\text{id}} = 0 \tag{11.38}$$

The volume of an ideal solution is thus equal to the sum of the volumes of the pure components. Figure 11.6 shows the variation with composition of the molar volume of an ideal binary solution.

At any composition the values of the partial molar volumes $\bar{V}_A^M$ and $\bar{V}_B^M$ are obtained as the intercepts of the tangent to the V-composition line with the respective axes. As the molar volume of the solution is a linear function of composition when the solution is ideal, then, trivially, the tangent at any point coincides with the straight line such that

$$\bar{V}_A = V_A^0 \quad \text{and} \quad \bar{V}_B = V_B^0$$

The Heat of Formation of an Ideal Solution

For a component in solution, the Gibbs-Helmholtz equation [Eq. (5.37)] is

$$\left[\frac{\partial(\bar{G}_i/T)}{\partial T}\right]_{P,\,\text{comp}} = -\frac{\bar{H}_i}{T^2} \tag{11.39}$$

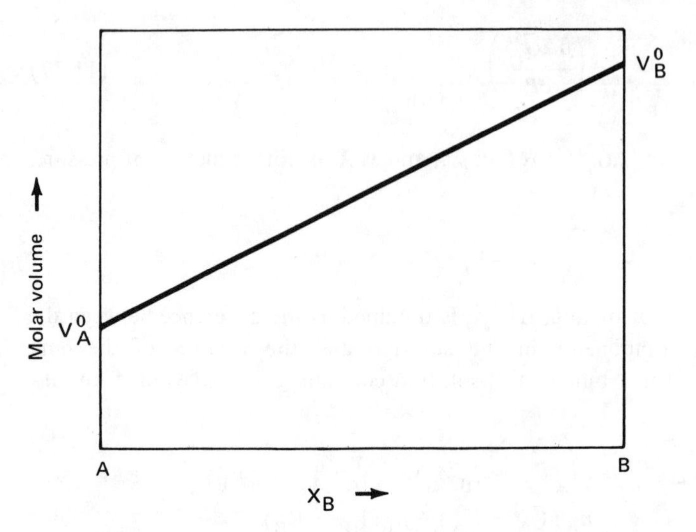

Fig. 11.6. The variation with composition of the molar volume of a Raoultian solution.

and for the pure component is

$$\left[\frac{\partial(G_i^0/T)}{\partial T}\right]_{P,\,\text{comp}} = -\frac{H_i^0}{T^2} \tag{11.40}$$

where $\bar{H}_i$ and H_i^0 are, respectively, the partial molar enthalpy of i in the solution and the standard molar enthalpy of i. Subtraction of Eq. (11.40) from Eq. (11.39) gives

$$\left[\frac{\partial\left(\dfrac{\bar{G}_i - G_i^0}{T}\right)}{\partial T}\right]_{P,\,\text{comp}} = -\frac{(\bar{H}_i - H_i^0)}{T^2}$$

or

$$\left(\frac{\partial\,(\Delta\bar{G}_i^M/T)}{\partial T}\right)_{P,\,\text{comp}} = -\frac{\Delta\bar{H}_i^M}{T^2} \tag{11.41}$$

where $\Delta\bar{H}_i^M$ is the partial molar heat of solution of i.

For an ideal solution $\Delta\bar{G}_i^M = RT\ln X_i$, substitution of which into Eq. (11.41) gives

$$\frac{d(R \ln X_i)}{dT} = -\frac{\Delta \bar{H}_i^M}{T^2}$$

and as X_i is independent of temperature, it is seen that, for a component i in an ideal solution,

$$\Delta \bar{H}_i^M = \bar{H}_i - H_i^0 = 0$$

or

$$\bar{H}_i = H_i^0 \tag{11.42}$$

The heat of formation of a solution (or heat of mixing of the components) is obtained as the difference between the heats or enthalpies of the components in solution and the heats or enthalpies of the components before mixing; i.e., for the mixing of n_A moles of A and n_B moles of B

$$\begin{aligned}\Delta H'^M &= (n_A \bar{H}_A + n_B \bar{H}_B) - (n_A H_A^0 + n_B H_B^0)\\ &= n_A(\bar{H}_A - H_A^0) + n_B(\bar{H}_B - H_B^0)\\ &= n_A \Delta \bar{H}_A^M + n_B \Delta \bar{H}_B^M\end{aligned}$$

As, for an ideal solution, $\Delta \bar{H}_i^M = 0$, it is seen that the heat of formation of an ideal solution, $\Delta H^{M,\mathrm{id}}$, is zero; i.e.,

$$\Delta H^{M,\mathrm{id}} = 0 \tag{11.43}$$

The Entropy of Formation of an Ideal Solution

The fundamental equation, Eq. (5.25), gives

$$\left(\frac{\partial G}{\partial T}\right)_{P,\,\mathrm{comp}} = -S$$

Thus for the formation of a solution

$$\left(\frac{\partial \Delta G^M}{\partial T}\right)_{P,\,\mathrm{comp}} = -\Delta S^M$$

For an ideal solution

$$\Delta G^{M,\mathrm{id}} = RT(X_A \ln X_A + X_B \ln X_B)$$

and hence

$$\Delta S^{M,\text{id}} = -\left(\frac{\partial \Delta G^{M,\text{id}}}{\partial T}\right)_{P,\,\text{comp}} = -R(X_A \ln X_A + X_B \ln X_B) \tag{11.44}$$

Equation (11.44) shows that the entropy of formation of an ideal binary solution is independent of the temperature of the solution.

Equation (4.17) gave, for the mixing of N_A particles of A with N_B particles of B,

$$\begin{aligned}\Delta S'_{\text{conf}} &= k \ln \frac{(N_A + N_B)!}{N_A!N_B!}\\ &= k[\ln (N_A + N_B)! - \ln N_A! - \ln N_B!] \qquad (4.17)\end{aligned}$$

Application of Stirling's theorem* gives

$$\begin{aligned}\Delta S'_{\text{conf}} &= k[(N_A + N_B) \ln (N_A + N_B) - (N_A + N_B)\\ &\quad -N_A \ln N_A + N_A - N_B \ln N_B + N_B]\\ &= -k\left[N_A \ln\left(\frac{N_A}{N_A + N_B}\right) + N_B \ln\left(\frac{N_B}{N_A + N_B}\right)\right]\end{aligned}$$

Now

$$\frac{N_A}{N_A + N_B} = \frac{n_A}{n_A + n_B} = X_A$$

and, similarly,

$$\frac{N_B}{N_A + N_B} = X_B$$

Also

$$N_A \text{ particles of A} = \frac{N_A}{\mathcal{A}} \text{ moles of A} = n_A \text{ moles of A}$$

and

*Stirling's theorem is $m! \doteq \sqrt{2\pi m}\, m^m e^m$; thus $\ln m! = \frac{1}{2} \ln (2\pi m) + m \ln m - m$, which, for large values of m, can be written as $\ln m! \doteq m \ln m - m$.

$$N_{\mathrm{B}} \text{ particles of B} = \frac{N_{\mathrm{B}}}{\mathcal{A}} \text{ moles of B} = n_{\mathrm{B}} \text{ moles of B}$$

where $\mathcal{A}$ is Avogadro's number. Thus

$$\Delta S'_{\mathrm{conf}} = -k\mathcal{A}(n_{\mathrm{A}} \ln X_{\mathrm{A}} + n_{\mathrm{B}} \ln X_{\mathrm{B}})$$

But as Boltzmann's constant (k) times Avogadro's number ($\mathcal{A}$) equals the Gas Constant (R),

$$\Delta S'_{\mathrm{conf}} = -R(n_{\mathrm{A}} \ln X_{\mathrm{A}} + n_{\mathrm{B}} \ln X_{\mathrm{B}})$$

Division by $n_{\mathrm{A}} + n_{\mathrm{B}}$, the total number of moles, gives

$$\Delta S_{\mathrm{conf}} = -R(X_{\mathrm{A}} \ln X_{\mathrm{A}} + X_{\mathrm{B}} \ln X_{\mathrm{B}}) \tag{11.45}$$

which is identical with Eq. (11.44). Thus the entropy increase due to the formation of an ideal solution is a measure of the increase in the number of spatial configurations which become available to the system as a result of the solution process. This is dependent only on the numbers of moles of the solution components present and is independent of the temperature. Figure 11.7 shows

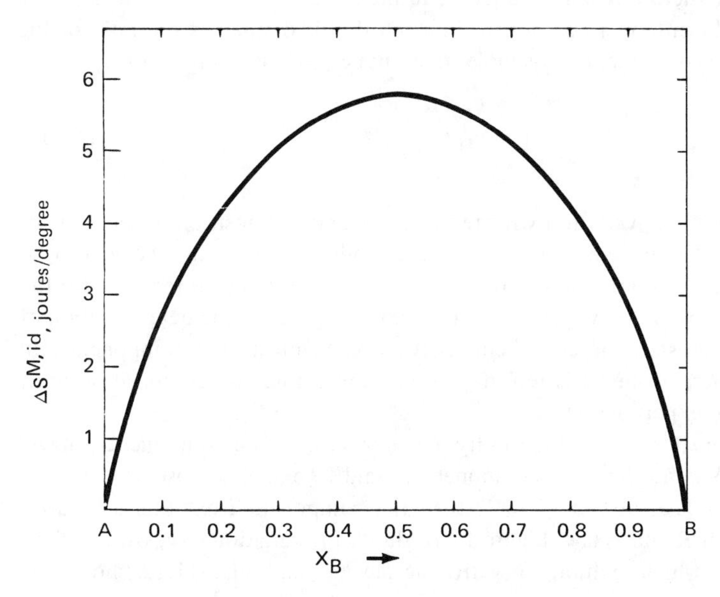

Fig. 11.7. The variation with composition of the entropy of formation of a binary Raoultian solution.

the variation of $\Delta S^{M,\text{id}}$ with composition in a binary A–B solution. As

$$\Delta S^M = X_A \Delta \bar{S}_A^M + X_B \Delta \bar{S}_B^M$$

it is seen that in an ideal solution,

$$\Delta \bar{S}_A^M = -R \ln X_A \quad \text{and} \quad \Delta \bar{S}_B^M = -R \ln X_B$$

For any solution

$$\Delta G^M = \Delta H^M - T\Delta S^M$$

and for an ideal solution, as $\Delta H^{M,\text{id}} = 0$, then

$$\Delta G^{M,\text{id}} = -T\Delta S^{M,\text{id}}$$

11.7 NONIDEAL SOLUTIONS

A nonideal solution is one in which the activities of the components are not equal to their mole fractions. However, in view of the convenience of the concept of activity and the simplicity of Raoult's law, it is convenient to define an additional thermodynamic function termed the *activity coefficient*, γ. The activity coefficient of a component in a solution is defined as the ratio of the activity of the component to its mole fraction; i.e., for the component i,

$$\gamma_i = \frac{a_i}{X_i} \tag{11.46}$$

A knowledge of the variation with temperature and composition of the value of γ_i is of central importance in solution thermodynamics, as the value of γ_i is required for the determination of the value of a_i, which in turn is required for the determination of $\Delta \bar{G}_i^M$, which again in turn is required for the determination of the equilibrium state of any chemical reaction involving the component i in solution. Generally the variation of γ_i with temperature and composition must be determined experimentally.

γ_i can be greater or less than unity, (if $\gamma_i = 1$, then the component behaves ideally). If $\gamma_i > 1$, then the component is said to exhibit a positive deviation from Raoult's law, and if $\gamma_i < 1$, then the component i exhibits a negative deviation from Raoult's law. Figure 11.10 shows the variation of a_i with X_i for a component i which exhibits negative deviations, and Fig. 11.12 shows the variation of a_i with X_i for a component which exhibits positive deviation. Figures 11.11 and 11.13 show the corresponding variations of γ_i with X_i.

If γ_i varies with temperature, then $\Delta\bar{H}_i^M$ has a nonzero value; i.e., from Eq. (11.41),

$$\frac{\partial(\Delta\bar{G}_i^M/T)}{\partial T} = -\frac{\Delta\bar{H}_i^M}{T^2}$$

and

$$\Delta\bar{G}_i^M = RT \ln a_i = RT \ln \gamma_i + RT \ln X_i$$

Thus

$$\frac{\partial(\Delta\bar{G}_i^M/T)}{\partial T} = \frac{\partial(R \ln \gamma_i)}{\partial T} = -\frac{\Delta\bar{H}_i^M}{T^2}$$

and as

$$d\left(\frac{1}{T}\right) = -\frac{dT}{T^2}$$

then

$$\frac{\partial(R \ln \gamma_i)}{\partial(1/T)} = \Delta\bar{H}_i^M \tag{11.47}$$

Generally an increase in the temperature of a nonideal solution results in a decrease in the extent to which its components deviate from ideal behavior; i.e., if $\gamma_i > 1$, then an increase in temperature decreases the value of γ_i toward unity; and if $\gamma_i < 1$, then an increase in temperature increases the value of γ_i toward unity. Thus in a solution, the components of which exhibit positive deviations from ideality, the values of the activity coefficients decrease as the temperature increases, and hence, from Eq. (11.47), the partial molar heats of solution of the components are positive quantities. Thus ΔH^M, the molar heat of formation of the solution, is a positive quantity, indicating that the solution process is endothermic. ΔH^M is the quantity of heat absorbed, per mole of the solution, from the thermostating heat reservoir surrounding the solution at the temperature T. Conversely, in a solution, the components of which exhibit negative deviations from ideality, the activity coefficients increase with increasing temperature, and hence the individual values of $\Delta\bar{H}_i^M$ and the value of ΔH^M are negative. Such a solution forms exothermically, and ΔH^M is the heat absorbed by the thermostating heat reservoir, per mole of solution formed, at the temperature T.

In the case of an A–B binary solution, exothermic mixing indicates a tendency toward compound formation between the two components. In this case the A-B attractions are greater than either the A-A or B-B attractions, and there is a tendency toward "ordering" in the solution; i.e., A atoms attempt to have only B atoms as nearest neighbors, and the B atoms attempt to have only A atoms as nearest neighbors. Endothermic mixing, on the other hand, indicates a tendency toward phase separation or "clustering" in the solution. The A-A and B-B attractions are greater than the A-B attractions, and hence the A atoms attempt to have only A atoms as nearest neighbors, and the B atoms attempt to have only B atoms as nearest neighbors. In both cases the equilibrium configuration of the solution is reached as a compromise between the enthalpy factors which, being determined by the magnitudes of the atomic interactions, attempt to either completely order or completely unmix the solution, and the entropy factor which attempts to maximize the randomness of mixing of the atoms in the solution.

11.8 APPLICATION OF THE GIBBS–DUHEM RELATIONSHIP TO ACTIVITY DETERMINATION

As was stated in Sec. 11.4, it is very often found in binary solutions that the variation with composition of the activity of only one component can be experimentally determined. In such cases the variation of the activity of the other component can be determined by means of application of the Gibbs-Duhem equation [Eq. (11.20)],

$$\sum_i X_i d\bar{Q}_i = 0$$

For a binary A–B solution, Eq. (11.20), in terms of $\Delta\bar{G}_i^M$, is given as,

$$X_A d\Delta\bar{G}_A^M + X_B d\Delta\bar{G}_B^M = 0 \tag{11.48}$$

and, as $\Delta\bar{G}_i^M = RT \ln a_i$, then

$$X_A\, d \ln a_A + X_B d \ln a_B = 0 \tag{11.49}$$

or

$$d \log a_A = -\frac{X_B}{X_A} d \log a_B \tag{11.50}$$

If the variation of a_B with composition is known, then by integration of Eq. (11.50) the value of $\log a_A$ at the composition $X_A = X_A$ is obtained as

$$\log a_A\Big|_{X_A=X_A} = -\int_{\log a_B \text{ at } X_A=1}^{\log a_B \text{ at } X_A=X_A} (X_B/X_A)\, d \log a_B \qquad (11.51)$$

As an analytical expression for the variation of a_B (and hence $\log a_B$) with composition is not usually computed, Eq. (11.51) is solved by graphical integration.

Figure 11.8 illustrates a typical variation of $\log a_B$ with composition in an A-B solution. The value of $\log a_A$ at $X_A = X_A$ is given by the shaded area under the curve. Two points are to be noticed in Fig. 11.8

1. As $X_B \rightarrow 1$, $a_B \rightarrow 1$, $\log a_B \rightarrow 0$, and $X_B/X_A \rightarrow \infty$. Thus the curve exhibits a tail to infinity as $X_B \rightarrow 1$
2. As $X_B \rightarrow 0$, $a_B \rightarrow 0$, and $\log a_B \rightarrow -\infty$. Thus the curve exhibits a tail to minus infinity as $X_B \rightarrow 0$

Of these two points, point 2 is the more serious, as the calculation of $\log a_A$ at

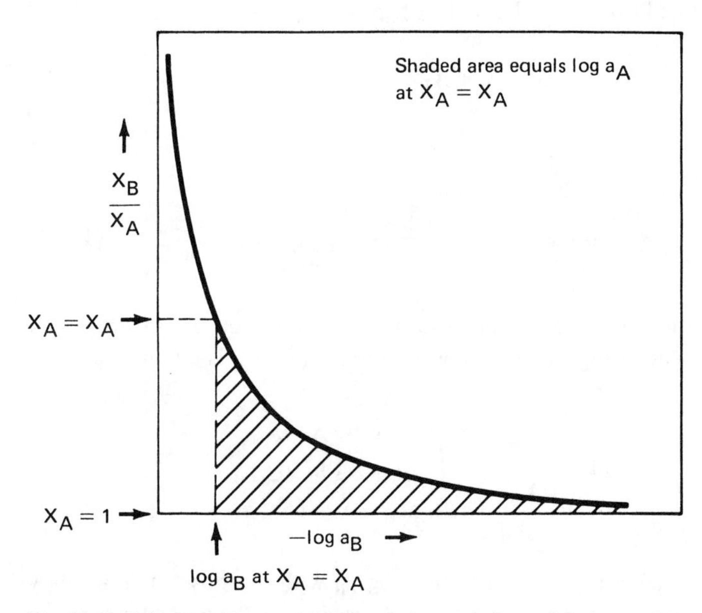

Fig. 11.8. Schematic representation of the variation of $\log a_B$ with X_B/X_A in a binary solution, and illustration of the application of the Gibbs-Duhem equation to calculation of the activity of component A.

any composition involves the evaluation of the area under a curve which tails to minus infinity. This introduces an uncertainty into the calculation.

The tail to minus infinity as $X_B \to 0$ can be avoided by considering activity coefficients instead of activities in the Gibbs-Duhem equation. In a binary A-B solution

$$X_A + X_B = 1$$

Thus

$$dX_A + dX_B = 0 \tag{11.52}$$

Therefore

$$X_A \frac{dX_A}{X_A} + X_B \frac{dX_B}{X_B} = 0$$

or

$$X_A \, d \log X_A + X_B \, d \log X_B = 0 \tag{11.53}$$

Subtraction of Eq. (11.53) from Eq. (11.49) gives

$$X_A d \log \gamma_A + X_B d \log \gamma_B = 0$$

whence

$$d \log \gamma_A = - \frac{X_B}{X_A} d \log \gamma_B \tag{11.54}$$

Thus if the variation of γ_B with composition is known, then by integration of Eq. (11.54) the value of $\log \gamma_A$ at the composition $X_A = X_A$ is obtained as

$$\log \gamma_A \Big|_{X_A = X_A} = - \int_{\log \gamma_B \text{ at } X_A=1}^{\log \gamma_B \text{ at } X_A=X_A} (X_B/X_A) d \log \gamma_B \tag{11.55}$$

Figure 11.9 illustrates a typical variation of $\log \gamma_B$ with composition in a binary A-B solution. The value of $\log \gamma_A$ at $X_B = X_B$ is given as the shaded area under the curve between the limits $\log \gamma_B$ at $X_B = X_B$ and $\log \gamma_B$ at $X_B = 0$.

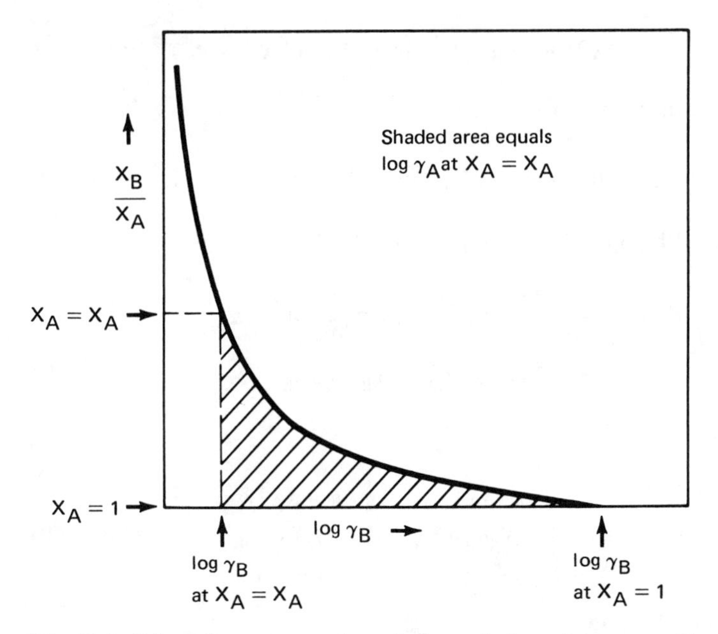

Fig. 11.9. Schematic representation of the variation of log γ_B with X_B/X_A in a binary solution, and illustration of the application of the Gibbs-Duhem equation to calculation of the activity coefficient of component A.

The α-Function

As a further aid to the integration of the Gibbs-Duhem equation, the α-function is introduced.* For the component i

$$\alpha_i = \frac{\ln \gamma_i}{(1 - X_i)^2} \tag{11.56}$$

The α-function is always finite by virtue of the fact that $\gamma_i \to 1$ as $X_i \to 1$. For the components of a binary A–B solution,

$$\alpha_A = \frac{\ln \gamma_A}{X_B^2} \quad \text{and} \quad \alpha_B = \frac{\ln \gamma_B}{X_A^2}$$

or

*L. S. Darken and R. W. Gurry, "Physical Chemistry of Metals," p. 264, McGraw-Hill Book Company, New York, 1953.

$$\ln \gamma_A = \alpha_A X_B^2 \quad \text{and} \quad \ln \gamma_B = \alpha_B X_A^2 \tag{11.57}$$

Differentiation of $\ln \gamma_B = \alpha_B X_A^2$ gives

$$d \ln \gamma_B = 2\alpha_B X_A dX_A + X_A^2 \, d\alpha_B \tag{11.58}$$

and substitution of Eq. (11.58) into Eq. (11.54) gives

$$\begin{aligned} d \ln \gamma_A &= -\frac{X_B}{X_A} 2\alpha_B X_A dX_A - \frac{X_B}{X_A} X_A^2 \, d\alpha_B \\ &= -2X_B \alpha_B dX_A - X_B X_A d\alpha_B \end{aligned} \tag{11.59}$$

Integration of Eq. (11.59) gives

$$\ln \gamma_A = -\int_{X_A=1}^{X_A=X_A} 2X_B \alpha_B \, dX_A - \int_{\alpha_B \text{ at } X_A=1}^{\alpha_B \text{ at } X_A=X_A} X_B X_A d\alpha_B \tag{11.60}$$

By virtue of the identity

$$\int d(xy) = \int y dx + \int x dy$$

the second integral on the right-hand side of Eq. (11.60) can be written as

$$\int X_B X_A d\alpha_B = \int d(X_B X_A \alpha_B) - \int \alpha_B d(X_B X_A)$$

substitution of which into Eq. (11.60) gives

$$\begin{aligned} \ln \gamma_A &= -\int 2X_B \alpha_B dX_A - \int d(X_B X_A \alpha_B) + \int \alpha_B d(X_B X_A) \\ &= -\int 2X_B \alpha_B dX_A - X_B X_A \alpha_B + \int \alpha_B X_B dX_A + \int \alpha_B X_A dX_B \\ &= -\int 2X_B \alpha_B dX_A - X_B X_A \alpha_B + \int \alpha_B X_B dX_A - \int \alpha_B X_A dX_A \\ &= -X_B X_A \alpha_B - \int (2X_B - X_B + X_A) \alpha_B dX_A = -X_B X_A \alpha_B - \int_{X_A=1}^{X_A=X_A} \alpha_B dX_A \end{aligned} \tag{11.61}$$

Thus ln γ_A at $X_A = X_A$ is obtained as $-X_B X_A \alpha_B$ minus the area under the plot of α_B versus X_A from $X_A = X_A$ to $X_A = 1$; and as α_B is everywhere finite, this integration does not involve a tail to infinity.

Figure 11.10 shows the variation of a_{Ni} with composition in the Fe–Ni system at 1600°C as determined by Zellars et al.,* and Fig. 11.11 shows the corresponding variation of γ_{Ni} with composition. Extrapolation of γ_{Ni} to $X_{Ni} =$ 0 in Fig. 11.11 gives the value of the Henry's law constant [k in Eq. (11.14)] as 0.66 for Ni in Fe at 1600°C. This, then, is the slope of the Henry's law line for Ni in Fe drawn in Fig. 11.10. The variation of γ_{Fe} with composition, shown in Fig. 11.11, is determined from consideration of either Fig. 11.14 or Fig. 11.16. Figure 11.14 shows the variation of log γ_{Ni} with X_{Ni}/X_{Fe}, graphical integration of which, according to Eq. (11.55), gives the variation of log γ_{Fe} with composition. In the graphical integration, as log γ_{Ni} increases with increasing X_{Ni}/X_{Fe}, the integrated area under the curve between $X_{Ni} = X_{Ni}$ and $X_{Ni} = 0$

*G. R. Zellars, S. L. Payne, J. P. Morris, and R. L. Kipp, The Activities of Iron and Nickel in Liquid Fe–Ni Alloys, *Trans. AIME*, **215**:181 (1959).

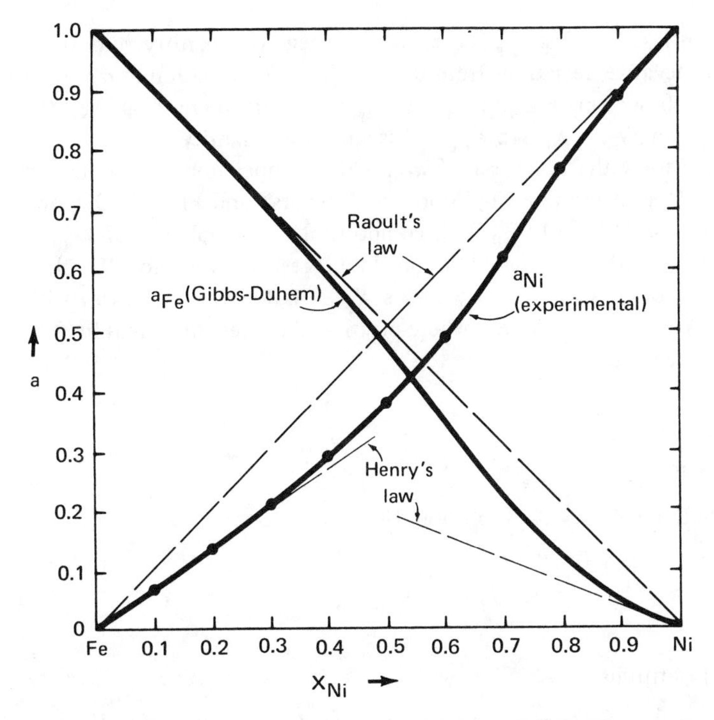

Fig. 11.10. Activities in the system iron-nickel at 1600°C. *(From G. R. Zellars, S. L. Payne, J. P. Morris, and R. L. Kipp, The Activities of Iron and Nickel in Liquid Fe-Ni Alloys, Trans. AIME, 215:181 [1959].)*

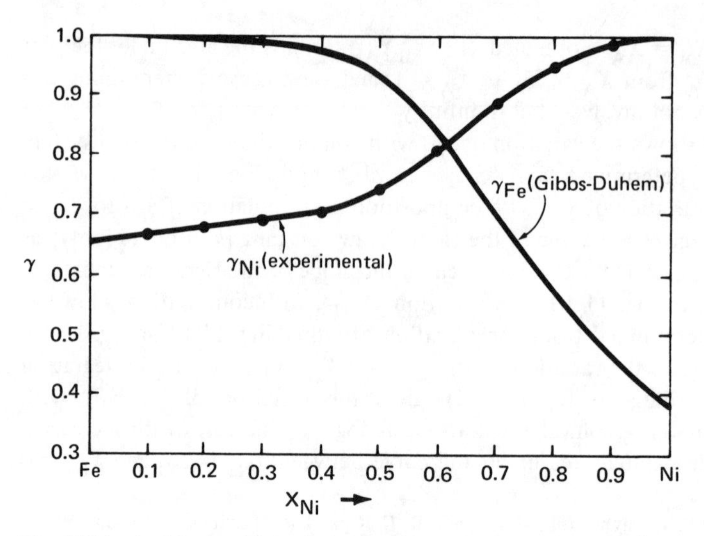

Fig. 11.11. Activity coefficients in the system iron-nickel at 1600°C.

is a positive quantity. Thus $\log \gamma_{Fe}$ is everywhere a negative quantity, and so Fe, like Ni, exhibits negative deviations from Raoult's law. The variation of α_{Ni} with composition is shown in Fig. 11.16. As α_{Ni} is everywhere negative, the integrated area from $X_{Fe} = X_{Fe}$ to $X_{Fe} = 1$ is a positive quantity.

Figure 11.12 shows the variation of a_{Cu} with composition in the Fe-Cu system at 1550°C as determined by Morris and Zellars* and Fig. 11.13 shows the corresponding variation of γ_{Cu} with composition. Extrapolation of γ_{Cu} to $X_{Cu} = 0$ gives $k_{Cu} = 10.1$. Figures 11.15 and 11.17, respectively, show the plots of $\log \gamma_{Cu}$ versus X_{Cu}/X_{Fe} and α_{Cu} versus X_{Fe}. As $\log \gamma_{Cu}$ decreases with increasing X_{Cu}/X_{Fe}, and α_{Cu} is everywhere positive, the integrated areas in both figures are negative quantities.

The Relationship between Henry's and Raoult's Laws

For the solute B in a binary A-B solution, Henry's law gives

$$a_B = k_B X_B$$

or, in terms of logarithms,

*J. P. Morris and G. R. Zellars, Vapor Pressure of Liquid Copper and Activities in Liquid Fe-Cu Alloys, *Trans. AIME,* **206**:1086 (1956).

$$\log a_B = \log k_B + \log X_B$$

differentiation of which gives,

$$d \log a_B = d \log X_B$$

Inserting this into the Gibbs-Duhem equation [Eq. (11.49)] gives

$$d \ln a_A = -\frac{X_B}{X_A} d \ln X_B = -\frac{X_B}{X_A}\frac{dX_B}{X_B} = -\frac{dX_B}{X_A}$$

$$= \frac{dX_A}{X_A} = d \ln X_A$$

Integration gives

$$\ln a_A = \ln X_A + \ln \text{constant}$$

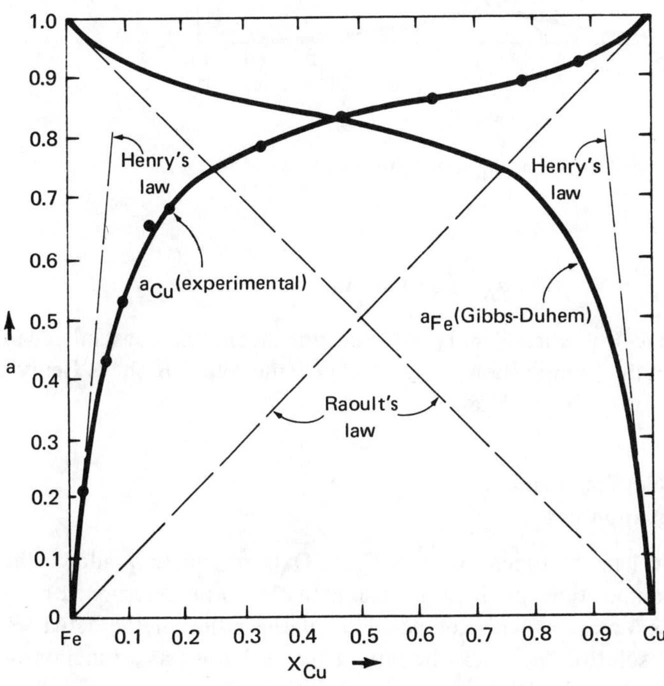

Fig. 11.12. Activities in the system iron-copper at 1550°C. *(From J. P. Morris and G. R. Zellars, Vapor Pressure of Liquid Copper and Activities in Liquid Fe-Cu Alloys, Trans. AIME, **206**:1086 [1956].)*

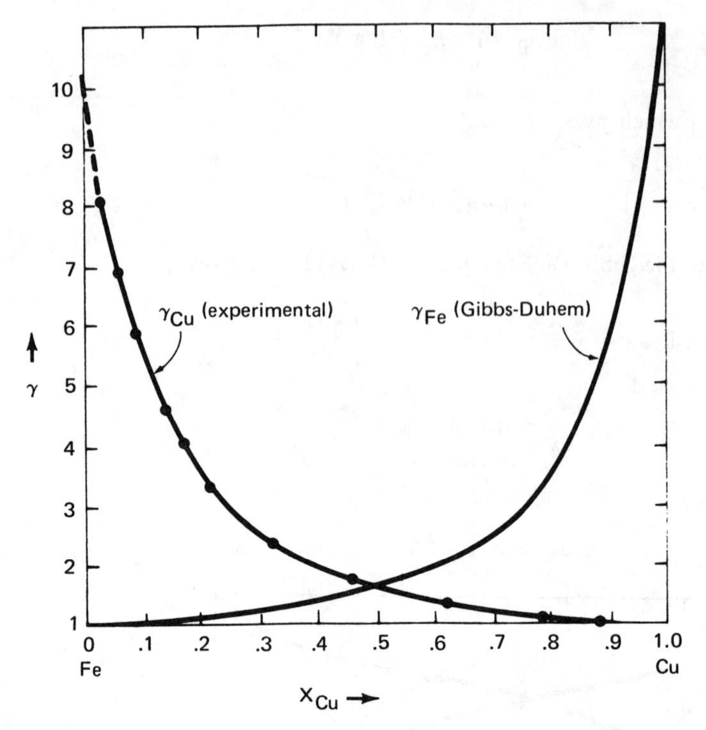

Fig. 11.13. Activity coefficients in the system iron-copper at 1550°C.

or

$$a_A = \text{constant } X_A$$

But by definition, $a_i = 1$ when $X_i = 1$, and hence the integration constant equals unity. Thus, over the composition range in which the solute B obeys Henry's law, the solvent A obeys Raoult's law.

The Belton-Fruehan Treatment of the Gibbs-Duhem Equation

The Belton-Fruehan treatment of the Gibbs-Duhem equation allows the variation, with composition, of the individual activities to be determined from knowledge of the variation, with composition, of the ratios of the activities. Consider an A-B solution in which the ratio a_A/a_B is known as a function of composition. Equation (11.49) gives

$$X_A d \ln a_A + X_B d \ln a_B = 0$$

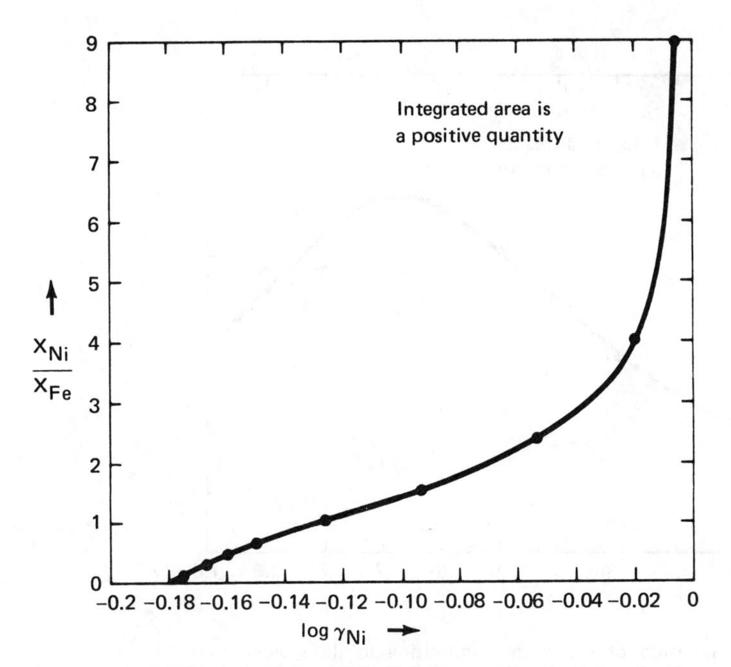

Fig. 11.14. Application of the Gibbs-Duhem equation to determination of the activities of iron in the system iron-nickel.

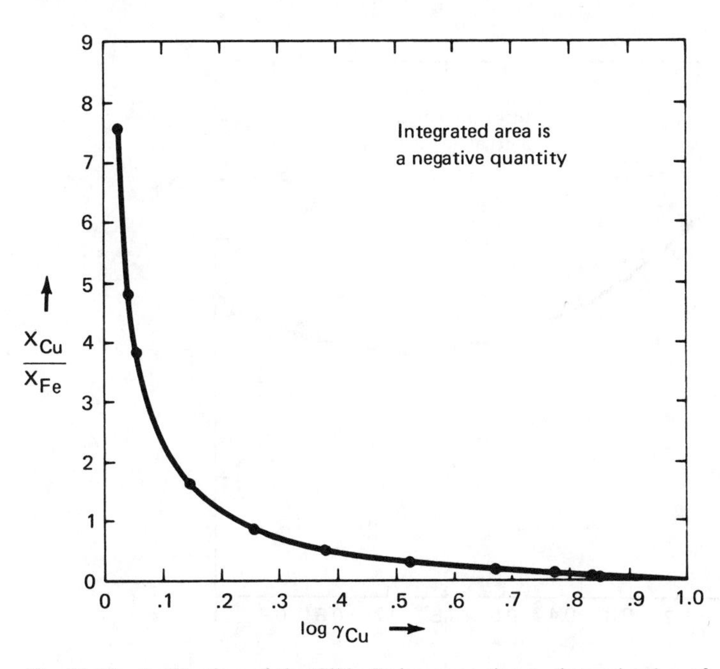

Fig. 11.15. Application of the Gibbs-Duhem equation to determination of the activities of iron in the system iron-copper.

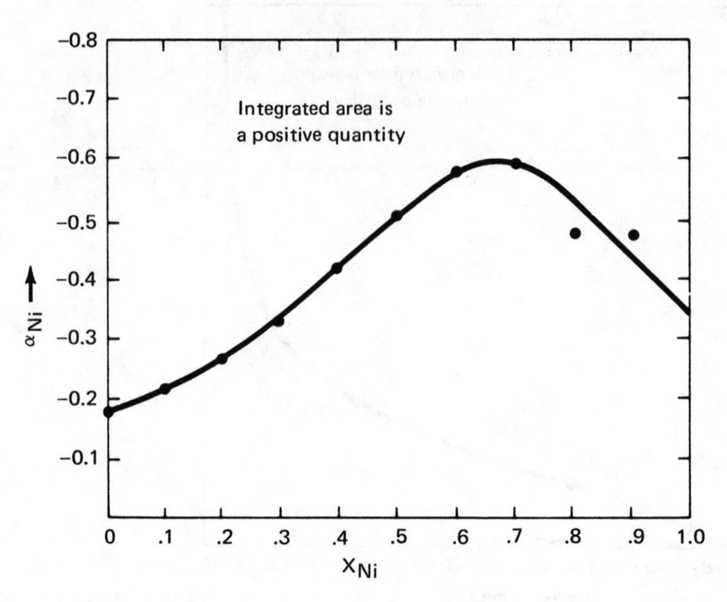

Fig. 11.16. The variation of α_{Ni} with composition in the system iron-nickel.

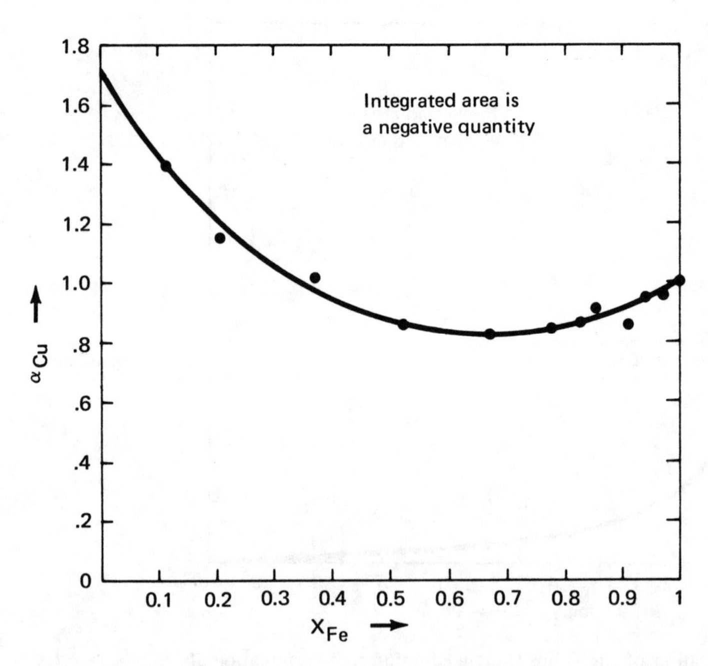

Fig. 11.17. The variation of α_{Cu} with composition in the system iron-copper.

Subtracting $d \ln a_B$ from both sides gives

$$X_A d \ln a_A + X_B d \ln a_B - d \ln a_B = -d \ln a_B$$

and as $X_B = (1 - X_A)$, then

$$X_A d \ln a_A + d \ln a_B - X_A d \ln a_B - d \ln a_B = -d \ln a_B$$

or

$$d \ln a_B = -X_A d \ln\left(\frac{a_A}{a_B}\right) \tag{11.62}$$

Equation (11.53) gives

$$X_A d \ln X_A + X_B d \ln X_B = 0$$

and subtracting $d \ln X_B$ from both sides and rearranging as above gives

$$d \ln X_B = -X_A d \ln\left(\frac{X_A}{X_B}\right) \tag{11.63}$$

Subtracting Eq. (11.63) from Eq. (11.62) gives

$$d \ln a_B - d \ln X_B = -X_A \left[d \ln\left(\frac{a_A}{a_B}\right) - d \ln\left(\frac{X_A}{X_B}\right)\right]$$

or

$$d \ln \gamma_B = -X_A d \left[\ln\left(\frac{a_A}{a_B}\right) - \ln\left(\frac{X_A}{X_B}\right)\right]$$

whence

$$\log \gamma_B\big|_{X_B} = -\int_{X_B=1}^{X_B=X_B} X_A d \left[\log\left(\frac{a_A}{a_B}\right) - \log\left(\frac{X_A}{X_B}\right)\right] \tag{11.64}$$

Figure 11.18 shows the variation of $\log(a_{Fe}/a_{Ni})$ with composition as determined experimentally by Belton and Fruehan* at 1600°C. Included in Fig. 11.18 is the corresponding variation of $\log(X_{Fe}/X_{Ni})$ with composition. Fig. 11.19 shows the variation of $[\log(a_{Fe}/a_{Ni}) - \log(X_{Fe}/X_{Ni})]$ with X_{Fe},

*G. R. Belton and R. J. Fruehan, The Determination of Activities by Mass Spectrometry, I: The Liquid Metallic Systems Iron-Nickel and Iron-Cobalt, *J. Phys. Chem.*, **71**:1403 (1967).

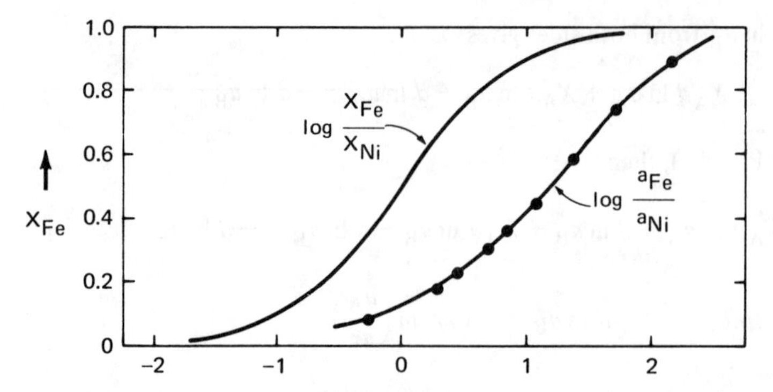

Fig. 11.18. The variations of $\log(X_{Fe}/X_{Ni})$ and $\log(a_{Fe}/a_{Ni})$ with composition in the system iron-nickel at 1600°C.

graphical integration of which from $[\log(a_{Fe}/a_{Ni}) - \log(X_{Fe}/X_{Ni})]$ at $X_{Fe} = X_{Fe}$ to $[\log(a_{Fe}/a_{Ni}) - \log(X_{Fe}/X_{Ni})]$ at $X_{Fe} = 0$ gives the value of $-\log \gamma_{Ni}$ at $X_{Ni} = X_{Ni}$. As the term $[\log(a_{Fe}/a_{Ni}) - \log(X_{Fe}/X_{Ni})]$ everywhere increases with increasing X_{Fe}, the integrated area is always positive, and hence γ_{Ni} is everywhere less than unity. The variation of $\log \gamma_{Fe}$ with composition is obtained by graphical integration of the equation

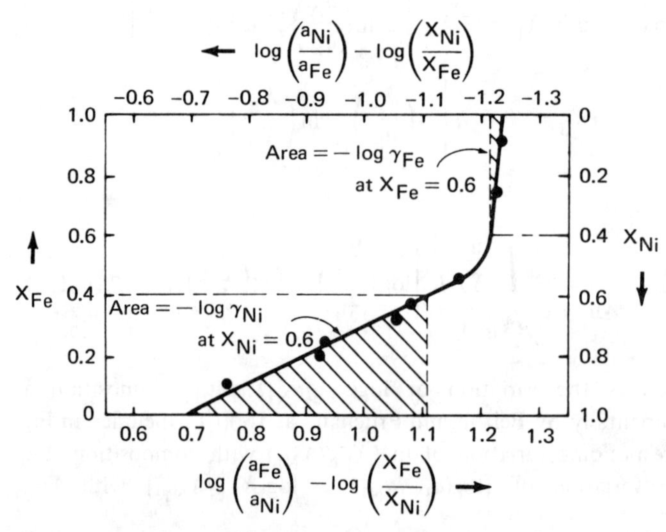

Fig. 11.19. The Belton-Fruehan plot for determination of the activities of Fe and Ni in iron-nickel solutions. *(From G. R. Belton and R. J. Fruehan, The Determination of Activities by Mass Spectrometry, I: The Liquid Metallic Systems Iron-Nickel and Iron-Cobalt, J. Phys. Chem., 71:1403 [1967].)*

$$\log \gamma_{Fe} = -\int_{X_{Fe}=1}^{X_{Fe}=X_{Fe}} X_{Ni} d\left[\log\left(\frac{a_{Ni}}{a_{Fe}}\right) - \log\left(\frac{X_{Ni}}{X_{Fe}}\right)\right]$$

and as, in Fig. 11.19, the term $[\log(a_{Ni}/a_{Fe}) - \log(X_{Ni}/X_{Fe})]$ everywhere increases with X_{Ni}, the integrated area is always positive; and hence γ_{Fe} is everywhere less than unity. Figure 11.20 shows the derived variation, with composition, of a_{Ni} and a_{Fe}.

Figure 11.21 shows the variation of $\log(a_{Co}/a_{Fe})$ with composition in Co-Fe melts at 1590°C as determined by Belton and Fruehan, and Fig. 11.22 shows the corresponding plot of $[\log(a_{Co}/a_{Fe}) - \log(X_{Co}/X_{Fe})]$ versus X_{Co}. With increasing X_{Co} the integrated area in Fig. 11.22 is initially a positive quantity, increasing to the composition $X_{Co} = 0.35$, whereafter it rapidly decreases, becoming zero at $X_{Co} = 0.44$, and thereafter becomes an increasingly negative quantity. Thus, as is seen in Fig. 11.23, γ_{Fe} is less than unity in the range $X_{Fe} = 0.56$ to $X_{Fe} = 1$, and is greater than unity in the range $X_{Fe} = 0.56$ to $X_{Fe} = 0$. The derived variation of a_{Fe} with composition is shown in Fig. 11.24. Integration of the equation

$$\log \gamma_{Co} = -\int_{X_{Co}=1}^{X_{Co}=X_{Co}} X_{Fe} d\left[\log\left(\frac{a_{Fe}}{a_{Co}}\right) - \log\left(\frac{X_{Fe}}{X_{Co}}\right)\right]$$

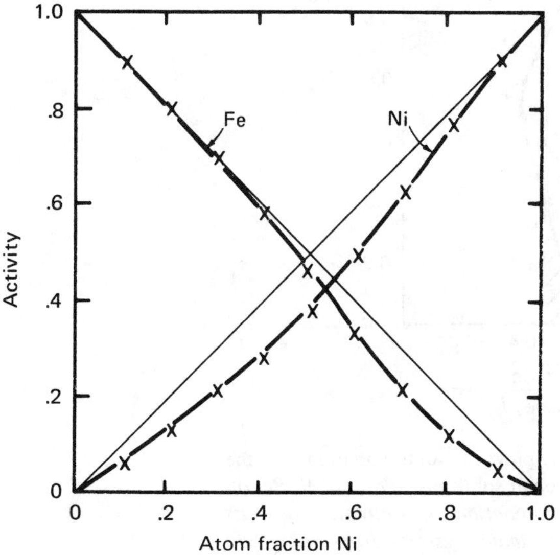

Fig. 11.20. The activities in the system iron-nickel (Belton and Fruehan–*full line*; Zellars et al.–*crosses*).

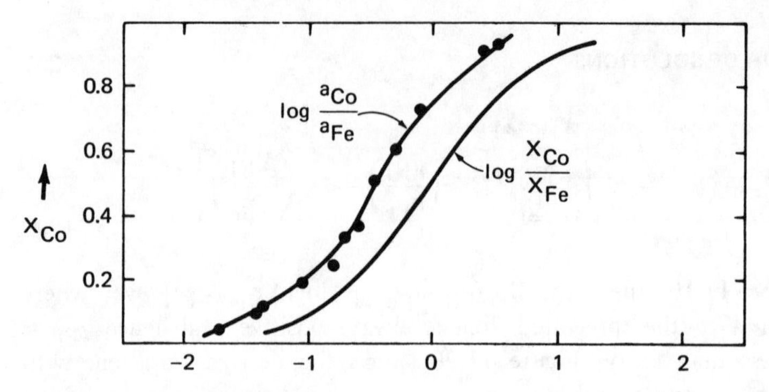

Fig. 11.21. The variations with composition of $\log(a_{Co}/a_{Fe})$ and $\log(X_{Co}/X_{Fe})$ in the system iron-cobalt at 1600°C.

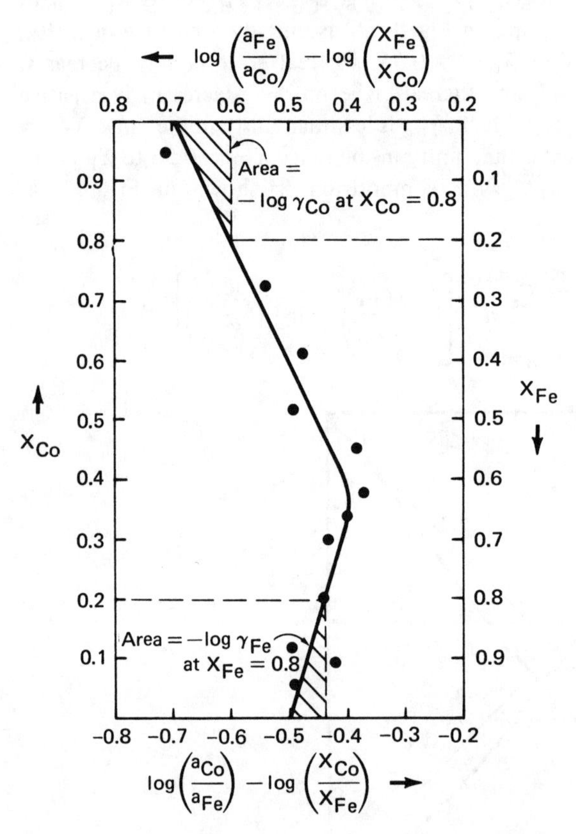

Fig. 11.22. The Belton-Fruehan plot for determination of the activities of Fe and Co in iron-cobalt solutions. *(From G. R. Belton and R. J. Fruehan, The Determination of Activities by Mass Spectrometry, I: The Liquid Metallic Systems Iron-Nickel and Iron-Cobalt, J. Phys. Chem., 71:1403 [1967].)*

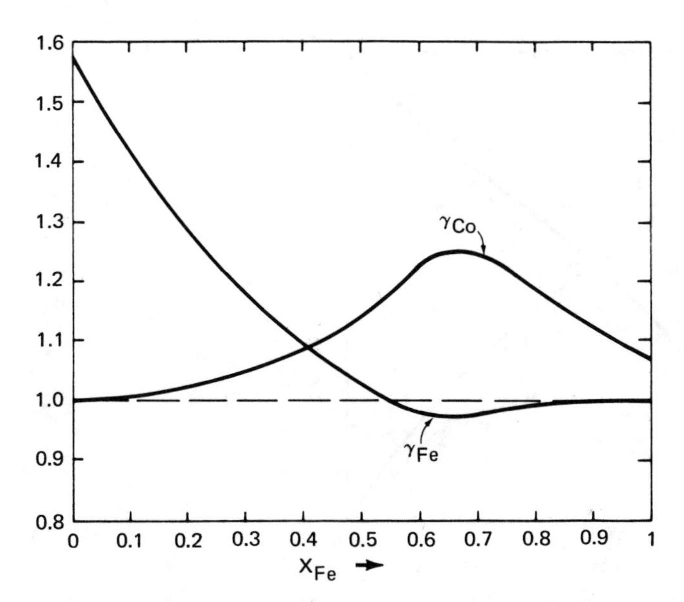

Fig. 11.23. The activity coefficients in the system iron-cobalt at 1600°C. *(From G. R. Belton and R. J. Fruehan, The Determination of Activities by Mass Spectrometry, I: The Liquid Metallic Systems Iron-Nickel and Iron-Cobalt, J. Phys. Chem., 71:1403 [1967].)*

shows that, as $\log(a_{Fe}/a_{Co}) - \log(X_{Fe}/X_{Co})$ decreases with increasing X_{Fe} up to $X_{Fe} = 0.65$ and thereafter increases with increasing X_{Fe} the integrated area passes through a maximum negative value at $X_{Fe} = 0.65$. Thus the value of γ_{Co}, which is everywhere greater than unity, passes through a maximum at $X_{Co} = 0.35$. This is shown in Fig. 11.23, and the corresponding variation of a_{Co} with composition is shown in Fig. 11.24.

Direct Calculation of the Integral Free Energy of the Solution

Equation (11.33*b*) gave

$$\Delta \bar{G}_A^M = \Delta G^M + \frac{X_B d\Delta G^M}{dX_A}$$

Rearranging and dividing by X_B^2 gives

$$\frac{\Delta \bar{G}_A^M dX_A}{X_B^2} = \frac{X_B d\Delta G^M - \Delta G^M dX_B}{X_B^2} = d\left(\frac{\Delta G^M}{X_B}\right)$$

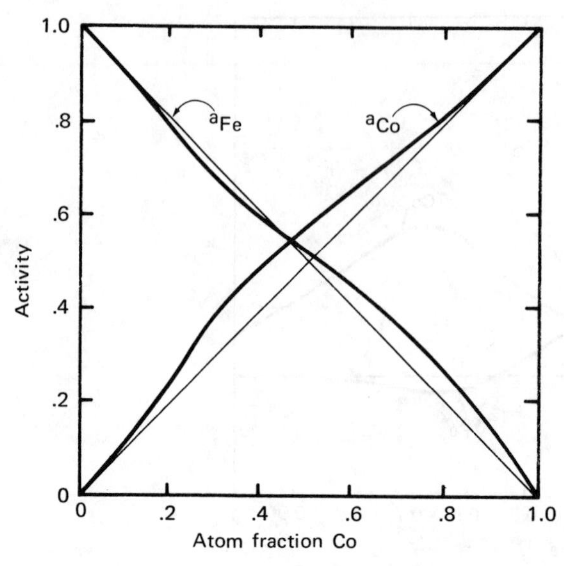

Fig. 11.24. The activities in the system iron-cobalt at 1600°C. *(From G. R. Belton and R. J. Fruehan, The Determination of Activities by Mass Spectrometry, I: The Liquid Metallic Systems Iron-Nickel and Iron-Cobalt, J. Phys. Chem., 71:1403 [1967].)*

or

$$d\left(\frac{\Delta G^M}{X_B}\right) = \frac{\Delta \bar{G}_A^M}{X_B^2}\, dX_A$$

Integrating between $X_A = X_A$ and $X_A = 0$ gives

$$\Delta G^M = X_B \int_0^{X_A} \frac{\Delta \bar{G}_A^M}{X_B^2}\, dX_A \tag{11.65}$$

and as $\Delta \bar{G}_A^M = RT \ln a_A$, the integral free energy of mixing A and B can be obtained directly from the variation of a_A with composition as

$$\Delta G^M = RTX_B \int_0^{X_A} \frac{\ln a_A}{X_B^2}\, dX_A \tag{11.66}$$

For the experimentally determined activities of Ni in Fe and Cu in Fe illustrated in Figs. 11.10 and 11.12,

$$\Delta G^M_{\text{(in the system Ni}-\text{Fe)}} = RTX_{\text{Fe}} \int_0^{X_{\text{Ni}}} \frac{\ln a_{\text{Ni}}}{X_{\text{Fe}}^2} \, dX_{\text{Ni}}$$

and

$$\Delta G^M_{\text{(in the system Cu}-\text{Fe)}} = RTX_{\text{Fe}} \int_0^{X_{\text{Cu}}} \frac{\ln a_{\text{Cu}}}{X_{\text{Fe}}^2} \, dX_{\text{Cu}}$$

The graphical integrations of these equations are illustrated in Fig. 11.25, where line (a) is $(\ln a_{\text{Cu}})/X_{\text{Fe}}^2$ versus X_{Cu}, and line (c) is $(\ln a_{\text{Ni}})/X_{\text{Fe}}^2$ versus X_{Ni}. Line (b) shows the variation of $(\ln X_i)/(1-X_i)^2$ with X_i, which is the variation of the function for a component i which exhibits Raoultian behavior. As is seen, some uncertainty is introduced into the integration by virtue of the fact that the function $(\ln a_i)/(1-X_i)^2 \to -\infty$ as $X_i \to 0$. In Fig. 11.25 the shaded area (which is the value of the integral between $X_{\text{Cu}} = 0.5$ and $X_{\text{Cu}} = 0$), multiplied by the factor 2.303 X 8.3144 X 1823 X 0.5 equals ΔG^M in the system at $X_{\text{Fe}} = 0.5$.

The ΔG^M-composition curves obtained from the graphical integrations are shown in Fig. 11.26. With respect to the Raoultian solution (line b), it is seen that

$$\begin{aligned}
\Delta G^M &= RT(1-X_i) \int_0^{X_i} \frac{\ln X_i}{(1-X_i)^2} \, dX_i \\
&= RT(1-X_i) \left[\frac{X_i \ln X_i}{1-X_i} + \ln(1-X_i) \right] \\
&= RT(X_i \ln X_i + (1-X_i) \ln(1-X_i))
\end{aligned}$$

in agreement with Eq. (11.34).

The uncertainty due to the infinite tail as $X_i \to 0$ can be eliminated if the equation is used to calculate the integral excess free energy (see Sec. 11.9).

Equation (11.65) is a general equation which relates the integral and partial molar values of any extensive thermodynamic function; e.g.,

$$\Delta H^{\text{M}} = X_{\text{B}} \int_0^{X_{\text{A}}} \frac{\Delta \bar{H}_{\text{A}}^{\text{M}}}{X_{\text{B}}^2} \, dX_{\text{A}} \tag{11.67}$$

and

$$\Delta S^{\text{M}} = X_{\text{B}} \int_0^{X_{\text{A}}} \frac{\Delta \bar{S}_{\text{A}}^{\text{M}}}{X_{\text{B}}^2} \, dX_{\text{A}} \tag{11.68}$$

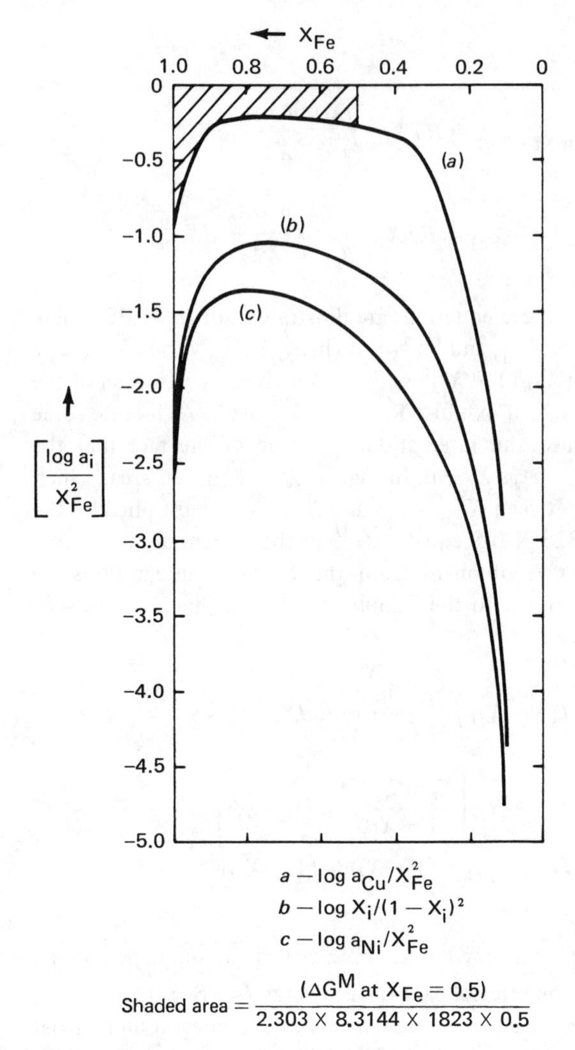

Fig. 11.25. Illustration of the direct calculation of the integral free energies of mixing in the systems iron-copper at 1550°C and iron-nickel at 1600°C.

11.9 REGULAR SOLUTIONS

Thus far two classes of solution have been distinguished:

1. Ideal or Raoultian solutions in which $a_i = X_i$, $\Delta \overline{H}_i^M = 0$, $\Delta \overline{V}_i^M = 0$, and $\Delta \overline{S}_i^M = -R \ln X_i$
2. Nonideal solutions in which $a_i \neq X_i$ and $\Delta \overline{H}_i^M \neq 0$

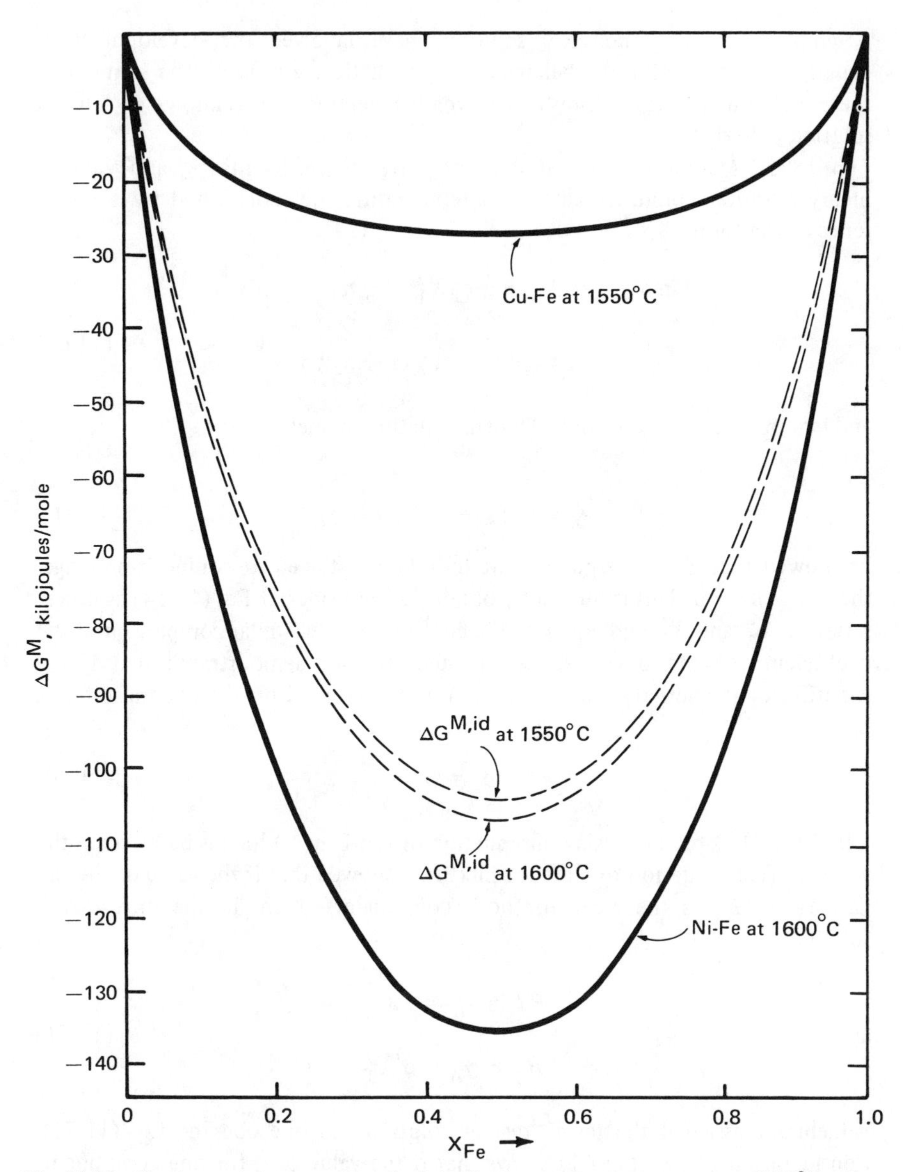

Fig. 11.26. The integral free energies of mixing in the systems iron-copper at 1550°C and iron-nickel at 1600°C.

Attempts to classify nonideal solutions have involved the development of equations that describe the behavior of hypothetical solutions. The simplest of these mathematical formalisms is that which generates what is known as "regular solution behavior."

In 1895* Margules suggested that the activity coefficients γ_A and γ_B of a binary solution could, at any given temperature, be represented by a power series of the form

$$\ln \gamma_A = \alpha_1 X_B + \frac{1}{2}\alpha_2 X_B^2 + \frac{1}{3}\alpha_3 X_B^3 + \cdots$$

$$\ln \gamma_B = \beta_1 X_A + \frac{1}{2}\beta_2 X_A^2 + \frac{1}{3}\beta_3 X_A^3 + \cdots \qquad (11.69)$$

and by application of the Gibbs-Duhem equation, namely,

$$X_A d \ln \gamma_A = - X_B d \ln \gamma_B \qquad (11.54)$$

he showed that if these equations are to hold over the entire composition range, then $\alpha_1 = \beta_1 = 0$. This is proved by obtaining both sides of Eq. (11.54) as power series of X_A and X_B and equating the coefficients. By similar comparison of the coefficients of the power series, Margules further demonstrated that if the variation of the activity coefficients can be represented by the quadratic terms only, then

$$\alpha_2 = \beta_2 \qquad (11.70)$$

In 1929 Hildebrand,† using an equation of van Laar‡ which is based upon the van der Waals equation of state for mixtures, showed that if the value of the van der Waals "*b*" is the same for both components, then, in the binary A-B solution,

$$RT \ln \gamma_B = \alpha' X_A^2$$

and

$$RT \ln \gamma_A = \alpha' X_B^2 \qquad (11.71)$$

Hildebrand assigned the term "regular solution" to one obeying Eq. (11.71). Consideration of Eq. (11.61) shows that if the value of α for one component, say component B, is independent of composition, then

*M. Margules, Uber die Zusammensetzung der gesattigten Dampfe von Mischungen, *Sitzungsberichte. Akad. Wiss. Vienna*, **104**:1243 (1895).

†J. H. Hildebrand, Solubility XII, Regular Solutions, *J. Am. Chem. Soc.*, **51**:66 (1929).

‡J. J. van Laar, The Vapor Pressure of Binary Mixtures, *Z. Physik. Chem.*, **72**:723 (1910).

$$\begin{aligned}\ln \gamma_A &= -X_A X_B \alpha_B - \alpha_B (X_A - 1)\\ &= -X_A X_B \alpha_B + \alpha_B X_B\\ &= \alpha_B X_B (1 - X_A)\\ &= \alpha_B X_B^2\end{aligned}$$

But as Eq. (11.57) gave $\ln \gamma_A = \alpha_A X_B^2$, it is seen that

$$\alpha_A = \alpha_B = \alpha$$

From Eq. (11.71), α for a regular solution is an inverse function of temperature, i.e.,

$$\alpha = \frac{\alpha'}{RT} \tag{11.72}$$

Hildebrand defined a regular solution as one which has a nonzero heat of formation and an ideal entropy of formation; i.e., for the component i of a regular solution,

$$\Delta \bar{H}_i^M \neq 0 \quad \text{and} \quad \Delta \bar{S}_i^M = \Delta \bar{S}_i^{M,\text{id}} = -R \ln X_i$$

The properties of a regular solution are best examined via the concept of excess functions. The excess value of an extensive thermodynamic solution property is simply the difference between its actual value and the value it would have were the solution ideal; e.g., in terms of the free energy of the solution,

$$G = G^{\text{id}} + G^{\text{xs}} \tag{11.73}$$

where G = the molar free energy of the solution
G^{id} = the molar free energy which the solution would have were it ideal
G^{xs} = the excess molar free energy of the solution

Subtraction of the free energy of the unmixed components, $(X_A G_A^0 + X_B G_B^0)$, from both sides of Eq. (11.73) gives

$$\Delta G^M = \Delta G^{M,\text{id}} + G^{\text{xs}} \tag{11.74}$$

As for any solution,

$$\Delta G^M = \Delta H^M - T\Delta S^M$$

and, for an ideal solution,

$$\Delta G^{M,\mathrm{id}} = -T\Delta S^{M,\mathrm{id}}$$

then
$$G^{\mathrm{xs}} = \Delta G^M - \Delta G^{M,\mathrm{id}} = \Delta H^M - T(\Delta S^M - \Delta S^{M,\mathrm{id}}) \tag{11.75}$$

As, for a regular solution, $\Delta S^M = \Delta S^{M,\mathrm{id}}$, then

$$G^{\mathrm{xs}} = \Delta H^M \tag{11.76}$$

Now
$$\begin{aligned}\Delta G^M &= RT(X_A \ln a_A + X_B \ln a_B)\\ &= RT(X_A \ln X_A + X_B \ln X_B) + RT(X_A \ln \gamma_A + X_B \ln \gamma_B)\end{aligned}$$

and as
$$\Delta G^{M,\mathrm{id}} = RT(X_A \ln X_A + X_B \ln X_B)$$

then
$$G^{\mathrm{xs}} = RT(X_A \ln \gamma_A + X_B \ln \gamma_B) \tag{11.77}$$

For a regular solution, $\ln \gamma_A = \alpha X_B^2$ and $\ln \gamma_B = \alpha X_A^2$, substitution of which into Eq. (11.77) gives

$$G^{\mathrm{xs}} = RT\alpha X_A X_B \tag{11.78}$$

or, from Eq. (11.72), gives

$$G^{\mathrm{xs}} = \alpha' X_A X_B \tag{11.79}$$

It is thus seen that G^{xs} for a regular solution is independent of temperature. That G^{xs} (and hence ΔH^M) for a regular solution are independent of temperature can also be demonstrated as follows.

$$\left(\frac{\partial G^{\mathrm{xs}}}{\partial T}\right)_{P,\,\mathrm{comp}} = -S^{\mathrm{xs}}$$

and as S^{xs} for a regular solution is zero, then G^{xs} is independent of temperature. From Eqs. (11.78) and (11.79), at any given composition,

$$\bar{G}_A^{\mathrm{xs}} = RT_1 \ln \gamma_{A(T_1)} = RT_2 \ln \gamma_{A(T_2)} = \alpha' X_B^2$$

and hence, for a regular solution,

$$\frac{\ln \gamma_A \text{ at the temperature } T_2}{\ln \gamma_A \text{ at the temperature } T_1} = \frac{T_1}{T_2} \tag{11.80}$$

Equation (11.80) is of considerable practical use in converting activity data at one temperature to activity data at another temperature.

Figures 11.27 and 11.28, respectively, show the symmetrical variations with composition of the activities and activity coefficients in the tin-thallium system at three temperatures as determined experimentally by Hildebrand and Sharma,* and Fig. 11.29 shows the linear variations of log γ_{Tl} with X_{Sn}^2, the slopes of which equal α at the given temperatures. Inasmuch as the variation of γ_i with X_i at a given temperature is concerned, this system exhibits the behavior of a regular solution; but Fig. 11.30 shows that αT, which, for strict adherence to the model, should be independent of T, decreases slowly as the temperature is increased. Figure 11.31 shows the variation with composition of ΔG^M, ΔH^M and $-T\Delta S^M$ for the Tl-Sn system at 414°C. It is to be noted that a parabolic relationship for ΔH^M or G^{xs} should not be taken as being a demonstration that the solution is regular, as it is frequently found that G^{xs} or ΔH^M can be

*J. H. Hildebrand and J. N. Sharma, The Activities of Molten Alloys of Thallium with Tin and Lead, *J. Am. Chem. Soc.*, **51**:462 (1929).

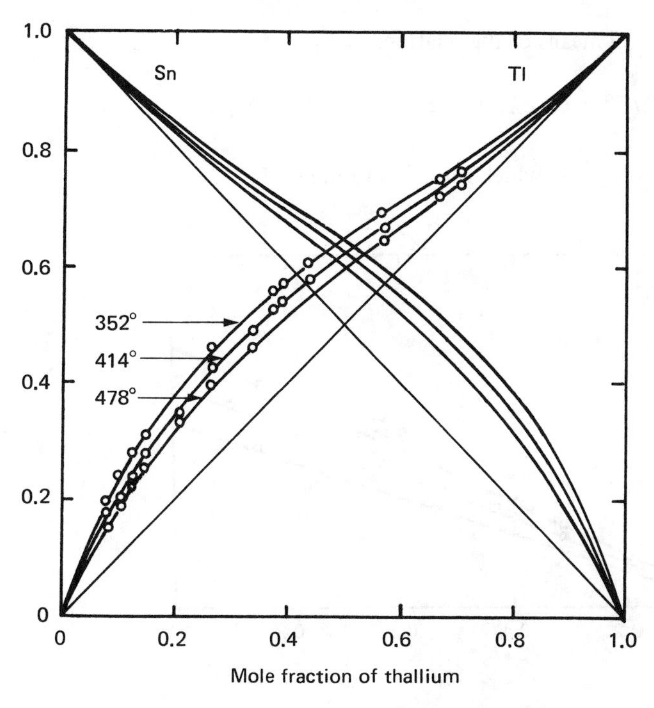

Fig. 11.27. Activities in the system thallium-tin. *(From J. H. Hildebrand and J. N. Sharma, The Activities of Molten Alloys of Thallium with Tin and Lead, J. Am. Chem. Soc., 51:462 [1929].)*

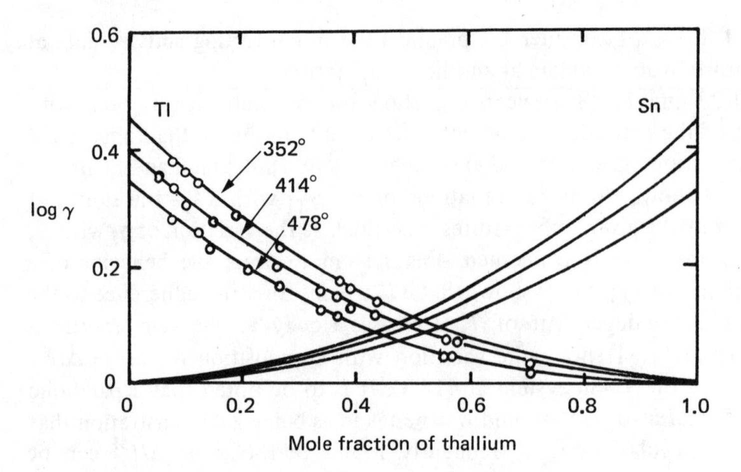

Fig. 11.28. Activity coefficients in the system thallium-tin. *(From J. H. Hildebrand and J. N. Sharma, The Activities of Molten Alloys of Thallium with Tin and Lead, J. Am. Chem. Soc., 51:462 [1929].)*

adequately expressed by means of the relations

$$\Delta H^M = bX_A X_B \qquad \text{or} \qquad G^{xs} = b' X_A X_B$$

where b and b' are unequal, in which case, from Eq. (11.75),

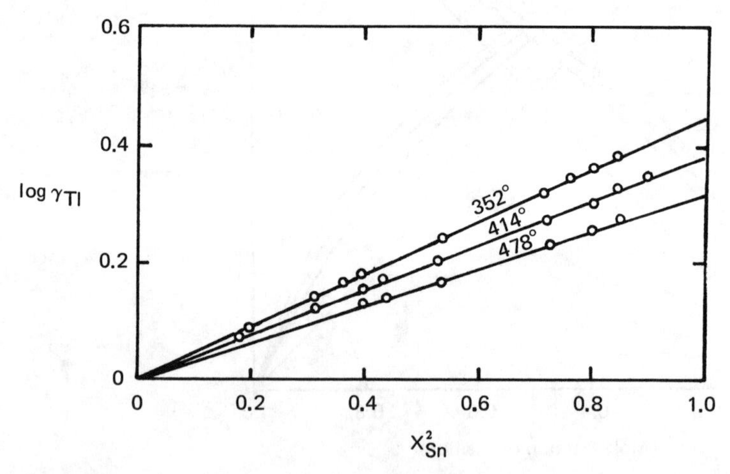

Fig. 11.29. Log γ_{Tl} versus X_{Sn}^2 in the system thallium-tin. *(From J. H. Hildebrand and J. N. Sharma, The Activities of Molten Alloys of Thallium with Tin and Lead, J. Am. Chem. Soc., 51:462 [1929].)*

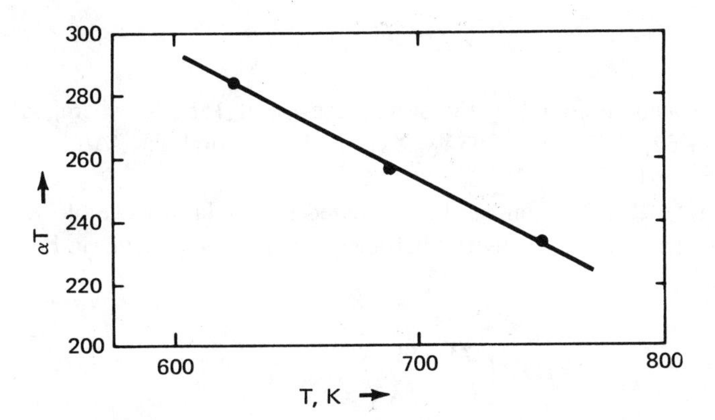

Fig. 11.30. The product αT versus T for the system thallium-tin.

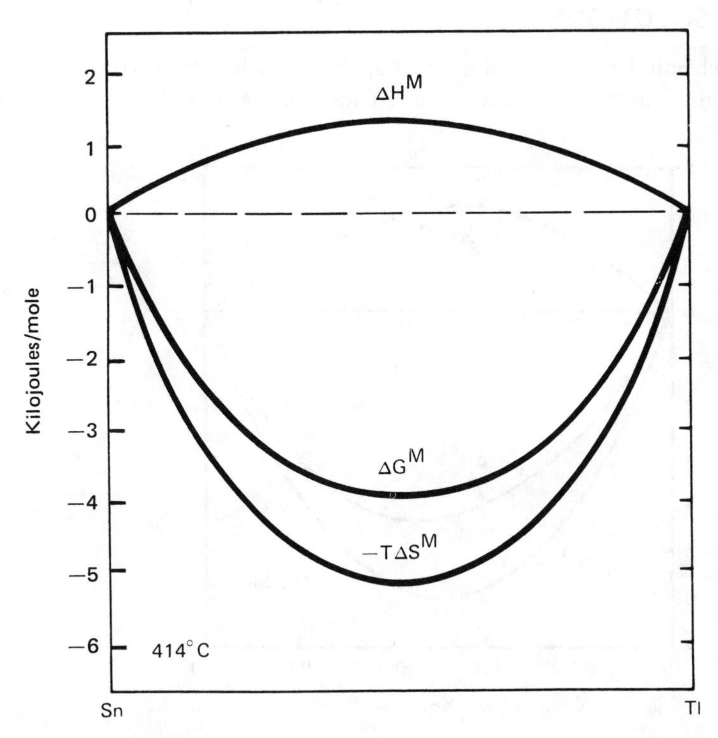

Fig. 11.31. The enthalpy, entropy, and free energy of mixing of thallium and tin at 414°C.

$$\Delta S^M \neq \Delta S^{M,\text{id}}$$

This type of behavior is found in the system Au-Cu at 1550 K, where, as shown in Fig. 11.32, $G^{xs} = -24{,}063 X_{Cu} X_{Au}$ joules is parabolic, ΔH^M is asymmetric, and $S^{xs} \neq 0$.

Having defined G^{xs}, this function can be calculated from a knowledge of the composition dependence of the activity coefficient of one component via Eq. (11.66), written as

$$G^{xs} = RTX_B \int_0^{X_A} \frac{\ln \gamma_A}{X_B^2}\, dX_A$$

Thus, for a Raoultian ideal solution, as $\gamma_A = 1$, $G^{xs} = 0$; and for a regular solution, as $(\ln \gamma_A)/X_B^2 = \alpha$, $G^{xs} = RT\alpha X_A X_B$.

11.10 THE QUASI-CHEMICAL MODEL OF SOLUTIONS

The quasi-chemical model of solutions is applied to solutions of components which are considered to have equal molar volumes in the pure state and which

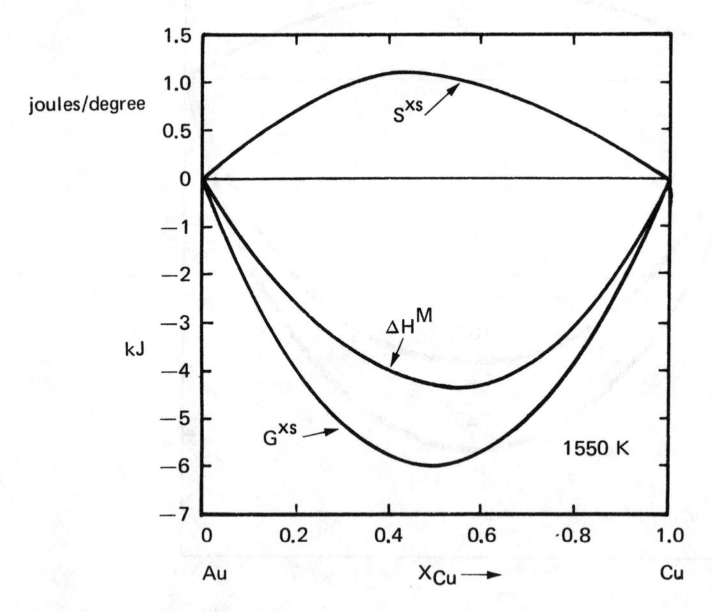

Fig. 11.32. The molar excess free energy, enthalpy, and excess entropy of mixing of gold and copper at 1550 K.

have zero volume change on mixing. Furthermore the interatomic forces are only significant over short distances, such that only nearest neighbor interactions need be considered. The energy of the solution is thus calculated by summing the atom-atom bond energies.

Consider 1 mole of a crystal containing n_A atoms of A and n_B atoms of B such that

$$X_A = \frac{n_A}{n_A + n_B} = \frac{n_A}{\mathcal{A}} \quad \text{and } X_B = \frac{n_B}{\mathcal{A}}$$

where $\mathcal{A}$ is Avogadro's number. This solution contains three types of atomic bonds:

1. A–A bonds the energy of each of which is E_{AA}
2. B–B bonds the energy of each of which is E_{BB}
3. A–B bonds the energy of each of which is E_{AB}

By considering the zero of energy as being that when the atoms are infinitely far apart, E_{AA}, E_{BB}, and E_{AB} are negative quantities. Let z be the coordination number of an atom in the crystal; i.e., each atom has z nearest neighbors. If, in the solution, there are P_{AA} A–A bonds, P_{BB} B–B bonds, and P_{AB} A–B bonds, then the energy of the solution, E, is obtained as the linear combination,

$$E = P_{AA}E_{AA} + P_{BB}E_{BB} + P_{AB}E_{AB} \tag{11.81}$$

and the problem of calculating E becomes one of calculating the values of P_{AA}, P_{BB}, and P_{AB}.

The number of A atoms × the number of bonds per atom

= the number of A-B bonds + the number of A-A bonds × 2

(The factor 2 arises from the fact that each A–A bond involves 2 A atoms.) Thus

$$n_A z = P_{AB} + 2P_{AA}$$

or

$$P_{AA} = \frac{n_A z}{2} - \frac{P_{AB}}{2} \tag{11.82}$$

Similarly, for B, $n_{\mathrm{B}}z = P_{\mathrm{AB}} + 2P_{\mathrm{BB}}$

or

$$P_{\mathrm{BB}} = \frac{n_{\mathrm{B}}z}{2} - \frac{P_{\mathrm{AB}}}{2} \tag{11.83}$$

Substitution of Eqs. (11.82) and (11.83) into Eq. (11.81) gives

$$\begin{aligned} E &= \left(\frac{n_{\mathrm{A}}z}{2} - \frac{P_{\mathrm{AB}}}{2}\right)E_{\mathrm{AA}} + \left(\frac{n_{\mathrm{B}}z}{2} - \frac{P_{\mathrm{AB}}}{2}\right)E_{\mathrm{BB}} + P_{\mathrm{AB}}E_{\mathrm{AB}} \\ &= \tfrac{1}{2}zn_{\mathrm{A}}E_{\mathrm{AA}} + \tfrac{1}{2}zn_{\mathrm{B}}E_{\mathrm{BB}} + P_{\mathrm{AB}}(E_{\mathrm{AB}} - \tfrac{1}{2}(E_{\mathrm{AA}} + E_{\mathrm{BB}})) \end{aligned} \tag{11.84}$$

Consider now the energies of the unmixed components. For n_{A} atoms in pure A,

the number of A–A bonds × 2 = the number of atoms
× the number of bonds per atom

i.e.,
$$P_{\mathrm{AA}} = \tfrac{1}{2}n_{\mathrm{A}}z$$

and similarly, for n_{B} atoms in pure B,

$$P_{\mathrm{BB}} = \tfrac{1}{2}n_{\mathrm{B}}z$$

Thus ΔE^M = (the energy of the solution) − (the energy of the unmixed components)

$$= P_{\mathrm{AB}}(E_{\mathrm{AB}} - \tfrac{1}{2}(E_{\mathrm{AA}} + E_{\mathrm{BB}}))$$

For the mixing process, from Eq. (5.10),

$$\Delta H^M = \Delta E^M - P\Delta V^M$$

and, as it has been stipulated that $\Delta V^M = 0$, then

$$\Delta H^M = \Delta E^M = P_{\mathrm{AB}}(E_{\mathrm{AB}} - \tfrac{1}{2}(E_{\mathrm{AA}} + E_{\mathrm{BB}})) \tag{11.85}$$

Equation (11.85) indicates that for given values of E_{AA}, E_{BB}, and E_{AB}, ΔH^M depends on P_{AB}, and further that, for the solution to be ideal, i.e., for $\Delta H^M = 0$,

$$E_{AB} = \frac{(E_{AA} + E_{BB})}{2} \tag{11.86}$$

Thus, contrary to the preliminary discussion in Sec. 11.2 which suggested that for ideal mixing it was necessary that $E_{AA} = E_{BB} = E_{AB}$, it is seen that a sufficient condition is that E_{AB} be the average of E_{AA} and E_{BB}. If $|E_{AB}| > |(E_{AA} + E_{BB})/2|$, then, from Eq. (11.85), ΔH^M is a negative quantity, corresponding to negative departures from Raoultian ideal behavior; and if $|E_{AB}| < |(E_{AA} + E_{BB})/2|$, then ΔH^M is a positive quantity, corresponding to positive departures from Raoultian ideality.

If $\Delta H^M = 0$, then the mixing of the n_A atoms of A with the n_B atoms of B is random, in which case Eq. (11.45) gives

$$\Delta S^M = \Delta S^{M,\mathrm{id}} = -R(X_A \ln X_A + X_B \ln X_B)$$

In solutions which do not depart too greatly from ideality, that is, $|\Delta H^M| \leqslant RT$, approximately the same random distribution of atoms as occurs in an ideal solution may be assumed, in which case P_{AB} in Eq. (11.85) may be calculated as follows.

Consider two neighboring lattice sites, labeled 1 and 2 in the mole of A–B crystal. The probability that site 1 is occupied by an A atom equals

$$\frac{\text{the number of A atoms in the crystal}}{\text{the number of lattice sites in the crystal}} = \frac{n_A}{\mathcal{O}} = X_A$$

and similarly, the probability that site 2 is occupied by a B atom is X_B. The probability that site 1 is occupied by an A atom and site 2 is simultaneously occupied by a B atom is thus $X_A X_B$. But the probability that site 1 is occupied by a B atom and site 2 is simultaneously occupied by an A atom is also $X_A X_B$. Thus the probability that a neighboring pair of sites contains an A–B pair is $2X_A X_B$. By a similar argument, the probability that the neighboring sites contain an A–A pair is X_A^2 and that the neighboring sites contain a B–B pair is X_B^2. As the mole of crystal contains $\frac{1}{2} z\mathcal{O}$ pairs of lattice sites, then

the number of A–B pairs = the number of pairs of sites
× the probability of an A–B pair

i.e.,

$$P_{AB} = \tfrac{1}{2} z\mathcal{O} \times 2X_A X_B = z\mathcal{O} X_A X_B \tag{11.87}$$

Similarly $$P_{AA} = \tfrac{1}{2} z\mathfrak{N} \times X_A^2 = \tfrac{1}{2} z\mathfrak{N} X_A^2$$

and $$P_{BB} = \tfrac{1}{2} z\mathfrak{N} \times X_B^2 = \tfrac{1}{2} z\mathfrak{N} X_B^2$$

Substituting Eq. (11.87) into Eq. (11.85) gives

$$\Delta H^M = z\mathfrak{N} X_A X_B [E_{AB} - \tfrac{1}{2}(E_{AA} + E_{BB})]$$

and if $$\Omega = z\mathfrak{N} [E_{AB} - \tfrac{1}{2}(E_{AA} + E_{BB})]$$

then $$\Delta H^M = \Omega X_A X_B \tag{11.88}$$

which indicates that ΔH^M is a parabolic function of composition. As random mixing is assumed, then the quasi-chemical model corresponds to the regular solution model; i.e.,

$$\Delta H^M = G^{xs} = \Omega X_A X_B = RT\alpha X_A X_B \tag{11.89}$$

and hence

$$\alpha = \frac{\Omega}{RT} \tag{11.90}$$

Application of Eq. (11.33*a*) to the heat of mixing gives

$$\Delta \bar{H}_A^M = \Delta H^M + X_B \frac{\partial \Delta H^M}{\partial X_A}$$

and from Eq. (11.89)

$$\frac{\partial \Delta H^M}{\partial X_A} = \Omega(X_B - X_A)$$

Thus

$$\Delta \bar{H}_A^M = \Omega X_A X_B + X_B \Omega (X_B - X_A) = \Omega X_B^2 \tag{11.91a}$$

and similarly

$$\Delta \bar{H}_B^M = \Omega X_A^2 \tag{11.91b}$$

As the mixing is random, then

$$\Delta \bar{S}_A^M = -R \ln X_A \quad \text{and} \quad \Delta \bar{S}_B^M = -R \ln X_B$$

and hence

$$\Delta \bar{G}_A^M = \Delta \bar{H}_A^M - T\Delta \bar{S}_A^M$$
$$= \Omega X_B^2 + RT \ln X_A \tag{11.92}$$

But

$$\Delta \bar{G}_A^M = RT \ln a_A$$
$$= RT \ln \gamma_A + RT \ln X_A \tag{11.93}$$

comparison of which with Eq. (11.92) indicates that

$$\ln \gamma_A = \frac{\Omega}{RT} X_B^2 = \alpha X_B^2 \tag{11.94}$$

The value of γ thus depends on the value of Ω, which, in turn, depends on the relative values of the bond energies E_{AA}, E_{BB}, and E_{AB}. If Ω is negative, then $\gamma_A < 1$; and if Ω is positive, then $\gamma_A > 1$.

Henry's law gives that ln γ_A approaches constancy as X_B approaches unity. Thus as $X_B \to 1$, $\ln \gamma_A \to \ln \gamma_A^0 = \Omega/RT$, with this limiting value being approached asymptotically. Similarly, via the relation between Henry's and Raoult's laws, Raoult's law is approached asymptotically by the component i as $X_i \to 1$.

The applicability of the quasi-chemical model to actual solutions decreases as the magnitude of Ω increases; i.e., if the magnitude of E_{AB} is significantly greater or less than the average of E_{AA} and E_{BB}, then random mixing of the A and B atoms cannot be assumed. The equilibrium configuration of a solution at constant T and P is that which minimizes the free energy G, where $G = H - TS$ is measured relative to the unmixed components. As has been seen, minimization of G occurs as a compromise between minimization of H and maximization of S. If $|E_{AB}| > |\frac{1}{2}(E_{AA} + E_{BB})|$, then minimization of H corresponds to maximization of the number of A–B pairs (complete ordering of the solution). On the other hand, maximum S corresponds to completely random mixing. Minimization of G thus occurs as a compromise between maximization of P_{AB} (the tendency toward which increases with increasingly negative values of Ω) and random mixing (the tendency toward which increases with increasing temperature). The critical parameters are thus Ω and T; and if Ω is appreciably negative and the temperature is not too high, then $P_{AB(actual)} > P_{AB(random)}$, in which case the assumption of random mixing is invalid. Similarly, if $|E_{AB}| < |\frac{1}{2}(E_{AA} + E_{BB})|$, then minimization of H corresponds to minimization of the number of A–B pairs (complete clustering in the solution), and minimization of G occurs as a compromise between minimization of P_{AB} (the tendency toward which increases with increasingly positive values of Ω) and random mixing. Thus if Ω is appreciably positive and the temperature is not too high, then

$P_{AB(actual)} < P_{AB(random)}$, in which case the assumption of random mixing is again invalid.

In order for the quasi-chemical model to be applicable, it is necessary that the above-mentioned compromise be such that the equilibrium solution configuration be not too distant from random mixing. As the entropy contribution to the free energy is temperature-dependent, then (1) for any value of Ω, more-nearly random mixing occurs as the temperature is increased; and (2) for any given temperature, more-nearly random mixing occurs with smaller values of Ω.

The preceding discussion can be illustrated qualitatively by Fig. 11.33*a* and *b*. In these figures the *x*-axis represents the range of spatial configurations available to the atoms in a 50% A–50% B solution. The extreme configurations are complete ordering and complete clustering (immiscibility between A and B) with the random configuration occurring between these extremes. The relationship between the entropy of mixing ΔS^M, and the solution configuration is given by curve I. This has a maximum value of $\Delta S^{M,\text{id}}$ in the random configuration; and as the randomness decreases by moving toward ordering or clustering, ΔS^M decreases. In both extreme configurations, ΔS^M is zero. Thus the $-T\Delta S^M$ is drawn as curve II. Line III represents the variation of the heat of formation of the solution, ΔH^M, with configuration, which in Fig. 11.33*a* is drawn for a solution which forms exothermically and in Fig. 11.33*b* is drawn for an endothermically forming solution. The sum of the ΔH^M and $-T\Delta S^M$ curves

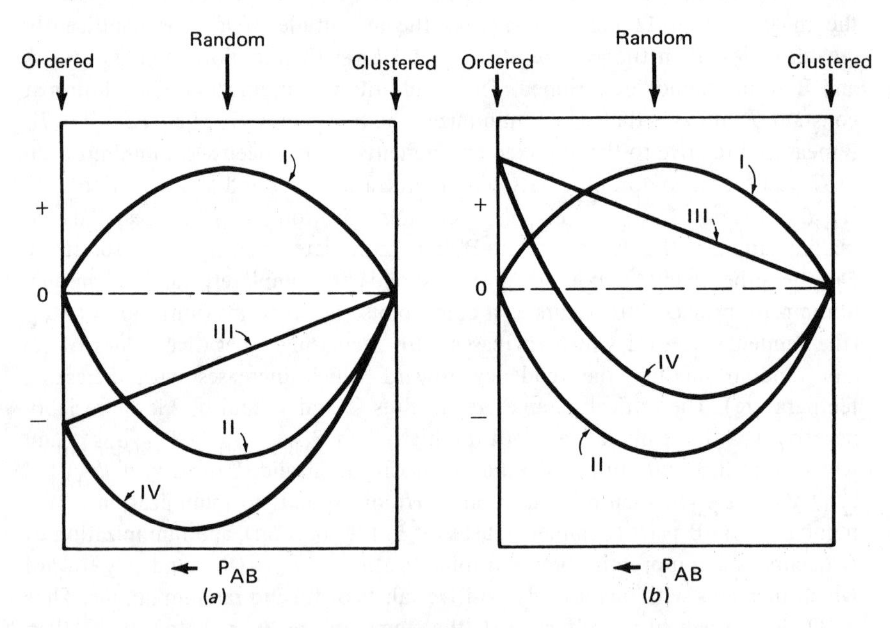

Fig. 11.33. Illustration of origins of deviations from regular solution behavior.

gives the ΔG^M curve (curve IV), and the minimum in this curve occurs at the equilibrium configuration. Figure 11.33*a* shows that the equilibrium configuration of exothermically forming solutions lies between the ordered and random configurations, and Fig. 11.33*b* shows that the equilibrium configuration of endothermically forming solutions lies between the clustered and random configurations. It is seen that the random configuration is the equilibrium configuration only when $\Delta H^M = 0$ and that as the magnitude of $|\Omega|$ for the A–B system increases, then, at a constant temperature, the position of the minimum in the ΔG^M curve moves further away from the random configuration. Similarly for any given system (of fixed Ω), as T and hence $|T\Delta S^M|$ increases, the position of the minimum in the ΔG^M curve moves toward the random configuration. Fig. 11.33*a* and *b* also illustrates that both extreme configurations are physically unrealizable as, in order to have the minimum in the ΔG^M curve coincide with either extreme, infinite values of ΔH^M would be required (negative for complete ordering, and positive for complete clustering). Similarly, with a nonzero ΔH^M the random configuration only becomes the equilibrium configuration at infinite temperature.

If $V_A^0 \neq V_B^0$, then the lattice parameter of the crystal will vary with composition; and hence, as the interatomic distances vary with composition, then so will the values of the bond energies be composition-dependent. This and other refinements to the model have been discussed by Wagner.*

11.11 SUMMARY

1. Raoult's law is $p_i = X_i p_i^0$, and a component of a solution that conforms with this law is said to exhibit Raoultian behavior. In all solutions, the behavior of the component i approaches Raoult's law as $X_i \to 1$.

2. Henry's law is $p_i = k' X_i$, and a component of a solution that conforms with this equation is said to exhibit Henrian behavior. In all solutions, the behavior of the component i approaches Henry's law as $X_i \to 0$. In a binary solution, Henry's law is obeyed by the solute in that composition range in which Raoult's law is obeyed by the solvent.

3. The activity of the component i in a solution, with respect to a given standard state, is the ratio of the vapor pressure of i (strictly, the fugacity of i) exerted by the solution to the vapor pressure (the fugacity) of i in the given standard state. If the standard state is chosen as being pure i, then $a_i = p_i/p_i^0$. An activity is thus a ratio, and its introduction effects a normalization of the vapor pressure exerted by the component i in the solution. In terms of activity, Raoult's law is $a_i = X_i$, and Henry's law is $a_i = kX_i$.

*C. Wagner, "Thermodynamics of Alloys," p. 36, Addison-Wesley Publishing Co., Reading, Mass., 1952.

4. The difference between the value of an extensive thermodynamic property per mole of i in a solution, and the value of the property per mole of i in its standard state, is termed the partial molar property change of i for the solution process; i.e., if Q is any extensive thermodynamic property, the property change due to solution of 1 mole of i is $\Delta \bar{Q}_i^M = \bar{Q}_i - Q_i^0$. In the case of free energy, $\Delta \bar{G}_i^M = \bar{G}_i - G_i^0$. This free energy difference is related to the activity of i in solution with respect to the standard state as $\Delta \bar{G}_i^M = RT \ln a_i$, and $\Delta \bar{G}_i^M$ is termed the partial molar free energy of solution of i.

The free energy change resulting from the formation of 1 mole of solution from the pure components i (termed the integral free energy change) is $\Delta G^M = \Sigma_i X_i \Delta \bar{G}_i^M$, so that, for the binary A-B, $\Delta G^M = X_A \Delta \bar{G}_A^M + X_B \Delta \bar{G}_B^M$. As $\Delta \bar{G}_A^M = RT \ln a_A$, then $\Delta G^M = RT(X_A \ln a_A + X_B \ln a_B)$. For a Raoultian solution, as $a_i = X_i$, then $\Delta G^M = RT(X_A \ln X_A + X_B \ln X_B)$. For any general extensive thermodynamic property Q, $\Delta Q^M = \Sigma_i X_i \Delta \bar{Q}_i^M$.

5. A Raoultian solution has the properties $a_i = X_i$, $\bar{V}_i = V_i^0$ (i.e., there is no change in volume on mixing the components), $\bar{H}_i = H_i^0$ (i.e., there is zero heat of mixing), and $\Delta G^{M,\text{id}} = RT(X_A \ln X_A + X_B \ln X_B)$. As $\Delta S^{M,\text{id}} = -(d\Delta G^{M,\text{id}}/dT)$, $\Delta S^{M,\text{id}} = -R\Sigma X_i \ln X_i$, so that in a Raoultian solution, $\Delta \bar{S}_i^M = -R \ln X_i$. $\Delta S^{M,\text{id}}$ is thus independent of temperature and is simply an expression for the maximum number of spatial configurations available to the system.

6. The thermodynamics of non-Raoultian solutions is dealt with by introducing the activity coefficient γ, which, for the component i, is defined as $\gamma_i = a_i/X_i$. γ_i, which can have values of greater or less than unity, thus quantifies the deviation of i from Raoultian behavior. As $\ln a_i = \ln X_i + \ln \gamma_i$, $d \ln a_i/d(1/T) = \Delta \bar{H}_i^M/R = d \ln \gamma_i/d(1/T)$. Thus if $d\gamma_i/dT$ is positive, $\Delta \bar{H}_i^M$ is negative and vice versa. The magnitude of the heat of formation of a nonideal solution is determined by the magnitude of the deviations of the solution components from Raoultian behavior. Non-Raoultian components approach Raoultian behavior with increasing temperature. Thus if $\gamma_i < 1$, then $d\gamma_i/dT$ is positive; and if $\gamma_i > 1$, $d\gamma_i/dT$ is negative. Solutions, the components of which exhibit negative deviations for Raoult's law, form exothermically; that is, $\Delta H^M < 0$, and vice versa.

7. The Gibbs-Duhem relationship is $\Sigma_i X_i d\bar{Q}_i = 0$ at constant temperature and pressure, where $\bar{Q}_i$ is the partial molar value of the general extensive thermodynamic function Q of the solution component i.

8. The excess value of an extensive thermodynamic property of a solution is the difference between the actual value and the value that the property would have if the components obeyed Raoult's law. Thus for the general function Q, $Q^{xs} = Q - Q^{\text{id}}$; or for the free energy, either $G^{xs} = G - G^{\text{id}}$, or $G^{xs} = \Delta G^M - \Delta G^{M,\text{id}}$. As $\gamma_i = a_i/X_i$, then $G^{xs} = RT\Sigma X_i \ln \gamma_i$.

9. A regular solution is one which has an ideal entropy of formation and a

nonzero heat of formation from its pure components. The activity coefficients of the components of a regular solution are given by the expression $RT \ln \gamma_i = \alpha'(1 - X_i)^2$, where α' is a temperature-independent constant, the value of which is characteristic of the particular solution. Thus $\ln \gamma_i$ varies inversely with temperature; and as $\bar{G}_i^{xs} = RT \ln \gamma_i$, then $\bar{G}^{xs} = \Delta \bar{H}_i^M$ is independent of temperature. Furthermore the heat of formation of a regular solution, being equal to G^{xs}, is a parabolic function of composition; that is, $\Delta H^M = G^{xs} = RT\alpha X_A X_B = \alpha' X_A X_B$.

10. Regular solution behavior is predicted by the quasi-chemical model of solutions. In this model, random mixing of the components is assumed, and the energy of the solution is the sum of the individual atom-atom bond energies in the solution. Random mixing can only be assumed if, in the system A-B, the A–B bond energy is not significantly different from the average of the A–A and B–B bond energies in the pure components. For any given such deviation, the validity of the assumption of random mixing increases with increasing temperature. The quasi-chemical model predicts tendency toward Raoultian behavior and Henrian behavior as $X_i \rightarrow 1$ and as $X_i \rightarrow 0$, respectively.

PROBLEMS

11.1 One mole of solid Cr at 1600°C is added to a large quantity of Fe-Cr liquid solution (in which $X_{Fe} = 0.8$) which is also at 1600°C. If Fe and Cr form Raoultian solutions, calculate the heat and entropy changes in the solution resulting from the addition. Assume that the heat capacity difference between solid and liquid Cr is negligible.

11.2 The free energies of formation, ΔG^M, of liquid Sn–Cu alloys at 1400 K, as a function of composition, are listed below. From this data calculate the composition dependencies of a_{Sn} and a_{Cu} at 1400 K.

X_{Sn}	0	0.1	0.2	0.3	0.4	0.5	0.6	0.7	0.8	0.9	1
ΔG^M, joules/mole	0	−8042	−11,860	−13,370	−13,810	−13,250	−12,090	−10,390	−8104	−5004	0

11.3 The variation, with composition, of G^{xs} for liquid Au–Cu alloys at 1550 K, shown in Fig. 11.32 is

X_{Cu}	0.1	0.2	0.3	0.4	0.5	0.6	0.7	0.8	0.9
G^{xs}, joules/mole	−2170	−3850	−5050	−5770	−6010	−5770	−5050	−3850	−2170

Calculate:

a. $\bar{G}_{Au}^{xs}$ and $\bar{G}_{Cu}^{xs}$ at $X_{Cu} = 0.3$

b. ΔG^M at $X_{Cu} = 0.3$

c. The partial pressures of Cu and Au exerted by the $X_{Cu} = 0.3$ alloy at 1550 K

11.4 The partial pressures of A exerted by A–B alloys at 1000 K are

X_A	1	0.9	0.8	0.7	0.6	0.5	0.4	0.3	0.2
$p_A \times 10^6$	5	4.4	3.75	2.9	1.8	1.1	0.8	0.6	0.4

Determine (*a*) the composition range over which Henry's law is obeyed by the solute A, and (*b*) the value of the Henry's law constant at 1000 K. If the temperature variation of the Henry's law constant is given as

$$\log k_A = -\frac{109.3}{T} - 0.2886$$

(*c*) calculate $\Delta \bar{H}_A^M$ in the composition range over which A obeys Henry's law, and (*d*) write an equation for the variation of ΔH^M with composition over the same composition range.

11.5 The activities of Cu in liquid Fe–Cu alloys at 1550°C have been determined as

X_{Cu}	1	0.883	0.792	0.626	0.467	0.328	0.217	0.171	0.142	0.088	0.061	0.0442	0.0265
a_{Cu}	1	0.923	0.888	0.870	0.821	0.786	0.729	0.687	0.660	0.521	0.424	0.325	0.216

Using, separately, Eqs. (11.55) and (11.61), determine the Gibbs-Duhem calculated variation of a_{Fe} with composition in this system at 1550°C.

11.6 The activities of Ni in liquid Fe–Ni alloys at 1600°C have been determined as

X_{Ni}	1	0.9	0.8	0.7	0.6	0.5	0.4	0.3	0.2	0.1
a_{Ni}	1	.89	.766	.62	.485	.374	.283	.207	.136	.067

Using, separately, Eqs. (11.55) and (11.61), determine the Gibbs-Duhem calculated variation of a_{Fe} with composition in this system at 1600°C.

11.7 At 473°C the Pb-Sn system exhibits regular solution behavior with the activity coefficient of Pb being given by

$$\log \gamma_{Pb} = -0.32(1 - X_{Pb})^2$$

Write the corresponding equation for the variation of γ_{Sn} with composition at 473°C. If 1 mole of lead at 25°C is added to a large quantity of liquid alloy of composition $X_{Pb} = 0.5$ which is thermostated at 473°C, calculate (*a*) the heat flow from the thermostat into the liquid alloy, and (*b*) the entropy increase in the surroundings resulting from the process. Calculate (*c*) a_{Pb} in the $X_{Pb} = 0.5$ alloy at 746 K and 1000 K.

11.8 Demonstrate that if the activity coefficients of a binary solution can be expressed as

$$\ln \gamma_A = \alpha_1 X_B + \tfrac{1}{2}\alpha_2 X_B^2 + \tfrac{1}{3}\alpha_3 X_B^3 + \cdots$$

and

$$\ln \gamma_B = \beta_1 X_A + \tfrac{1}{2}\beta_2 X_A^2 + \tfrac{1}{3}\beta_3 X_A^3 + \cdots$$

over the entire composition range, then $\alpha_1 = \beta_1 = 0$, and that if the variation of the activity coefficients can be represented by the quadratic terms alone, then

$$\alpha_2 = \beta_2$$

11.9 The activity coefficient of Zn in liquid Cd–Zn alloys at 435°C can be represented by

$$\log \gamma_{Zn} = 0.38\, X_{Cd}^2 - 0.13\, X_{Cd}^3$$

Calculate the corresponding expression for the composition dependence of $\log \gamma_{Cd}$, and hence calculate a_{Cd} in the $X_{Cd} = 0.5$ alloy at 435°C.

CHAPTER TWELVE

FREE ENERGY-COMPOSITION AND PHASE DIAGRAMS OF BINARY SYSTEMS

12.1 INTRODUCTION

Before making use of the concept of activity in the determination of equilibrium in reaction systems containing components in condensed solutions, it is instructive to examine the relationship between free energy (activity) and phase stability (as is normally represented by isobaric phase diagrams using temperature and composition as the variables). When a liquid solution is cooled, a liquidus temperature is eventually reached, at which point a stable solid phase begins to separate from the liquid solution. This solid phase could be a pure component, a solid solution of the same or different composition from the liquid, or a chemical compound comprising two or more of the components. In all possible cases it is to be expected that the variation of the free energy-composition relationship with temperature will predict the phase change at the liquidus temperature. If, over the entire composition range, liquid solutions are stable, then the free energies of all the liquid states are lower than those of any possible solid states; and, conversely, if the temperature of the system is lower than the lowest solidus temperature, then the free energies of the solid states are everywhere lower than the free energies of the liquid states. At intermediate temperatures the free energy-composition relationship would be expected to show composition ranges over which liquid states are stable, composition ranges over which solid states are stable, and intermediate composition ranges over which solid and liquid phases coexist in equilibrium with each other. Thus, by virtue of the facts that (1) the state of lowest free energy is the stable state, and (2) when phases coexist in equilibrium, $\bar{G}_i$ has the same value in all of the coexisting phases, there must exist a quantitative correspondence between free energy–composition diagrams and "phase diagrams." This correspondence is examined in this chapter, where it will be seen that "normal" phase diagrams are generated by, and are simply representations of, free energy–composition diagrams.

12.2 FREE ENERGY AND ACTIVITY

The free energy of mixing of the components A and B, to form a mole of solution, is given as

$$\Delta G^M = RT(X_A \ln a_A + X_B \ln a_B)$$

and ΔG^M is the difference between the free energy of a mole of homogeneous solution and the free energy of the corresponding numbers of moles of unmixed components. If the solution is ideal, i.e., if $a_i = X_i$, then the curve of the free energy of mixing, given as

$$\Delta G^{M,\mathrm{id}} = RT(X_A \ln X_A + X_B \ln X_B)$$

has the characteristic shape shown, at the temperature T, as curve I in Fig. 12.1. As $\Delta H^{M,\mathrm{id}} = 0$, then $\Delta G^{M,\mathrm{id}} = -T\Delta S^{M,\mathrm{id}}$, and hence curve I in Fig. 12.1 is obtained as $-T \times$ (the curve drawn in Fig. 11.7). It is thus seen that the shape of the curve of $\Delta G^{M,\mathrm{id}}$ versus composition is dependent only on temperature.

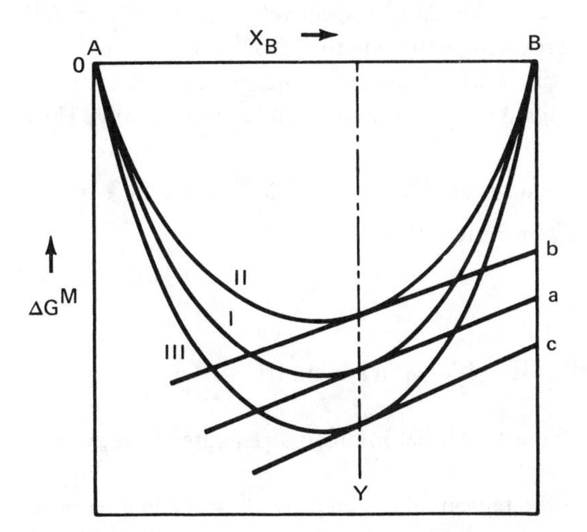

Fig. 12.1. The integral free energies of mixing in binary systems exhibiting ideal behavior (I), positive deviation from ideal behavior (II), and negative deviation from ideal behavior (III).

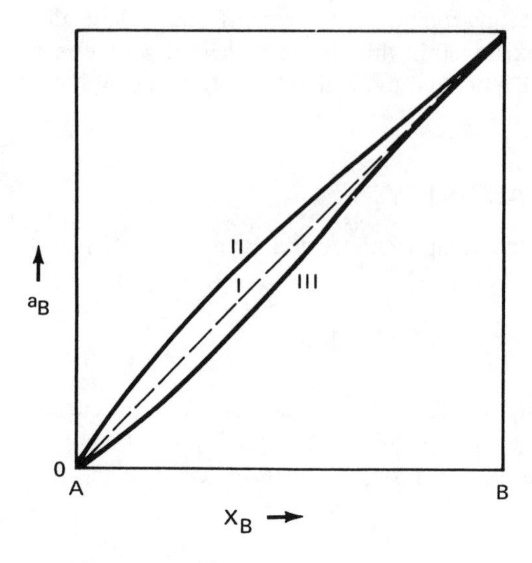

Fig. 12.2. The activities of component B obtained from lines I, II, and III in Fig. 12.1.

If the solution exhibits slight positive deviation from ideal mixing, i.e., if $\gamma_i > 1$ and $a_i > X_i$, then, at the temperature T, the curve of the free energy of mixing is typically as shown by curve II in Fig. 12.1; and if the solution shows slight negative deviation, i.e., if $\gamma_i < 1$ and $a_i < X_i$, then, at temperature T, the curve of the free energy of mixing is typically as shown by curve III in Fig. 12.1. From Eq. (11.33*a*) the tangent drawn to the ΔG^M curve at any composition intersects the $X_A = 1$ and $X_B = 1$ axes at $\Delta \bar{G}_A^M$ and $\Delta \bar{G}_B^M$ respectively, and, as $\Delta \bar{G}_i^M = RT \ln a_i$, a correspondence is provided between the ΔG^M-composition and activity-composition curves. In Fig. 12.1, at the composition Y, tangents drawn to curves I, II, and III intersect the $X_B = 1$ axis at a, b, and c, respectively. Thus

$$|Bb = \Delta \bar{G}_B^M = RT \ln a_B \text{ (in system II)}| < |Ba = \Delta \bar{G}_B^M = RT \ln X_B|$$
$$< |Bc = \Delta \bar{G}_B^M = RT \ln a_B \text{ (in system III)}|$$

from which it is seen that

$$\gamma_B \text{ in system II} > 1 > \gamma_B \text{ in system III}$$

Variation, with composition, of the tangential intercepts generates the a_i versus X_i curves shown in Fig. 12.2.

As $X_i \to 0$, $a_i \to 0$, and hence the tangential intercept $\Delta \bar{G}_i^M = RT \ln a_i \to -\infty$, which indicates that all ΔG^M versus composition curves have vertical tangents at their extremities. Similarly the $\Delta S^{M,\text{id}}$ versus composition curve in Fig. 11.7 has vertical tangents at its extremities.

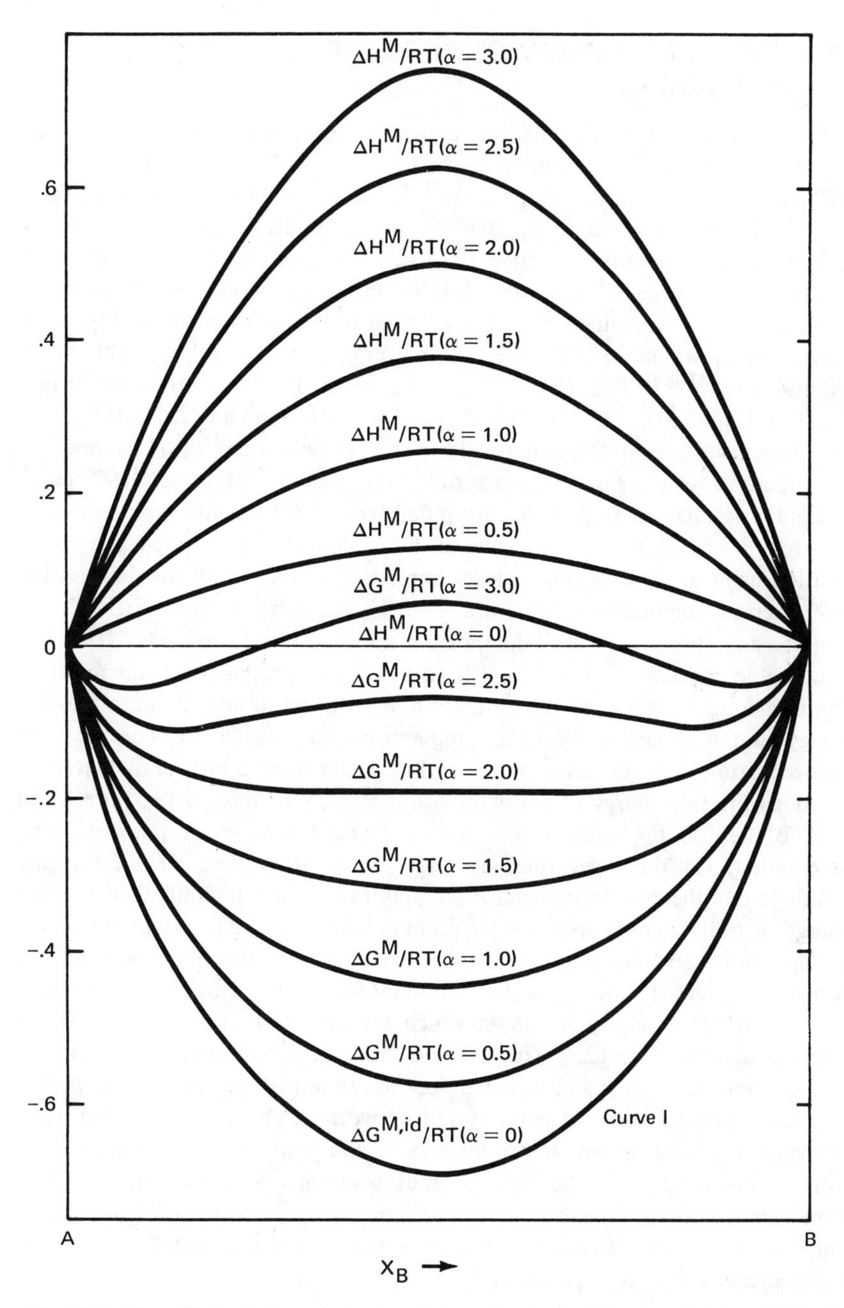

Fig. 12.3. The effect of the magnitude of α on the integral heats and integral free energies of mixing of the components of a binary regular solution.

12.3 THE FREE ENERGY OF REGULAR SOLUTIONS

If curves II and III in Fig. 12.1 are drawn for regular solutions, then deviation of ΔG^M from $\Delta G^{M,\text{id}}$ is due only to the nonzero heat of mixing, and the difference between the two curves, $\Delta G^{M,\text{id}} - \Delta G^M = -G^{\text{xs}} = -RT\alpha X_A X_B = -\Delta H^M$. For curve II, $|\Delta G^M| < |\Delta G^{M,\text{id}}|$, and thus ΔH^M is a positive quantity (α is positive). Similarly for curve III, $|\Delta G^M| > |\Delta G^{M,\text{id}}|$, and thus ΔH^M is a negative quantity (α is negative). It is of interest to consider the effect of increasingly positive values of α on the shape of the ΔG^M curve. In Fig. 12.3, curve I is drawn as $-\Delta S^{M,\text{id}}/R = (X_A \ln X_A + X_B \ln X_B)$. This curve represents $\Delta G^{M,\text{id}}/RT$. $\Delta H^M/RT = \alpha X_A X_B$ curves are drawn for $\alpha = 0, +0.5, +1.0, +1.5, +2.0, +2.5$, and $+3.0$, and the corresponding $\Delta G^M/RT$ curves are drawn as the sum of the particular $\Delta H^M/RT$ and $-\Delta S^{M,\text{id}}/R$ curves. It is seen that, as α increases in magnitude, the shape of the $\Delta G^M/RT$ curve continuously changes from a shape typified by $\alpha = 0$ to a form typified by $\alpha = 3$. Before discussing the consequences of this change of shape of the curve on the solution behavior, it is pertinent to examine the significance of the shape of the ΔG^M versus composition curve. In Fig. 12.4*a*, curve I from Fig. 12.1 is reproduced. This curve, at all compositions, is "convex downwards." Thus the most stable configuration of A and B, mixed in any proportion, is the resultant homogeneous solution, as the formation of this solution minimizes the free energy of the system at the fixed temperature and pressure. Consider, further, two separate solutions, say *a* and *b* in Fig. 12.4*a*. Before mixing of these two solutions the free energy of the two-solution system, with respect to pure A and pure B, lies on the straight line joining *a* and *b* (the exact position being determined, via the lever rule, by the relative proportions of the separate solutions.) If the solutions *a* and *b* are present in equal amount, then the free energy of the system is given by the point *c*. When the two solutions are mixed, a single homogeneous solution is formed as, thereby, the free energy of the system is decreased from *c* to *d*, the minimum free energy which the system may have. Consider now Fig. 12.4*b* in which the $\Delta G^M/RT$ curve for $\alpha = +3$ is reproduced from Fig. 12.3. This curve is "convex downwards" only between A and *n* and between *p* and B and is "convex upwards" between *n* and *p*. The free energy of a system of composition between *m* and *q* is minimized when the system occurs as two solutions, one of composition *m* and the other of composition *q*; e.g., if the homogeneous solution *r* separates into the two coexisting solutions *m* and *q*, then the free energy is decreased from *r* to *s*. For the equilibrium coexistence of two separate solutions, at the temperature *T* and pressure *P*, it is required that,

$$\bar{G}_A(\text{in solution } m) = \bar{G}_A(\text{in solution } q) \qquad (a)$$

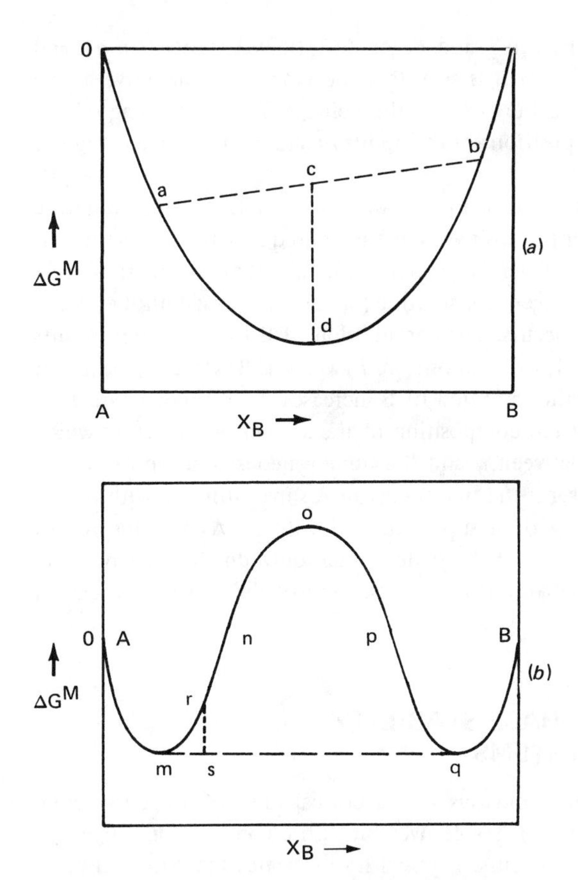

Fig. 12.4(*a*). The integral free energies of mixing of binary components that form a complete range of solutions. (*b*) The integral free energies of mixing of binary components in a system that exhibits a miscibility gap.

and $$\bar{G}_B(\text{in solution } m) = \bar{G}_B(\text{in solution } q) \qquad (b)$$

Subtracting G_A^0 from both sides of Eq. (*a*) gives

$$RT \ln a_{A(\text{in solution } m)} = RT \ln a_{A(\text{in solution } q)}$$

or $$a_A(\text{in solution } m) = a_A(\text{in solution } q) \qquad (c)$$

Similarly $$a_B(\text{in solution } m) = a_B(\text{in solution } q) \qquad (d)$$

Equations (*c*) and (*d*) are the criteria for equilibrium coexistence of two solutions (or phases) at constant T and P. As $\Delta\bar{G}_A^M(\text{in } m) = \Delta\bar{G}_A^M(\text{in } q)$, and $\Delta\bar{G}_B^M(\text{in } m) = \Delta\bar{G}_B^M(\text{in } q)$, then it is seen that the tangent to the curve at the point m is also the tangent to the curve at the point q. The positioning of this double tangent defines the positions of the points m and q on the free energy of mixing curve.

The A-B system, as represented in Fig. 12.4*b*, is such that, at the temperature T, the value of α is sufficiently positive that the consequent tendency towards clustering of like atoms is sufficiently great to bring about phase separation. As B is initially added to A, a homogeneous solution (phase I) is formed until at $X_B = m$, saturation of A with B occurs. Further addition of B to the system results in the appearance of phase II, of composition q (which is B saturated with A at the given temperature). Further addition of B increases the ratio of phase II to phase I present until the overall composition of the system reaches q, at which point phase I disappears. Between q and B a homogeneous solution of A in B occurs. The curve mn represents the free energy of A supersaturated with B, and qp represents the free energy of B supersaturated with A. As the line AmqB represents the equilibrium state of the system, then only this line has physical significance. AmqB is the isobaric, isothermal section of the system as it occurs in G-T-P-composition space.

12.4 CRITERION OF PHASE STABILITY IN REGULAR SYSTEMS

For a given value of T, it is obvious that a critical value of α occurs below which a homogeneous solution is stable over the entire composition range and above which phase separation occurs. Figure 12.5 illustrates the criteria used in the determination of this critical value. Figure 12.5*a*, *b*, and *c* illustrates the variations of ΔG^M, $\partial \Delta G^M/\partial X_B$, $\partial^2 \Delta G^M/\partial X_B^2$, and $\partial^3 \Delta G^M/\partial X_B^3$ with composition for $\alpha < \alpha_{\text{critical}}$, $\alpha = \alpha_{\text{critical}}$, and $\alpha > \alpha_{\text{critical}}$, respectively. The critical value of α is seen to be that which makes $\partial^2 \Delta G^M/\partial X_B^2$ and $\partial^3 \Delta G^M/\partial X_B^3$ simultaneously equal to zero at that composition at which immiscibility becomes imminent. For a regular solution,

$$\Delta G^M = RT(X_A \ln X_A + X_B \ln X_B) + RT\alpha X_A X_B$$

$$\frac{\partial \Delta G^M}{\partial X_B} = RT\left[\ln \frac{X_B}{X_A} + \alpha(X_A - X_B)\right]$$

$$\frac{\partial^2 \Delta G^M}{\partial X_B^2} = RT\left[\frac{1}{X_A} + \frac{1}{X_B} - 2\alpha\right]$$

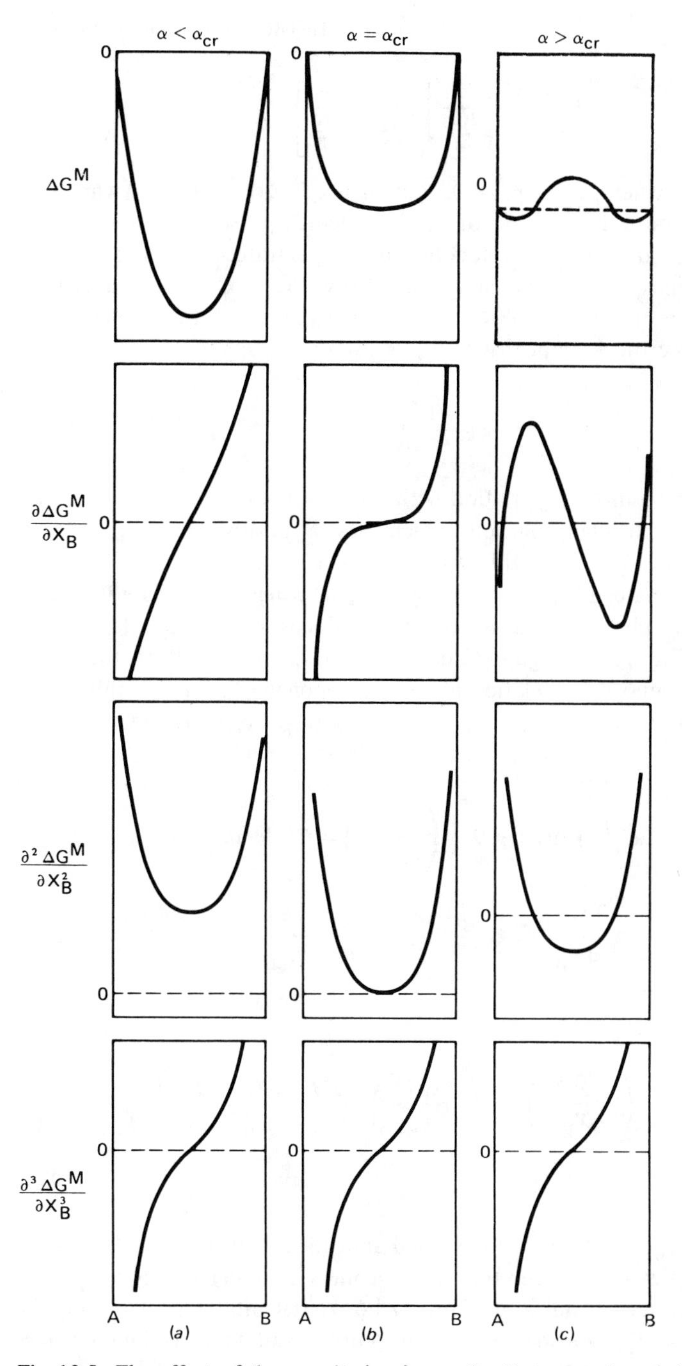

Fig. 12.5. The effect of the magnitude of α on the first, second, and third derivatives of the integral free energy of mixing with respect to composition.

$$\frac{\partial^3 \Delta G^M}{\partial X_B^3} = RT\left[\frac{1}{X_A^2} - \frac{1}{X_B^2}\right]$$

$\partial^3 \Delta G^M/\partial X_B^3 = 0$ when $X_A = X_B = 0.5$, and hence $\partial^2 \Delta G^M/\partial X_B^2 = 0$ when $\alpha = 2$, which is thus the critical value of α, above which phase separation occurs. From Eq. (11.72), α is an inverse function of temperature; and hence, for any given regular binary system which has a positive heat of mixing, a critical temperature occurs above which $\alpha < 2$ and below which $\alpha > 2$. Equation (11.90) indicates that this critical temperature, T_{cr}, is given as

$$T_{cr} = \frac{\Omega}{2R} \tag{12.1}$$

Figure 12.6*a* represents the variation with temperature of the ΔG^M-composition curve for a regular solution which has a positive heat of mixing ($\Omega = +16{,}630$ joules/mole). In this system $T_{cr} = 16{,}630/2R = 1000$ K. Figure 12.6*b* shows the phase diagram of this system, in which the miscibility curve bounding the two-phase region is simply the locus of the double tangent compositions in Fig. 12.6*a*. Figure 12.6*c* shows the variation, with temperature, of the activity-composition relationship for component B. At T_{cr} this curve exhibits a horizontal inflexion at $X_B = 0.5$, as is seen by the following. From Eq. (11.33*b*),

$$\Delta \bar{G}_B^M = \Delta G^M + X_A\left(\frac{\partial \Delta G^M}{\partial X_B}\right) = RT \ln a_B$$

Thus

$$\frac{\partial \Delta \bar{G}_B^M}{\partial X_B} = X_A \frac{\partial^2 \Delta G^M}{\partial X_B^2} = \frac{RT}{a_B}\frac{\partial a_B}{\partial X_B} \tag{12.2}$$

and

$$\frac{\partial^2 \Delta \bar{G}_B^M}{\partial X_B^2} = X_A\left(\frac{\partial^3 \Delta G^M}{\partial X_B^3}\right) - \left(\frac{\partial^2 \Delta G^M}{\partial X_B^2}\right) = \frac{RT}{a_B}\frac{\partial^2 a_B}{\partial X_B^2} - \frac{RT}{a_B^2}\left(\frac{\partial a_B}{\partial X_B}\right)^2 \tag{12.3}$$

As, at T_{cr} and $X_B = 0.5$, both the second and third derivatives of ΔG^M with respect to X_B are zero, it is seen that the first and second derivatives of a_B with respect to X_B are zero, indicating, thus, a horizontal inflexion in the a_B-X_B curve. When $T < T_{cr}$, the derived activity curve exhibits a maximum and a minimum, which occur at the spinodal compositions (where $\partial^2 \Delta G^M/\partial X_B^2$, and hence $\partial a_B/\partial X_B$ equal zero), e.g., the points *b* and *c* on the curve drawn for $T =$

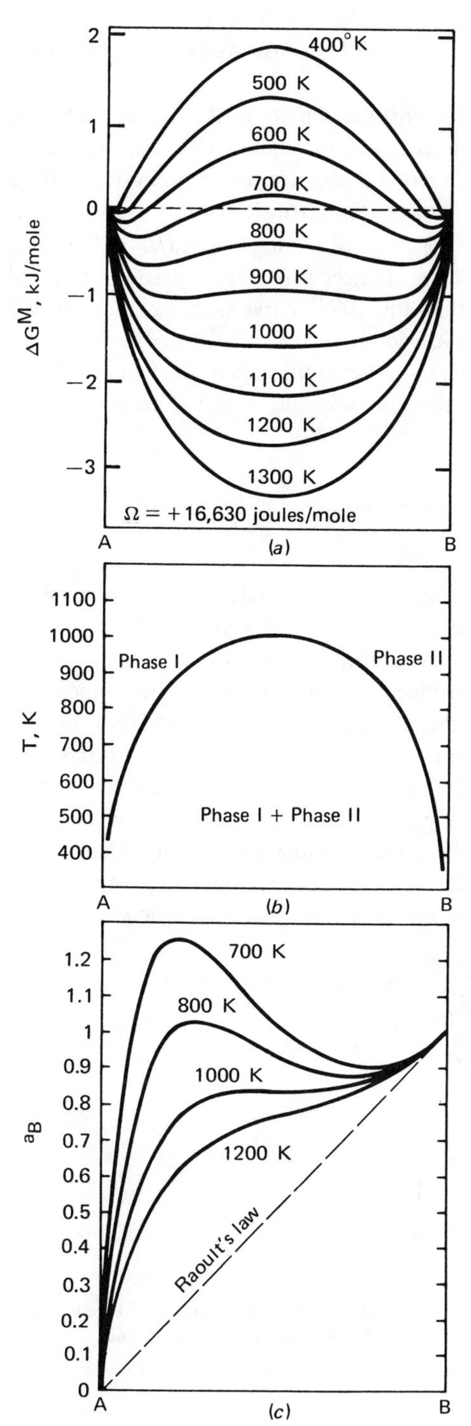

Fig. 12.6. (*a*) The effect of temperature on the integral free energies of mixing in a binary regular solution for which $\Omega = +16{,}630$ joules/mole. (*b*) The loci of the double tangent points in Fig. 12.6*a* with temperature (which generates the phase diagram of the system). (*c*) The activities of component B derived from Fig. 12.6*a*.

800 K in Fig. 12.7. The curve *ab* represents the activity of B in phase I, which is supersaturated with B; and the curve *cd* represents the activity of B in phase II, which is supersaturated with A. Between the compositions *b* and *c* the derived $\partial a_B/\partial X_B$ is negative. This situation violates the intrinsic stability criterion that $\partial a_i/\partial X_i$ be always positive (cf. $(\partial P/\partial V)_T > 0$ over the portion *JHF* in Fig. 8.7), and hence the derived activity curve between *b* and *c* has no physical significance, which also indicates that the ΔG^M curve between the spinodal compositions *m* and *q* in Fig. 12.4*b* has no physical significance. The horizontal line drawn between *a* and *d* in Fig. 12.7 represents the actual constant activity of B in the two-phase region, where the compositions *a* and *d* are the double tangent points on the ΔG^M curve.

12.5 LIQUID AND SOLID STANDARD STATES

Thus far the standard state of a component of a condensed system has been chosen at being the pure component in its stable state at the particular temperature and pressure of interest. At 1 atm pressure (the pressure normally considered), the stable state is determined by whether or not the temperature of interest is above or below the normal melting temperature of the component. Thus far, in the discussion of binary condensed solutions it has been tacitly assumed that the temperature of interest is above or below the melting temperatures of both components; e.g., Fig. 12.7 could refer equally well to liquid immiscibility, in which case the standard states are the pure liquids, or to solid immiscibility, in which case the standard states are the two pure solids. As the standard state of a component is simply a reference state against which the

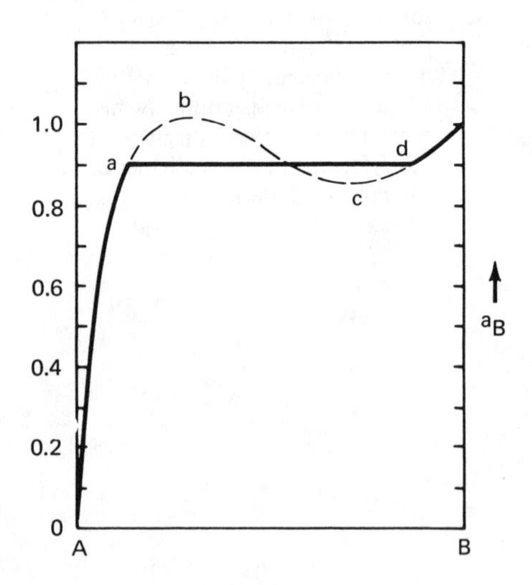

Fig. 12.7. The activity of B at 800 K derived from Fig. 12.6*a*.

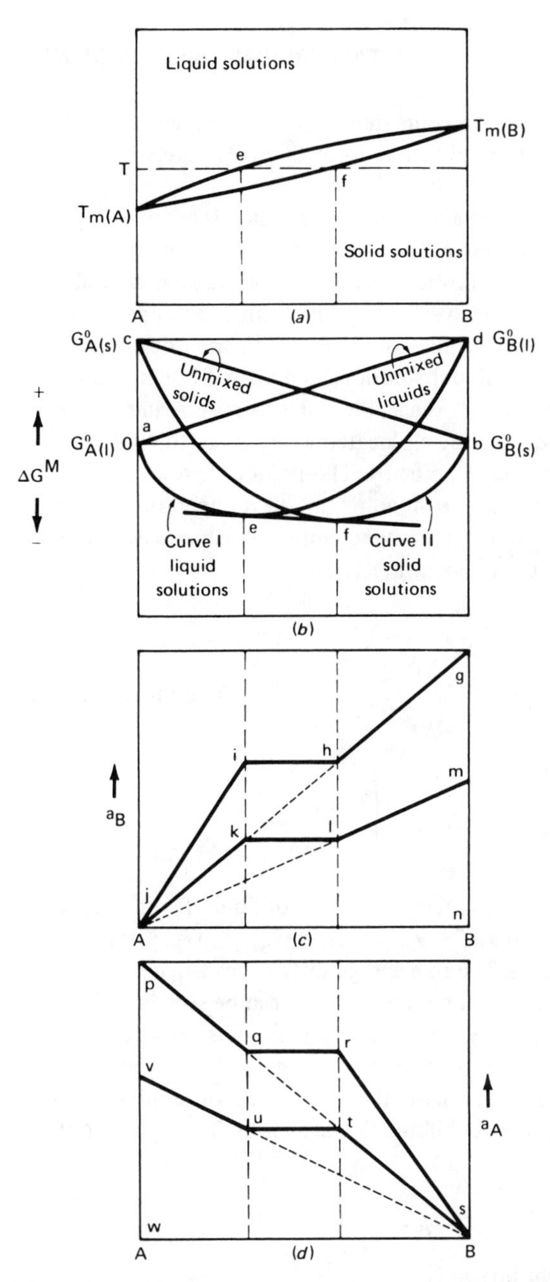

Fig. 12.8. (*a*) The phase diagram of the system A–B. (*b*) The integral free energies of mixing in the system A–B at the temperature *T*. (*c*) The activities of B at the temperature *T*, and comparison between the solid and liquid standard states of B. (*d*) The activities of A at the temperature *T*, and comparison between the solid and liquid standard states of A.

component in any other state can be compared, it follows that any state can be chosen as the standard state and this choice is normally made purely on the basis of convenience.

Consider the binary system A–B at a temperature T which is below $T_{m(\mathrm{B})}$, the melting temperature of B, and above $T_{m(\mathrm{A})}$, the melting temperature of A. Consider further that this system forms ideal liquid solutions and ideal solid solutions. The phase diagram of the system together with the temperature T is shown in Fig. 12.8*a*. Figure 12.8*b* shows the two ΔG^M-composition curves of interest, curve I being drawn for liquid solutions and curve II being drawn for solid solutions. At the temperature T, the stable states of pure A and pure B are located at $\Delta G^M = 0$, with liquid A being located at $X_\mathrm{A} = 1$ (the point *a*) and solid B being located at $X_\mathrm{B} = 1$ (the point *b*). The point *c* represents the molar free energy of solid A with respect to liquid A at the temperature T; and as $T > T_{m(\mathrm{A})}$, then $G^0_{\mathrm{A}(s)} - G^0_{\mathrm{A}(l)}$ is a positive quantity which equals minus the free energy of melting of A at the temperature T. That is,

$$G^0_{\mathrm{A}(s)} - G^0_{\mathrm{A}(l)} = -\Delta G^0_{m(\mathrm{A})} = -(\Delta H^0_{m(\mathrm{A})} - T\Delta S^0_{m(\mathrm{A})})$$

and if $c_{p,\mathrm{A}(s)} = c_{p,\mathrm{A}(l)}$, that is, if $\Delta H^0_{m(\mathrm{A})}$ and $\Delta S^0_{m(\mathrm{A})}$ are independent of temperature, then

$$\Delta G^0_{m(\mathrm{A})} = \Delta H^0_{m(\mathrm{A})} \left[\frac{T_{m(\mathrm{A})} - T}{T_{m(\mathrm{A})}}\right] \tag{12.4}$$

Similarly the point *d* represents the molar free energy of liquid B with respect to solid B at the temperature T; and as $T < T_{m(\mathrm{B})}$, then $G^0_{\mathrm{B}(l)} - G^0_{\mathrm{B}(s)}$ is a positive quantity, being equal to $\Delta G^0_{m(\mathrm{B})}$. The free energy differences $G^0_{\mathrm{B}(l)} - G^0_{\mathrm{B}(s)}$ and $G^0_{\mathrm{A}(s)} - G^0_{\mathrm{A}(l)}$, and their dependence on temperature, can be seen graphically in Figs. 7.1 and 7.2. The straight line joining *a* and *d* represents the free energy of unmixed liquid A and liquid B with reference to unmixed liquid A and solid B; the straight line joining *c* and *b* represents the free energy of unmixed solid A and solid B with reference to unmixed liquid A and solid B. The equation of the straight line *cb* is

$$\Delta G = -X_\mathrm{A} \Delta G^0_{m(\mathrm{A})}$$

and the equation of the straight line *ad* is

$$\Delta G = X_\mathrm{B} \Delta G^0_{m(\mathrm{B})}$$

At any composition, the formation of a homogeneous liquid solution from pure liquid A and pure solid B can be regarded as being a two-step process involving,

1. The melting of X_B moles of B, for which $\Delta G = X_B \Delta G^0_{m(B)}$
2. The mixing of X_B moles of liquid B and X_A moles of liquid A to form an ideal liquid solution, for which

$$\Delta G = \Delta G^{M,\mathrm{id}} = RT(X_A \ln X_A + X_B \ln X_B)$$

Thus the molar free energy of the formation of an ideal liquid solution, $\Delta G^M_{(l)}$, from liquid A and solid B is obtained as

$$\Delta G^M_{(l)} = RT(X_A \ln X_A + X_B \ln X_B) + X_B \Delta G^0_{m(B)} \tag{12.5}$$

which is thus the equation of curve I in Fig. 12.8*b*. Similarly, at any composition, the formation of an ideal solid solution from liquid A and solid B involves a free energy change of

$$\Delta G^M_{(s)} = RT(X_A \ln X_A + X_B \ln X_B) - X_A \Delta G^0_{m(A)} \tag{12.6}$$

which is the equation of curve II in Fig. 12.8*b*.

At the composition *e*, the tangent to the liquid solutions curve is also the tangent to the solid solutions curve at the composition *f*. Hence, at the temperature *T*, liquid *e* is in equilibrium with solid *f*; i.e., *e* is the liquidus composition and *f* is the solidus composition, as is seen in Fig. 12.8*a*. As the temperature is varied, say lowered, consideration of Figs. 7.1 and 7.2 shows that the magnitude of *ca* decreases and the magnitude of *db* increases. The consequent movement of curves I and II relative to one another is such that the double tangent positions *e* and *f* shift to the left. Correspondingly, if the temperature is increased, the relative movement of the free energy curves is such that *e* and *f* shift to the right. The loci of *e* and *f* with change in temperature trace out the liquidus and solidus lines respectively.

Specifically, for equilibrium between the solid and liquid phases,

$$\Delta \bar{G}^M_A \text{ (in the solid solution)} = \Delta \bar{G}^M_A \text{ (in the liquid solution)} \tag{12.7}$$

and

$$\Delta \bar{G}^M_B \text{ (in the solid solution)} = \Delta \bar{G}^M_B \text{ (in the liquid solution)} \tag{12.8}$$

At any temperature *T*, these two conditions fix the solidus and liquidus compositions, i.e., the double tangent points. From Eq. (12.5),

$$\frac{\partial \Delta G^M_{(l)}}{\partial X_{A(l)}} = RT\,[\ln X_{A(l)} - \ln X_{B(l)}] - \Delta G^0_{m(B)}$$

Thus,

$$X_{B(l)} \cdot \frac{\partial \Delta G^M_{(l)}}{\partial X_{A(l)}} = RT\,[X_{B(l)} \ln X_{A(l)} - X_{B(l)} \ln X_{B(l)}] - X_{B(l)}\, \Delta G^0_{m(B)} \tag{12.9}$$

From Eq. (11.33*a*),

$$\Delta \bar{G}^M_A \text{ (in liquid solutions)} = \Delta G^M_{(l)} + X_{B(l)} \cdot \frac{\partial \Delta G^M_{(l)}}{\partial X_{A(l)}}$$

and thus adding Eqs. (12.5) and (12.9) gives

$$\Delta \bar{G}^M_A \text{ (in liquid solutions)} = RT \ln X_{A(l)} \tag{12.10}$$

From Eq. (12.6),

$$\frac{\partial \Delta G^M_{(s)}}{\partial X_{A(s)}} = RT\,[\ln X_{A(s)} - \ln X_{B(s)}] - \Delta G^0_{m(A)}$$

and thus,

$$X_{B(s)} \cdot \frac{\partial \Delta G^M_{(s)}}{\partial X_{A(s)}} = RT\,[X_{B(s)} \ln X_{A(s)} - X_{B(s)} \ln X_{B(s)}] - X_{B(s)}\, \Delta G^0_{m(A)} \tag{12.11}$$

Adding Eqs. (12.6) and (12.11) gives

$$\Delta \bar{G}^M_A \text{ (in solid solutions)} = \Delta G^M_{(s)} + X_{B(s)} \cdot \frac{\partial \Delta G^M_{(s)}}{\partial X_{A(s)}} = RT \ln X_{A(s)} - \Delta G^0_{m(A)} \tag{12.12}$$

Thus, from Eqs. (12.7), (12.10), and (12.12),

$$RT \ln X_{A(l)} = RT \ln X_{A(s)} - \Delta G^0_{m(A)} \tag{12.13}$$

Similarly, from Eqs. (12.5) and (11.33*b*),

$$\Delta \bar{G}^M_B \text{ (in liquid solutions)} = \Delta G^M_{(l)} + X_{A(l)} \cdot \frac{\partial \Delta G^M_{(l)}}{\partial X_{B(l)}} = RT \ln X_{B(l)} + \Delta G^0_{m(B)} \tag{12.14}$$

and from Eqs. (12.6) and (11.33*b*),

$$\Delta \bar{G}_{\mathrm{B}}^{M} \text{ (in solid solutions)} = \Delta G_{(s)}^{M} + X_{\mathrm{A}(s)} \cdot \frac{\partial \Delta G_{(s)}^{M}}{\partial X_{\mathrm{B}(s)}} = RT \ln X_{\mathrm{B}(s)} \tag{12.15}$$

Thus, from Eqs. (12.8), (12.14), and (12.15),

$$RT \ln X_{\mathrm{B}(l)} + \Delta G_{m(\mathrm{B})}^{0} = RT \ln X_{\mathrm{B}(s)} \tag{12.16}$$

The solidus and liquidus compositions are thus determined by Eqs. (12.13) and (12.16) as follows.

Equation (12.13) can be written as

$$X_{\mathrm{A}(l)} = X_{\mathrm{A}(s)} \cdot \exp\left(\frac{-\Delta G_{m(\mathrm{A})}^{0}}{RT}\right) \tag{12.17}$$

and, noting that $X_{\mathrm{B}} = 1 - X_{\mathrm{A}}$, Eq. (12.16) can be written as

$$(1 - X_{\mathrm{A}(l)}) = (1 - X_{\mathrm{A}(s)}) \cdot \exp\left(\frac{-\Delta G_{m(\mathrm{B})}^{0}}{RT}\right) \tag{12.18}$$

Combining Eqs. (12.17) and (12.18) gives

$$X_{\mathrm{A}(s)} = \frac{1 - \exp(-\Delta G_{m(\mathrm{B})}^{0}/RT)}{\exp(-\Delta G_{m(\mathrm{A})}^{0}/RT) - \exp(-\Delta G_{m(\mathrm{B})}^{0}/RT)} \tag{12.19}$$

and

$$X_{\mathrm{A}(l)} = \frac{[1 - \exp(-\Delta G_{m(\mathrm{B})}^{0}/RT)] \exp(-\Delta G_{m(\mathrm{A})}^{0}/RT)}{\exp(-\Delta G_{m(\mathrm{A})}^{0}/RT) - \exp(-\Delta G_{m(\mathrm{B})}^{0}/RT)} \tag{12.20}$$

Thus, if $c_{p,i(s)} = c_{p,i(l)}$, such that

$$\Delta G_{m(i)}^{0} = \Delta H_{m(i)}^{0} \cdot \frac{T_{m(i)} - T}{T_{m(i)}} \tag{12.4}$$

it is seen that the phase diagram for a system that forms ideal solid and liquid solutions is determined only by the melting temperatures and the molar heats of fusion of the components. The system Mn_2SiO_4–Fe_2SiO_4 forms both ideal solid and ideal liquid solutions, and hence the phase diagram can be calculated from

Eqs. (12.19), (12.20), and (12.4) with $T_{m,Mn_2SiO_4} = 1620$ K, $T_{m,Fe_2SiO_4} = 1490$ K, $\Delta H^0_{m,Mn_2SiO_4} = 89.66$ kJ, and $\Delta H^0_{m,Fe_2SiO_4} = 92.17$ kJ. The calculated phase diagram is shown in Fig. 12.9.

Referring again to Figs. 12.8*a* and 12.8*b*, when $T < T_{m(A)}$, both $\Delta G^0_{m(A)}$ and $\Delta G^0_{m(B)}$ are positive quantities, and so the liquid solutions curve lies everywhere above the solid solutions curve, in accord with the fact that the solid solution is everywhere stable. Similarly, when $T > T_{m(B)}$, both $\Delta G^0_{m(A)}$ and $\Delta G^0_{m(B)}$ are negative quantities, and so the solid solutions curve lies everywhere above the liquid solutions curve. When either $T > T_{m(B)}$ or $T < T_{m(A)}$, the most convenient choice of standard state is obvious; but when $T_{m(A)} < T < T_{m(B)}$, then either the solid or the liquid can conveniently be chosen as the standard state.

Figure 12.8*c* illustrates the activity-composition relationships of component B. These relationships are again obtained from the variation of the intercept to the free energy of mixing curve *aefb* with the $X_B = 1$ axis. However, as two standard states are available—the point *b* for solid B, and the point *d* for liquid B—then the lengths of the tangential intercepts either can be measured from *b*, in which case the activities of B are obtained with respect to solid B as the standard state, or the lengths of the intercepts can be measured from *d*, in which case the activities are obtained with respect to liquid B as the standard state.

If pure solid B is chosen as the standard state and is located at the point *g* in Fig. 12.8*c*, then the length of *gn* is, by definition, unity, and this defines the

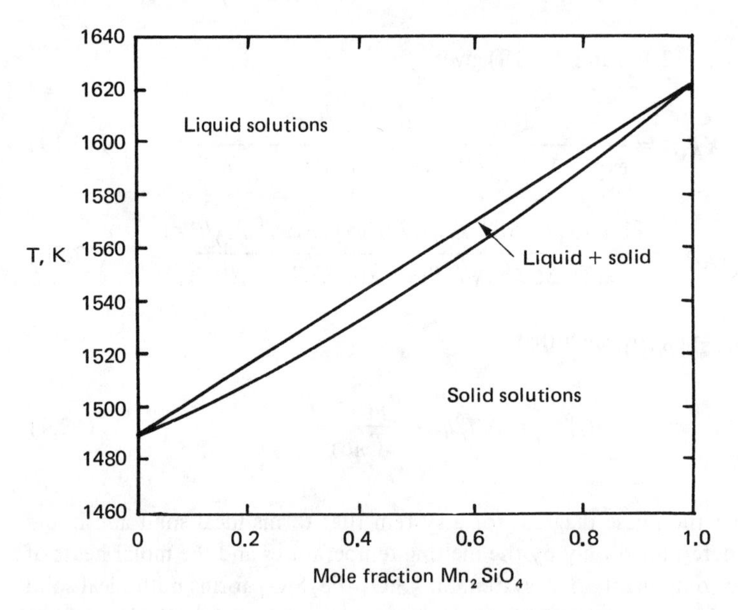

Fig. 12.9. The phase diagram for the system Fe_2SiO_4–Mn_2SiO_4.

solid standard state activity scale. The line *ghij* then represents a_B in the solutions with respect to solid B having unit activity at *g*. This line is obtained from the variation of the tangential intercepts from the curve *aefb* with the $X_B = 1$ axis measured from the point *b*. On this activity scale, Raoult's law is given by *jg*, and the points *i* and *h* respectively represent the activity of B in the coexisting liquid solution *e* and solid solution *f*. The point *m* represents the activity of liquid B and, as is seen, the activity of liquid B with respect to solid B is less than unity, being given by the ratio *mn*/*gn*. This can be examined quantitatively as follows. For B in any state along the curve *aefb*, in which state its partial molar free energy is $\bar{G}_B$,

$$\bar{G}_B = G^0_{B(l)} + RT \ln (a_B \text{ with respect to liquid B})$$

and

$$\bar{G}_B = G^0_{B(s)} + RT \ln (a_B \text{ with respect to solid B})$$

Thus

$$G^0_{B(l)} - G^0_{B(s)} = \Delta G^0_{m(B)} = RT \ln \left(\frac{a_B \text{ with respect to solid B}}{a_B \text{ with respect to liquid B}} \right) \tag{12.21}$$

As $T < T_{m(B)}$, $\Delta G^0_{m(B)}$ is a positive quantity; and hence, in any state,

$$a_B(\text{with respect to solid B}) > a_B(\text{with respect to liquid B})$$

where both activities are measured on the same activity scale. For pure B, $a_{B(s)} > a_{B(l)}$; that is, $gn > mn$ in Fig. 12.8c. And if $gn = 1$, then $mn = \exp(-\Delta G^0_{m(B)}/RT)$. Equation (12.21) simply states that

(The tangential intercept from any point on the curve *aefb*, measured from *b*)
+ (the length *bd*)
= (the tangential intercept from the same point on the curve measured from *d*)

i.e.,

$$RT \ln (a_B \text{ with respect to solid B}) + (G^0_{B(s)} - G^0_{B(l)}) = RT \ln (a_B \text{ with respect to liquid B})$$

or

$$\Delta G^0_{m(B)} = RT \ln \left(\frac{a_B \text{ with respect to solid B}}{a_B \text{ with respect to liquid B}} \right)$$

If pure liquid B is chosen as the standard state and is located at the point m, then the length mn is, by definition, unity, and this defines the liquid standard state activity scale. Raoult's law on this scale is given by the line jm, and the activities of B in the solutions with respect to pure liquid B having unit activity are represented by the line $mlkj$. The activity of solid B, located at g, is greater than unity on the liquid standard state activity scale, being equal to $\exp(\Delta G^0_{m(\mathrm{B})}/RT)$. When measured on one or the other of the two scales, the lines $jihg$ and $jklm$ vary in the constant ratio $\exp(\Delta G^0_{m(\mathrm{B})}/RT)$, but $jihg$ measured on the solid standard state activity scale is identical with $jklm$ measured on the liquid standard state activity scale.

Figure 12.8*d* illustrates the variation of a_A with composition in the solutions at the temperature T. In this case, as $T > T_{m(\mathrm{A})}$, $\Delta G^0_{m(\mathrm{A})}$ is a negative quantity; and hence, from Eq. (12.3) applied to component A,

$$a_\mathrm{A}(\text{with respect to liquid A}) > a_\mathrm{A}(\text{with respect to solid A})$$

when measured on the same activity scale.

If pure liquid A is chosen as the standard state and is located at the point p, then the length of pw is, by definition, unity, and the line $pqrs$ represents a_A in the solutions with respect to the liquid standard state. On the liquid standard state activity scale, the activity of solid A, located at the point v, has the value $\exp(\Delta G^0_{m(\mathrm{A})}/RT)$. If, on the other hand, pure solid A is chosen as the standard state, then the length of vw is, by definition, unity, and Raoult's law is given by vs. The line $vuts$ represents the activities of A in the solutions with respect to pure solid A. On the solid standard state activity scale, liquid A, located at the point p, has the value $\exp(-\Delta G^0_{m(\mathrm{A})}/RT)$. Again the two lines, measured on one or the other of the two activity scales, vary in the constant ratio $\exp(\Delta G^0_{m(\mathrm{A})}/RT)$ and, when measured on their respective activity scales, are identical.

If the temperature of the system is decreased below the value of T indicated in Fig. 12.8*a*, then the length of ac, being equal to $\Delta G^0_{m(\mathrm{A})}$ at the temperature of interest, decreases in magnitude; and, correspondingly, the magnitude of $\Delta G^0_{m(\mathrm{B})}$, and so the length of bd, increases. The resultant change in the relative positions of the curves I and II in Fig. 12.8*b* is such that the double tangent points e and f shift to the left towards A. The effect of this on the activity curves is as follows. In the case of both components

$$\frac{a_i \text{ with respect to solid } i}{a_i \text{ with respect to liquid } i} = \exp\left(\frac{\Delta G^0_{m(i)}}{RT}\right)$$

which, from Eq. (12.4),

$$= \exp\left[\Delta H^0_{m(i)}\left(\frac{T_{m(i)} - T}{RTT_{m(i)}}\right)\right]$$

With respect to component B, if the temperature, which is less than $T_{m(\mathrm{B})}$, is lowered, then the ratio $a_{\mathrm{B(solid)}}/a_{\mathrm{B(liquid)}}$, which is greater than unity, increases. Hence in Fig. 12.8*c*, the ratio *gn*/*mn* increases. With respect to component A, if the temperature, which is greater than $T_{m(\mathrm{A})}$, is lowered, then the ratio $a_{\mathrm{A(solid)}}/a_{\mathrm{A(liquid)}}$, which is less than unity, increases. Hence the ratio *vw*/*pw* in Fig. 12.8*d* increases. When the temperature is lowered to $T_{m(\mathrm{A})}$, then, as solid A and liquid A coexist in equilibrium, $\Delta G^0_{m(\mathrm{A})}$ is zero, and the points *p* and *v* coincide. Similarly, if the temperature is increased, then the ratio *vw*/*pw* decreases and the ratio *mn*/*gn* increases until, at $T = T_{m(\mathrm{B})}$, the points *m* and *g* coincide.

12.6 PHASE DIAGRAMS, FREE ENERGY, AND ACTIVITY

Complete solid solubility of the components A and B requires that A and B have the same crystal structure and be of comparable atomic size, electronegativity, and valency, If any of these conditions are not met, then a miscibility gap will occur in the solid state. Consider the system A-B, the phase diagram of which is shown in Fig. 12.10*a*, in which A and B have differing crystal structures. Two terminal solid solutions, α and β, occur. The free energy of mixing curves, at the temperature T_1, are shown in Fig. 12.10*b*. In this figure, *a* and *c*, located at $\Delta G^M = 0$, respectively represent pure solid A and pure liquid B; and *b* and *d* respectively represent pure liquid A and pure solid B. The curve *aeg* (curve I) is the free energy of mixing of solid A and solid B to form homogeneous α solid solution which has the same crystal structure as has A. This curve intersects the $X_\mathrm{B} = 1$ axis at the value of *G*, relative to liquid B, which solid B would have if it had the same crystal structure as A. In principle this value can be calculated using the methods of quantum mechanics. Similarly the curve *dh* (curve II) represents the free energy of mixing of solid A and solid B to form homogeneous β solid solution which has the same crystal structure as B. This curve intersects the $X_\mathrm{A} = 1$ axis at the value of *G*, relative to stable solid A, which solid A would have if it had the same crystal structure as B. The curve *bfc* (curve III) represents the free energy of mixing liquid A and liquid B to form a homogeneous liquid solution. As curve II lies everywhere above curve III, solid solution β has no stable existence at the temperature T_1. Figure 12.10*c* shows the activity-composition relationships of the components at T_1, drawn with respect to solid as the standard state for A and liquid as the standard state for B. These relationships are drawn in accordance with the assumption that the liquid solutions exhibit Raoultian ideality and the solid solutions show positive deviations from Raoult's law.

As the temperature decreases below T_1, the length *ab* increases and the length

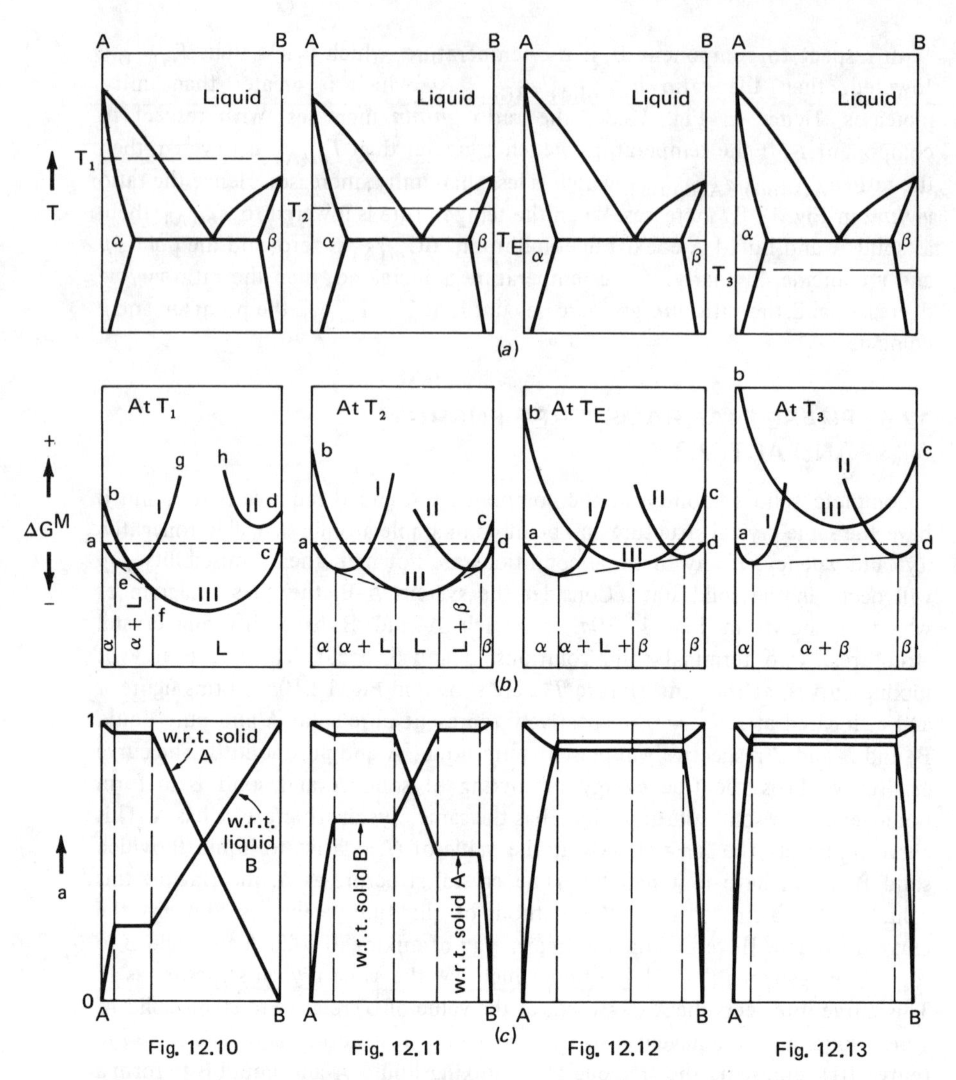

Fig. 12.10 Fig. 12.11 Fig. 12.12 Fig. 12.13

Figs. 12.10–12.13. The effect of temperature on the free energies of mixing and the activities of the components of the system A–B.

cd decreases until, at $T = T_{m(\mathrm{B})}$, the points *c* and *d* coincide at $\Delta G^M = 0$. At $T_2 < T_{m(\mathrm{B})}$, the point *d* (solid B) lies at $\Delta G^M = 0$, and the point *c* (liquid B) lies above *d*. As curve II now lies partially below curve III, two double tangents can be drawn, one to the curves I and III and the other to the curves II and III. The ΔG^M-composition curves at the temperature T_2 are shown in Fig. 12.11*b*, and the corresponding activity-composition curves are shown in Fig. 12.11*c*, in which the solid is the standard state for both components.

At $T = T_E$, the eutectic temperature, curve III has changed its position relative to curves I and II such that the two double tangents in Fig. 12.11*b* coincide and become a triple tangent to the three curves. Figure 12.12*b* and *c* shows respectively the ΔG^M and activity curves at T_E. Figure 12.13*b* and *c* shows the same relationships at $T_3 < T_E$.

If the ranges of solid solubility in the α and β phases are immeasurably small, then, as an approximation, it is normally said that A and B are insoluble in one another in the solid state. The phase diagram of such a system could be as in Fig. 12.14*a*. As the ΔG^M-composition curves of any system have vertical tangents at their extremities, then any pure substance presents an infinite chemical sink to any other substance, or conversely, it is impossible to obtain an absolutely pure substance; i.e., no one substance is absolutely insoluble in any other substance. As the solid solubility range in Fig. 12.14*a* is so small that it may be neglected on the scale of Fig. 12.14*a*, then also the α and β solid solution curves (curves I and II in Figs. 12.10 to 12.13) are so compressed toward the X_A and X_B axes, respectively, that on the scale of Figs. 12.10 to 12.13, they coincide with the vertical axes. The sequence in Fig. 12.15 illustrates how, as the solid solubility of B in A decreases, the α solid solution free energy of mixing curve is compressed against the X_A axis. Figure 12.14*b* shows the free energy-composition relationships for the A–B system of Fig. 12.14*a* at the temperature T. The "double tangent" to the α solid solution and liquid solution curves is reduced to the tangent drawn to the liquid solutions curve from the point on the X_A axis which represents pure solid A. Figure 12.14*c* shows the activity-composition relationships of the two components at the temperature T. Again these are drawn in accordance with the supposition that the liquid A–B solutions are ideal. In Fig. 12.14*c*, *pqr* is a_A with respect to solid A having unit activity at *p*, *s* is the activity of liquid A with respect to solid A at *p*, *str* is a_A with respect to liquid A having unit activity at *s*, and *Auvw* is a_B with respect to liquid B having unit activity at *w*.

In a case where a binary system exhibits complete miscibility in the liquid state and virtually complete immiscibility in the solid state (e.g., Fig. 12.14*a*), the activity-composition of a component in the liquid solutions can be obtained from consideration of the liquidus curve of the component. At any temperature T, (Fig. 12.14*a*), the system of composition between pure A and the liquidus composition comprises pure solid A in equilibrium with a liquid solution of the liquidus composition. Thus, at T,

$$G^\circ_{A(s)} = \bar{G}_{A(l)}$$

$$= G^\circ_{A(l)} + RT \ln a_A$$

where a_A is with respect to liquid A as the standard state.

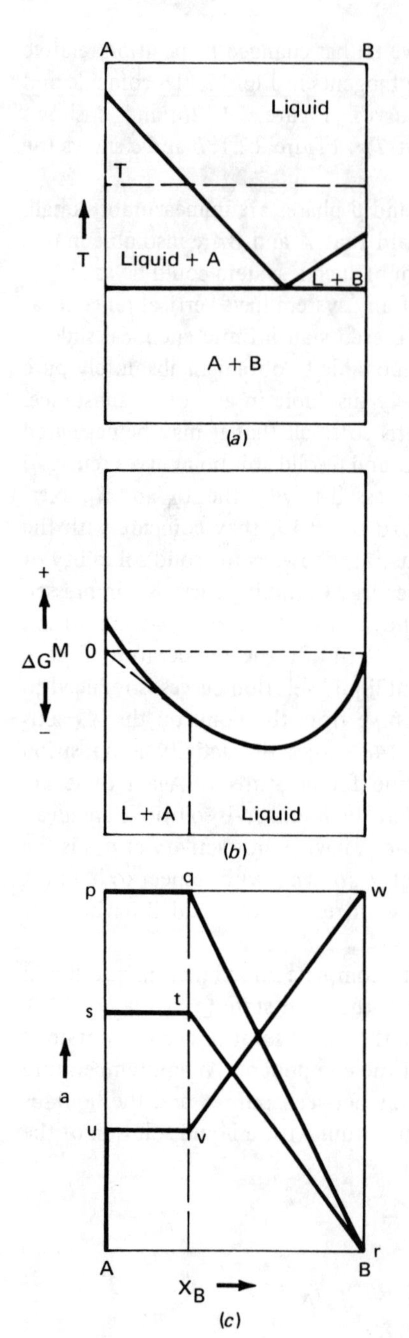

Fig. 12.14. The integral free energies of mixing and the activities in a binary eutectic system that exhibits complete liquid miscibility and virtual complete solid immiscibility.

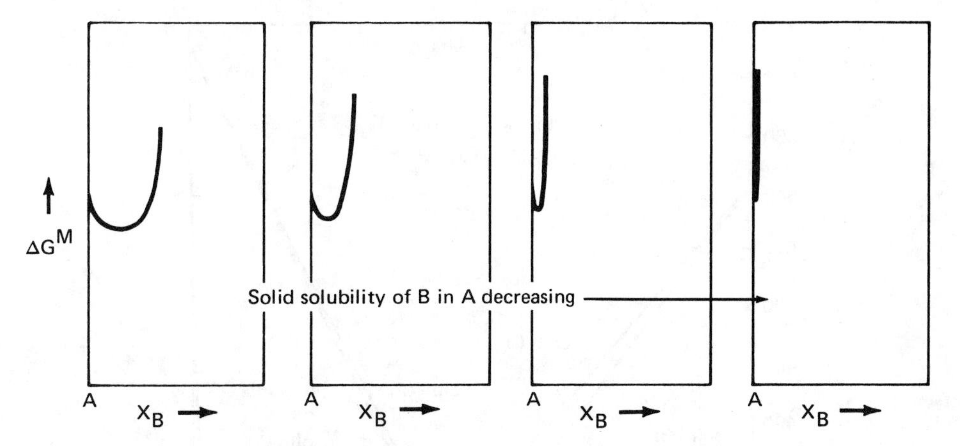

Fig. 12.15. The effect of decreasing solid solubility on the integral free energy of mixing curve.

Thus

$$\Delta G^0_{m(\mathrm{A})} = -RT \ln a_{\mathrm{A}} \tag{12.22}$$

or, if the liquid solutions are Raoultian,

$$\Delta G^0_{m(\mathrm{A})} = -RT \ln X_{\mathrm{A}(l)} \tag{12.23}$$

It can be noted that Eq. (12.23) is simply Eq. (12.17) with $X_{\mathrm{A}(s)} = 1$.

Consider the application of Eq. (12.23) to calculation of the liquidus lines in a binary eutectic system. The phase diagram for the system Cd–Bi is shown in Fig. 12.16; cadmium is virtually insoluble in solid bismuth, and the maximum solubility of bismuth in solid cadmium is 2.75 mole percent at the eutectic temperature of 419 K. If the liquid solutions are ideal, the Bi liquidus is obtained from Eq. (12.23) as

$$\Delta G^0_{m(\mathrm{Bi})} = -RT \ln X_{\mathrm{Bi\,(liquidus)}}$$

$\Delta H^0_{m(\mathrm{Bi})} = 10{,}900$ joules at $T_{m,\mathrm{Bi}} = 544$ K, and thus

$$\Delta S^0_{m(\mathrm{Bi})} = \frac{10{,}900}{544} = 20.0 \text{ joules/degree at 544 K}$$

$$c_{p,\,\mathrm{Bi}(s)} = 18.8 + 22.6 \times 10^{-3}\,T \text{ joules/degree}$$

$$c_{p,\,\mathrm{Bi}(l)} = 20 + 6.15 \times 10^{-3}\,T + 21.1 \times 10^{5}\,T^{-2} \text{ joules/degree}$$

Thus,

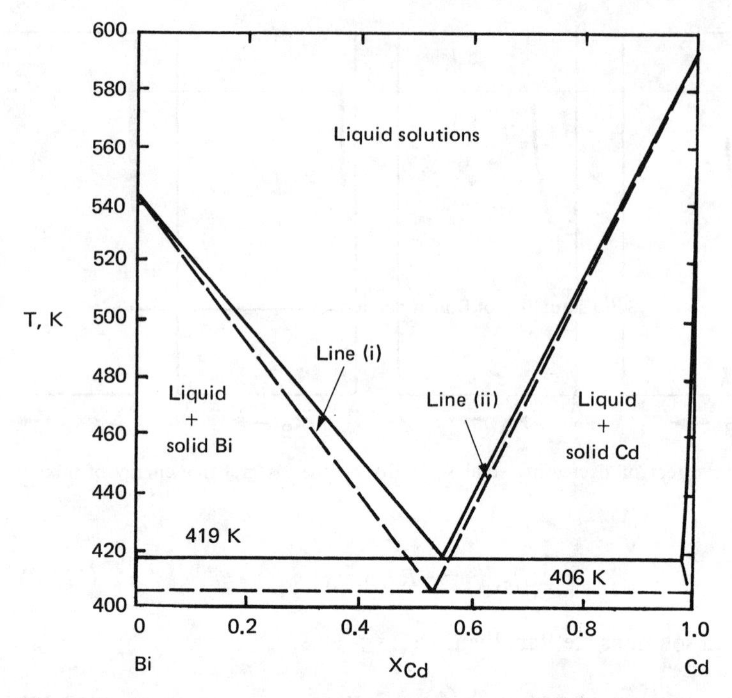

Fig. 12.16. Phase diagram for the system Bi–Cd. (Full lines–experimentally determined liquidus lines; Broken lines–calculated assuming no solid solution and ideal mixing in the liquid solutions.)

$$c_{p,\,\mathrm{Bi}(l)} - c_{p,\,\mathrm{Bi}(s)} = \Delta c_{p,\,\mathrm{Bi}}$$

$$= 1.2 - 16.45 \times 10^{-3}T + 21.1 \times 10^{5}T^{-2} \text{ joules/degree}$$

and

$$\Delta G^0_{m(\mathrm{Bi})} = \Delta H^0_{m(\mathrm{Bi}),544} + \int_{544}^{T} \Delta c_{p,\,\mathrm{Bi}}\,dT - T\left[\Delta S^0_{m(\mathrm{Bi}),544} + \int_{544}^{T} \frac{\Delta c_{p,\,\mathrm{Bi}}}{T}\,dT\right]$$

$$= 16{,}560 - 23.79T - 1.2T\ln T + 8.225 \times 10^{-3}T^2 - 10.55 \times 10^{5}T^{-1}$$

$$= -RT\ln X_{\mathrm{Bi\,(liquidus)}} \qquad (12.24)$$

or,

$$\ln X_{\mathrm{Bi\,(liquidus)}} = \frac{-1992}{T} + 2.861 + 0.144\ln T - 9.892 \times 10^{-4}T + 1.269 \times \frac{10^5}{T^2}$$

This equation is drawn as broken line (i) in Fig. 12.16.

Similarly, if the small solid solubility of Bi in Cd is ignored,

$$\Delta G^{0}_{m(\text{Cd})} = -RT \ln X_{\text{Cd (liquidus)}}$$

$\Delta H^{0}_{m(\text{Cd})} = 6400$ joules at $T_{m,\text{Cd}} = 594$ K, and thus $\Delta S^{0}_{m(\text{Cd})} = 6400/594 = 10.77$ joules/degree at 594 K and

$$c_{p,\text{Cd}(s)} = 22.2 + 12.3 \times 10^{-3} T \text{ joules/degree}$$

$$c_{p,\text{Cd}(l)} = 29.7 \text{ joules/degree}$$

Thus

$$c_{p,\text{Cd}(l)} - c_{p,\text{Cd}(s)} = \Delta c_{p,\text{Cd}} = 7.5 - 12.3 \times 10^{-3} T \text{ joules/degree}$$

$$\Delta G^{0}_{m(\text{Cd})} = \Delta H^{0}_{m(\text{Cd}),594} + \int_{594}^{T} \Delta c_{p,\text{Cd}}\, dT - T\left[\Delta S^{0}_{m(\text{Cd}),594} + \int_{594}^{T} \frac{\Delta c_{p,\text{Cd}}}{T}\, dT\right]$$

$$= 4115 + 37.32T - 7.5T \ln T + 6.15 \times 10^{-3} T^{2} \text{ joules}$$

$$= -RT \ln X_{\text{Cd (liquidus)}} \tag{12.25}$$

or $$\ln X_{\text{Cd (liquidus)}} = \frac{-495}{T} - 4.489 + 0.90 \ln T - 7.397 \times 10^{-4} T$$

which is drawn as the broken line (ii) in Fig. 12.16. Lines (i) and (ii) intersect at the composition of the Raoultian liquid which is simultaneously saturated with Cd and Bi, and 406 K, which would be the eutectic temperature if the liquids were ideal. The actual liquidus lines lie above those calculated and the actual eutectic temperature is 419 K. From Eq. (12.24), $\Delta G^{0}_{m(\text{Bi}),419\text{ K}} = 2482$ joules, and from Eq. (12.25), $\Delta G^{0}_{m(\text{Cd}),419\text{K}} = 1898$ joules. Thus, from Eq. (12.22), in the actual eutectic melt,

$$a_{\text{Bi}} = \exp\left(\frac{-2482}{8.3144 \times 419}\right) = 0.49$$

and $$a_{\text{Cd}} = \exp\left(\frac{-1898}{8.3144 \times 419}\right) = 0.58$$

As the actual eutectic composition is $X_{\text{Cd}} = 0.55$, $X_{\text{Bi}} = 0.45$,

$$\gamma_{\mathrm{Bi}} = \frac{0.49}{0.45} = 1.09$$

and
$$\gamma_{\mathrm{Cd}} = \frac{0.58}{0.55} = 1.05$$

in the eutectic melt. Thus positive deviations from Raoultian ideality raise the liquidus temperatures.

It is now of interest to examine what happens to the liquidus lines as the magnitude of the positive deviation from Raoultian behavior in the liquids increases, i.e., as G^{xs} becomes increasingly positive. Assuming regular solution behavior, Eq. (12.22), written in the form

$$-\Delta G^{0}_{m(\mathrm{A})} = RT \ln X_{\mathrm{A}} + RT \ln \gamma_{\mathrm{A}}$$

becomes
$$-\Delta G^{0}_{m(\mathrm{A})} = RT \ln X_{\mathrm{A}} + RT\alpha(1 - X_{\mathrm{A}})^2$$

or, from Eq. (11.90),

$$-\Delta G^{0}_{m(\mathrm{A})} = RT \ln X_{\mathrm{A}} + \Omega(1 - X_{\mathrm{A}})^2 \tag{12.26}$$

Consider a hypothetical system A-B in which $\Delta H^{0}_{m(\mathrm{A})} = 7690$ joules at $T_{m,\mathrm{A}} = 1996$ K. Thus, for this system,

$$-7690 + 3.85T = RT \ln X_{\mathrm{A}} + \Omega(1 - X_{\mathrm{A}})^2$$

where X_{A} is the A-liquidus composition at the temperature T. Fig. 12.17 shows the A-liquidus drawn for $\Omega = 0$, 10, 20, 23.45, 30, 40, and 50 kilojoules. As Ω exceeds some critical value (which is 23.45 kilojoules in this case), the form of the liquidus line changes from a monotonic decrease in liquidus temperature with decreasing X_{A} to a form that contains a maximum and a minimum. At the critical value of Ω the maximum and minimum coincide at $X_{\mathrm{A}} = 0.5$ to produce a horizontal inflexion in the liquidus curve. It is apparent that, when Ω exceeds the critical value, isothermal tie-lines cannot be drawn between pure solid A and all points on the liquidus lines, which, necessarily, means that the calculated liquidus lines are impossible.

From Eq. (12.21),

$$\ln a_{\mathrm{A}} = \frac{-\Delta H^{0}_{m(\mathrm{A})}}{RT} + \frac{\Delta H^{0}_{m(\mathrm{A})}}{RT_{m(\mathrm{A})}}$$

Thus
$$d \ln a_{\mathrm{A}} = \frac{da_{\mathrm{A}}}{a_{\mathrm{A}}} = \frac{\Delta H^{0}_{m(\mathrm{A})}}{RT^2}\, dT$$

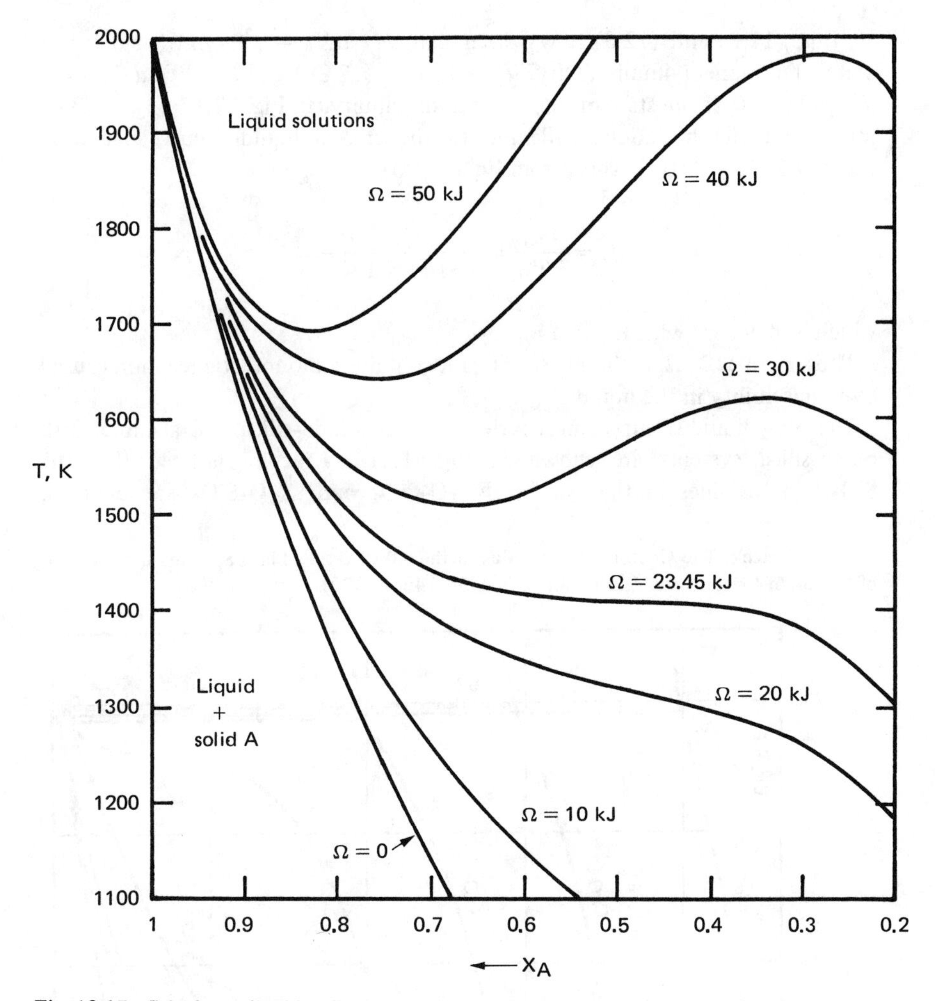

Fig. 12.17. Calculated liquidus lines assuming regular solution behavior in the liquid solutions and no solid solubility of B in A.

or

$$\frac{dT}{dX_A} = \frac{RT^2}{\Delta H^0_{m(A)} \cdot a_A} \cdot \frac{da_A}{dX_A} \tag{12.28}$$

and also

$$\frac{d^2T}{dX_A^2} = \frac{2RT}{\Delta H^0_{m(A)} \cdot a_A} \cdot \frac{da_A}{dX_A} \cdot \frac{dT}{dX_A} - \frac{RT^2}{\Delta H^0_{m(A)} \cdot a_A^2} \left(\frac{da_A}{dX_A}\right)^2 + \frac{RT^2}{\Delta H^0_{m(A)} \cdot a_A} \cdot \frac{d^2a_A}{dX_A^2} \tag{12.29}$$

In Eqs. (12.2) and (12.3) it was seen that $da_A/dX_A = d^2a_A/dX_A^2 = 0$ at the state of imminent immiscibility. Thus in Eqs. (12.28) and (12.29), $dT/dX_A = d^2T/dX_A^2 = 0$ at the state of imminent immiscibility. In Fig. 12.17, $\Omega_{cr} = 23.45$ joules and the horizontal inflexion in the critical liquidus curve occurs at $X_A = 0.5$, $T = 1410$ K. Thus, from Eq. (11.90),

$$\alpha_{cr} = \frac{\Omega_{cr}}{RT_{cr}} = \frac{23{,}450}{8.3144 \times 1410} = 2$$

which is in accord with Eq. (12.1).

Thus, with $\Omega > \Omega_{cr}$, the phase diagram contains a monotectic reaction caused by immiscibility in the liquid.

The SiO_2-liquidus curves in a series of alkali oxide-silica and alkaline earth oxide-silica systems are shown in Fig. 12.18. Kracek* noticed that the SiO_2-liquidus lines in the systems Rb_2O-SiO_2 and Cs_2O-SiO_2 are identical,

*F. C. Kracek, The Cristobalite Liquidus in the Alkali Oxide-Silica Systems and the Heat of Fusion of Cristobalite, *J. Am. Chem. Soc.* 52:1436 (1930).

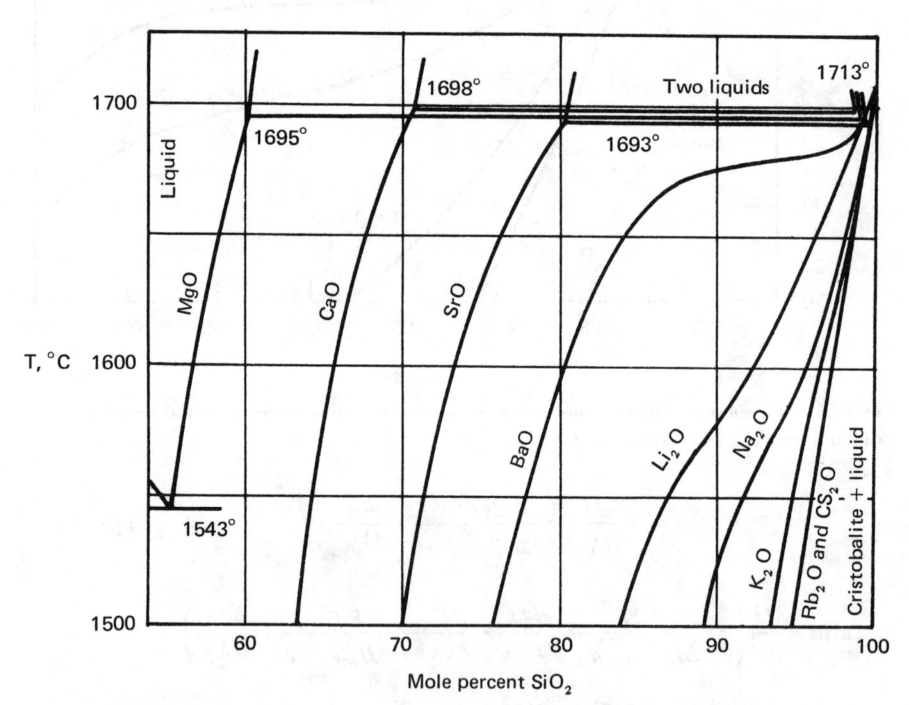

Fig. 12.18. The silica liquidus lines in several alkali oxide-silica and alkaline earth oxide-silica systems.

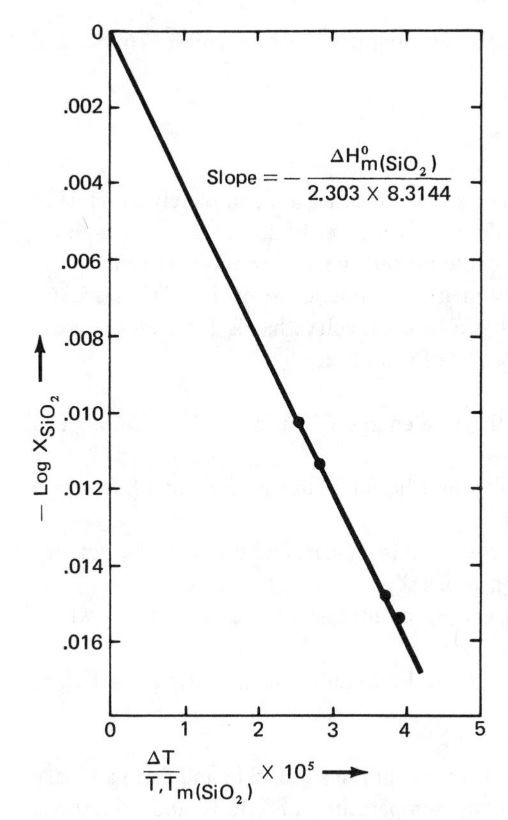

Fig. 12.19. The heat of fusion of silica calculated from the silica liquidus lines in the systems Rb_2O-SiO_2 and Cs_2O-SiO_2. *(From F. C. Kracek, The Cristobalite Liquidus in the Alkali Oxide-Silica Systems and the Heat of Fusion of Cristobalite, J. Am. Chem. Soc.,* **52:***1436 [1930].)*

which led him to assume that these two systems form ideal liquid solutions. For this to be the case, a plot of ln $X_{SiO_2\,(\text{liquidus})}$ in the two systems against

$$\frac{T_{m(SiO_2)} - T_{\text{liquidus}}}{T_{m(SiO_2)} \cdot T_{\text{liquidus}}}$$

should be a straight line with a slope of $\Delta H^0_{m(SiO_2)}/R$. Such a plot is shown in Fig. 12.19, and the slope of the line, being −401.5, gives the molar heat of fusion of SiO_2 (cristobalite) as 7690 joules (which is the value used in the previously discussed hypothetical system A-B). From Fig. 12.18 it is seen that the magnitudes of the positive deviations from Raoultian ideality in the liquid solutions increase in the order K_2O-SiO_2, Na_2O-SiO_2, Li_2O-SiO_2, BaO-SiO_2 SrO-SiO_2, CaO-SiO_2, MgO-SiO_2. The magnitude of the positive deviation in the system BaO-SiO_2 is such that liquid immiscibility almost occurs, and in the systems SrO-SiO_2, CaO-SiO_2, and MgO-SiO_2 the deviations are sufficient to cause liquid immiscibility with monotectic temperatures of 1693°, 1698°, and 1695°C.

PROBLEMS

12.1 The Fe–S phase diagram indicates that Fe and S are completely miscible in the liquid state and are virtually immiscible in the solid state. The liquidus composition at 1200°C is 62 mole percent Fe and 38 mole percent S. The heat of fusion of iron at its melting temperature of 1535°C is 15,360 joules/mole, and liquid iron has a heat capacity that is 1.3 joules/degree·mole higher than the heat capacity of solid iron.

a. Derive an expression for the free energy of fusion of iron, ΔG_m^0, as a function of temperature.
b. Calculate the activity of iron in a liquid solution of Fe and S in which $X_{Fe} = 0.62$ at 1200°C.
c. Assume that the liquid Fe–S system is regular, and calculate the activity of iron in the liquid alloy in which $X_{Fe} = 0.62$ at 1535°C.
d. Calculate the excess free energy of mixing of the solution in which $X_{Fe} = 0.62$ at 1535°C.
e. Calculate the composition of the Fe liquidus in the system Fe–FeS at 1200°C.

12.2 Gold and silicon are mutually insoluble in the solid state and form a binary eutectic system with a eutectic temperature of 636 K and a eutectic composition of $X_{Si} = 0.186$, $X_{Au} = 0.814$. Calculate the free energy of the eutectic melt relative to (*a*) unmixed liquid Au and liquid Si, and (*b*) unmixed solid Au and solid Si.

12.3 Cesium and rubidium form complete ranges of solid and liquid solutions and a minimum occurs in the solidus and liquidus curves at $X_{Cs} = 0.5$ and 282 K. If it is assumed that the liquid solutions are ideal, calculate the free energy of mixing of solid Cs and Rb to form the solid solution of $X_{Cs} = 0.5$ at 282 K.

12.4 FeO and MnO form ideal liquid and solid solutions. Calculate the temperature at which equilibrium melting begins when the equimolar solid solution is heated. Calculate the composition of the first-formed liquid and the temperature at which equilibrium melting is complete.

CHAPTER THIRTEEN

REACTION EQUILIBRIA IN SYSTEMS CONTAINING COMPONENTS IN CONDENSED SOLUTION

13.1 INTRODUCTION

Dissolving the pure component i in a condensed solution that is in contact with a vapor phase causes a decrease in the vapor pressure exerted by i from the value p_i^0 (exerted by pure i) to the value p_i (exerted by i when it exists in solution). This decrease in the equilibrium vapor pressure corresponds, via Eq. (8.9), to a decrease of $RT \ln (p_i/p_i^0)$ in the partial molar free energy of i in the vapor phase. As phase equilibrium is maintained between the coexisting vapor phase and condensed solution phase, the partial molar free energy of i in the solution is $RT \ln (p_i/p_i^0)$ less than the molar free energy of pure condensed i at the temperature T. As the activity, a_i, of i in the solution with respect to pure i is defined as p_i/p_i^0, then the partial molar free energy of i in the condensed solution is less than the molar free energy of pure i by the quantity $RT \ln a_i$. The value of p_i (and hence a_i) depends on the composition and the nature of the components of the solution and on the temperature; and inasmuch that the solution of i affects the value of $\bar{G}_i$, it thus necessarily affects the equilibrium state of any chemical reaction system in which the component i is involved.

As an example, consider the equilibrium between silica, silicon, and oxygen gas,

$$SiO_{2(s)} = Si_{(s)} + O_{2(g)}$$

From Eq. (9.4), the criterion for reaction equilibrium at any temperature and total pressure is

$$\bar{G}_{SiO_2} = \bar{G}_{Si} + \bar{G}_{O_2}$$

If the SiO_2 and Si present in the system are both pure, and if the pure state is chosen as the standard state, then

$$G^0_{SiO_2} = G^0_{Si} + \bar{G}_{O_2}$$

or

$$G^0_{SiO_2} = G^0_{Si} + G^0_{O_2} + RT \ln p_{O_2(eq,\, T)}$$

It has been seen in Chap. 10 that, as the values of G_i^0 are dependent only on T, then at the temperature T there exists a unique oxygen partial pressure, $p_{O_2(eq,T)}$ at which equilibrium occurs in the system. This unique oxygen pressure is easily calculated as

$$p_{O_2(eq,\, T)} = \exp\left(\frac{1}{RT}(G^0_{SiO_2} - G^0_{Si} - G^0_{O_2})\right)$$

and if it is required to reduce pure SiO_2 to pure Si at the temperature T, then the oxygen pressure in the system must be less than $p_{O_2(eq,T)}$, such that

$$G^0_{SiO_2} > G^0_{Si} + \bar{G}_{O_2}$$

and the required reduction proceeds spontaneously with the required decrease in free energy.

Suppose now that the silica to be reduced occurs at the activity a_{SiO_2} in an Al_2O_3-SiO_2 solution. The criterion that equilibrium occurs between SiO_2, Si, and O_2 is still that

$$\bar{G}_{SiO_2} = \bar{G}_{Si} + \bar{G}_{O_2}$$

but now

$$\bar{G}_{SiO_2} = G^0_{SiO_2} + RT \ln a_{SiO_2}$$

and hence, in terms of standard free energies,

$$G^0_{SiO_2} + RT \ln a_{SiO_2} = G^0_{Si} + G^0_{O_2} + RT \ln p'_{O_2(eq,\, T)}$$

Thus, for a given a_{SiO_2}, there now exists a new unique equilibrium oxygen pressure, $p'_{O_2(eq,T)}$, which is given as

$$p'_{O_2(eq,\, T)} = p_{O_2(eq,\, T)} a_{SiO_2}$$

and so, if it is required to reduce SiO_2 from an Al_2O_3-SiO_2 solution to pure Si, the oxygen pressure in the system must now be less than $p'_{O_2(eq,T)}$.

It is thus seen that the possibility of reducing SiO_2 from an Al_2O_3-SiO_2 solution to pure Si with a gas of given oxygen partial pressure and at a given temperature depends entirely on the solution thermodynamics of the system Al_2O_3-SiO_2. Generally, the calculation of the equilibrium state of any reaction system involving components in condensed solution requires a knowledge of the thermodynamic properties of the various solutions present in the system. The influence of solution thermodynamics on reaction equilibria is examined in this chapter.

13.2 REACTION EQUILIBRIUM CRITERIA IN SYSTEMS CONTAINING COMPONENTS IN CONDENSED SOLUTION

Consider the general reaction,

$$aA + bB = cC + dD$$

occurring at the temperature T and the pressure P. If none of the reactants or reaction products occurs in its standard state, then the free energy difference between the products and reactants is

$$\Delta G = c\bar{G}_C + d\bar{G}_D - a\bar{G}_A - b\bar{G}_B \tag{13.1}$$

If, however, all the reactants and products are present in their standard states, then the free energy difference between the products and the reactants is the standard free energy change, ΔG^0, given as

$$\Delta G^0 = cG_C^0 + dG_D^0 - aG_A^0 - bG_B^0 \tag{13.2}$$

Subtraction of Eq. (13.2) from Eq. (13.1) gives

$$\Delta G - \Delta G^0 = c(\bar{G}_C - G_C^0) + d(\bar{G}_D - G_D^0) - a(\bar{G}_A - G_A^0) - b(\bar{G}_B - G_B^0) \tag{13.3}$$

For a component i occurring in some state other than its standard state, Eq. (11.28) gives

$$\bar{G}_i = G_i^0 + RT \ln a_i$$

where a_i is the activity of i with respect to the standard state, and thus Eq. (13.3) can be written as

$$\Delta G - \Delta G^0 = c(RT\ln a_C) + d(RT\ln a_D) - a(RT\ln a_A) - b(RT\ln a_B)$$

$$= RT\ln\left[\frac{a_C^c a_D^d}{a_A^a a_B^b}\right] = RT\ln Q \qquad (13.4)$$

where $Q = a_C^c a_D^d / a_A^a a_B^b$ is termed the activity quotient. Reaction equilibrium is established when the reaction has proceeded to such an extent that

$$a\bar{G}_A + b\bar{G}_B = c\bar{G}_C + d\bar{G}_D$$

i.e., that the free energy of the system at the fixed T and P has been minimized or that ΔG for the reaction is zero. Hence, at equilibrium,

$$\Delta G^0 = -RT\ln Q^{eq} \qquad (13.5)$$

where Q^{eq} is the value of the activity quotient at reaction equilibrium. From Eq. (9.8),

$$\Delta G^0 = -RT\ln K$$

and hence

$$Q^{eq} = K$$

i.e., at reaction equilibrium the activity quotient is numerically equal to the equilibrium constant K.

Consider the oxidation of the pure solid metal M by gaseous oxygen to form the pure solid metal oxide MO_2.

$$M_{(s)} + O_{2\,(g)} = MO_{2\,(s)} \qquad \text{(i)}$$

at the temperature T and the pressure P. For this reaction

$$Q = \frac{a_{MO_2}}{a_M a_{O_2}}$$

As M and MO_2 are pure, i.e., occur in their standard states, then $a_M = a_{MO_2} = 1$; and from the formal definition of activity, the activity of oxygen in the gas phase is given as

$$a_{O_2} = \frac{\text{the pressure of oxygen in the gas phase}}{\text{the pressure of oxygen in its standard state}}$$

As the standard state for gaseous species has been chosen as being 1 atm pressure at the temperature of interest, then the activity of oxygen in the gas phase is simply equal to its partial pressure (assuming ideal behavior of the gas).

Thus $$Q = \frac{1}{p_{O_2}} \quad \text{and} \quad Q^{eq} = \frac{1}{p_{O_2(eq,\, T)}} = K$$

Now consider that the metal M is being oxidized from a solution, in which it occurs at the activity a_M, to form the pure metal oxide, at the same temperature T. In this case,

$$Q^{eq} = \frac{1}{a_M p_{O_2(eq,\, T)}} = K$$

and as K is dependent only on temperature, and $a_M < 1$, it is seen that the oxygen pressure required to maintain equilibrium between M in solution and pure MO_2 is greater than that required for equilibrium between pure M and pure MO_2 at the same temperature. Similarly, if the pure metal M is being oxidized to form the metal oxide which occurs in solution at the activity a_{MO_2}, then

$$Q^{eq} = \frac{a_{MO_2}}{p_{O_2(eq,\, T)}} = K$$

in which case it is seen that the oxygen pressure required for equilibrium between pure M and MO_2 in solution is less than that required for equilibrium between pure M and pure MO_2.

In Fig. 13.1, ab is drawn as the variation of the standard free energy change, ΔG^0, with temperature, for the oxidation

$$M_{(s)} + O_{2(g,\, 1\ atm)} = MO_{2(s)} \tag{i}$$

At the temperature T, $\Delta G^0 = cd$, and the oxygen pressure for equilibrium between pure solid M and pure solid MO_2 is drawn as the point e on the oxygen pressure nomographic scale. Consider now the reaction

$$M_{(\text{in solid solution at } a_M)} + O_{2(g,\, 1\ atm)} = MO_{2\,(s)} \tag{ii}$$

for which, at the temperature T, the free energy change is $\Delta G_{(ii)}$. Reaction (ii) can be written as the sum of

$$M_{(\text{in solution at } a_M)} \rightarrow M_{(s,\, \text{pure})} \tag{iii}$$

for which

$$\Delta G_{(\text{iii})} = G^0_{\text{M}} - \bar{G}_{\text{M}} = -\Delta \bar{G}^{\text{M}}_{\text{M}} = -RT \ln a_{\text{M}}$$

and
$$M_{(s)} + O_{2(g,\,1\,\text{atm})} = MO_{2\,(s)} \qquad \text{(i)}$$

for which
$$\Delta G_{(\text{i})} = \Delta G^0 \text{ at the temperature } T$$

Thus,

$$\begin{aligned}\Delta G_{(\text{ii})} &= \Delta G_{(\text{i})} + \Delta G_{(\text{iii})} \\ &= \Delta G^0 - RT \ln a_{\text{M}}\end{aligned}$$

At the temperature T, ΔG^0 is a negative quantity, and as $a_{\text{M}} < 1$, then $\Delta G_{(\text{ii})}$ is a smaller negative quantity, drawn in Fig. 13.1 as *cf*. The effect, on the oxidation of M, of the solution of M is thus seen to be an anticlockwise rotation of the free energy line *ab* about the point *a* (ΔG^0 at $T = 0$), with the extent of this rotation being such that, at the temperature T, the vertical separation from the standard free energy line equals $RT \ln a_{\text{M}}$. The extent of the rotation is determined by the value of a_{M}, and the oxygen pressure required for equilibrium between M in solution and pure MO_2 is increased from *e* to *g* in accordance with

$$\begin{aligned}K_T &= \frac{1}{p_{O_2(T,\,\text{eq M}_{(s)}/\text{MO}_{2\,(s)})}} \\ &= \frac{1}{a_{\text{M}} p_{O_2(T,\,\text{eq M at } a_{\text{M}})/\text{MO}_{2\,(s)})}}\end{aligned}$$

Consider the reaction

$$M_{(s,\,\text{pure})} + O_{2\,(g,\,1\,\text{atm})} = MO_{2\ (\text{in solid solution at } a_{\text{MO}_2})} \qquad \text{(iv)}$$

for which, at the temperature T, the free energy change is $\Delta G_{(\text{iv})}$. Reaction (iv) can be written as the sum of

$$MO_{2(s,\,\text{pure})} \longrightarrow MO_{2(\text{in solution at } a_{\text{MO}_2})} \qquad \text{(v)}$$

and reaction (i), i.e.,

$$\Delta G_{(\text{iv})} = \Delta G^0 + RT \ln a_{\text{MO}_2}$$

At the temperature T, ΔG^0 is a negative quantity; and as $a_{\text{MO}_2} < 1$, then $\Delta G_{(\text{iv})}$ is a larger negative quantity, drawn in Fig. 13.1 as *ch*. The effect on the oxidation of M of the solution of MO_2 is thus seen to be a clockwise rotation of

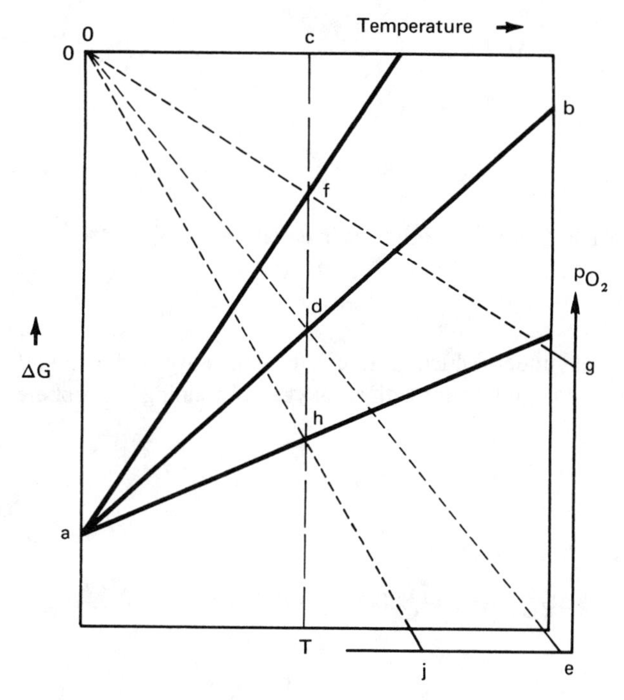

Fig. 13.1. The effect of nonunit activities of the reactants and products of a reaction on the ΔG-T relationship for the reaction.

the free energy line about the point a with the extent of the rotation being such that, at the temperature T, the vertical separation from the standard free energy line equals $RT \ln a_{MO_2}$. The extent of the rotation is determined by the value of a_{MO_2}, and the oxygen pressure required for equilibrium between MO_2 in solution and pure M is decreased from e to j in accordance with

$$K_T = \frac{1}{p_{O_2(T,\,\text{eq}\ M_{(s)}/MO_{2(s)})}} = \frac{a_{MO_2}}{p_{O_2(T,\,\text{eq}\ M_{(s)}/MO_{2\,(\text{at}\ a_{MO_2})})}}$$

In the general case,

$$M_{(\text{in solution at}\ a_M)} + O_{2(g,\,\text{at}\ p_{O_2})} = MO_{2\,(\text{in solution at}\ a_{MO_2})} \qquad \text{(vi)}$$

for which, at the temperature T, the free energy change is $\Delta G_{(\text{vi})}$

$$\Delta G_{(\text{vi})} = \Delta G^0 - RT \ln a_M - RT \ln p_{O_2} + RT \ln a_{MO_2}$$

or
$$= \Delta G^0 + RT \ln \frac{a_{MO_2}}{a_M p_{O_2}}$$
$$= \Delta G^0 + RT \ln Q$$

At equilibrium a_M, a_{MO_2} and p_{O_2} are such that $\Delta G_{(vi)} = 0$, and hence

$$\Delta G^0 = -RT \ln Q^{eq} = -RT \ln K \text{ as in equation (13.5)}$$

Example

Examine the conditions under which a liquid Fe-Mn alloy and a liquid FeO-MnO solution can be in equilibrium with an oxygen-containing atmosphere at 1800°C.

For

$$Mn_{(l)} + \tfrac{1}{2}O_{2(g)} = MnO_{(l)} \qquad (a)$$
$$\Delta G^0_{(a)} = -344{,}800 + 55.90T \text{ joules}$$

For

$$Fe_{(l)} + \tfrac{1}{2}O_{2(g)} = FeO_{(l)} \qquad (b)$$
$$\Delta G^0_{(b)} = -232{,}700 + 45.31T \text{ joules}$$

The pertinent equilibrium is

$$FeO_{(l)} + Mn_{(l)} = MnO_{(l)} + Fe_{(l)} \qquad (c)$$

for which

$$\Delta G^0_{(c),2073} = \Delta G^0_{(a),2073} - \Delta G^0_{(b),2073}$$
$$= -228{,}900 + 138{,}800$$
$$= -90{,}100 \text{ joules}$$
$$= -8.3144 \times 2073 \ln K_{(c),2073} \text{ joules}$$

Therefore

$$K_{(c),2073} = 186 = \frac{(a_{MnO})[a_{Fe}]}{(a_{FeO})[a_{Mn}]}$$

where (a_{MnO}) = the activity of MnO in the liquid oxide phase with respect to pure liquid MnO

(a_{FeO}) = the activity of FeO in the liquid oxide phase with respect to iron-saturated liquid iron oxide (see Sec. 13.7)

$[a_{Mn}]$ = the activity of Mn in the alloy phase with respect to pure liquid Mn

$[a_{Fe}]$ = the activity of Fe in the liquid alloy phase with respect to pure liquid Fe

As both the liquid metal phase and the liquid oxide solution phase exhibit Raoultian behavior, then the condition for equilibrium at 2073 K is that

$$\frac{(X_{MnO})[X_{Fe}]}{(X_{FeO})[X_{Mn}]} = 186 \qquad (d)$$

or

$$\frac{[X_{Fe}]}{[X_{Mn}]} = 186 \times \frac{(X_{FeO})}{(X_{MnO})}$$

Consider now the influence of the partial pressure of oxygen in the gaseous atmosphere

$$\Delta G^0_{(a)} = -RT \ln K_{(a)} = -RT \ln \frac{(a_{MnO})}{[a_{Mn}]p^{\frac{1}{2}}_{O_2}}$$

Thus, as

$$\Delta G^0_{(a),2073} = -228{,}900 \text{ joules}$$

$$K_{(a),2073} = 5.856 \times 10^5 = \frac{(a_{MnO})}{[a_{Mn}]p^{\frac{1}{2}}_{O_2}}$$

or

$$\frac{(a_{MnO})}{[a_{Mn}]} = 5.856 \times 10^5 \, p^{\frac{1}{2}}_{O_2} \qquad (e)$$

Also

$$\Delta G^0_{(b)} = -RT \ln K_{(b)} = -RT \ln \frac{(a_{FeO})}{[a_{Fe}]p^{\frac{1}{2}}_{O_2}}$$

and as

$$\Delta G^0_{(b),2073} = -138{,}800 \text{ joules}$$

then

$$K_{(b),2073} = 3.14 \times 10^3 = \frac{(a_{\mathrm{FeO}})}{[a_{\mathrm{Fe}}]p_{O_2}^{1/2}}$$

or

$$\frac{(a_{\mathrm{FeO}})}{[a_{\mathrm{Fe}}]} = 3.14 \times 10^3 p_{O_2}^{1/2} \qquad (f)$$

Thus at 2073 K and any oxygen pressure p_{O_2}, the individual ratios

$$\frac{(a_{\mathrm{FeO}})}{[a_{\mathrm{Fe}}]} \left(= \frac{(X_{\mathrm{FeO}})}{[X_{\mathrm{Fe}}]} \right) \quad \text{and} \quad \frac{(a_{\mathrm{MnO}})}{[a_{\mathrm{Mn}}]} \left(= \frac{(X_{\mathrm{MnO}})}{[X_{\mathrm{Mn}}]} \right)$$

are fixed by Eqs. (e) and (f), combination of which gives

$$\frac{(X_{\mathrm{MnO}})[X_{\mathrm{Fe}}]}{(X_{\mathrm{FeO}})[X_{\mathrm{Mn}}]} = \frac{5.856 \times 10^5}{3.14 \times 10^2} = 186$$

in accordance with Eq. (d). Establishment of equilibrium (c) requires that the ΔG-T lines for Fe oxidation and Mn oxidation intersect at 2073 K, i.e., that $\Delta G_{(c),2073} = 0$. For any oxidation $M + \frac{1}{2}O_2 = MO$, clockwise rotation of the ΔG-T line (e.g., the line ab in Fig. 13.1) about the point of its intersection with the $T = 0$ K axis, is effected by decreasing the ratio a_{MO}/a_M below unity and, conversely, anticlockwise rotation is effected by increasing the ratio a_{MO}/a_M above unity. Also, as the equilibrium constant K is a function only of temperature, then at any oxygen pressure, p_{O_2}, in the system M-MO-O_2 at the temperature T, the equilibrium a_{MO}/a_M ratio must be

$$\frac{a_{MO}}{a_M} = \frac{p_{O_2}^{\frac{1}{2}}}{p_{O_2(\mathrm{eq},\, T,\, \mathrm{pure\ M/pure\ MO})}^{\frac{1}{2}}}$$

where $p_{O_2(\mathrm{eq},T,\ \mathrm{pure\ M/pure\ MO})}$ is that unique oxygen pressure at the temperature T required for equilibrium between pure M and pure MO. Thus for any oxygen pressure, p_{O_2}, equilibrium (c) is established when Eqs. (e) and (f) are satisfied. Under these conditions the ΔG-T lines for Fe oxidation and Mn oxidation intersect at 1800°C.

Thus as a result of the ability to vary a_M and a_{MO}, equilibrium (c) can be established at any T and any p_{O_2}. This is in contrast to the situation illustrated in Figs. 10.4 and 10.5 where, if both metals and both oxides are present in their pure states, then an equilibrium such as (c) could only be achieved at the single unique state (unique T and unique p_{O_2}) given by the intersection of the ΔG^0-T lines for the two separate oxidation reactions. The restrictions on general multiphase multicomponent equilibria are discussed in Sec. 13.4.

13.3 ALTERNATIVE STANDARD STATES

Up to this point in the discussion of solution thermodynamics, the standard state of a solution component has been chosen as being the pure component in its stable state of existence at the temperature of interest. This is called the Raoultian standard state, and in Fig. 13.2 this standard state for the component B is located at the point r.

In the case where, at the temperature of interest, the pure component exists in a physical state which is different from that of the solution, the Henrian standard state may be more convenient than the Raoultian standard state. Such cases include the solution of a gas in a solid or liquid solvent or the solution of a solid in a liquid solvent. The Henrian standard state is obtained from consideration of Henry's law, which, strictly being a limiting law obeyed by the solute B at infinite dilution, is expressed as

$$\frac{a_B}{X_B} \rightarrow k_B \text{ as } X_B \rightarrow 0$$

where a_B is the activity of B in the solution with respect to the Raoultian standard state, and k_B is the Henry's law constant at the temperature T. Alternatively Henry's law can be written as

$$\frac{a_B}{X_B} \rightarrow \gamma_B^0 \text{ as } X_B \rightarrow 0 \qquad (13.6)$$

where γ_B^0 ($= k_B$) is the constant activity coefficient which quantifies the difference between Raoultian solution behavior of B and Henrian solution behavior of B. If the solute obeys Henry's law over a finite composition range, then, over this range,

$$a_B = \gamma_B^0 X_B$$

The Henrian standard state is obtained by extrapolating the Henry's law line to $X_B = 1$. This state (the point h in Fig. 13.2) represents pure B in the hypothetical, nonphysical state in which it would exist as a pure component if it obeyed Henry's law over the entire composition range, i.e., if pure, it behaved as it does in dilute solution. The activity of B in the Henrian standard state with respect to the Raoultian standard state having unit activity is given by Eq. (13.6) as

$$a_B = \gamma_B^0$$

Thus, in Fig. 13.2, if rb is unity, then $hb = \gamma_B^0$.

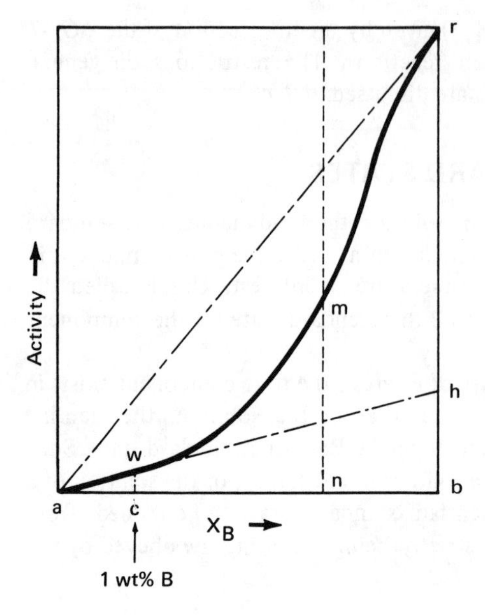

Fig. 13.2. Illustration of the Raoultian, Henrian, and 1 weight percent standard states for component B in a binary A–B system.

Having defined the Henrian standard state thus, the activity of B in a solution, with respect to the Henrian standard state having unit activity is given as

$$h_B = f_B X_B \tag{13.7}$$

where h_B = the Henrian activity
f_B = the Henrian activity coefficient

In that composition range over which the solute B obeys Henry's law, $f_B = 1$, and the solute exhibits Henrian ideality.

The mole fraction of B in an A–B solution is related to its concentration in weight percent by

$$X_B = \frac{\text{wt\% B}/MW_B}{\text{wt\% B}/MW_B + (100 - \text{wt\% B})/MW_A}$$

where MW_A and MW_B are, respectively, the molecular weights of A and B. Thus, in dilute solution, as the mole fraction of B is virtually proportional to the weight percentage of B, i.e.,

$$X_B \doteqdot \frac{\text{wt\% B} \times MW_A}{100 \times MW_B}$$

a third standard state can be introduced. This is the 1 weight percent standard state, and as use of this state eliminates the necessity of converting weight percentages, obtained by chemical analysis, to mole fractions for the purpose of thermodynamic calculation, it is of particular convenience for use in metallurgical systems containing dilute solutes. This standard state is formally defined as

$$\frac{h_{B(wt\%)}}{wt\%\ B} \rightarrow 1 \text{ as } wt\%\ B \rightarrow 0$$

and is located at that point on the Henry's law line which corresponds to a concentration of 1 weight percent B in A (the point w in Fig. 13.2). With respect to the 1 wt% standard state having unit activity, the activity of B, $h_{B(wt\%)}$, is given as

$$h_{B(wt\%)} = f_{B(wt\%)} wt\%\ B \tag{13.8}$$

where $f_{B(wt\%)}$ is the weight percent activity coefficient, and in that composition range over which the solute B obeys Henry's law, $f_{B(wt\%)}$ is unity, and hence

$$h_{B(wt\%)} = wt\%\ B$$

which is of considerable practical convenience.

From consideration of the similar triangles awc and ahb in Fig. 13.2, the activity of B in the 1 wt% standard state with respect to the Henrian standard state having unit activity is

$$\frac{wc}{hb} = \frac{ac}{ab} = \frac{MW_A}{100 MW_B}$$

and with respect to the Raoultian standard state having unit activity is

$$\frac{\gamma_B^0 MW_A}{100 MW_B}$$

The value of the equilibrium constant K for any reaction, being equal to the quotient of the activities of the reactants and products at reaction equilibrium, necessarily depends on the choice of standard states for the reaction components. Similarly the magnitude of ΔG^0 for the reaction depends on the choice of standard states, and hence, in order to convert from the use of one standard state to another, it is necessary that the free energy differences between the various standard states be known.

For the change of standard state,

$$B_{\text{(in the Raoultian standard state)}} \rightarrow B_{\text{(in the Henrian standard state)}}$$

$$\Delta G^0_B(R \to H) = G^0_{B(H)} - G^0_{B(R)} = RT \ln \frac{a_{B(\text{in the Henrian standard state})}}{a_{B(\text{in the Raoultian standard state})}}$$

where both activities are measured on the same activity scale. On either the Raoultian or the Henrian scales

$$\frac{a_{B(\text{in the Henrian standard state})}}{a_{B(\text{in the Raoultian standard state})}} = \frac{hb}{rb} = \gamma^0_B$$

and hence

$$\Delta G^0_{B(R \to H)} = RT \ln \gamma^0_B \tag{13.9}$$

where γ^0_B is the Henrian activity coefficient *at the temperature T*.

For the change of standard state,

$$B_{(\text{in the Henrian standard state})} \to B_{(\text{in the 1 wt\% standard state})}$$

$$\Delta G^0(H \to 1 \text{ wt\%}) = G^0_{(1 \text{ wt\%})} - G^0_{(H)} = RT \ln \frac{a_{B(\text{in the 1 wt\% standard state})}}{a_{B(\text{in the Henrian standard state})}}$$

where again both activities are measured on the same activity scale.

$$\frac{a_{B(\text{in the 1 wt\% standard state})}}{a_{B(\text{in the Henrian standard state})}} = \frac{wc}{hb} = \frac{ac}{ab} = \frac{MW_A}{100MW_B}$$

and hence

$$\Delta G^0_{B(H \to 1 \text{ wt\%})} = RT \ln \left(\frac{MW_A}{100MW_B} \right) \tag{13.10}$$

For the change of state, Raoultian → 1 wt%, combination of Eqs. (13.9) and (13.10) gives,

$$\Delta G^0_{B(R \to 1 \text{ wt\%})} = RT \ln \left(\frac{\gamma^0_B MW_A}{100MW_B} \right) \tag{13.11}$$

Using the subscript (*R*) to denote the Raoultian standard state, the subscript (*H*) to denote the Henrian standard state, and the subscript (wt%) to denote the 1 weight percent standard state, consider again the oxidation of metal M to the oxide MO_2 at the temperature *T*.

$$M_{(R)} + O_{2(g)} = MO_{2(R)}$$

For this equilibrium

$$\Delta G^0_{(R)} = -RT\ln K_{(R)} = -RT\ln \frac{a_{MO_2}}{a_M p_{O_2}}$$

If M occurs in dilute solution, in which case it may be more convenient to use the Henrian standard state for M, then for

$$M_{(H)} + O_{2(g)} = MO_{2(R)}$$

$$\Delta G^0_{(H)} = \Delta G^0_{(R)} - \Delta G^0_{M(R\rightarrow H)}$$

i.e.,

$$-RT\ln K_{(H)} = -RT\ln K_{(R)} - RT\ln \gamma^0_M$$

or

$$RT\ln \frac{a_{MO_2}}{h_M p_{O_2}} = RT\ln \frac{a_{MO_2}}{a_M p_{O_2}} + RT\ln \gamma^0_M$$

Thus

$$a_M = h_M \gamma^0_M \tag{13.12}$$

which relates the activity of M in solution with respect to the Raoultian standard state to the activity of M in solution with respect to the Henrian standard state, e.g., in the case of the composition m in Fig. 13.2,

$$a_B = \frac{mn}{rb} = \frac{mn}{hb}\frac{hb}{rb} = h_B \gamma^0_B$$

Similarly if it is convenient to use the 1 wt% standard state for M, then, for

$$M_{(1\ \mathrm{wt\%})} + O_{2(g)} = MO_{2(R)}$$

$$\Delta G^0_{(1\ \mathrm{wt\%})} = \Delta G^0_{(R)} - \Delta G^0_{M(R\rightarrow 1\ \mathrm{wt\%})}$$

or

$$-RT\ln K_{(\mathrm{wt\%})} = -RT\ln K_{(R)} - RT\ln \frac{\gamma^0_M MW_{\mathrm{solvent}}}{100 MW_M}$$

or

$$-RT\ln \frac{a_{MO_2}}{f_{M(\mathrm{wt\%})}\,\mathrm{wt\%\ M}\,p_{O_2}} = -RT\ln \frac{a_{MO_2}}{a_M p_{O_2}} - RT\ln \frac{\gamma^0_M MW_{\mathrm{solvent}}}{100 MW_M}$$

or

$$a_M = f_{M(\mathrm{wt\%})}\,\mathrm{wt\%\ M}\,\gamma^0_M \frac{MW_{\mathrm{solvent}}}{100 MW_M} \tag{13.13}$$

Example

The activity, a_{Si}, of silicon in binary Fe–Si liquid alloys is shown in Fig. 13.3 at two temperatures. As is seen, Si exhibits considerable negative deviation from Raoult's law, e.g., at $X_{Si} = 0.1$ and 1420°C, $a_{Si} = 0.00005$. In considering dilute solutions of Si in Fe, there is an advantage to using either the Henrian standard state or the 1 wt% in Fe standard state.

For the change of standard state from Raoultian to Henrian at the temperature T,

$$\Delta G^0_{Si(R \to H)} = RT \ln \gamma^0_{Si}$$

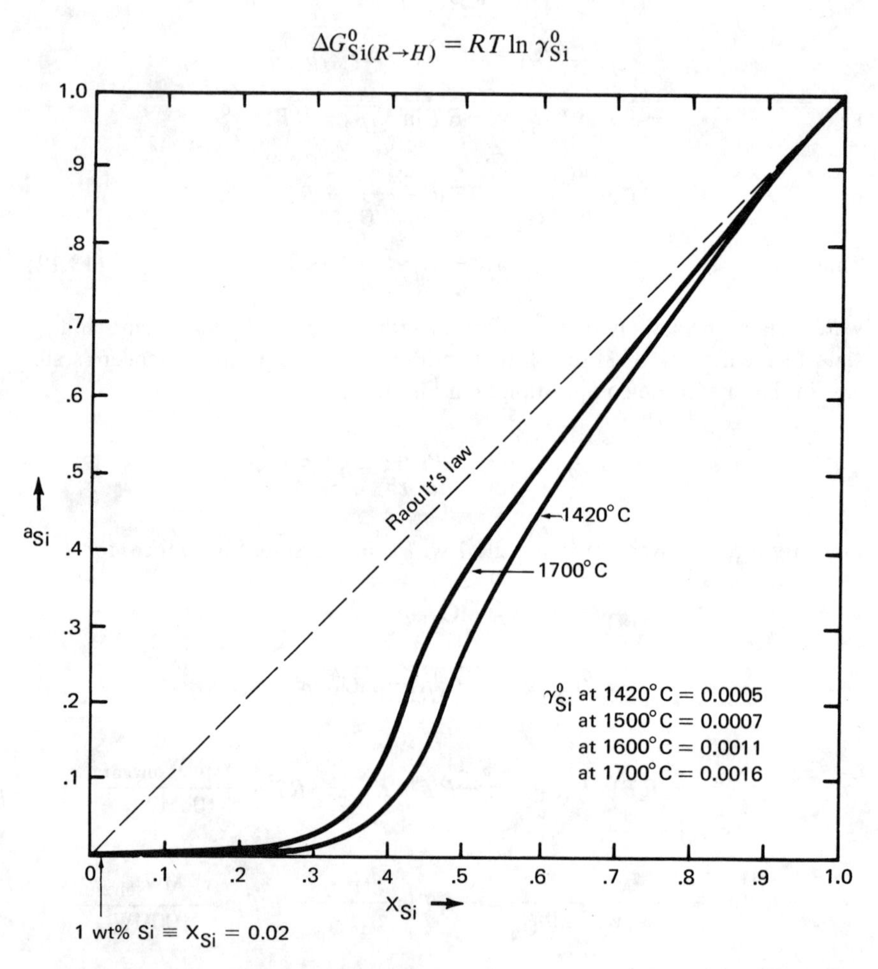

Fig. 13.3. The activities of silicon in iron-silicon melts at several temperatures. *(From J. F. Elliott, M. Gleiser, and V. Ramakrishna, "Thermochemistry for Steelmaking," Addison-Wesley Publishing Co., Reading, Mass., 1963.)*

and as the experimentally determined variation of log γ^0_{Si} with temperature is

$$\log \gamma^0_{Si} = -\frac{6230}{T} + 0.37$$

then

$$\Delta G^0_{Si(R \to H)} = 8.3144T \times 2.303 \log \gamma^0_{Si}$$
$$= -119{,}300 + 7.08T \text{ joules}$$

Also for the change of standard state from Henrian to 1 wt% in Fe at the temperature T,

$$\Delta G^0_{Si(H \to 1\text{ wt\%})} = RT \ln \frac{MW_{Fe}}{100MW_{Si}}$$
$$= RT \ln \left(\frac{55.85}{100 \times 28.09}\right)$$
$$= -32.6T \text{ joules}$$

so, for the change $Si_{(R)} \to Si_{(1\text{ wt\% in Fe})}$

$$\Delta G^0_{(R \to 1\text{ wt\%})} = \Delta G^0_{(R \to H)} + \Delta G^0_{(H \to 1\text{ wt\%})}$$
$$= -119{,}300 - 25.5T \text{ joules} \qquad \text{(i)}$$

Now, given that a liquid Fe–Si alloy is in equilibrium with a SiO_2-saturated FeO–SiO_2 melt and an oxygen-containing atmosphere (the significance of SiO_2-saturation being that $a_{SiO_2} = 1$ in the liquid oxide solution), calculate the relationship between the equilibrium weight percentage of Si in the Fe–Si alloy and the oxygen pressure in the gaseous atmosphere.

The pertinent equilibrium is

$$Si_{(1\text{ wt\%})} + O_{2(g)} = SiO_{2(s)} \qquad \text{(ii)}$$

For the reaction

$$Si_{(l)} + O_{2(g)} = SiO_{2(s)} \qquad \text{(iii)}$$

$\Delta G^0_{(iii)} = -952{,}700 + 204T$ joules in the temperature range 1700 to 2000 K. As $\Delta G^0_{(iii)}$ involves oxidation of Si in the Raoultian standard state, calculation of $\Delta G^0_{(ii)}$ requires knowledge of the free energy difference between Si in the Raoultian standard state and Si in the 1 wt% in Fe standard state. This has been determined to be

$$\Delta G^0_{(i)} = -119{,}300 - 25.5T \text{ joules}$$

and hence

$$\Delta G^0_{(ii)} = \Delta G^0_{(iii)} - \Delta G^0_{(i)} = -833{,}400 + 229.5T \text{ joules}$$

$$= -RT \ln \frac{a_{SiO_2}}{h_{Si(1\ wt\%)} p_{O_2}}$$

As $a_{SiO_2} = 1$, then

$$\ln h_{Si(1\ wt\%)} = -\frac{833{,}400}{8.3144T} + \frac{229.5}{8.3144} - \ln p_{O_2}$$

If it can be assumed that, over an initial composition range of Si in Fe, Si exhibits Henrian behavior, then, in this range, $h_{Si(1\ wt\%)} = \text{wt\% Si}$, and hence

$$\ln \text{wt\% Si} = -\frac{833{,}400}{8.3144T} + \frac{229.5}{8.3144} - \ln p_{O_2}$$

Thus, in order that the equilibrium weight percentage of Si in the metal phase be 1.0, p_{O_2} at 1600°C must be 5.57×10^{-12}.

The equilibrium oxygen pressure for any other weight percentage of Si at 1600°C is calculated as

$$p_{O_2} = \frac{5.57 \times 10^{-12}}{\text{wt\% Si}}$$

The error incurred in this calculation by the assumption that $h_{Si(1\ wt\%)} = \text{wt\%}$ Si over some initial composition range is demonstrated when this calculation is considered again in Sec. 13.9.

Example: The Activity of Carbon in Iron-Carbon Alloys

In Sec. 10.7 it was seen that the equilibrium

$$C_{(graphite)} + CO_{2(g)} = 2CO_{(g)} \qquad (a)$$

sets a lower limit on the CO_2/CO ratio which can be obtained in a CO_2-CO gas mixture at any temperature and fixed total pressure. For Eq. (*a*),

$$\Delta G^0_a = 170{,}700 - 174.5T \text{ joules} = -RT \ln K_a$$

where

$$K_a = \frac{p^2_{CO}}{p_{CO_2} a_{C(gr)}} \tag{b}$$

and $a_{C(gr)}$ is the activity of carbon with respect to solid graphite as the standard state. Thus when a CO-CO_2 gas mixture is in equilibrium with solid graphite at the temperature T and the pressure P, $a_{C(gr)} = 1$, and the p_{CO} and p_{CO_2} values are fixed by the conditions

$$P = p_{CO} + p_{CO_2} \quad \text{and} \quad K_a = \frac{p^2_{CO}}{p_{CO_2}}$$

However, if, at constant T and P, the p_{CO_2}/p_{CO} ratio in the CO_2-CO gas mixture is increased above the minimum limit set by the presence of solid carbon, then the activity of carbon (with respect to solid graphite) in the mixture falls below unity, being given, via Eq. (*b*), as

$$a_{C(gr)} = \frac{1}{K_a} \frac{p^2_{CO}}{p_{CO_2}}$$

If such a CO-CO_2 gas mixture is allowed to contact pure iron at the temperature T, equilibrium is established when carbon has dissolved in the iron (being transferred from the gas phase) to the extent that the activity of carbon in the iron equals the activity of carbon in the gas mixture. Thus by observing the variation of the equilibrium carbon content of iron with gas composition, the activity-composition relationship of carbon in iron can be obtained. Such an investigation has been conducted by Smith,* whose results at 1000°C are listed in the first three columns of Table 13.1.

Carbon has a limited solubility in iron and at 1000°C γ-iron (the austenite phase) is saturated with carbon at 1.5 weight percent C ($X_C = 0.066$); i.e., a 1.5 wt% C − 98.5 wt% Fe alloy is in equilibrium with pure solid graphite, and hence the activity of carbon in the 1.5 wt% C alloy, with respect to solid graphite, is unity at 1000°C. The activity of carbon in an undersaturated alloy is thus calculated as

$$a_{C(gr)} = \frac{(p^2_{CO}/p_{CO_2})_{\text{in the equilibrating gas}}}{(p^2_{CO}/p_{CO_2})_{\text{in equilibrium with graphite}}} \tag{c}$$

where, from Eq. (*b*), $(p^2_{CO}/p_{CO_2})_{\text{in eq with solid graphite}} = K_a$. From

*R. P. Smith, Equilibrium of Iron-Carbon Alloys with Mixtures of CO-CO_2 and CH_4-H_2, *J. Am. Chem. Soc.*, **68**:1163 (1946).

Table 13.1

p^2_{CO}/p_{CO_2}	wt% C	X_C	$a_{C(gr)}$	K'_d	$\log K'_d$	K'_e	$\log K'_e$
1.98	.0360	.00167	.0143	1186	3.074	55.0	1.740
2.49	.0487	.00226	.0180	1102	3.042	55.1	1.709
3.12	.0563	.00262	.0225	1190	3.076	55.4	1.744
4.21	.0740	.00344	.0304	1224	3.088	56.9	1.755
7.29	.133	.00616	.0526	1184	3.073	54.8	1.739
13.8	.242	.0112	.100	1235	3.092	57.0	1.756
27.4	.455	.0208	.198	1315	3.119	60.2	1.780
43.4	.655	.0298	.313	1458	3.164	66.3	1.821
56.2	.810	.0366	.405	1535	3.186	69.4	1.841
70.8	.963	.0433	.511	1635	3.214	73.5	1.866
84.1	1.081	.0484	.607	1738	3.240	77.8	1.891
99.4	1.206	.0538	.717	1849	3.267	82.4	1.916
113.3	1.321	.0587	.818	1932	3.286	85.8	1.933
130.2	1.462	.0646	.939	2016	3.305	89.1	1.950
131.7	1.466	.0648	.950	2034	3.309	89.8	1.954
132.4	1.471	.0649	.955	2043	3.310	90.0	1.954
138.6	1.500	.0662	1.00	2094	3.321	92.4	1.966

$\Delta G^0_{a,1273}$, $K_{a,1273} = 137.8$ and from direct measurement at 1000°C, Smith obtained $K_{a,1273} = 138.6$.

The activities of C in Fe, calculated from Eq. (*c*) are listed in column 4 of Table 13.1, and the $a_{C(gr)}$-X_C and $a_{C(gr)}$-wt% C relationships are shown in Fig. 13.4*a* and *b* respectively.

Activities of C in Fe–C alloys, with respect to the Henrian standard state, are calculated as follows. The equilibrium of interest is

$$C_{\text{(Henrian standard state)}} + CO_{2(g)} = 2CO_{(g)} \qquad (d)$$

for which

$$\Delta G^0_d = -RT \ln K_d = -RT \ln \frac{p^2_{CO}}{p_{CO_2} h_C}$$

Recalling that

$$h_C = f_C X_C$$

where $f_C \to 1$ as $X_C \to 0$ K_d can be written as

$$K_d = \frac{p^2_{CO}}{p_{CO_2} f_C X_C}$$

or

$$K'_d = K_d f_C = \frac{p^2_{CO}}{p_{CO_2} X_C}$$

where K'_d is termed the "apparent equilibrium constant." (Note that as $f_C \rightarrow 1$ as $X_C \rightarrow 0$, then $K'_d \rightarrow K_d$ as $X_C \rightarrow 0$.)

The values of K'_d and $\log K'_d$ for each composition are listed in columns 5 and 6 of Table 13.1. By plotting $\log K'_d$ against X_C (as in Fig. 13.5*a*) and extrapolating to $X_C = 0$, the value of $\log K_{d,1273}$ is obtained as 3.06.

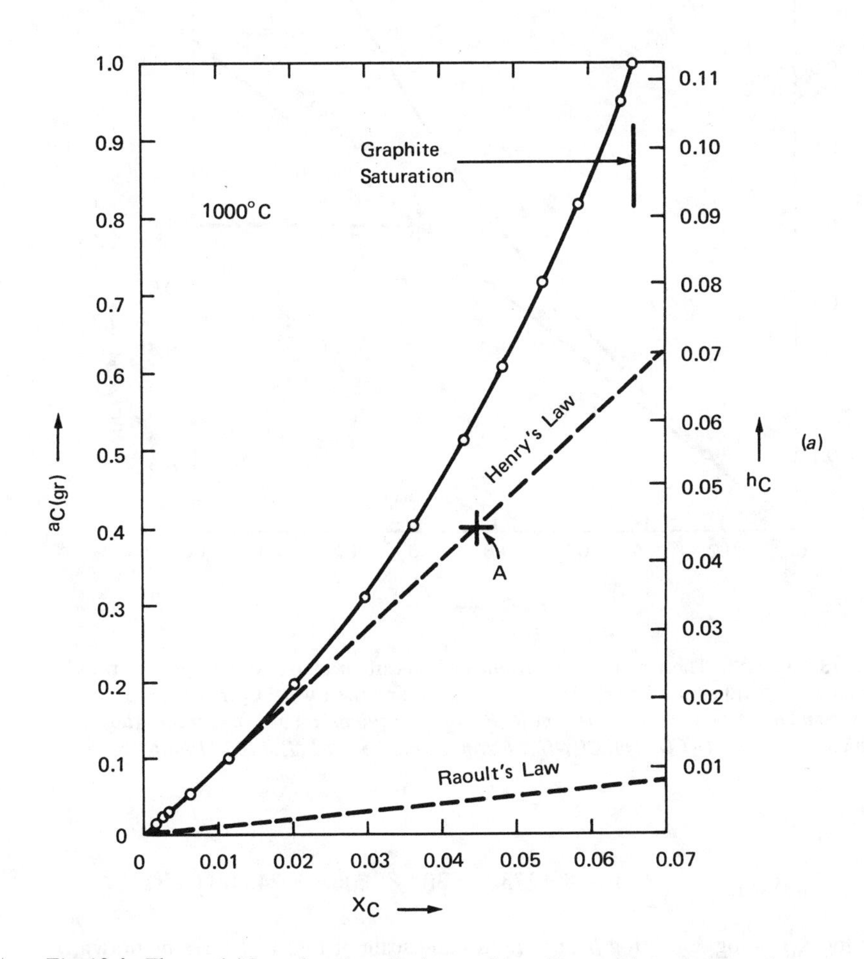

Fig. 13.4. The activities of carbon in iron-carbon alloys with respect to the graphite standard state, the Henrian standard state, and the 1 weight percent carbon in iron standard state at 1000°C. *(From R. P. Smith, Equilibrium of Iron-Carbon Alloys with Mixtures of CO-CO$_2$ and CH$_4$-H$_2$, J. Am. Chem. Soc., 68 (1):1163 [1946].)*

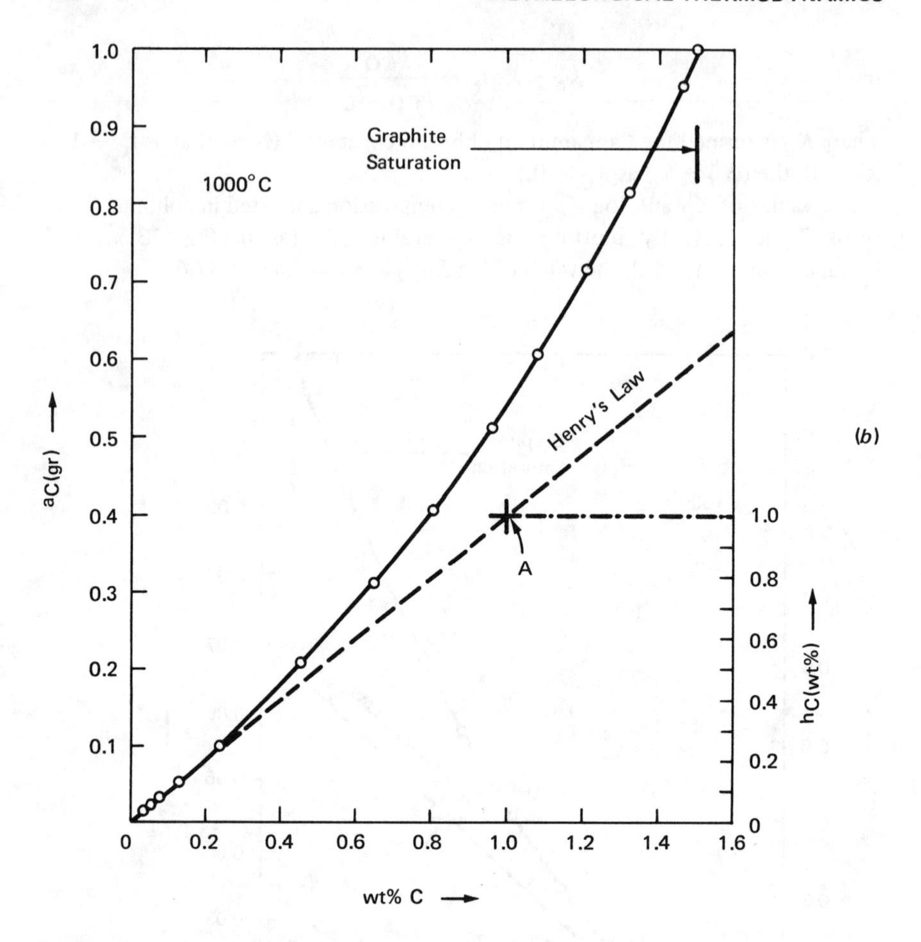

Fig. 13.4 (*Cont.*). The activities of carbon in iron-carbon alloys with respect to the graphite standard state, the Henrian standard state, and the 1 weight percent carbon in iron standard state at 1000°C. *(From R. P. Smith, Equilibrium of Iron-Carbon Alloys with Mixtures of CO-CO$_2$ and CH$_4$-H$_2$, J. Am. Chem. Soc., 68 (1):1163 [1946].)*

Thus

$$\Delta G^0_{d,1273} = -8.3144 \times 1273 \times 2.303 \times 3.06 = -74{,}600 \text{ joules}$$

As $\log K'_d = \log K_d + \log f_C$, the right-hand scale of Fig. 13.5*a* can be drawn to give the variation of $\log f_C$ with X_C.

Now from

$$C_{(gr)} + CO_{2(g)} = 2CO_{(g)} \quad \Delta G^0_{a,1273} = -51{,}440 \text{ joules}$$

and $$C_{(\text{Henrian})} + CO_{2(g)} = 2CO_{(g)} \quad \Delta G^0_{d,1273} = -74{,}600 \text{ joules}$$

the free energy change for the change of standard state, is obtained as

$$C_{(gr)} = C_{(\text{Henrian})}$$

$$\begin{aligned}\Delta G^0 &= -51{,}440 + 74{,}600 \text{ joules}\\ &= 23{,}160 \text{ joules}\\ &= RT \ln \gamma^0_C\\ &= 8.3144 \times 1273 \ln \gamma^0_C\end{aligned}$$

Thus $\gamma^0_C = 8.92$; i.e., the Henry's law line intersects the $X_C = 1$ axis at 8.92 on the Raoult's law activity scale. The Henry's law line thus passes through the point $a_{C(gr)} = 0.446$, $X_C = 0.05$, and hence can be drawn accordingly in Fig. 13.4*a*. The Henrian activity scale is drawn on the right-hand edge of Fig. 13.4*a*. Note that $h_C = 0.07$ at $X_C = 0.07$.

The 1 wt% C in Fe standard state is now located on the Henry's law line at the composition 1 wt% C. One wt% C is equivalent to $X_C = 0.045$; and thus, in Fig. 13.4*b*, the Henry's law line passes through the point A ($a_{C(gr)} = 0.4$, wt% C = 1), which corresponds to the point A in Fig. 13.4*a*.

Alternatively the 1 wt% standard state can be determined as follows. The equilibrium of interest is now

$$C_{(1 \text{ wt\% in Fe})} + CO_{2(g)} = 2CO_{(g)} \tag{e}$$

for which

$$K_e = \frac{p^2_{CO}}{p_{CO_2} h_{C(1 \text{ wt\%})}} = \frac{p^2_{CO}}{p_{CO_2} f_{C(1 \text{ wt\%})} \text{ wt\% C}}$$

The 1 wt% standard state is defined as $f_{(\text{wt\%})} \to 1$ as wt% $\to 0$ and hence a second "apparent equilibrium constant" can be defined as

$$K'_e = K_e f_{C(1 \text{ wt\%})} = \frac{p^2_{CO}}{p_{CO_2} \text{ wt\% C}}$$

where, again, $K'_e \to K_e$ as wt% C $\to 0$.

The values of K'_e and $\log K'_e$ are listed in columns 7 and 8 of Table 13.1, and plotting $\log K'_e$ against wt% C (as in Fig. 13.5*b*) and extrapolating to wt% C = 0 gives $\log K_{e,1273} = 1.729$

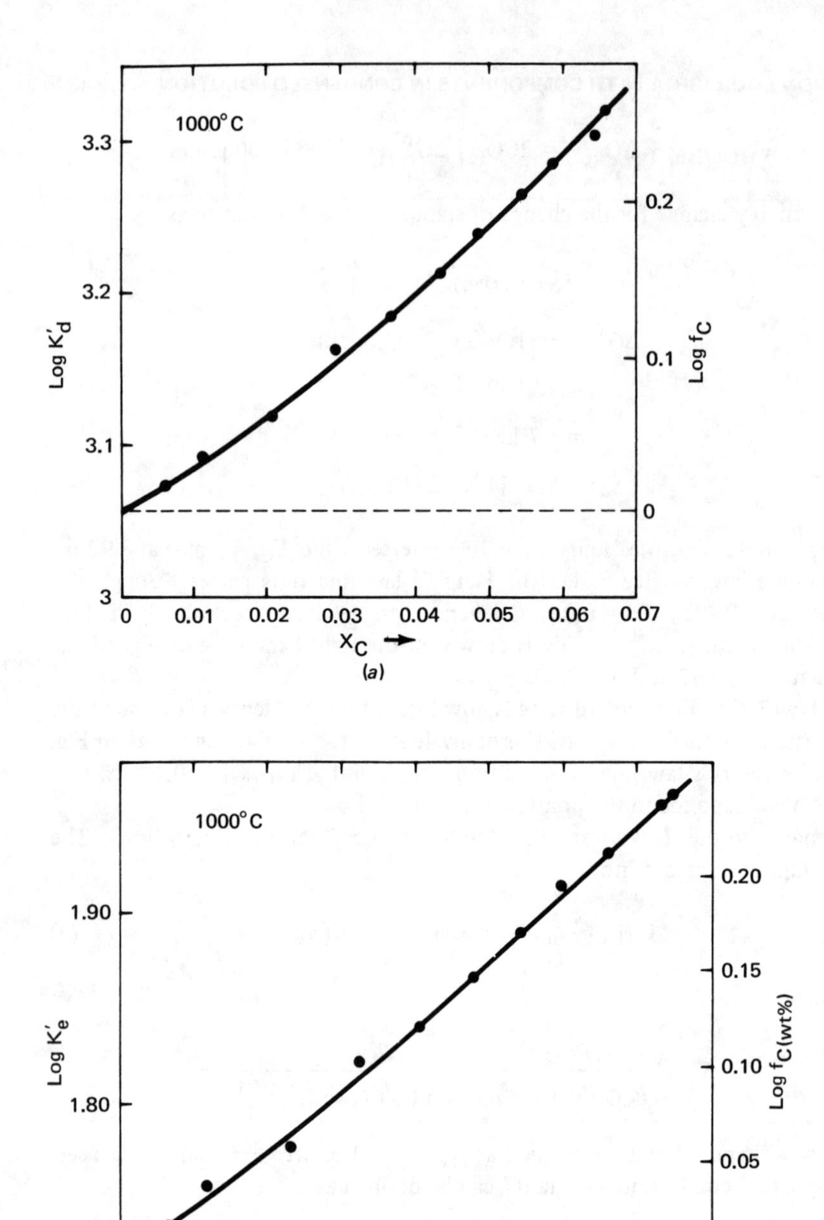

Fig. 13.5. (*a*) Log K'_d and log f_C versus the mole fraction of carbon in iron-carbon alloys at 1000°C. (*b*) Log K'_e and log $f_{C(1\ wt\%)}$ versus weight percent carbon in iron-carbon alloys at 1000°C.

$$\Delta G^0_{e,1273} = -8.3144 \times 1273 \times 2.303 \times 1.729$$
$$= -42{,}140 \text{ joules}$$

and so for

$$C_{(gr)} = C_{(1 \text{ wt\% in Fe})}$$
$$\Delta G^0 = \Delta G^0_a - \Delta G^0_e$$
$$= -51{,}440 + 42{,}140$$
$$= -9300 \text{ joules}$$
$$= RT \ln \frac{\gamma^0_C MW_{Fe}}{100 \times MW_C}$$
$$= 8.3144 \times 1273 \times \ln \frac{\gamma^0_C \times 55.85}{100 \times 12}$$

Thus $\gamma^0_C = 8.92$, in agreement with the previous derivation.

The variations of $a_{C(gr)}$ with both composition and temperature are illustrated by the isoactivity lines in Fig. 13.6.* The $a_{C(gr)} = 1$ line traces out the variation of the solubility limit of carbon in iron with temperature. The figure also includes (as the dashed lines) the Gibbs-Duhem calculated activities of Fe in austenite with respect to pure solid iron and, as the dash-dot-dash lines, the Gibbs-Duhem calculated activities of Fe in liquid alloys with respect to the liquid iron standard state.

13.4 THE GIBBS PHASE RULE

In Chap. 7 it was found that the number of degrees of freedom available to a one-component system at equilibrium was related to the number of phases present by means of a simple rule. This rule, the Gibbs phase rule, as applied to a one-component system, was easily derived by virtue of the simplicity of representation, by means of a pressure-temperature diagram, of the phase relationships occurring in the system. In a multicomponent system, however, the phase relationships cannot be as simply represented; and in such systems the phase rule is a powerful tool in the determination of possible equilibria and the restrictions on these equilibria. The general derivation of the phase rule is as follows.

Consider a system containing C chemical species, $i, j, k, \ldots$ (none of which enter into chemical reaction with one another), which occur in P phases

*M. G. Benz and J. F. Elliott, The Austenite Solidus and Revised Iron-Carbon Diagram, *Trans. Met. Soc. AIME,* **221**:323 (1961).

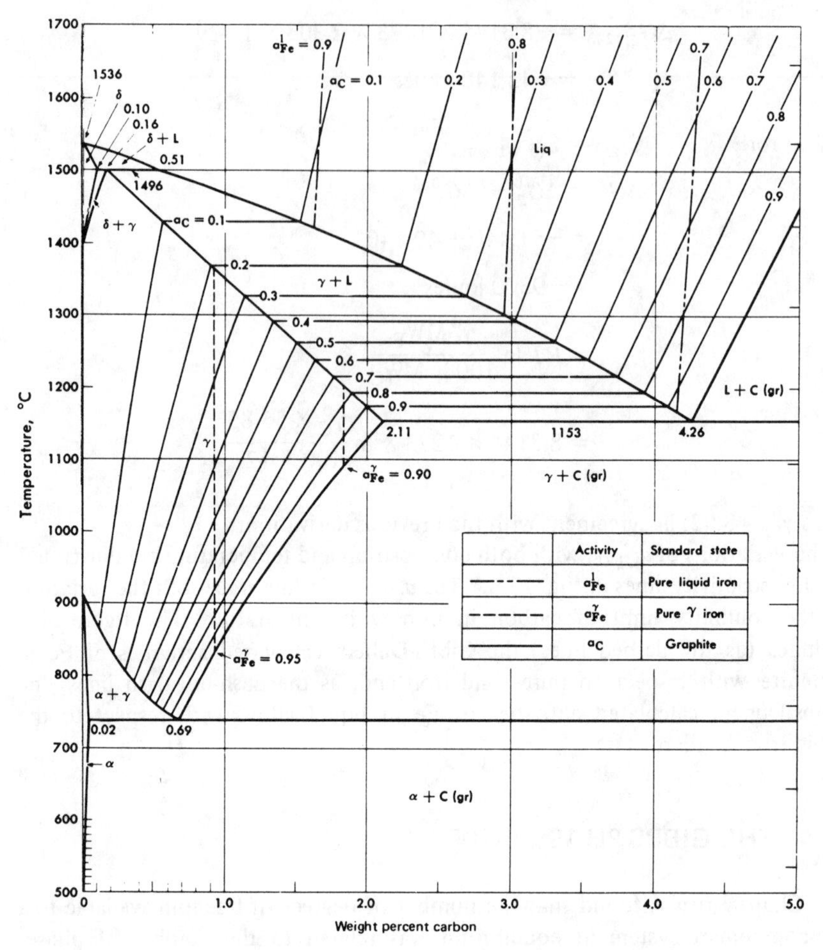

Fig. 13.6. The phase diagram of the iron-carbon system showing isoactivity lines. *(From M. G. Benz and J. F. Elliott, The Austenite Solidus and Revised Iron-Carbon Diagram, Trans. Met. Soc. AIME, 221:323 [1961].)*

$\alpha, \beta, \gamma, \ldots$. As the thermodynamic state of each of the P phases is determined by specification of its temperature, pressure, and composition (where composition is expressed in terms of $C - 1$ composition variables such as mole fractions or weight percentages), then the state of the entire system is specified when its $P(C + 1)$ variables are fixed. The conditions that the system be at complete equilibrium are

$$T_\alpha = T_\beta = T_\gamma = \ldots \qquad (P - 1) \text{ equalities of temperature}$$

$P_\alpha = P_\beta = P_\gamma = \ldots$ $(P-1)$ equalities of pressure

$a_{i(\alpha)} = a_{i(\beta)} = a_{i(\gamma)} = \ldots$ $(P-1)$ equalities of the activity of the species i

$a_{j(\alpha)} = a_{j(\beta)} = a_{j(\gamma)} = \ldots$ $(P-1)$ equalities of the activity of the species j

and so on for each of the C chemical species. Thus the total number of equilibrium conditions, given as the number of equations among the variables of the system, equals

$$(P-1)(C+2)$$

The number of degrees of freedom, F, which the equilibrium in the system may have is defined as the maximum number of variables which may be independently altered in value without disturbing the equilibrium in the system. This number F is obtained as the difference between the total number of variables available to the system and the minimum number of equations among these variables that is required for maintenance of the equilibrium;

i.e.,

$$\begin{aligned} F &= P(C+1) - (P-1)(C+2) \\ &= C + 2 - P \end{aligned} \tag{13.14}$$

In a system of nonreacting species, the number of species C equals the number of components in the system. However if some of the species enter into reaction with one another, such that the equilibrium of the system includes a number of reaction equilibria–in addition to the phase, temperature, and pressure equilibria–then the number of equations among the variables which must be specified is increased by R, the number of such independent reaction equilibria occurring in the system; e.g., if the species i and j react to form the species k, then the establishment of this reaction equilibrium requires that, in each of the P phases,

$$\bar{G}_i + \bar{G}_j = \bar{G}_k$$

which increases the number of equations among the variables by one. Thus if the system contains N species among which there are R independent reaction equilibria, then

$$\begin{aligned} F &= P(N+1) - (P-1)(N+2) - R \\ &= (N-R) + 2 - P \end{aligned}$$

In order that the phase rule, as given by Eq. (13.14), be generally applicable to both reactive and nonreactive systems, the number of components in the former is defined as

$$C = N - R$$

C can thus be determined either as

1. The minimum number of chemical species required to produce the system at equilibrium
2. The number of species in the system minus the number of independent reaction equilibria among these species

Consider the system M, MO, and O_2 in which the reaction equilibrium

$$M_{(s)} + \frac{1}{2}O_{2\,(g)} = MO_{(s)}$$

is established. This system has three phases (the condensed phases M and MO, and the gaseous oxygen phase) and has two components. The available variables are the temperature T and the total pressure P. As, at equilibrium, the species M and MO occur in fixed states, i.e., M saturated with oxygen and MO saturated with M, the activities of these two species are fixed. Thus the total pressure P equals the sum of the oxygen pressure in the system and the vapor pressures of the solid phases M and MO; and as the latter two are fixed at any given temperature, then any variation in P can only be effected by varying p_{O_2}. From the phase rule, $F = C + 2 - P = 2 + 2 - 3 = 1$; i.e., the system at equilibrium has only one degree of freedom. Thus either T may be arbitrarily fixed, in which case the equilibrium constant K_T and hence $p_{O_2(eq,T)}$ is fixed; or $P = p_{O_2} +$ (the vapor pressures of M and MO) can be arbitrarily fixed, in which case K_T and hence T is fixed. If an inert gas is added to the system, then $P = p_{O_2} + p_{\text{inert gas}}$ + vapor pressures of M and MO, and so p_{O_2} and $p_{\text{inert gas}}$ may be independently varied. The addition of the inert gas as the third component increases the number of degrees of freedom to two, but this additional degree of freedom is restricted to variation of $p_{\text{inert gas}}$; i.e., in addition to either T or p_{O_2} being independently variable, $p_{\text{inert gas}}$ may be independently varied.

Consider the equilibrium

$$M_{(s)} + CO_{2\,(g)} = MO_{(s)} + CO_{(g)}$$

This three-component, three-phase equilibrium has two degrees of freedom which may be selected from T, P, p_{CO} and p_{CO_2}. For example, fixing T and P uniquely fixes p_{CO} and p_{CO_2} via

$$K_T = \frac{p_{CO}}{p_{CO_2}} \quad \text{and} \quad P = p_{CO} + p_{CO_2}$$

This type of equilibrium is illustrated in Figs. 10.17 and 10.18.

If the system contains the solid carbide MC, then the three-component four-phase system (M + MO + MC + gas) has one degree of freedom which again can be selected from T, P, p_{CO}, and p_{CO_2}. As $R = N - C = 5 - 3 = 2$, the two independent reaction equilibria can be selected as

$$M_{(s)} + CO_{2(g)} = MO_{(s)} + CO_{(g)} \quad \text{(i)}$$

and

$$M_{(s)} + 2CO_{(g)} = MC_{(s)} + CO_{2(g)} \quad \text{(ii)}$$

Fixing T fixes

$$K_{T(\text{i})} = \frac{p_{CO}}{p_{CO_2}}$$

and

$$K_{T(\text{ii})} = \frac{p_{CO_2}}{p_{CO}^2}$$

which uniquely fixes p_{CO} and p_{CO_2}, and hence $P = p_{CO} + p_{CO_2}$.

If solid carbon is also present such that the system contains the phases M, MO, MC, C, and gaseous CO and CO_2, then the number of independent reaction equilibria is increased by one, e.g., the independent equilibria

$$M_{(s)} + CO_{2(g)} = MO_{(s)} + CO_{(g)} \quad \text{(i)}$$

$$M_{(s)} + 2CO_{(g)} = MC_{(s)} + CO_{2(g)} \quad \text{(ii)}$$

and

$$C_{(s)} + CO_{2(g)} = 2CO_{(g)} \quad \text{(iii)}$$

occur, and the number of phases present is increased by one. In this case $F = 0$ and the system is invariant, occurring at a unique T and unique values of p_{CO} and p_{CO_2}.

In a multiphase, multicomponent system in which several independent reaction equilibria occur, the number of such equilibria can be calculated as follows. First write a chemical equation for the formation of each species present from its constituent elements; e.g., in the previous example

$$M + \tfrac{1}{2}O_2 = MO \quad (a)$$

$$M + C = MC \quad (b)$$

$$C + \tfrac{1}{2}O_2 = CO \quad (c)$$

$$C + O_2 = CO_2 \quad (d)$$

Then combine these equations in such a manner that those elements not considered as being present in the system are eliminated. The resultant number of equations is then the number of independent reaction equilibria R. In the above system the species present are M, MO, MC, C, CO and CO_2. Hence, from (a),

$$\tfrac{1}{2}O_2 = MO - M$$

and thus in (c)

$$C + MO - M = CO \quad \text{or} \quad C + MO = CO + M \tag{iv}$$

in (d)

$$C + 2MO - 2M = CO_2 \quad \text{or} \quad C + 2MO = CO_2 + 2M \tag{v}$$

and in (b)

$$M + C = MC \tag{vi}$$

Thus three independent equilibria occur, combinations of which produce other equilibria which occur in the system; e.g.,

$$MC + 2MO = CO_2 + 3M \tag{vii}$$
$$MC + MO = CO + 2M \tag{viii}$$
$$M + CO_2 = MO + CO \tag{i}$$
$$M + 2CO = CO_2 + MC \tag{ii}$$
$$C + CO_2 = 2CO \tag{iii}$$

When any three of the equilibria (i) to (viii) are established, then the other five are established.

Example 1

Consider again the example discussed on p. 416, in which an examination was made of the conditions under which a liquid Fe-Mn solution and a liquid FeO-MnO solution can be in equilibrium with an oxygen-containing atmosphere.

This is a three-component system (Fe-Mn-O) existing in three phases (metal-oxide-gas) and hence, from the phase rule, the equilibrium has two degrees of freedom, which can be selected from the variables T, p_{O_2}, X_{Fe}, X_{Mn}, X_{FeO}, and X_{MnO}. With five species (O_2, Fe, Mn, FeO, MnO) and three components, there are two independent reaction equilibria, which can be selected as

$$Fe + \tfrac{1}{2}O_2 = FeO$$

for which

$$K_{(i),T} = \frac{a_{FeO}}{a_{Fe} \cdot p_{O_2}^{1/2}} = \frac{X_{FeO}}{X_{Fe} \cdot p_{O_2}^{1/2}} \qquad \text{(i)}$$

and

$$Mn + \tfrac{1}{2}O_2 = MnO$$

for which

$$K_{(ii),T} = \frac{a_{MnO}}{a_{Mn} \cdot p_{O_2}^{1/2}} = \frac{X_{MnO}}{X_{Mn} \cdot p_{O_2}^{1/2}} \qquad \text{(ii)}$$

(*a*) If T and p_{O_2} are chosen as the independent variables,

$$\frac{X_{FeO}}{X_{Fe}} = K_{(i),T} \cdot p_{O_2}^{1/2}$$

is fixed by Eq. (i), and

$$\frac{X_{MnO}}{X_{Mn}} = \frac{1 - X_{FeO}}{1 - X_{Fe}} = K_{(ii),T} \cdot p_{O_2}^{1/2}$$

is fixed by Eq. (ii). Thus X_{FeO} (and hence X_{MnO}) and X_{Fe} (and hence X_{Mn}) are fixed.

(*b*) If T and X_{Fe} are chosen as the independent variables, $X_{Mn} = 1 - X_{Fe}$ is automatically fixed.

$$\frac{X_{MnO}}{X_{FeO}} = \frac{1 - X_{FeO}}{X_{FeO}} = \frac{K_{(ii),T} \cdot X_{Mn}}{K_{(i),T} \cdot X_{Fe}}$$

is fixed by Eqs. (i) and (ii), which fixes X_{FeO} (and hence X_{MnO}), and

$$p_{O_2}^{1/2} = \frac{X_{FeO}}{X_{Fe} \cdot K_{(i),T}}$$

is fixed by Eq. (i).

(*c*) If p_{O_2} and X_{Fe} are chosen as the independent variables, from Eq. (i),

$$X_{FeO} = K_{(i),T} \cdot X_{Fe} \cdot p_{O_2}^{1/2} = \exp\left(\frac{-\Delta H^0_{(i)}}{RT}\right) \exp\left(\frac{\Delta S^0_{(i)}}{R}\right) \cdot X_{Fe} \cdot p_{O_2}^{1/2}$$

and from Eq. (ii),

$$X_{MnO} = 1 - X_{FeO} = K_{(ii),T} \cdot X_{Mn} \cdot p_{O_2}^{1/2} = \exp\left(\frac{-\Delta H^0_{(ii)}}{RT}\right) \exp\left(\frac{\Delta S^0_{(ii)}}{R}\right) X_{Mn} p_{O_2}^{1/2}$$

simultaneous solution of which fixes T and X_{FeO}. Thus the fixing of any two of the variables fixes the values of all the others. In the previous discussion of this example, $T = 1800°C$ and p_{O_2} were selected as the independent variables.

Example 2

Consider the equilibria which can occur among the solid stoichiometric phases V_2O_4, V_2O_5, and $VOSO_4$ and an SO_2-SO_3 gas phase.*

Firstly consider the criteria for equilibrium among all three solid phases and the SO_2-SO_3 gas phase. This system contains four phases and three components (V, O, and S) and so has $F = C + 2 - P = 1$ degree of freedom. Thus at any temperature the four-phase equilibrium occurs at fixed values of p_{SO_2} and p_{SO_3}. Such an equilibrium is said to be univariant. Now consider the various equilibria which can occur between two of the solid phases and the gas phase. There are three such equilibria, namely

$$V_2O_{4(s)} - V_2O_{5(s)} - SO_{2(g)} - SO_{3(g)} \quad (a)$$
$$V_2O_{4(s)} - VOSO_{4(s)} - SO_{2(g)} - SO_{3(g)} \quad (b)$$
$$V_2O_{5(g)} - VOSO_{4(s)} - SO_{2(g)} - SO_{3(g)} \quad (c)$$

For (*a*) the pertinent equilibrium is

$$V_2O_{4(s)} + SO_{3(g)} = V_2O_{5(s)} + SO_{2(g)}$$

for which

$$K_{(a)} = \frac{p_{SO_2}}{p_{SO_3}}$$

*See H. Flood and O. J. Kleppa, Investigations on the Equilibria in the System V_2O_4, V_2O_5, $VOSO_4$, SO_2, SO_3, *J. Am. Chem. Soc.*, 69:998 (1947).

For (*b*) the pertinent equilibrium is

$$2VOSO_{4(s)} = V_2O_{5(s)} + SO_{2(g)} + SO_{3(g)}$$

for which

$$K_{(b)} = p_{SO_2}p_{SO_3}$$

and for (*c*) the pertinent equilibrium is

$$2VOSO_{4(s)} = V_2O_{4(s)} + 2SO_{3(g)}$$

for which

$$K_{(c)} = p^2_{SO_3}$$

In each of (*a*), (*b*), and (*c*), $F = C + 2 - P = 2$ degrees of freedom, which, if T is fixed, leaves one degree of freedom to be chosen from p_{SO_2}, p_{SO_3}, or $P = p_{SO_2} + p_{SO_3}$. If P is chosen as the second degree of freedom, then in (*a*) the values of p_{SO_2} and p_{SO_3} are fixed by the requirements that

$$P = p_{SO_2} + p_{SO_3} \quad \text{and} \quad K_{(a)} = \frac{p_{SO_2}}{p_{SO_3}}$$

in (*b*) the values of p_{SO_2} and p_{SO_3} are fixed by the requirements that

$$P = p_{SO_2} + P_{SO_3} \quad \text{and} \quad K_{(b)} = p_{SO_2}p_{SO_3}$$

and in (*c*) p_{SO_2} and p_{SO_3} are fixed by the requirements that

$$P = p_{SO_2} + P_{SO_3} \quad \text{and} \quad p^2_{SO_3} = K_{(c)}$$

The conditions of phase equilibrium can thus be represented on an isothermal diagram, the coordinates of which are p_{SO_2} and p_{SO_3}.

For equilibrium (*a*), the allowed variation of p_{SO_2} and p_{SO_3} is that $p_{SO_2}/p_{SO_3} = K_{(a)}$. This variation is shown as the line *OA* in Fig. 13.7*a*; and *in the absence of considerations as to the stability of* $VOSO_4$, below *OA*, V_2O_4 is stable with respect to V_2O_5, and above *OA*, the reverse is the case.

For equilibrium (*b*), the allowed variation of p_{SO_2} and p_{SO_3} is that $p_{SO_2}p_{SO_3} = K_{(b)}$. This variation is shown as the line *BC* in Fig. 13.7*a*; and *in the absence of considerations as to the stability of* V_2O_4, above *BC*, $VOSO_4$ is stable with respect to V_2O_5, and below *BC* the reverse is the case. Lines *OA* and *BC* intersect at the point *P* which thus gives the unique values of p_{SO_2} and p_{SO_3} (at the particular temperature T) required for the four-phase equilibrium.

For equilibrium (*c*), p_{SO_3} is constant, being given as $p^2_{SO_3} = K_{(c)}$. This is

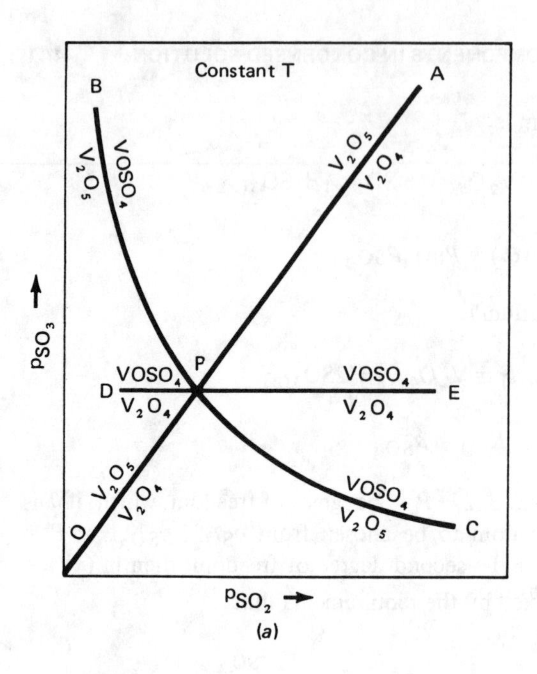

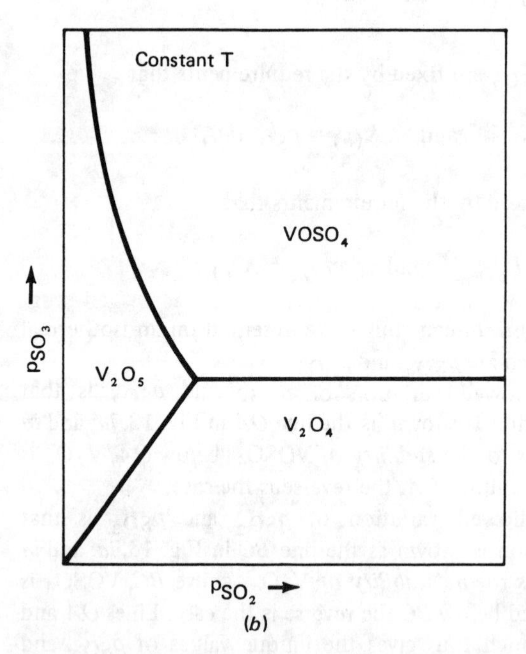

Fig. 13.7. (*a*) The V_2O_5–V_2O_4, $VOSC_4$–V_2O_5, and $VOSO_4$–V_2O_4 equilibria in SO_2–SO_3 atmospheres at constant temperature. (*b*) Regions of phase stability of $VOSO_4$, V_2O_5, and V_2O_4 in SO_2–SO_3 atmospheres at constant temperature.

shown as the line *DE* which passes through *P*, and *in the absence of considerations as to the stability of* V_2O_5, above *DE*, $VOSO_4$ is stable with respect to V_2O_4, and below *DE* the reverse is the case.

Figure 13.7*a* thus shows six lines radiating from the point *P*. As there are three fields of single-phase stability, then three of these lines represent stable equilibria and three represent metastable equilibria, and the problem is now to distinguish between the two types of equilibria. It is a property of such phase diagrams that the lines of metastable and stable equilibria radiate alternatively from a point such as *P* (see, for example, Figs. 7.10 and 7.13). Thus one set of lines is *PO-PB-PE* and the other is *PD-PC-PA*. Below the line *OPE*, V_2O_4 is stable; above the line *BPE*, $VOSO_4$ is stable; and to the left of the line *OPB*, V_2O_5 is stable.

No such similar scheme exists for the lines *DPC*, *CPA*, and *APD*, and thus the lines *PO-PE-PB* are the stable equilibrium lines with the equilibrium along *PC* being metastable with respect to V_2O_4, the equilibrium along *PA* being metastable with respect to $VOSO_4$, and the equilibrium along *PD* being metastable with respect to V_2O_5. The phase, or stability diagram is thus as shown in Fig. 13.7*b*.

Example 3: Phase Equilibria in the System Cu-O-S

The solid phases that can exist in this system are metallic Cu, the oxides Cu_2O and CuO, the sulfides Cu_2S and CuS, the sulfate $CuSO_4$, and the oxysulfate $CuO \cdot CuSO_4$, and the gas phase is an O_2-S_2 mixture. Just as the fixing of the activity of one component in a binary system fixes the activity of the other component, the fixing of the activities of two of the components of a ternary system fixes the activity of the third component. Thus when the partial pressures of O_2 and S_2 are fixed in the system Cu-O-S, the activity of Cu is fixed, and a definite equilibrium state occurs. Although the simplest choice of variables is p_{O_2}, p_{S_2}, and *T*, it is more convenient, for practical reasons, to consider p_{O_2}, p_{SO_2}, and *T* as the variables. Thus two-dimensional representation of the phase stability can be considered, (a) at constant temperature with p_{SO_2} and p_{O_2} as the variables, or (b) at constant p_{SO_2} (or p_{O_2}) with *T* and p_{O_2} (or p_{SO_2}) as the variables.

At fixed values of p_{O_2} and p_{S_2}, p_{SO_2} is fixed by the equilibrium

$$\tfrac{1}{2}S_2 + O_2 = SO_2$$

Application of the phase rule to a three-component system indicates that an equilibrium among five phases has no degrees of freedom. Thus, if a gas phase is always present,

1. Four condensed phases can be in equilibrium with one another and a gas phase at an invariant state,
2. Three condensed phases can be in equilibrium with one another and a gas phase at an arbitrarily chosen temperature,
3. Two condensed phases can be in equilibrium with one another and a gas phase at an arbitrarily chosen temperature and an arbitrarily chosen value of p_{O_2} *or* p_{SO_2}, and
4. One condensed phase can be in equilibrium with the gas phase at an arbitrarily chosen temperature and arbitrarily chosen values of p_{O_2} *and* p_{SO_2}.

As seven solid phases can exist, there are $(7 \times 6 \times 5)/(3 \times 2) = 35$ possible equilibria involving three condensed phases and the gas phase, $(7 \times 6)/2 = 21$ possible equilibria involving two condensed phases and the gas phase, and seven possible equilibria involving a single condensed phase and the gas phase. However, most of the 35 four-phase and the 21 three-phase equilibria are metastable.

Consider construction of the stability diagram for the system Cu-O-S at 700°C (973 K) using $\log p_{SO_2}$ and $\log p_{O_2}$ as the variables. At this temperature, the standard free energies of interest are

	$\Delta G^0_{973\text{ K}}$ (kJ)	
$2Cu_{(s)} + \frac{1}{2}O_{2(g)} = Cu_2O_{(s)}$	-94	(i)
$Cu_{(s)} + \frac{1}{2}O_{2(g)} = CuO_{(s)}$	-69	(ii)
$2Cu_{(s)} + \frac{1}{2}S_{2(g)} = Cu_2S_{(s)}$	-101	(iii)
$Cu_{(s)} + \frac{1}{2}S_{2(g)} = CuS_{(s)}$	-37	(iv)
$2Cu_{(s)} + \frac{1}{2}S_{2(g)} + \frac{5}{2}O_{2(g)} = CuO \cdot CuSO_{4(s)}$	-483	(v)
$Cu_{(s)} + \frac{1}{2}S_{2(g)} + 2O_{2(g)} = CuSO_{4(s)}$	-407.5	(vi)
$\frac{1}{2}S_{2(g)} + O_{2(g)} = SO_{2(g)}$	-293	(vii)

A. The Equilibrium Cu-Cu₂O-Gas Phase

From Eq. (i), $2Cu + \frac{1}{2}O_2 = Cu_2O$,

$$\Delta G^0_{973} = -94{,}000 \text{ joules}$$

$$= -RT \ln \frac{1}{p_{O_2}^{1/2}}$$

$$= 8.3144 \times 973 \times 2.303 \log p_{O_2}^{1/2}$$

or $\log p_{O_2} = -10.1$. Thus, at 973 K, the equilibrium Cu-Cu_2O-gas requires $\log p_{O_2} = -10.1$, and this equilibrium, which is drawn as line (A) in Fig. 13.8*a*, is independent of p_{SO_2}. At lower values of p_{O_2}, Cu is stable relative to Cu_2O, and at higher values of p_{O_2}, Cu_2O is stable relative to Cu.

B. The Equilibrium Cu-Cu_2S-Gas Phase

From Eqs. (iii) and (vii),

$$2Cu + \tfrac{1}{2}S_2 = Cu_2S \qquad \Delta G^0_{973\,K} = -101{,}000 \text{ joules}$$

and

$$SO_2 = \tfrac{1}{2}S_2 + O_2 \qquad \Delta G^0_{973\,K} = 293{,}000 \text{ joules}$$

summation of which gives

$$2Cu + SO_2 = Cu_2S + O_2$$

$$\Delta G^0_{973\,K} = 192{,}000 \text{ joules}$$

$$= -RT \ln \frac{p_{O_2}}{p_{SO_2}}$$

$$= -8.3144 \times 973 \times 2.303 \log \frac{p_{O_2}}{p_{SO_2}} \qquad \text{(viii)}$$

Therefore, $\log (p_{O_2}/p_{SO_2}) = -10.31$, or $\log p_{SO_2} = \log p_{O_2} + 10.31$, which, drawn as line (B) in Fig. 13.8*a*, is the variation of p_{O_2} with p_{SO_2} required for the equilibrium Cu-Cu_2S-gas at 973 K. Above line (B), Cu_2S is stable relative to Cu, and below the line, Cu is stable relative to Cu_2S.

C. The Equilibrium Cu_2O-Cu_2S-Gas Phase

Lines (A) and (B) intersect at $\log p_{O_2} = -10.1$, $\log p_{SO_2} = 0.21$, which, at 973 K, are the values of $\log p_{O_2}$ and $\log p_{SO_2}$ at which the four-phase equilibrium Cu-Cu_2O-Cu_2S-gas occurs. The variation of $\log p_{O_2}$ with $\log p_{SO_2}$ required for the equilibrium Cu_2S-Cu_2O-gas must pass through this point.

From Eqs. (i) and (viii),

$$2Cu + \tfrac{1}{2}O_2 = Cu_2O \qquad \Delta G^0_{973} = -94{,}000 \text{ joules}$$

and

$$Cu_2S + O_2 = 2Cu + SO_2 \qquad \Delta G^0_{973} = -192{,}000 \text{ joules}$$

summation of which gives

$$Cu_2S + \tfrac{3}{2}O_2 = Cu_2O + SO_2$$

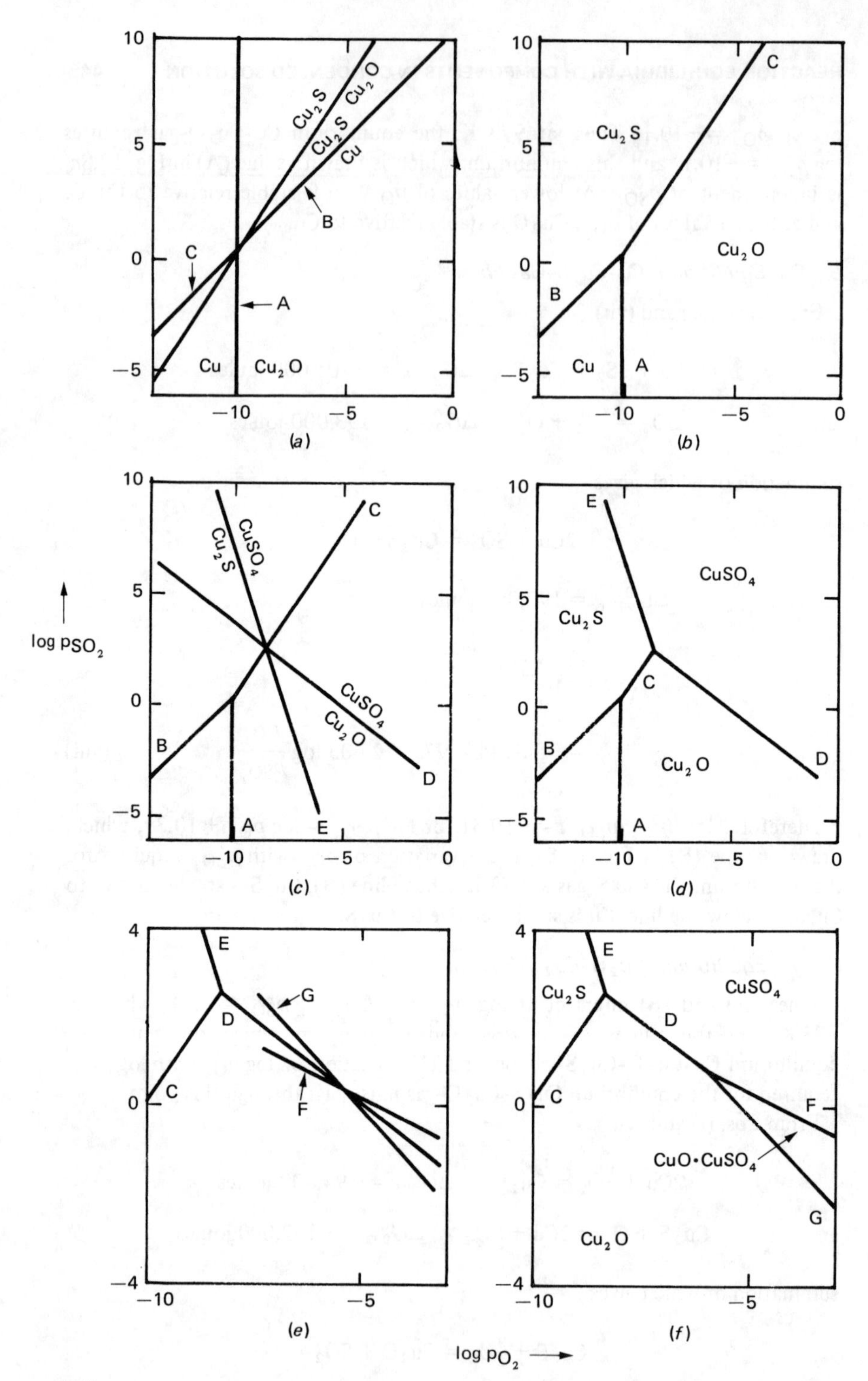

Cu
Cu₂O
Cu₂S
CuSO₄
CuO·CuSO₄
A
B
C
D
E
F
G
(a)
(b)
(c)
(d)
(e)
(f)
log p_{SO_2}
log p_{O_2}

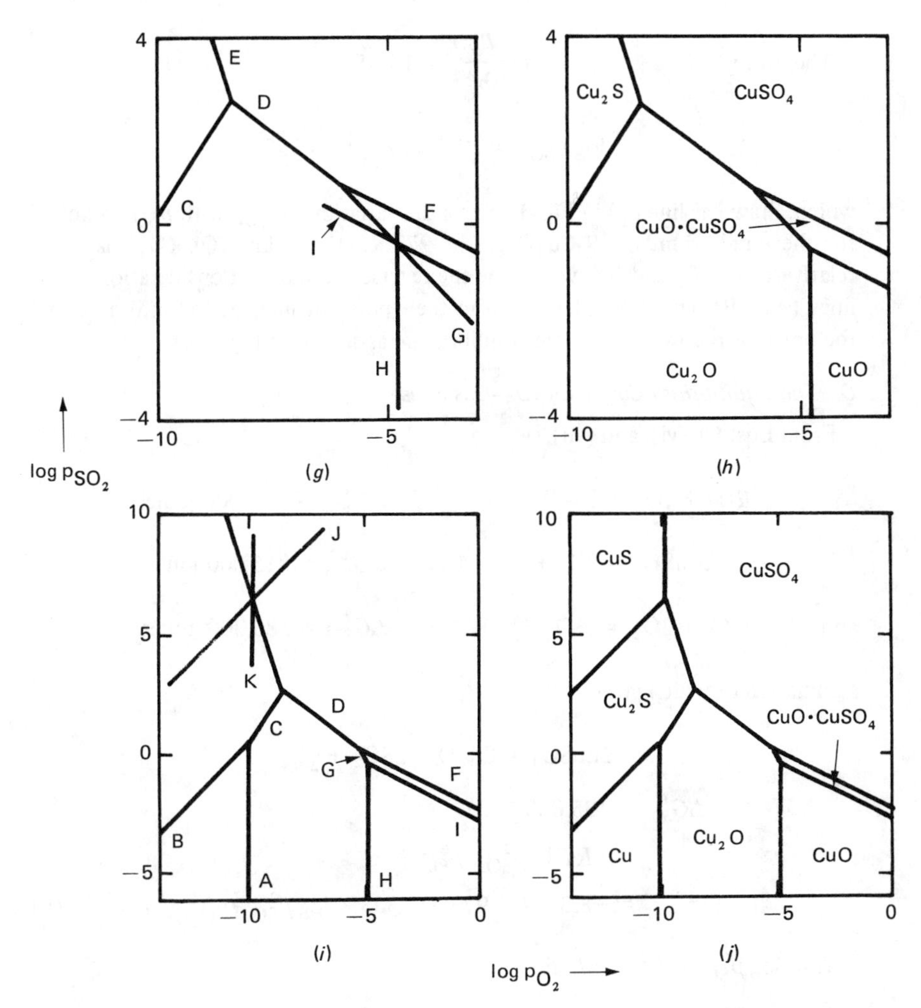

Fig. 13.8. Construction of the stability diagram for the system copper-sulfur-oxygen at 700°C.

$$\Delta G^0_{973} = -286{,}000 \text{ joules}$$

$$= -RT \ln \frac{p_{SO_2}}{p_{O_2}^{3/2}}$$

$$= -8.3144 \times 973 \times 2.303 \log \frac{p_{SO_2}}{p_{O_2}^{3/2}} \qquad \text{(ix)}$$

Therefore,
$$\log \frac{p_{SO_2}}{p_{O_2}^{3/2}} = 15.35$$

or
$$\log p_{SO_2} = 1.5 \log p_{O_2} + 15.35$$

which, drawn as line (C) in Fig. 13.8*a*, is the variation of p_{O_2} with p_{SO_2} required for the equilibrium Cu_2S-Cu_2O-gas at 973 K. Above line (C), Cu_2S is stable relative to Cu_2O, and below the line the reverse is the case. Consideration of the lines (A), (B), and (C) radiating from their point of intersection indicates that the portions representing stable equilibria are as shown in Fig. 13.8*b*.

D. The Equilibrium Cu_2O-$CuSO_4$-Gas Phase

From Eqs. (i), (vi), and (vii),

$$2Cu + \tfrac{1}{2}O_2 = Cu_2O \qquad \Delta G^0_{973} = -94{,}000 \text{ joules}$$

$$2CuSO_4 = 2Cu + S_2 + 4O_2 \qquad \Delta G^0_{973} = 815{,}000 \text{ joules}$$

and
$$S_2 + 2O_2 = 2SO_2 \qquad \Delta G^0_{973} = 586{,}000 \text{ joules}$$

summation of which gives

$$2CuSO_4 = Cu_2O + 2SO_2 + \tfrac{3}{2}O_2$$

$$\begin{aligned} \Delta G^0_{973} &= 135{,}000 \text{ joules} \\ &= -RT \ln p_{SO_2}^2 \cdot p_{O_2}^{3/2} \\ &= -8.3144 \times 973 \times 2.303 \times \log p_{SO_2}^2 \cdot p_{O_2}^{3/2} \end{aligned} \qquad \text{(x)}$$

Thus $\log p_{SO_2}^2 \cdot p_{O_2}^{3/2} = -7.25$

or
$$\log p_{SO_2} = -0.75 \log p_{O_2} - 3.62$$

This, drawn as line (D) in Fig. 13.8*c*, is the variation of p_{SO_2} with p_{O_2} required for the equilibrium Cu_2O-$CuSO_4$-gas. Above this line, $CuSO_4$ is stable relative to Cu_2O, and below the line, Cu_2O is stable relative to $CuSO_4$.

E. The Equilibrium Cu_2S-$CuSO_4$-Gas Phase

Lines (C) and (D) in Fig. 13.8*c* intersect at the point $\log p_{O_2} = -8.43$, $\log p_{SO_2} = 2.70$, which are thus the values of $\log p_{O_2}$ and $\log p_{SO_2}$ required for the four-phase equilibrium Cu_2O-Cu_2S-$CuSO_4$-gas at 973 K. From Eqs. (ix) and (x),

$$Cu_2S + \tfrac{3}{2}O_2 = Cu_2O + SO_2 \qquad \Delta G^0_{973} = -286{,}000 \text{ joules}$$

and
$$Cu_2O + 2SO_2 + \tfrac{3}{2}O_2 = 2CuSO_4 \qquad \Delta G^0_{973} = -135{,}000 \text{ joules}$$

summation of which gives

$$Cu_2S + 3O_2 + SO_2 = 2CuSO_4$$

$$\Delta G^0_{973} = -421{,}000 \text{ joules}$$

$$= -RT \ln \frac{1}{p^3_{O_2} \cdot p_{SO_2}}$$

$$= +8.3144 \times 973 \times 2.303 \log p^3_{O_2} \cdot p_{SO_2} \qquad \text{(xi)}$$

Thus $\log p^3_{O_2} \cdot p_{SO_2} = -22.60$, or $\log p_{SO_2} = -3 \log p_{O_2} - 22.60$, which, drawn as line (E) in Fig. 13.8*c*, is the variation of p_{O_2} with p_{SO_2} required, at 973 K, for the equilibrium Cu_2S-$CuSO_4$-gas. Above this line, $CuSO_4$ is stable relative to Cu_2S, and below the line the reverse is the case. As line (C), radiating from the point of intersection of lines (A), (B), and (C) in Fig. 13.8*b*, is stable, the lines radiating from the point of intersection of (C), (D), and (E), which represent stable equilibria, are as shown in Fig. 13.8*d*.

F. The Equilibrium $CuSO_4$-$CuO \cdot CuSO_4$-Gas Phase

From Eqs. (v), (vi), and (vii),

$$2Cu + \tfrac{1}{2}S_2 + \tfrac{5}{2}O_2 = CuO \cdot CuSO_4 \qquad \Delta G^0_{973} = -483{,}000 \text{ joules}$$

$$2CuSO_4 = 2Cu + S_2 + 4O_2 \qquad \Delta G^0_{973} = +815{,}000 \text{ joules}$$

$$\tfrac{1}{2}S_2 + O_2 = SO_2 \qquad \Delta G^0_{973} = -293{,}000 \text{ joules}$$

summation of which gives

$$2CuSO_4 = CuO \cdot CuSO_4 + \tfrac{1}{2}O_2 + SO_2$$

$$\Delta G^0_{973} = 39{,}000$$

$$= -RT \ln p^{1/2}_{O_2} \cdot p_{SO_2}$$

$$= -8.3144 \times 973 \times 2.303 \log p^{1/2}_{O_2} \cdot p_{SO_2}$$

Therefore, $$\log p_{O_2}^{1/2} \cdot p_{SO_2} = -2.09$$

or $$\log p_{SO_2} = -\tfrac{1}{2} \log p_{O_2} - 2.09$$

This is drawn as line (B) in Fig. 13.8*e* and is the variation of p_{O_2} with p_{SO_2} required for the equilibrium $CuSO_4$-$CuO \cdot CuSO_4$-gas at 973 K. Above the line, $CuSO_4$ is stable relative to $CuO \cdot CuSO_4$, and below the line the reverse is the case. Lines (D) and (F) intersect at $\log p_{O_2} = -6.13$, $\log p_{SO_2} = 0.97$, which is the point at which the four-phase equilibrium Cu_2O-$CuSO_4$-$CuO \cdot CuSO_4$-gas occurs.

G. The Equilibrium Cu_2O-$CuO \cdot CuSO_4$-Gas Phase

From Eqs. (i), (v), and (vii),

$$2Cu + \tfrac{1}{2}O_2 = Cu_2O \qquad \Delta G^0_{973} = -94{,}000 \text{ joules}$$

$$CuO \cdot CuSO_4 = 2Cu + \tfrac{1}{2}S_2 + \tfrac{5}{2}O_2 \qquad \Delta G^0_{973} = 483{,}000 \text{ joules}$$

$$\tfrac{1}{2}S_2 + O_2 = SO_2 \qquad \Delta G^0_{973} = -293{,}000 \text{ joules}$$

summation of which gives

$$CuO \cdot CuSO_4 = Cu_2O + O_2 + SO_2$$

$$\begin{aligned}\Delta G^0_{973} &= 96{,}000 \text{ joules}\\ &= -RT \ln p_{O_2} \cdot p_{SO_2}\\ &= -8.3144 \times 973 \times 2.303 \log p_{O_2} \cdot p_{SO_2}\end{aligned}$$

Thus $\log p_{O_2} \cdot p_{SO_2} = -5.15$, or $\log p_{SO_2} = -\log p_{O_2} - 5.15$. This is drawn as line (G) in Fig. 13.8*e* and is the variation of p_{O_2} with p_{SO_2} required for the equilibrium Cu_2O-$CuO \cdot CuSO_4$-gas at 973 K. Consideration of lines (D), (F), and (G) radiating from their point of intersection indicates that the lines representing stable equilibria are as drawn in Fig. 13.8*f*.

H. The Equilibrium Cu_2O-CuO-Gas Phase

From Eqs. (i) and (ii),

$$2Cu + \tfrac{1}{2}O_2 = Cu_2O \qquad \Delta G^0_{973} = -94{,}000 \text{ joules}$$

$$2CuO = 2Cu + O_2 \qquad \Delta G^0_{973} = 138{,}000 \text{ joules}$$

summation of which gives

$$2CuO = Cu_2O + \tfrac{1}{2}O_2$$

$$\Delta G^0_{973} = 44{,}000 \text{ joules}$$
$$= -RT \log p_{O_2}^{1/2}$$
$$= -8.3144 \times 973 \times 2.303 \log p_{O_2}^{1/2}$$

Thus $\log p_{O_2} = -4.72$. This is drawn as the line (H) in Fig. 13.8*g* and is the oxygen pressure at 973 K required for the equilibrium between Cu_2O and CuO. At higher oxygen pressures, CuO is stable relative to Cu_2O, and at lower oxygen pressures Cu_2O is stable relative to CuO.

Lines (G) and (H) intersect at the values of $\log p_{O_2}$ and $\log p_{SO_2}$ required for the four-phase equilibrium Cu_2O-CuO-CuO·$CuSO_4$-gas.

I. The Equilibrium CuO-CuO·CuSO$_4$-Gas Phase

From combination of Eqs. (ii), (v), and (vii),

$$2CuO + \tfrac{1}{2}O_2 + SO_2 = CuO \cdot CuSO_4$$

for which $\Delta G^0_{973} = -52{,}000$ joules

$$= -8.3144 \times 973 \times 2.303 \times \log \frac{1}{p_{O_2}^{1/2} \cdot p_{SO_2}}$$

Thus
$$\log p_{O_2}^{1/2} \cdot p_{SO_2} = -2.79$$

or
$$\log p_{SO_2} = -\tfrac{1}{2} \log p_{O_2} - 2.79$$

which, drawn as line (I) in Fig. 13.8*g*, is the variation of p_{O_2} with p_{SO_2} required for the equilibrium CuO-CuO·$CuSO_4$-gas at 973 K. The lines radiating from the intersection of (G), (H), and (I) which represent stable equilibria are as shown in Fig. 13.8*h*.

J. The Equilibrium Cu$_2$S-CuS-Gas Phase

From Eqs. (iii), (iv), and (vii),

$$2CuS + O_2 = Cu_2S + SO_2$$

$$\Delta G^0_{973} = -320{,}000 \text{ joules}$$

$$= -8.3144 \times 973 \times 2.303 \log \frac{p_{SO_2}}{p_{O_2}} \qquad \text{(xii)}$$

Thus $$\log p_{SO_2} = \log p_{O_2} + 17.18$$

This, drawn as line (J) in Fig. 13.8*i*, intersects line (E) at $\log p_{O_2} = -9.95$, $\log p_{SO_2} = 7.23$, which is the state at which the equilibrium Cu_2S-CuS-$CuSO_4$-gas phase occurs at 973 K.

K. The Equilibrium CuS-$CuSO_4$-Gas Phase

From Eqs. (xi) and (xii),

$$2CuS + 4O_2 = 2CuSO_4$$

for which
$$\Delta G^0_{973} = -741{,}000 \text{ joules}$$
$$= -18{,}631 \log p^4_{O_2}$$

Thus $\log p_{O_2} = -9.95$, which, drawn as the line (K) in Fig. 13.8*i*, is the oxygen pressure required for the equilibrium CuS-$CuSO_4$-gas at 973 K.

Consideration of the lines radiating from the intersection of (E), (J), and (K) gives the lines representing stable equilibria as shown in Fig. 13.8*j* and hence the complete stability diagram is as shown in Fig. 13.9. As is seen, only five of the possible 35 equilibria involving three condensed phases and a gas phase are stable, and only 11 of the possible 21 equilibria involving two condensed phases and a gas phase are stable.

The partial pressure of S_2 is obtained from Eq. (vii) as

$$-293{,}000 = -RT \ln \frac{p_{SO_2}}{p^{1/2}_{S_2} \cdot p_{O_2}} = -18{,}631 \log \frac{p_{SO_2}}{p^{1/2}_{S_2} \cdot p_{O_2}}$$

or
$$\log p_{SO_2} = \log p_{O_2} + 15.73 + \tfrac{1}{2} \log p_{S_2}$$

Lines of constant p_{S_2} thus have unit positive slope in Fig. 13.9. Similarly, lines of constant p_{SO_3} are obtained from consideration of the equilibrium

$$SO_2 + \tfrac{1}{2}O_2 = SO_3$$

for which $\Delta G^0_{973} = -7610$ joules. Thus,

$$-7610 = -18{,}631 \log \frac{p_{SO_3}}{p_{SO_2} \cdot p^{1/2}_{O_2}}$$

or
$$\log p_{SO_2} = \tfrac{1}{2} \log p_{O_2} - 0.41 + \log p_{SO_3}$$

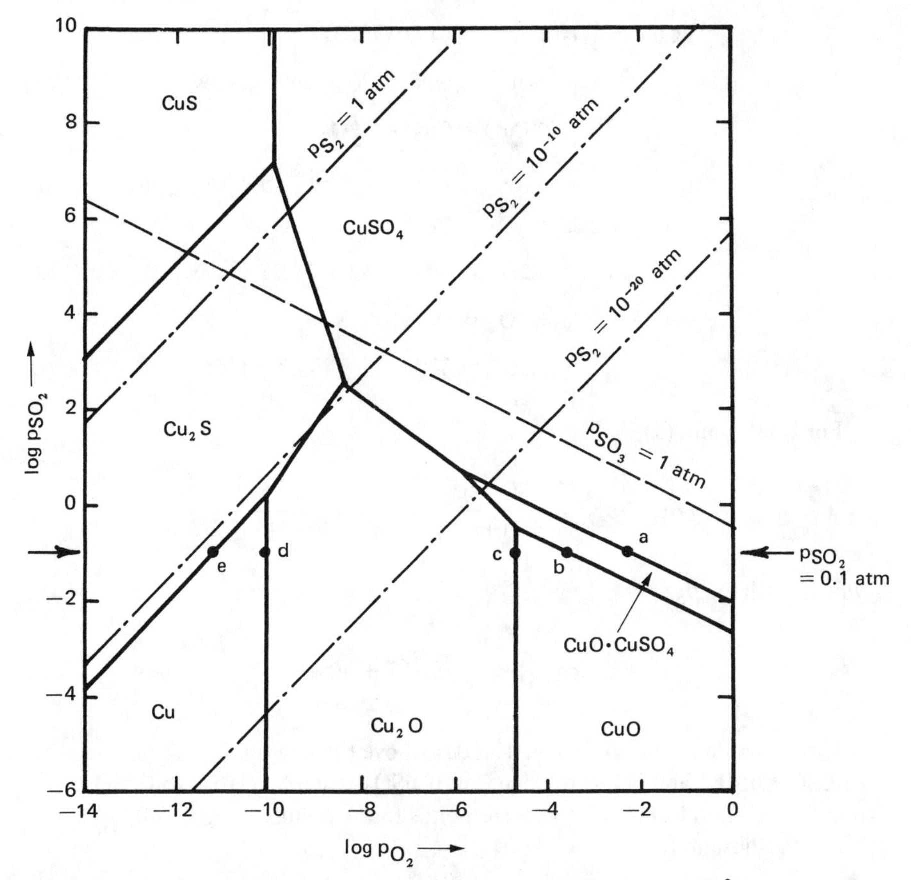

Fig. 13.9. The stability diagram for the system copper–sulfur–oxygen at 700°C.

which shows that lines of constant p_{SO_3} have a slope of $-\frac{1}{2}$. Figure 13.9 is an isothermal section of the three-dimensional stability diagram in T-log p_{SO_2}-log p_{O_2} space.

Consider now a section of the three-dimensional diagram at $p_{SO_2} = 0.1$ atm using log p_{O_2} and $1/T$ as the variables. In Fig. 13.9 it is seen that, with $p_{SO_2} = 0.1$ atm, five equilibria involving two condensed phases and a gas phase occur at 973 K, namely $CuSO_4$-$CuO \cdot CuSO_4$-gas at *a*, $CuO \cdot CuSO_4$-CuO-gas at *b*, CuO-Cu_2O-gas at *c*, Cu_2O-Cu at *d*, and Cu-Cu_2S-gas at *e*. The standard free energies of interest are

$$CuO \cdot CuSO_4 + SO_2 + \tfrac{1}{2}O_2 = 2CuSO_4$$

$$\Delta G^0 = -304{,}370 + 272T \text{ joules} \qquad \text{(a)}$$

$$2CuO + \tfrac{1}{2}O_2 + SO_2 = CuO{\cdot}CuSO_4$$
$$\Delta G^0 = -305{,}290 + 260T \text{ joules} \qquad \text{(b)}$$
$$2CuO = Cu_2O + \tfrac{1}{2}O_2$$
$$\Delta G^0 = 139{,}890 - 98.5T \text{ joules} \qquad \text{(c)}$$
$$2Cu + \tfrac{1}{2}O_2 = Cu_2O$$
$$\Delta G^0 = -166{,}840 + 74.7T \text{ joules} \qquad \text{(d)}$$
$$2Cu + SO_2 = O_2 + Cu_2S$$
$$\Delta G^0 = 230{,}670 - 40.6T \text{ joules} \qquad \text{(e)}$$

For equilibrium (a),

$$\log p_{SO_2} \cdot p_{O_2}^{1/2} = \frac{-304{,}370}{8.3144 \times 2.303T} + \frac{272}{8.3144 \times 2.303}$$

which, with $\log p_{SO_2} = -1$, gives

$$\log p_{O_2} = \frac{-31{,}790}{T} + 30.4 \qquad \text{(A)}$$

This is drawn as line A in Fig. 13.10*a*. Above the line $CuSO_4$ is stable relative to $CuO{\cdot}CuSO_4$, and below the line $CuO{\cdot}CuSO_4$ is stable relative to $CuSO_4$. The point *a* on line A in Fig. 13.10*a* corresponds to the point *a* in Fig. 13.9.

For equilibrium (b),

$$\log p_{SO_2} \cdot p_{O_2}^{1/2} = \frac{-305{,}285}{8.3144 \times 2.303T} + \frac{260}{8.3144 \times 2.303}$$

which, with $\log p_{SO_2} = -1$, gives

$$\log p_{O_2} = \frac{-31{,}886}{T} + 29.1 \qquad \text{(B)}$$

This is drawn as line B in Fig. 13.10*a*. Above line B $CuO{\cdot}CuSO_4$ is stable relative to CuO, and below the line the reverse is the case. The point *b* in Fig. 13.10*a* corresponds to the point *b* in Fig. 13.9.

For equilibrium (c),

$$\log p_{O_2} = \frac{-14{,}610}{T} + 10.29 \qquad \text{(C)}$$

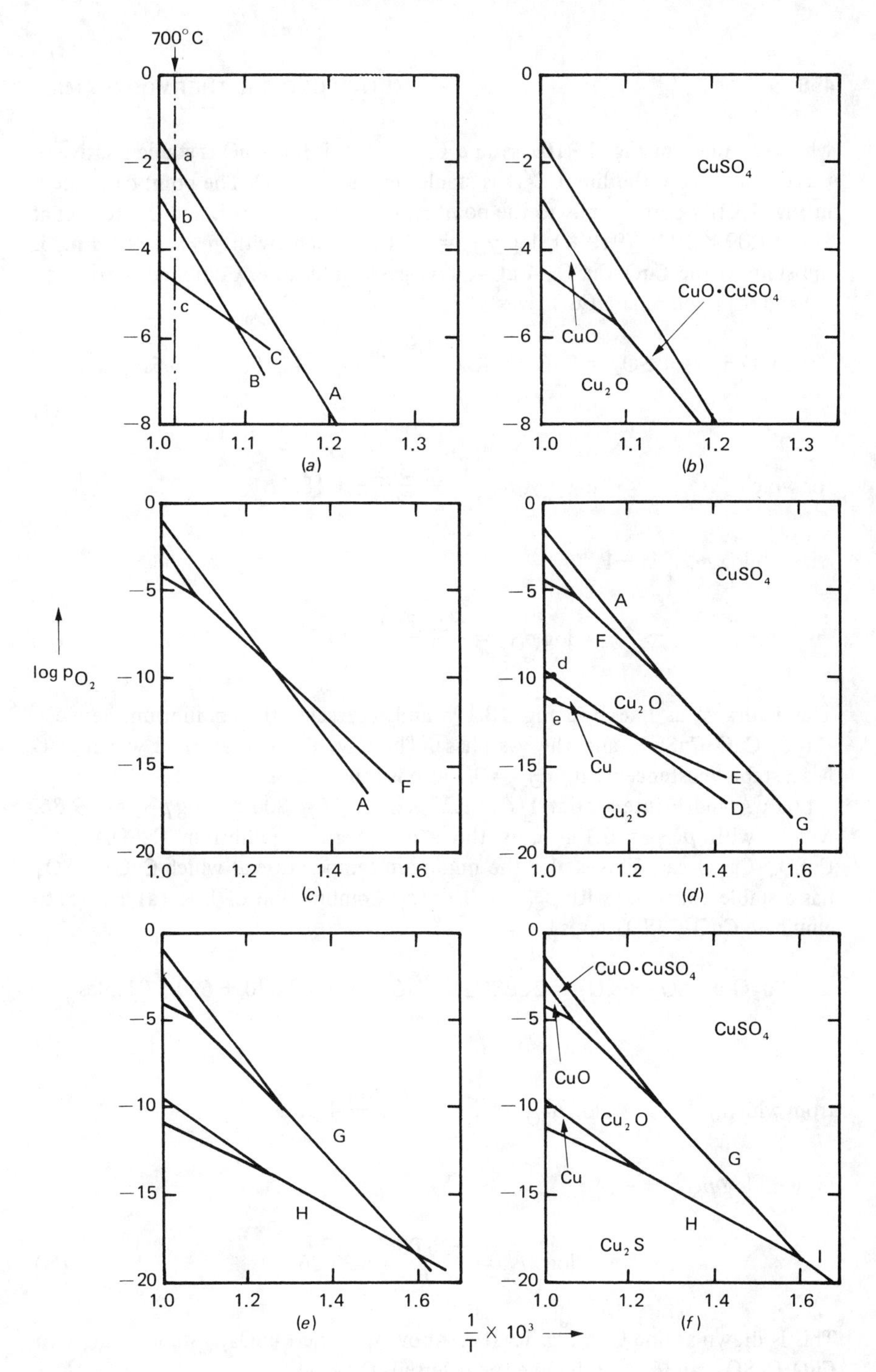

Fig. 13.10. Construction of the log p_{O_2} vs. $1/T$ diagram for the system copper–sulfur–oxygen at $p_{SO_2} = 0.1$ atm.

which is drawn in Fig. 13.10*a* as line C. Above this line CuO is stable relative to Cu_2O, and below the line Cu_2O is stable relative to CuO. The point *c* on line C in Fig. 13.10*a* corresponds to the point *c* in Fig. 13.9. Lines B and C intersect at $1/T = 1.09 \times 10^{-3}$ (918 K), $\log p_{O_2} = -5.63$, which, with $p_{SO_2} = 0.1$ atm, is the state of the $CuO \cdot CuSO_4$-CuO-Cu_2O-gas equilibrium. Combination of Eqs. (b) and (c) to eliminate CuO gives

$$Cu_2O + O_2 + SO_2 = CuO \cdot CuSO_4 \qquad \Delta G^0 = -445{,}180 + 358.5T \text{ joules} \tag{f}$$

for which $$\log p_{O_2} p_{SO_2} = \frac{-23{,}250}{T} + 18.72$$

or, with $\log p_{SO_2} = -1$,

$$\log p_{O_2} = \frac{-23{,}250}{T} + 19.72 \tag{F}$$

This is drawn as line F in Fig. 13.10*b* and represents the equilibrium between Cu_2O, $CuO \cdot CuSO_4$, and the gas phase. The lowest temperature at which CuO has a stable existence with $p_{SO_2} = 0.1$ atm is thus 918 K.

Lines A and F intersect at $1/T = 1.25 \times 10^{-3}$ ($T = 800$ K), $\log p_{O_2} = -9.36$, which, with $p_{SO_2} = 0.1$ atm, is the state of the equilibrium $CuSO_4$-$CuO \cdot CuSO_4$-Cu_2O-gas. This is also the minimum temperature at which $CuO \cdot CuSO_4$ has a stable existence with $p_{SO_2} = 0.1$ atm. Combination of Eqs. (a) and (f) to eliminate $CuO \cdot CuSO_4$ gives

$$Cu_2O + 2SO_2 + \tfrac{3}{2}O_2 = 2CuSO_4 \qquad \Delta G^0 = -749{,}540 + 630.5T \text{ joules} \tag{g}$$

from which, $$\log p_{SO_2}^2 \cdot p_{O_2}^{3/2} = \frac{-39{,}144}{T} + 32.9$$

or, with $\log p_{SO_2} = -1$,

$$\log p_{O_2} = \frac{-26{,}096}{T} + 23.26 \tag{G}$$

This is drawn as line G in Fig. 13.10*d*. Above this line $CuSO_4$ is stable relative to $CuO \cdot CuSO_4$, and below the line the reverse is the case.

For equilibrium (d),

$$\log p_{O_2} = \frac{-17{,}425}{T} + 7.80 \qquad \text{(D)}$$

This is drawn as line D in Fig. 13.10*d* and the point *d* on this line corresponds to the point *d* in Fig. 13.9. Above this line Cu_2O is stable relative to Cu, and below the line Cu is stable relative to Cu_2O.

For equilibrium (e),

$$\log \frac{p_{O_2}}{p_{SO_2}} = \frac{-12{,}047}{T} + 2.12$$

which, with $\log p_{SO_2} = -1$, gives

$$\log p_{O_2} = \frac{-12{,}047}{T} + 1.12 \qquad \text{(E)}$$

This is drawn as line E in Fig. 13.10*d* and the point *e* on this line corresponds to the point *e* in Fig. 13.9. Above the line Cu is stable relative to Cu_2S, and below the line Cu_2S is stable relative to Cu. Lines D and E intersect at $1/T = 1.24 \times 10^{-3}$ ($T = 805$ K), $\log p_{O_2} = -13.84$, which, with $p_{SO_2} = 0.1$ atm, is the lowest temperature at which metallic Cu can exist. Combination of Eqs. (d) and (e) to eliminate Cu gives

$$Cu_2S + \tfrac{3}{2}O_2 = SO_2 + Cu_2O \qquad \Delta G^0 = -397{,}510 + 115.3T \text{ joules} \qquad \text{(h)}$$

Thus,

$$\log \frac{p_{O_2}^{3/2}}{p_{SO_2}} = \frac{-20{,}759}{T} + 6.02$$

or, with $\log p_{SO_2} = -1$,

$$\log p_{O_2} = \frac{-13{,}839}{T} + 3.35 \qquad \text{(H)}$$

This is drawn as line H in Fig. 13.10*e*. Lines G and H intersect at $1/T = 1.16 \times 10^{-3}$ ($T = 615$ K), $\log p_{O_2} = -19.13$, which is thus the minimum temperature at which Cu_2O can exist with $p_{SO_2} = 0.1$ atm.

Combination of Eqs. (g) and (h) to eliminate Cu_2O gives

$$Cu_2S + SO_2 + 3O_2 = 2CuSO_4 \qquad \Delta G^0 = -1{,}147{,}050 + 745.8T \text{ joules}$$

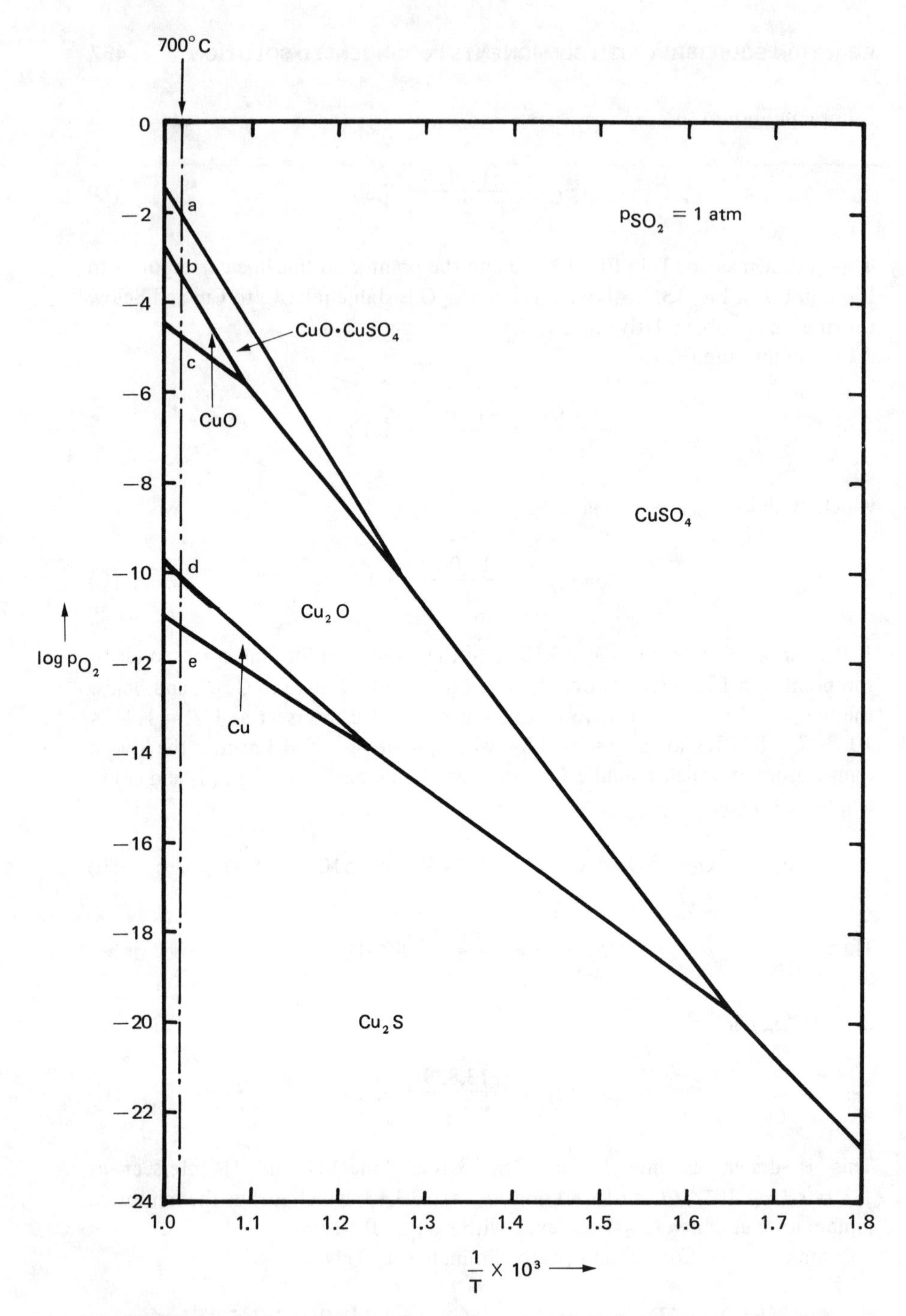

Fig. 13.11. The fields of phase stability in the system copper-sulfur-oxygen with p_{SO_2} = 0.1 atm.

With $\log p_{SO_2} = -1$, this gives

$$\log p_{O_2} = \frac{-19{,}967}{T} + 13.32$$

which is drawn as line I in Fig. 13.10*f*. The complete stability diagram with $p_{SO_2} = 0.1$ atm is thus as shown in Fig. 13.11.

13.5 BINARY SYSTEMS INVOLVING COMPOUND FORMATION

The phase relationships in a two-component system can be represented on an isobaric temperature-composition phase diagram. Such diagrams are the binary phase diagrams normally encountered in metallurgy. If the two components enter into chemical reaction with one another to form new species, e.g., compounds, then, in such systems, chemical reaction equilibria and phase equilibria are synonymous.

Consider the binary system A–B, the phase diagram of which is shown in Fig. 13.12. In the solid state the negative departures from ideality are sufficiently great that compound formation occurs, with there being negligible solubility of A in B, or B in A, and negligible range of nonstoichiometry of the compounds AB_3, AB, and A_3B. The system contains the three equilibria

$$3A + B = A_3B \qquad A + B = AB \qquad A + 3B = AB_3$$

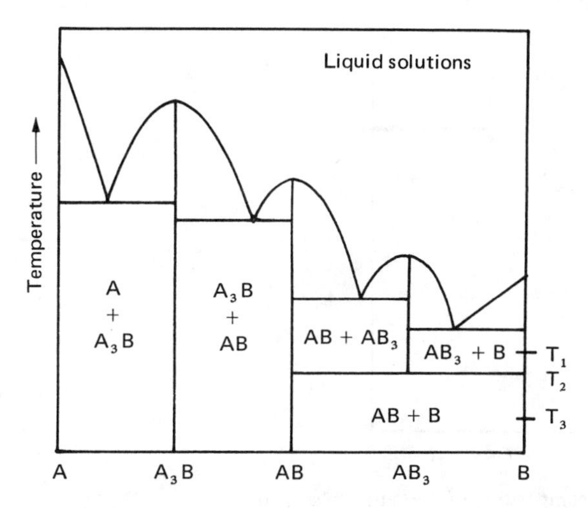

Fig. 13.12. The phase diagram for the system A–B in which are formed three stoichiometric compounds.

and if one of the elemental components (B) is appreciably volatile and the other (A) is not, then the thermodynamics of the system can be determined from a knowledge of the variation of $P = p_B$ across the system. Figure 13.13 shows the variation of $P = p_B$ with composition at the temperature T_1. In the composition range B-AB_3, virtually pure B is in equilibrium with AB_3 (saturated with B); and as pure B exists, the pressure exerted by the system is p_B^0, the saturated vapor pressure of B at the temperature T_1. In the composition range between AB (saturated with B) and AB_3 (saturated with A), the constant pressure exerted by the system is p_B'; in the range between A_3B (saturated with B) and AB (saturated with A) it is p_B''; and in the range between A (saturated with B) and A_3B (saturated with A) it is p_B'''. For each of these composition ranges, the two-component, three-phase equilibrium has one degree of freedom which is used once T_1 is specified, hence the constancy of $P = p_B$ within these ranges at fixed temperature. The activity of B in the system, defined as p_B/p_B^0, is thus

$$\frac{p_B^0}{p_B^0} = 1 \text{ in the range } AB_3\text{—B}$$

$$\frac{p_B'}{p_B^0} \quad \text{in the range } AB\text{—}AB_3$$

$$\frac{p_B''}{p_B^0} \quad \text{in the range } A_3B\text{—}AB$$

and

$$\frac{p_B'''}{p_B^0} \quad \text{in the range } A\text{—}A_3B$$

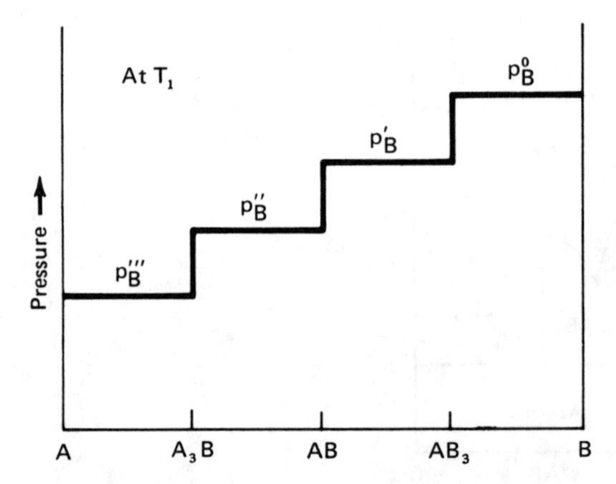

Fig. 13.13. The variation with composition of the vapor pressure of component B in the system shown in Fig. 13.10 at the temperature T_1.

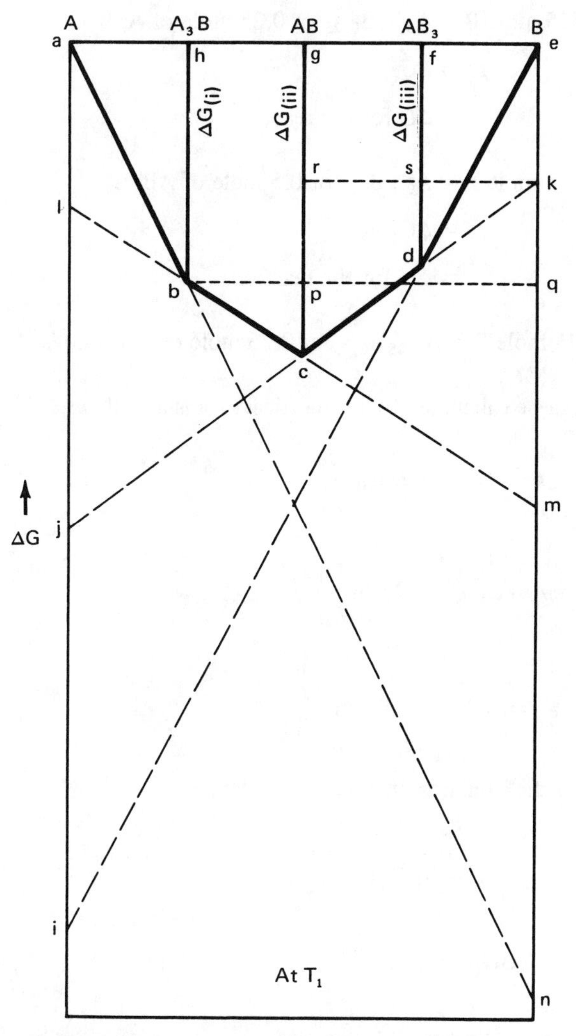

Fig. 13.14. The integral free energies in the system shown in Fig. 13.12 at the temperature T_1.

Thus as $\Delta\bar{G}_B^M = \bar{G}_B - G_B^0 = RT \ln p_B/p_B^0$, the free energy–composition diagram at the temperature T_1 is as shown in Fig. 13.14.

As Fig. 13.14 is drawn for 1 mole of the system, then

$$hb = \Delta G_{(i)}$$

$$= \Delta G \text{ for the reaction}$$

$$0.75 \text{ mole A} + 0.25 \text{ mole B} = \text{A}_{0.75}\text{B}_{0.25} \text{ or } 0.25 \text{ mole of A}_3\text{B}$$

$$gc = \Delta G_{(ii)}$$
$$= \Delta G \text{ for the reaction}$$

$$0.5 \text{ mole A} + 0.5 \text{ mole B} = \text{A}_{0.5}\text{B}_{0.5} \text{ or } 0.5 \text{ mole of AB}$$

$$fd = \Delta G_{(iii)}$$
$$= \Delta G \text{ for the reaction}$$

$$0.25 \text{ mole A} + 0.75 \text{ mole B} = \text{A}_{0.25}\text{B}_{0.75} \text{ or } 0.25 \text{ mole of AB}_3$$

These three free energies can be calculated by geometrical means as follows.

$$ek = \Delta \bar{G}_B'^M = RT \ln \frac{p_B'}{p_B^0}$$

$$em = \Delta \bar{G}_B''^M = RT \ln \frac{p_B''}{p_B^0}$$

and

$$en = \Delta \bar{G}_B'''^M = RT \ln \frac{p_B'''}{p_B^0}$$

Thus, from consideration of the similar triangles *ahb* and *aen*,

$$\frac{\Delta G_{(i)}}{\Delta \bar{G}_B'''^M} = \frac{1}{4}$$

and hence

$$\Delta G_{(i)} = \frac{1}{4} \Delta \bar{G}_B'''^M$$

Consideration of the similar triangles *bpc* and *bqm* gives

$$\frac{pc}{qm} = \frac{bp}{bq} = \frac{1}{3}$$

But

$$pc = gc - gp = \Delta G_{(ii)} - \Delta G_{(i)}$$

and

$$qm = em - eq = \Delta \bar{G}_B''^M - \Delta G_{(i)}$$

thus

$$3(\Delta G_{(ii)} - \Delta G_{(i)}) = (\Delta \bar{G}_B''^M - \Delta G_{(i)})$$

or
$$\Delta G_{(\mathrm{ii})} = \tfrac{1}{3}\,\Delta \bar{G}_{\mathrm{B}}^{\prime\prime\mathrm{M}} + \tfrac{2}{3}\,\Delta G_{(\mathrm{i})}$$

Consideration of the similar triangles *rck* and *sdk* gives

$$\frac{rc}{sd} = \frac{rk}{ks} = 2$$

But
$$rc = gc - gr = \Delta G_{(\mathrm{ii})} - \Delta \bar{G}_{\mathrm{B}}^{\prime\mathrm{M}}$$

and
$$sd = fd - fs = \Delta G_{(\mathrm{iii})} - \Delta \bar{G}_{\mathrm{B}}^{\prime\mathrm{M}}$$

such that
$$\Delta G_{(\mathrm{iii})} = \tfrac{1}{2}\,\Delta G_{(\mathrm{ii})} + \tfrac{1}{2}\,\Delta \bar{G}_{\mathrm{B}}^{\prime\mathrm{M}}$$

Thus,

$$\Delta G_{(\mathrm{i})} = \tfrac{1}{4}\,RT\ln\frac{p_{\mathrm{B}}^{\prime\prime\prime}}{p_{\mathrm{B}}^{0}}$$

$$\Delta G_{(\mathrm{ii})} = \tfrac{1}{3}\,RT\ln\frac{p_{\mathrm{B}}^{\prime\prime}}{p_{\mathrm{B}}^{0}} + \tfrac{1}{6}\,RT\ln\frac{p_{\mathrm{B}}^{\prime\prime\prime}}{p_{\mathrm{B}}^{0}} \quad \text{and}$$

$$\Delta G_{(\mathrm{iii})} = \tfrac{1}{6}\,RT\ln\frac{p_{\mathrm{B}}^{\prime\prime}}{p_{\mathrm{B}}^{0}} + \tfrac{1}{12}\,RT\ln\frac{p_{\mathrm{B}}^{\prime\prime\prime}}{p_{\mathrm{B}}^{0}} + \tfrac{1}{2}\,RT\ln\frac{p_{\mathrm{B}}^{\prime}}{p_{\mathrm{B}}^{0}}$$

and hence for

$$3\mathrm{A} + \mathrm{B} = \mathrm{A_3B} \qquad \Delta G^0_{(\mathrm{i})} = 4\Delta G_{(\mathrm{i})}$$
$$\mathrm{A} + \mathrm{B} = \mathrm{AB} \qquad \Delta G^0_{(\mathrm{ii})} = 2\Delta G_{(\mathrm{ii})} = \tfrac{2}{3}\,\Delta \bar{G}_{\mathrm{B}}^{\prime\prime\mathrm{M}} + \tfrac{4}{3}\,\Delta G_{(\mathrm{i})}$$
$$\mathrm{A} + 3\mathrm{B} = \mathrm{AB_3} \qquad \Delta G^0_{(\mathrm{iii})} = 4\Delta G_{(\mathrm{iii})} = 2\Delta G_{(\mathrm{ii})} + 2\Delta \bar{G}_{\mathrm{B}}^{\prime\mathrm{M}}$$

Figure 13.12 shows that below the temperature T_2 the compound AB_3 is unstable with respect to AB and B, with the invariant equilibrium

$$\mathrm{AB}_{(s)} + 2\mathrm{B}_{(s)} = \mathrm{AB}_{3(s)}$$

occurring at the temperature T_2. The standard free energy for the above reaction is calculated as $-\Delta G^0_{(\mathrm{ii})} + \Delta G^0_{(\mathrm{iii})}$, which equals $2RT\ln p'_{\mathrm{B}}/p^0_{\mathrm{B}}$.

Thus above T_2 $\quad \Delta G^0_{(\mathrm{iii})} - \Delta G^0_{(\mathrm{ii})} < 0 \quad$ and $\quad p'_{\mathrm{B}} < p^0_{\mathrm{B}}$

at T_2 $\quad \Delta G^0_{(\mathrm{iii})} - \Delta G^0_{(\mathrm{ii})} = 0 \quad$ and $\quad p'_{\mathrm{B}} = p^0_{\mathrm{B}}$

and below T_2 $\quad \Delta G^0_{(iii)} - \Delta G^0_{(ii)} > 0 \quad$ and $\quad p'_B > p^0_B$

The free energy–composition relationships at T_2 and T_3 are shown in Fig. 13.15*a* and *b*, respectively, which graphically illustrate that, at T_2, $\Delta G_{(iii)} = \frac{1}{2}\Delta G_{(ii)}$ and, at T_3, $\Delta G_{(iii)} < \frac{1}{2}\Delta G_{(ii)}$, such that, in the composition range B–AB, below T_2, the system occurring either as AB + AB_3 or AB_3 + B is metastable with respect to its occurrence as AB + B.

In considering the thermodynamics of a system such as shown in Fig. 13.12, two approaches are available, namely:

(i) The consideration that the compounds are ordered solid solutions

(ii) The consideration that the compounds form as the result of a chemical reaction between A and B

(i) Consider the compound AB_3 to be an ordered solid solution of A and B in the molar ratio 1/3. Then, in Fig. 13.14

$$fd = \Delta G^M = RT(X_A \ln a_A + X_B \ln a_B) = RT(0.75 \ln a_B + 0.25 \ln a_A)$$
$$= RT \ln a_B^{0.75} a_A^{0.25} \qquad \text{(i)}$$

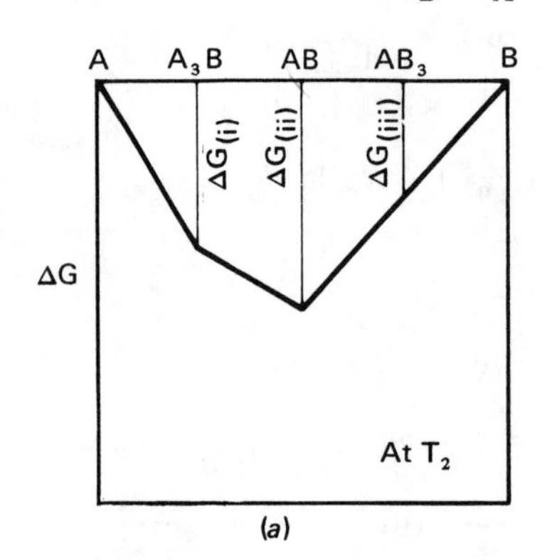

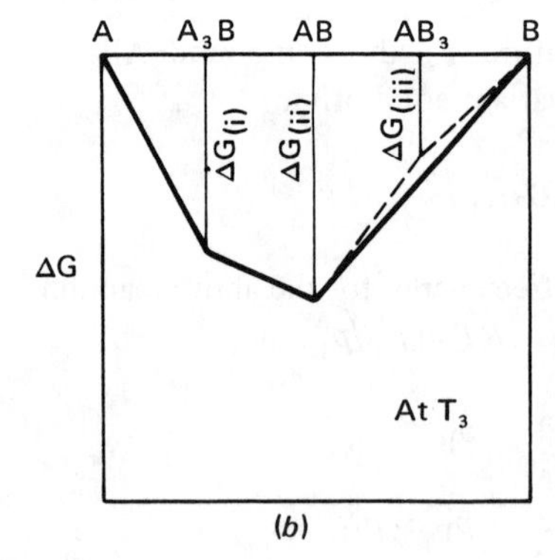

Fig. 13.15. (*a*) The integral free energies in the system shown in Fig. 13.12 at the temperature T_2. (*b*) The integral free energies in the system shown in Fig. 13.12 at the temperature T_3.

(ii) Consider the compound AB_3 to form as a result of the chemical reaction

$$A + 3B = AB_3$$

for which the standard free energy change is $\Delta G^0_{(iii)}$. Then

$$fd = \tfrac{1}{4}\,\Delta G^0_{(iii)} = -RT\ln\left(\frac{a_{AB_3}}{a_A a_B^3}\right)^{0.25}$$

$$= RT\ln\left(\frac{a_A^{0.25}a_B^{0.75}}{a_{AB_3}^{0.25}}\right) \qquad \text{(ii)}$$

As AB_3 is a "line compound," i.e., has a negligible range of nonstoichiometry, it is of fixed composition and hence exists in a fixed state. If this fixed state is taken as being the standard state in which $a_{AB_3} = 1$, then Eq. (ii) becomes

$$fd = \tfrac{1}{4}\,\Delta G^0_{(iii)} = RT\ln a_A^{0.25}a_B^{0.75}$$

which is identical with Eq. (i). In both Eqs. (i) and (ii) the standard states of A and B are the pure elements at the temperature T.

The variations of the activities of A and B in the compound AB_3 are limited by the separation of B and AB; e.g., when AB_3 is in equilibrium with B, $a_B = 1$ and hence $a_A = \exp(4\Delta G_{(iii)}/RT) = \exp(\Delta G^0_{(iii)}/RT)$ ($RT \ln a_A = ai$ in Fig. 13.14). If the activity of B in AB_3 is decreased below unity, then AB_3 is no longer saturated with B, and a_A increases in accordance with Eq. (i). The minimum activity of B in AB_3 is determined by saturation of AB_3 with A, at which point the compound AB appears. The minimum activity which B may have is obtained, from Fig. 13.14, as $RT \ln a_B = ek$, and the corresponding maximum activity of A is obtained from $RT \ln a_A = aj$. Nonsaturation of AB_3 with either A or B occurs when the pressure of B above the compound lies between the limits p'_B and p^0_B.

Similar considerations can be made with respect to the compounds AB and A_3B; e.g., in Fig. 13.14

$$gc = \Delta G^M = \tfrac{1}{2}\Delta G^0_{(ii)} = RT\ln a_A^{0.5}a_B^{0.5}$$

and

$$hb = \Delta G^M = \tfrac{1}{4}\,\Delta G^0_{(i)} = RT\ln a_A^{0.75}a_B^{0.25}$$

In the case where A and B exhibit a measurable solubility in one another and the compounds A_2B and AB_2 (labeled as phases β and γ, respectively) have measurable nonstoichiometry ranges, the phase diagram is as shown in Fig. 13.16*a*. Again if B is appreciably volatile and A is not, then the pressure-composition diagram at the temperature T_1 is as shown in Fig. 13.16*b*, and the corresponding free energy-composition diagram is as shown in Fig. 13.16*c*.

Example

Calculate the phase diagram for the binary uranium–carbon system in the range of temperature 1000–2500 K, given

$$U_{(s)} + C_{(s)} = UC_{(s)} \qquad \Delta G^0_{UC} = -105{,}000 + 6.28T \text{ joules}$$

$$2U_{(s)} + 3C_{(s)} = U_2C_{3(s)} \qquad \Delta G^0_{U_2C_3} = -236{,}800 + 25.1T \text{ joules}$$

$$U_{(s)} + 2C_{(s)} = UC_{2(s)} \qquad \Delta G^0_{UC_2} = -115{,}900 + 10.9T \text{ joules}$$

For the purpose of this calculation it will be assumed that uranium and carbon are insoluble in one another and that the carbides UC, U_2C_3, and UC_2 are stoichiometric compounds.

From the phase rule, at constant total pressure in a two-component system, a maximum of three condensed phases can be in equilibrium with one another, and as five condensed phases exist (U, UC, U_2C_3, UC_2, and C), there are $5 \times 4 \times 3/3! = 10$ possible three-phase equilibria and $5 \times 4/2 = 10$ possible two-phase equilibria. The simplest approach to determining the stable equilibria is to draw and examine the ΔG^M lines for a series of temperatures. From the above free energy equations,

$\Delta G^M_{U_{0.5}C_{0.5}}$ (the free energy of formation of one mole of the binary system of composition $X_C = 0.5) = \frac{1}{2}\,\Delta G^0_{UC} = -52{,}500 + 3.14T$ joules (a)

$\Delta G^M_{U_{0.4}C_{0.6}}$ (the free energy of formation of one mole of the binary system of composition $X_C = 0.6) = \frac{1}{5}\,\Delta G^0_{U_2C_3} = -47{,}360 + 5.02T$ (b)

and

$\Delta G^M_{U_{0.33}C_{0.67}}$ (the free energy of formation of one mole of the binary system of composition $X_C = \frac{2}{3}) = \frac{1}{3}\,\Delta G^0_{UC_2} = -38{,}633 + 3.63T$ (c)

Thus the values of ΔG^M at chosen temperatures are

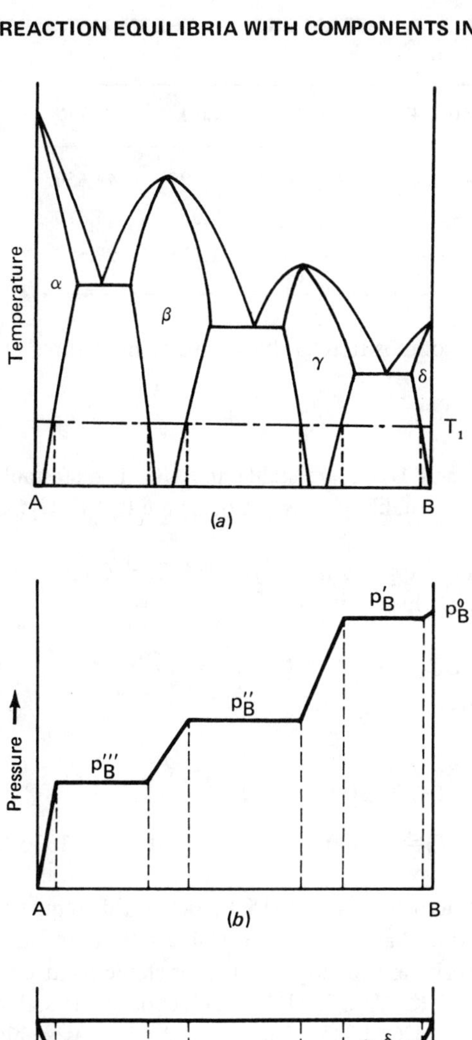

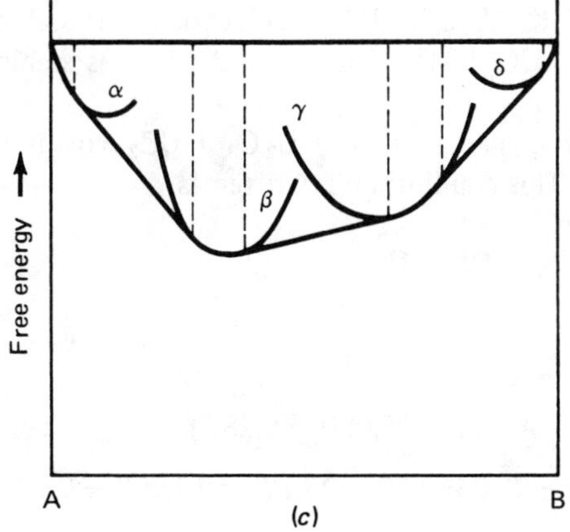

Fig. 13.16. (*a*) The phase diagram for a system A–B. (*b*) The vapor pressures of B at the temperature T_1. (*c*) The integral free energies at the temperature T_1.

	500 K	1000 K	1500 K	2000 K	2500 K
$\Delta G^M_{U_{0.5}C_{0.5}}$ (joules)	−50,930	−49,360	−47,790	−46,220	−44,650
$\Delta G^M_{U_{0.4}C_{0.6}}$ (joules)	−44,850	−42,340	−39,830	−37,320	−34,810
$\Delta G^M_{U_{0.33}C_{0.67}}$ (joules)	−36,818	−35,003	−33,188	−31,373	−29,558

and the variations of ΔG^M with composition at these five temperatures are shown in Fig. 13.17.

Examination of Fig. 13.17 shows that

(i) At 500 K and 1000 K, UC and U_2C_3 are stable and UC_2 is metastable relative to $U_2C_3 + C$. Thus the stable phases occurring are U, UC, U_2C_3, and C.

(ii) At 1500 K, UC is stable and it appears that $U_2C_3 + UC_2 + C$ exist in a stable three-phase equilibrium. This equilibrium is written as

$$2UC_2 = U_2C_3 + C$$

for which

$$\Delta G^0 = \Delta G^0_{U_2C_3} - 2\,\Delta G^0_{UC_2} = (-236{,}800 + 25.1T) - 2(-115{,}900 + 10.9T) = -5000 + 3.3T = 0 \quad \text{at} \quad T = 1515 \text{ K}$$

Thus the three-phase equilibrium occurs at 1515 K. Below this temperature ΔG^0 is negative, indicating that $U_2C_3 + C$ is stable relative to UC_2, and above 1515 K ΔG^0 is positive, showing that UC_2 is stable relative to $U_2C_3 + C$. Thus, below 1515 K, U, UC, U_2C_3, and C occur as stable phases and, above 1515 K, U, UC, U_2C_3, UC_2, and C occur as stable phases.

(iii) At 2000 K, UC is stable and it appears that $UC + U_2C_3 + UC_2$ exist in a stable three-phase equilibrium. This equilibrium is written as

$$U_2C_3 = UC + UC_2$$

for which

$$\Delta G^0 = \Delta G^0_{UC} + \Delta G^0_{UC_2} - \Delta G^0_{U_2C_3} = (-105{,}000 + 6.28T) + (-115{,}900 + 10.9T) - (-236{,}800 + 25.1T) = 15{,}900 - 7.92T$$

$$= 0 \quad \text{at} \quad T = 2007 \text{ K}$$

Thus the three-phase equilibrium occurs at 2007 K. Below this temperature ΔG^0 is positive, showing that U_2C_3 is stable relative to $UC + UC_2$ and, above 2007 K, ΔG^0 is negative, indicating that $UC + UC_2$ is stable relative to U_2C_3. Hence, below 2007 and above 1515 K, the stable phases are U, UC, U_2C_3, UC_2, and C, and above 2007 K the stable phase are U, UC, UC_2, and C.

(iv) At 2500 K, UC is stable and both U_2C_3 and UC_2 are metastable relative

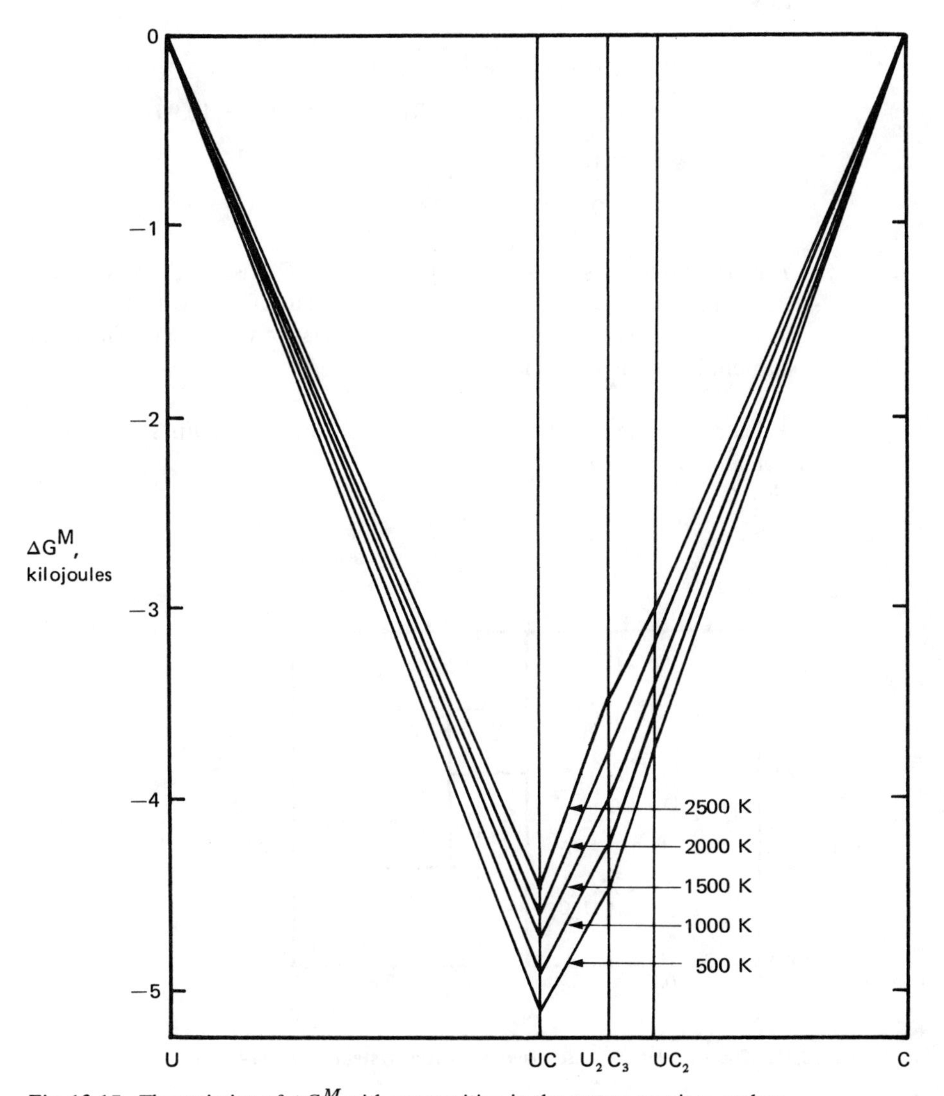

Fig. 13.17. The variation of ΔG^M with composition in the system uranium–carbon.

to UC + C. In view of the fact that it has been found that U, UC, UC_2, and C are stable above 2007 K, and the three-phase equilibrium UC + UC_2 + C must exist between 2007 and 2500 K. The pertinent equilibrium is

$$UC_2 = UC + C$$

for which

$$\begin{aligned} \Delta G^0 &= \Delta G^0_{UC} - \Delta G^0_{UC_2} \\ &= (-105{,}000 + 6.28T) - (-115{,}900 + 10.9T) \\ &= 10{,}900 - 4.62T \\ &= 0 \quad \text{at} \quad T = 2359\text{ K} \end{aligned}$$

Thus the three-phase equilibrium UC_2 + UC + C exists at 2359 K. Above this temperature, ΔG^0 is negative, showing that UC + C is stable relative to UC_2, and below 2359 K, UC_2 is stable relative to UC + C. The phase diagram is thus as shown in Fig. 13.18.

Now calculate the variations, with composition, of the activities of uranium and carbon at 1750 K.

From Eq. (a),

$$\Delta G^M_{U_{0.5}C_{0.5},1750\text{ K}} = -47{,}000 \text{ joules} \tag{d}$$

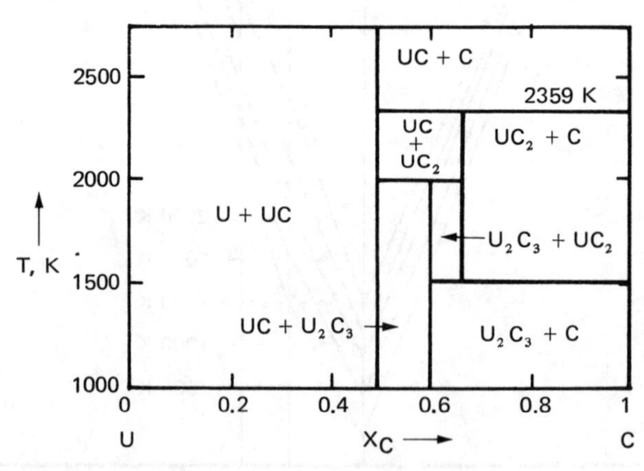

Fig. 13.18. The calculated phase diagram for the system uranium–carbon.

From Eq. (b), $$\Delta G^M_{U_{0.4}C_{0.6}, 1750 K} = -38{,}580 \text{ joules} \tag{e}$$

and from Eq. (c), $$\Delta G^M_{U_{0.33}C_{0.67}, 1750 K} = -32{,}280 \text{ joules} \tag{f}$$

Between the compositions U and UC, i.e., with $0 < X_C \leq 0.5$, the equation of the linear variation of ΔG^M with composition across the two-phase U + UC region is obtained from

$$\Delta G^M = 0 \quad \text{at} \quad X_C = 0$$

and $$\Delta G^M = -47{,}000 \text{ joules} \quad \text{at} \quad X_C = 0.5$$

as $$\Delta G^M = -94{,}000\, X_C \text{ joules}$$

This straight line intersects the $X_C = 0$ axis at $\Delta G^M = \Delta \bar{G}^M_U = 0$, and the $X_C = 1$ axis at $\Delta G^M = \Delta \bar{G}^M_C = -94{,}000$ joules. Thus, in equilibrated U + UC at 1750 K,

$$RT \ln a_U = \Delta \bar{G}^M_U = 0 \quad \text{or} \quad a_U = 1$$

and

$$RT \ln a_C = \Delta \bar{G}^M_C = -94{,}000 \quad \text{or} \quad a_C = 1.56 \times 10^{-3} \; (\log a_C = -2.81)$$

As a check, with $X_C = X_U = 0.5$,

$$\begin{aligned} \Delta G^M &= RT\,(X_U \ln a_U + X_C \ln a_C) \\ &= 8.3144 \times 1750(0.5 \ln 1 + 0.5 \ln 1.56 \times 10^{-3}) \\ &= -47{,}000 \text{ joules} \end{aligned}$$

in agreement with Eq. (d).

Between the compositions UC and U_2C_3, the equation of the linear variation of ΔG^M with composition is obtained from

$$\Delta G^M = -47{,}000 \text{ joules} \quad \text{at} \quad X_C = 0.5$$

and $$\Delta G^M = -38{,}580 \text{ joules} \quad \text{at} \quad X_C = 0.6$$

as $$\Delta G^M = 84{,}200 X_C - 89{,}100 \text{ joules}$$

This line intersects the $X_C = 0$ axis at $\Delta G^M = \Delta \bar{G}^M_U = -89{,}100$ joules, and the

$X_C = 1$ axis at $\Delta G^M = \Delta \bar{G}_C^M = -4900$ joules. Thus, in equilibrated $UC + U_2C_3$ at 1750 K,

$$\Delta \bar{G}_U^M = RT \ln a_U = -89{,}100 \text{ joules}$$

or

$$a_U = 2.19 \times 10^{-3} \quad (\log a_U = -2.66)$$

$$\Delta \bar{G}_C^M = RT \ln a_C = -4900 \text{ joules}$$

or

$$a_C = 0.714 \quad (\log a_C = -0.15)$$

As a check, with $X_C = 0.5$,

$$\Delta G^M = 8.3144 \times 1750 \,(0.5 \ln 2.19 \times 10^{-3} + 0.5 \ln 0.714)$$

$$= -47{,}000 \text{ joules, in agreement with Eq. (d)}$$

and, with $X_C = 0.6$,

$$\Delta G^M = 8.3144 \times 1750 \,(0.4 \ln 2.19 \times 10^{-3} + 0.6 \ln 0.714)$$

$$= -38{,}580 \text{ joules, in agreement with Eq. (e)}$$

Between the compositions U_2C_3 and UC_2, the equation of the variation of ΔG^M with composition is obtained from

$$\Delta G^M = -38{,}580 \text{ joules} \quad \text{at} \quad X_C = 0.6$$

and

$$\Delta G^M = -32{,}280 \text{ joules} \quad \text{at} \quad X_C = \tfrac{2}{3}$$

as

$$\Delta G^M = 94{,}500 X_C - 95{,}280 \text{ joules}$$

This line intersects the $X_C = 0$ axis at $-95{,}280$ joules, and the $X_C = 1$ axis at -780 joules. Thus,

$$\Delta \bar{G}_U^M = RT \ln a_U = -95{,}280 \text{ joules}$$

or

$$a_U = 1.43 \times 10^{-3} \quad (\log a_U = -2.84)$$

and

$$\Delta \bar{G}_C^M = RT \ln a_C = -780 \text{ joules}$$

or

$$a_C = 0.948 \quad (\log a_C = -0.023)$$

Thus, in U_2C_3, the activity of uranium can vary from 0.00219 to 0.00143 and the activity of carbon can vary, correspondingly, from 0.714 to 0.948, as the state of U_2C_3 is varied from that of being in equilibrium with UC to that of being in equilibrium with UC_2.

Between the compositions UC_2 and C, the variation of ΔG^M with composition is

$$\Delta G^M = 96{,}840 X_C - 96{,}840 \text{ joules}$$

Thus, in equilibrated UC_2 + C at 1750 K,

$$\Delta \bar{G}_U^M = RT \ln a_U = -96{,}840 \text{ joules}$$

or

$$a_U = 1.29 \times 10^{-3} \quad (\log a_U = -2.89)$$

and

$$\Delta \bar{G}_C^M = RT \ln a_C = 0 \quad \text{or} \quad a_C = 1$$

In UC_2 the activity of uranium varies from 0.00143 to 0.00129 and the activity of carbon varies from 0.948 to 1 as the state of the UC_2 varies from that in equilibrium with U_2C_3 to saturation with carbon. The variations of log a_U and log a_C with composition are shown in Fig. 13.19.

13.6 THE SOLUBILITY OF GASES IN METALS

If it is considered that the diatomic gas X_2 dissolves in the metal M to form molecules in solution, the appropriate equilibrium would be written as

$$X_{2(g)} = [X_2] \text{ (in metal)} \tag{a}$$

On the other hand, if it is considered that the gas dissolves to form atoms in solution, the equilibrium would be written as

$$\tfrac{1}{2} X_{2(g)} = [X] \text{ (in metal)} \tag{b}$$

where, in both cases, the square brackets denote solution in the metal phase. For equilibrium (*a*),

$$K_{H(a)} = \frac{[h_{X_2}]}{p_{X_2}}$$

and for equilibrium (*b*),

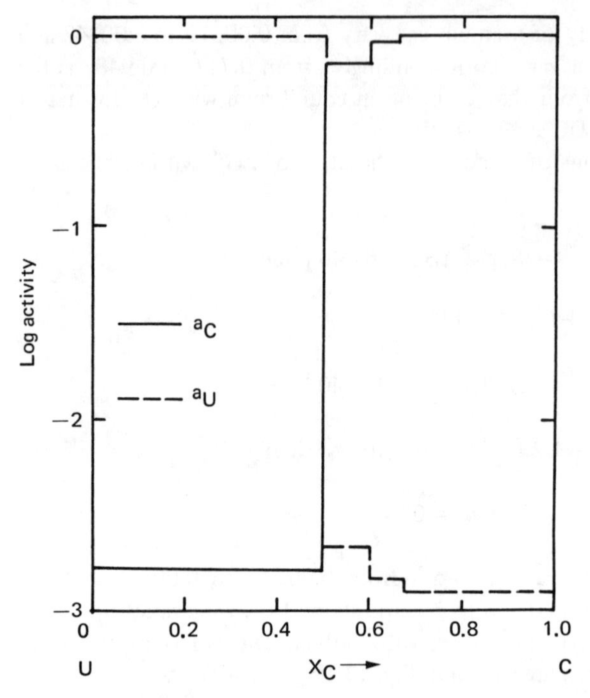

Fig. 13.19. The activities of uranium and carbon in the system U–C at 1750 K.

$$K_{H(b)} = \frac{[h_X]}{p_{X_2}^{\frac{1}{2}}} \tag{13.15}$$

If the dissolved gas is sufficiently dilute that it obeys Henry's law, then

$$K_{1\ \mathrm{wt\%}(a)} = \frac{[\mathrm{wt\%}\ X_2]}{p_{X_2}}$$

and

$$K_{1\ \mathrm{wt\%}(b)} = \frac{[\mathrm{wt\%}\ X]}{p_{X_2}^{\frac{1}{2}}} \tag{13.16}$$

Whether the gas dissolves atomically or molecularly can be ascertained experimentally by determining whether $K_{(a)}$ or $K_{(b)}$ is constant. It is invariably found that $K_{(b)}$ is constant, indicating thus that atomic solution occurs and that (*b*) is the correct expression for the equilibrium. Equations (13.15) and (13.16) are expressions of Sievert's law, and if the solute gas obeys Henry's law, then $K_{1\ \mathrm{wt}\%}$ is obtained as the solubility of the gas in weight percent under a pressure of 1 atm at the temperature of interest. For the solubility of N_2 in liquid iron at 1606°C, Pehlke and Elliot* obtained the variation of wt% N_2 with $p_{N_2}^{1/2}$ shown in Fig. 13.20 from which it is seen that for

$$\tfrac{1}{2}N_{2(g)} = [N]_{(1\ \mathrm{wt}\%\ \mathrm{in\ Fe})}$$

$K_{1879\ (\mathrm{wt}\%)} = 0.045$. Variation of the solubility of the gas under a pressure of

*R. Pehlke and J. F. Elliott, Solubility of Nitrogen in Liquid Iron Alloys, I: Thermodynamics, *Trans. Met. Soc. AIME,* **218:**1088 (1960).

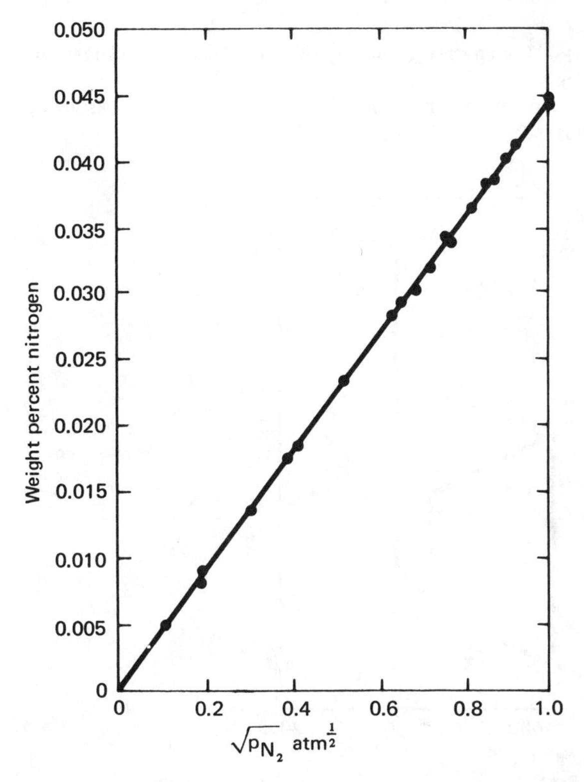

Fig. 13.20. The solubility of nitrogen in liquid iron as a function of $p_{N_2}^{1/2}$ at 1606°C. *(From R. Pehlke and J. F. Elliott, Solubility of Nitrogen in Liquid Iron Alloys: I, Thermodynamics, Trans. Met. Soc. AIME, 218:1088[1960].)*

1 atm, with temperature, describes the variation of $K_{(\mathrm{wt}\%)}$ with temperature. For nitrogen in iron this is shown in Fig. 13.21.*

The solubility of sulfur in liquid iron has been studied by Sherman, Elvander, and Chipman,† who determined the equilibrium sulfur content of liquid iron under a gas of known p_{H_2S}/p_{H_2} ratio; i.e., they studied the equilibrium

$$H_{2(g)} + [S]_{(1\ \mathrm{wt}\%)} = H_2S_{(g)}$$

for which

$$K = \frac{p_{H_2S}}{p_{H_2}} \frac{1}{[h_{S\,(\mathrm{wt}\%)}]} = \frac{p_{H_2S}}{p_{H_2}} \frac{1}{f_{S(\mathrm{wt}\%)}[\mathrm{wt}\%\ S]}$$

This can be written as

*L. S. Darken and R. W. Gurry, "The Physical Chemistry of Metals," p. 372, McGraw-Hill Book Company, New York, 1953.

†C. W. Sherman, H. I. Elvander, and J. Chipman, The Thermodynamic Properties of Sulfur in Molten Iron-Sulfur Alloys, *Trans. AIME*, **188**:334 (1950).

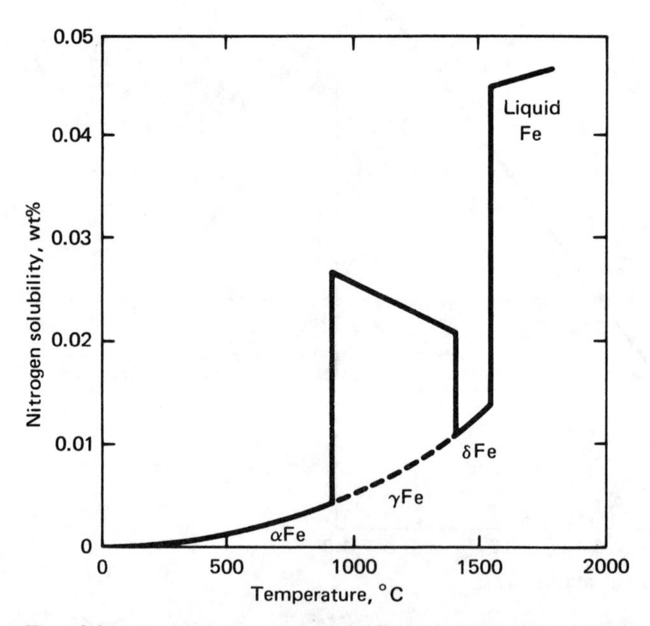

Fig. 13.21. The variation of the solubility of nitrogen in iron under a pressure of 1 atm of nitrogen with temperature. *(From L. S. Darken and R. W. Gurry, "Physical Chemistry of Metals," McGraw-Hill Book Company, New York, 1953.)*

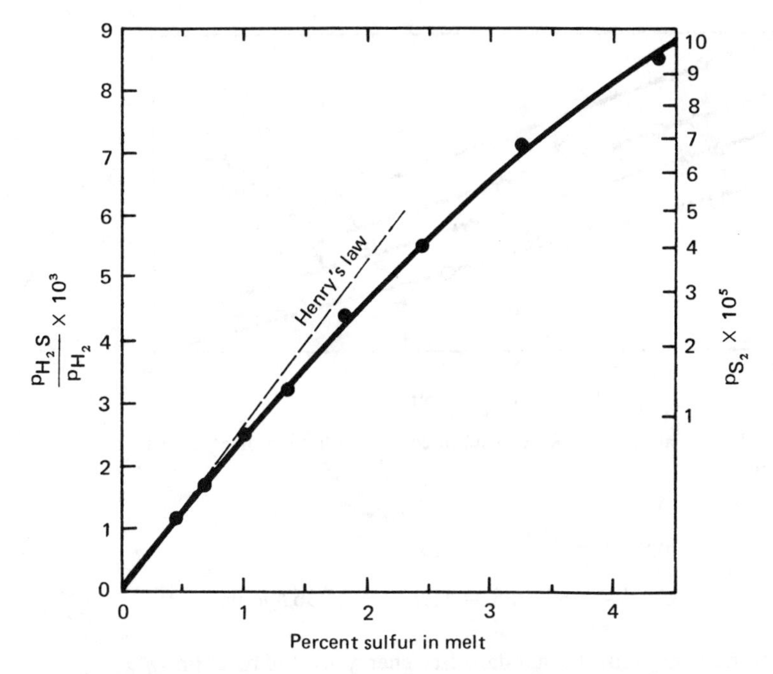

Fig. 13.22. The solubility of sulfur in liquid iron as a function of sulfur pressure. *(From C. W. Sherman, H. I. Elvander, and J. Chipman, Thermodynamic Properties of Sulfur in Molten Iron-Sulfur Alloys, Trans. AIME, 188:334 [1950].)*

$$K' = \frac{p_{H_2S}}{p_{H_2}} \frac{1}{[\text{wt\% S}]}$$

where K' is the "apparent equilibrium constant" which is given as $Kf_{S\,(\text{wt\%})}$.

The nonlinear variation of dissolved weight percent sulfur with p_{H_2S}/p_{H_2}, shown in Fig. 13.22, indicates that K' is not a true constant independent of composition; i.e., $f_{S\,(\text{wt\%})}$ is not unity and, at any value of [wt% S], the value of K' is obtained as the tangent to the curve in Fig. 13.22. From the definition of the 1 wt% standard state,

$$f_{S\,(\text{wt\%})} \to 1 \text{ as wt\% S} \to 0$$

the slope of the experimental curve in Fig. 13.22 increases toward the value K as [wt% S] $\to 0$. The variation of $\log K'$ with [wt% S], at four temperatures, is shown in Fig. 13.23, and extrapolation of these lines to [wt% S] $= 0$ gives the value of K for the particular temperature. The variation of $\log K$ with temperature then gives the required thermodynamics of the system; i.e.,

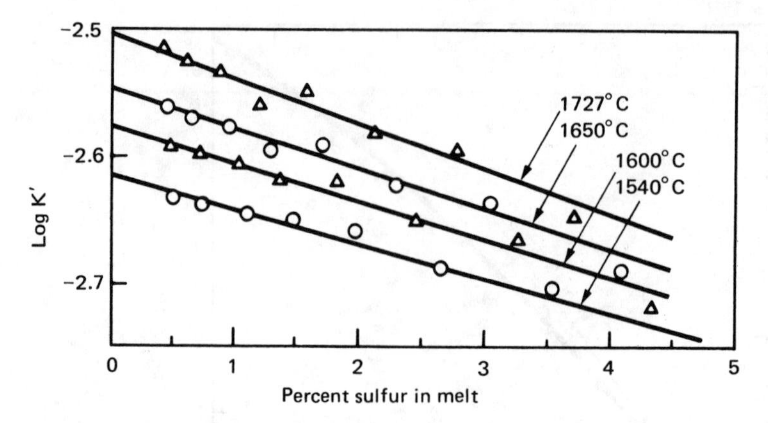

Fig. 13.23. The variation of log K' with sulfur content of liquid iron at several temperatures.

$$\log K = -\frac{2150}{T} - 1.429$$

or

$$\Delta G^0 = -RT \ln K = 41{,}170 + 27.36T \text{ joules}$$

Combination of this with the standard free energy for the reaction

$$H_2S_{(g)} = H_{2(g)} + \tfrac{1}{2}S_{2(g)}$$

namely

$$\Delta G^0 = 90{,}290 - 49.39T \text{ joules}$$

gives

$$\Delta G^0 = -131{,}460 + 22.03T \text{ joules}$$

for the change of state

$$\tfrac{1}{2}S_{2(g)} = [S]_{1\text{ wt\% in Fe}}$$

i.e., for the isothermal transfer of $\frac{1}{2}$ mole of sulfur gas at 1 atm to a 1 wt% solution of sulfur in iron.

Using a similar experimental technique, Floridis and Chipman* studied the thermodynamics of oxygen solution in liquid iron by measuring the equilibrium

$$H_{2(g)} + [O]_{(1\text{ wt\% in Fe})} = H_2O_{(g)}$$

The variations of [wt% O] with p_{H_2O}/p_{H_2} in the gas phase at 1500° and 1600°C

*T. P. Floridis and J. Chipman, Activity of Oxygen in Liquid Iron Alloys, *Trans. AIME*, 212:549 (1958).

are shown in Fig. 13.24. The curvature of the lines, which is similar to that in Fig. 13.23, indicates that Henry's law is not obeyed, and, again, K for the equilibrium is obtained as the limiting slope of the line as [wt% O] → 0. The variation of K with T in the range 1536°-1700°C gives

$$\Delta G^0 = -134{,}700 + 61.21T \text{ joules}$$

which, together with

$$\Delta G^0 = -246{,}000 + 54.8T \text{ joules}$$

for

$$H_{2(g)} + \tfrac{1}{2}O_{2(g)} = H_2O_{(g)}$$

gives

$$\Delta G^0 = -111{,}300 - 6.41T \text{ joules}$$

for the change of state

$$\tfrac{1}{2}O_{2(g)} = [O]_{(1\text{ wt\%})}$$

The deviations of the full lines from Henrian behavior (indicated by the broken lines) in Fig. 13.24 gives

$$\log f_{O(1\text{ wt\%})} = -0.20\ [\text{wt\% O}]$$

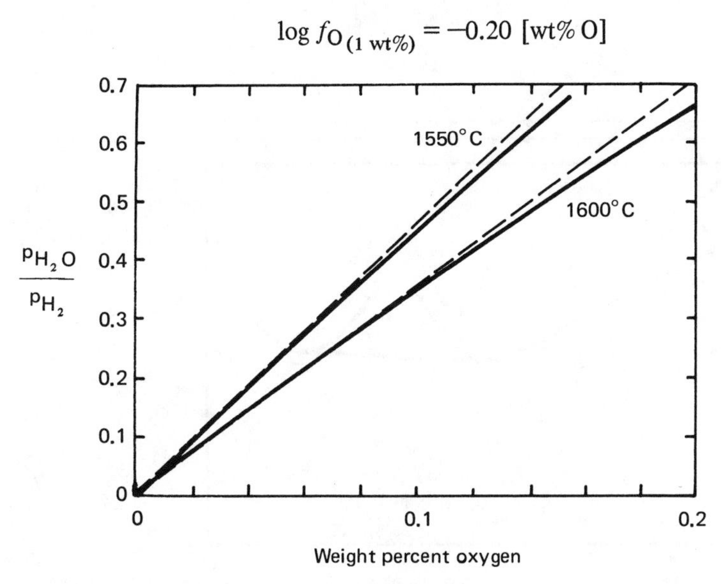

Fig. 13.24. The solubility of oxygen in liquid iron as a function of p_{H_2O}/p_{H_2} in the gas phase at 1550° and 1600°C (Henrian behavior is shown as the broken lines).

13.7 THE FORMATION OF OXIDE PHASES OF VARIABLE COMPOSITION

Figure 13.25 shows the free energy-composition relationship at some temperature T for the metal M-oxygen system in which measurable solubility of oxygen in M occurs and where the oxides "MO" and "M_3O_4" are of variable composition. Starting with pure M and increasing the pressure of oxygen above the system moves the molar free energy along the curve *fi* until, at $p_{O_2} = p_{O_2(M/MO)}$, saturation of the metal occurs and the metal-saturated "MO" phase of composition M_bO_a appears. If pure metal M and oxygen gas at 1 atm pressure are chosen as the standard states, then

$$\Delta G^M_{(\text{metal-saturated "MO"})} = jk = RT(b \ln a_M + a \ln p_{O_2}^{\frac{1}{2}})$$

$$= RT \ln a_M^b p_{O_2}^{\frac{1}{2}a}$$

or for the reaction

$$bM_{(s)} + \tfrac{1}{2}aO_{2\,(g)} = M_bO_{a(s)}$$

$$\Delta G^0 = jk = RT \ln \left(\frac{a_M^b p_{O_2}^{\frac{1}{2}a}}{a_{M_bO_a}} \right)$$

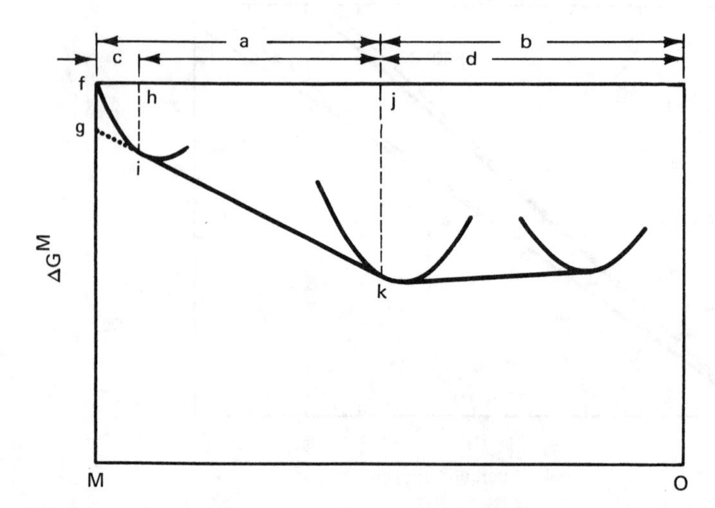

Fig. 13.25. The integral free energies of the system M–O which forms oxide phases of variable composition and which shows a significant solubility of oxygen in metallic M.

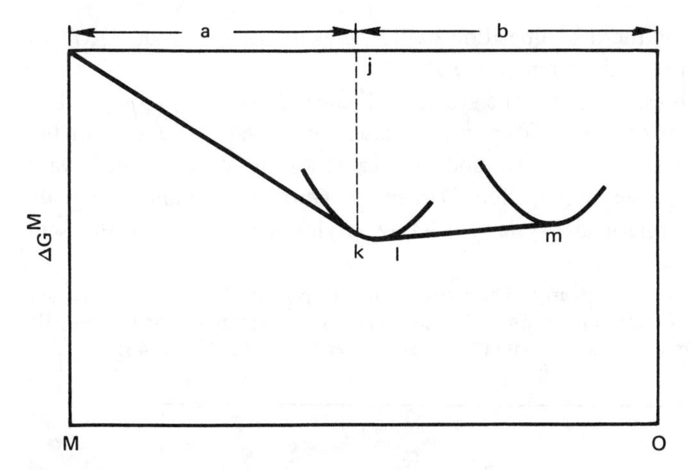

Fig. 13.26. The integral free energies of the system M–O which forms oxide phases of variable composition and which shows a negligible solubility of oxygen in metallic M.

which, if the metal-saturated oxide composition is chosen as the standard state, is identical with the above expression.

It is convenient to write the oxidation reaction such that it involves the consumption of an integral number of gram-atoms of oxygen. For example, for consumption of 1 gram-atom

$$yM + \tfrac{1}{2}O_2 = M_yO \qquad \text{where} \qquad y = \frac{b}{a}$$

$$\Delta G^0 = RT \ln \left(\frac{a_M^y p_{O_2}^{\frac{1}{2}}}{a_{M_yO}} \right) = \frac{jk}{a}$$

or for the consumption of 1 gram mole of oxygen

$$\Delta G^0 = RT \ln \left(\frac{a_M^{2y} p_{O_2}}{a_{M_yO}^2} \right)$$

If the solubility of oxygen in the metal M is virtually zero, then *fg* in Fig. 13.25 shrinks to a point, and Fig. 13.25 is redrawn as in Fig. 13.26. In this case choosing pure M, oxygen gas at 1 atm pressure, and the oxide of composition M_yO as the standard states gives, for the oxidation,

$$yM + \tfrac{1}{2}O_2 = M_yO$$

$$\Delta G^0 = RT \ln p_{O_2(M/MO)}^{1/2}$$

where $p_{O_2(M/MO)}$ is the decomposition oxygen pressure of the oxide "MO" to produce the metal M at the temperature T.

If the oxygen pressure above the system is increased above $p_{O_2(M/MO)}$, then the oxygen content of the "MO" phase increases, the molar free energy of the system moves along the line kl, and the activities of M and "MO" vary accordingly. In a classic investigation, Darken and Gurry* determined the phase relationships occurring in the system Fe-O by varying the oxygen pressure and

*L. S. Darken and R. W. Gurry: The System Iron-Oxygen, I: The Wustite Field and Related Equilibria, *J. Am. Chem. Soc.*, **67**:1398 (1945); The System Iron-Oxygen, II: Equilibria and Thermodynamics of Liquid Oxide and Other Phases, **68**:798 (1946).

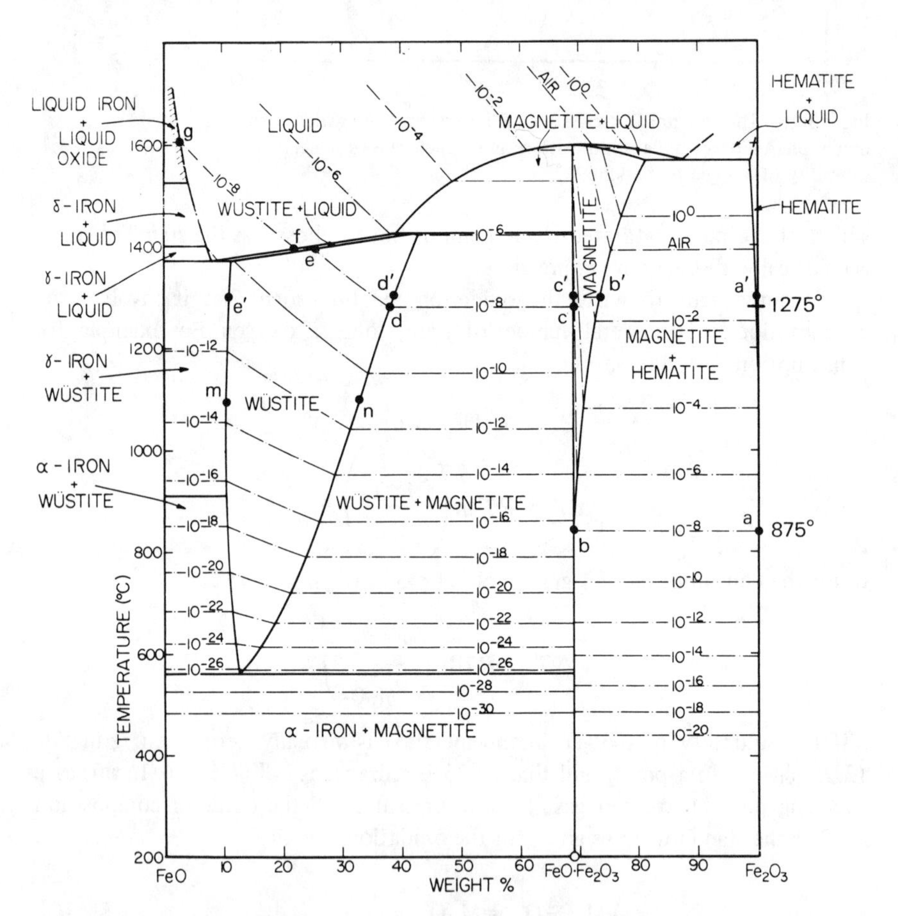

Fig. 13.27. The phase diagram of the system FeO-Fe_2O_3 showing the positions of the oxygen isobars. *(From A. Muan and E. F. Osborn, "Phase Equilibria among Oxides in Steelmaking," Addison-Wesley Publishing Co., Reading, Mass., 1965.)*

temperature and observing the consequential composition and phase changes. Their phase diagram, drawn for the components FeO and Fe_2O_3, is shown in Fig. 13.27. Consider the wustite ("FeO") phase field which, at 1100°C, extends from the composition m to the composition n. The variation of a_{Fe} in the wustite phase can be calculated from the experimentally determined variation of wustite composition with oxygen pressure via the Gibbs-Duhem equation

$$X_{Fe} d \ln a_{Fe} + X_O d \ln a_O = 0$$

i.e.,

$$\log a_{Fe} = -\int \frac{X_O}{X_{Fe}} d \log a_O = -\int \frac{X_O}{X_{Fe}} d \ln p_{O_2}^{\frac{1}{2}}$$

where the upper limit of integration is the oxygen pressure in equilibrium with the wustite composition of interest, and the lower limit is $p_{O_2(\text{Fe/"FeO"})}$ (the oxygen pressure at which wustite of composition m is in equilibrium with oxygen-saturated metallic iron) at which composition (as the solubility of oxygen in solid Fe is negligible) a_{Fe} is unity. Having thus determined the a_{Fe}-composition relationship, the $a_{\text{"FeO"}}$-composition relationship is determined as follows. If the standard state of oxygen is selected as being $p_{O_2(\text{Fe/"FeO"})}$ at the temperature of interest, then for

$$y\text{Fe}_{(s)} + \tfrac{1}{2}\text{O}_{2(g,\text{ at } p_{O_2(\text{Fe/"FeO"})})} = \text{Fe}_y\text{O}_{(s)}$$

as the standard states are in equilibrium with each other, $\Delta G^0 = 0$, and hence $K = 1$; i.e.,

$$a_{\text{"FeO"}} = a_{Fe} a_O \qquad \text{where} \qquad a_O = \left[\frac{p_{O_2}}{p_{O_2(\text{Fe/"FeO"})}}\right]^{1/2}$$

or

$$\log a_{\text{"FeO"}} = \log a_{Fe} + \log a_O$$

$$= -\int \frac{X_O}{X_{Fe}} d \log a_O + \int d \log a_O$$

$$= -\int \left(\frac{X_O}{X_{Fe}} - 1\right) d \log p_{O_2}^{\frac{1}{2}}$$

where the integration limits are the same as before.

The variations of a_{Fe}, $a_{\text{"FeO"}}$, and a_O across the wustite phase field at 1100, 1200, and 1300°C are shown in Fig. 13.28, and from these variations the free energy curve for wustite, kl in Fig. 13.26, can be determined.

For a fixed composition, the partial molar heats of solution of metal and oxygen in the wustite can be obtained from the Gibbs-Helmholtz relationship as

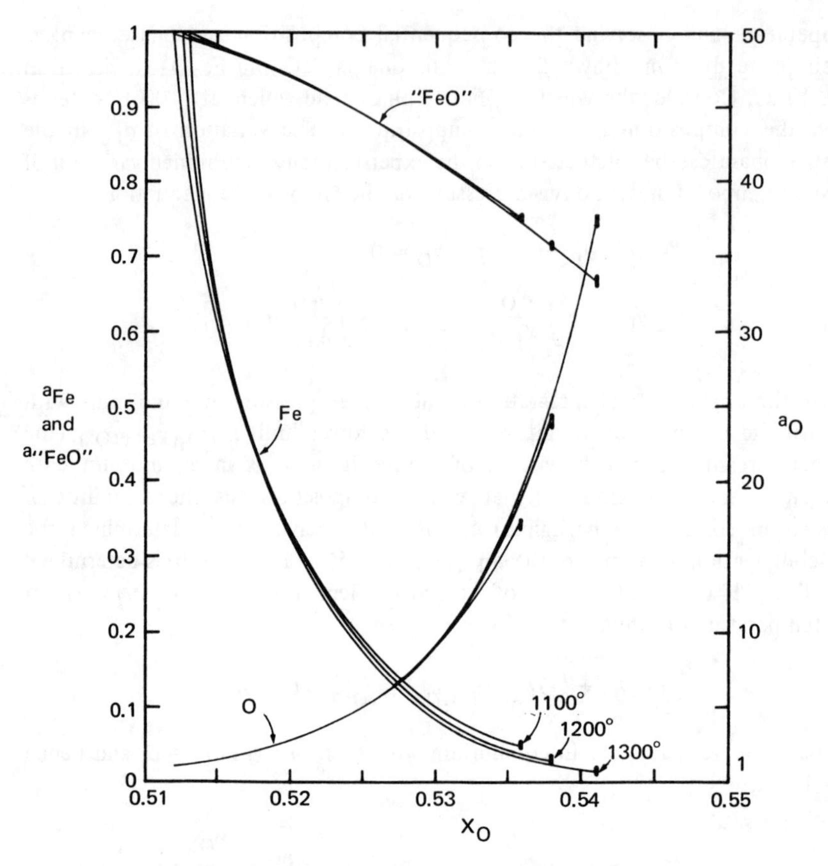

Fig. 13.28. The activities of iron, oxygen, and iron-saturated wustite in the wustite phase field at several temperatures.

$$\Delta \bar{H}_{\mathrm{O}}^{M} = \frac{R \partial \ln p_{\mathrm{O}_2}^{1/2}}{\partial (1/T)}$$

and

$$\Delta \bar{H}_{\mathrm{Fe}}^{M} = \frac{R \partial \ln a_{\mathrm{Fe}}}{\partial (1/T)}$$

Figure 13.29 shows the variations of $\Delta \bar{H}_{\mathrm{O}}^{M}$ and $\Delta \bar{H}_{\mathrm{Fe}}^{M}$ with composition.

At the temperature T, the limit of increase of p_{O_2} above homogeneous stable wustite is $p_{\mathrm{O}_2(\text{"FeO"}/\mathrm{Fe}_3\mathrm{O}_4)}$, the oxygen pressure at which wustite of composition l is in equilibrium with magnetite of composition m (in Fig. 13.26), and if these compositions are chosen as the standard states, then for

$$3\ \text{"FeO"}_{(\text{saturated with oxygen})} + \tfrac{1}{2}O_{2(g)} = Fe_3O_{4(\text{saturated with Fe})}$$

$$\Delta G^0 = RT \ln p^{1/2}_{O_2}(\text{"FeO"}/Fe_3O_4)$$

Figure 13.27 shows that the composition of wustite in equilibrium with magnetite varies significantly with temperature. Thus the heat of formation of magnetite from wustite cannot be calculated by applying the Gibbs-Helmholtz relationship to the variation of $p_{O_2}(\text{"FeO"}/Fe_3O_4)$ with temperature (the Gibbs-Helmholtz partial differential is for constant total pressure and constant composition). However, as the composition of magnetite in equilibrium with wustite is independent of temperature, the heat of the reaction

$$3\ Fe_{(s)} + 2\ O_{2\,(g)} = Fe_3O_{4(\text{ saturated with Fe})}$$

can be calculated using the Gibbs-Helmholtz relationship; i.e.,

$$K = \frac{1}{a^3_{Fe}p^2_{O_2}}$$

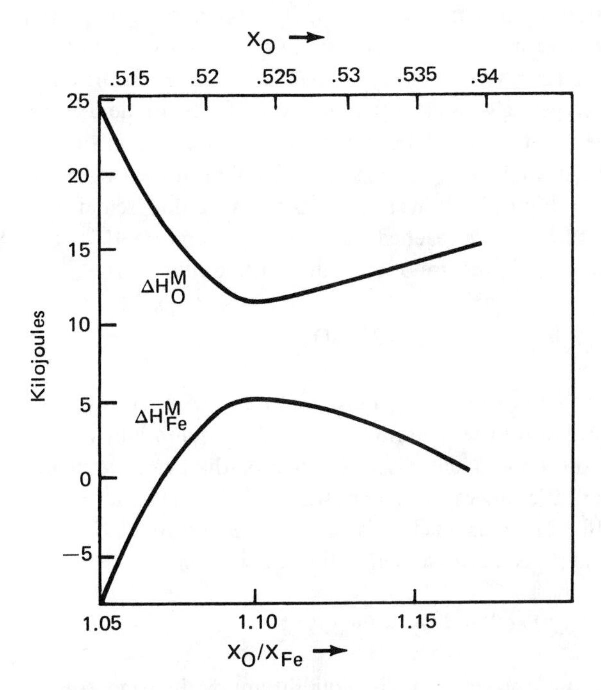

Fig. 13.29. The partial molar heats of solution of iron and oxygen in wustite. *(From L. S. Darken and R. W. Gurry, The System Iron-Oxygen, J. Am. Chem. Soc., 67:1398 [1945]; 68:798 [1946].)*

and hence

$$\Delta H^{o} = \frac{R d \ln K}{d\left(\frac{1}{T}\right)} = \frac{R d}{d\left(\frac{1}{T}\right)}\left[\ln\left(\frac{1}{a_{Fe}^{3}}\right) + \ln\left(\frac{1}{p_{O_2}^{2}}\right)\right]$$

In this expression, a_{Fe} and p_{O_2} are the respective values for equilibrium between wustite and magnetite and are obtained from the data in Fig. 13.28.

Below 550°C, homogeneous wustite is metastable with respect to iron and magnetite. This situation corresponds directly to Fig. 13.15*b*, in that, below 550°C, the tangent drawn from pure Fe to the magnetite free energy curve lies below the wustite curve. At 550°C this tangent is also the tangent to the wustite curve, and the two-component, four-phase equilibrium is invariant.

13.8 GRAPHICAL REPRESENTATION OF PHASE EQUILIBRIA

The temperature-composition diagram at 1 atm total pressure shown in Fig. 13.27 has, superimposed on it, oxygen isobars which trace the loci of variation of equilibrium composition with temperature under a fixed oxygen partial pressure in the system. For example, consider a small quantity of hematite at room temperature held in a gas reservoir of $p_{O_2} = 10^{-8}$ atm, and of sufficiently large volume that any oxygen given off by reduction of the oxide has an insignificant effect on the magnitude of the oxygen partial pressure in the gas reservoir. Let the oxide be heated slowly enough that equilibrium with the gas phase is maintained. From Fig. 13.27, it is seen that the oxide remains as homogeneous hematite until 875°C is reached, at which temperature 10^{-8} atm is the invariant equilibrium partial pressure of oxygen in the equilibrium

$$2\,Fe_3O_4 + \tfrac{1}{2}O_2 = 3Fe_2O_3$$

At 875°C magnetite of composition *b* is in equilibrium with hematite of composition *a*, and any increase in temperature upsets this equilibrium toward the magnetite side, with the consequent disappearance of the hematite phase. Further increase in temperature moves the composition of the oxide across the magnetite phase field until 1275°C is reached, at which temperature 10^{-8} is the invariant equilibrium partial pressure of oxygen in the equilibrium

$$3\,\text{"FeO"} + \tfrac{1}{2}O_2 = Fe_3O_4$$

At 1275°C, wustite of composition *d* is in equilibrium with magnetite of composition *c*. Further increase of temperature results in the disappearance of the magnetite phase, and solid homogeneous wustite exists until the melting temperature of 1400°C is reached, where solid wustite of composition *e* melts to give liquid oxide of composition *f* at $p_{O_2} = 10^{-8}$ Continued temperature increase takes the composition of the liquid oxide to saturation with iron at the temperature 1635°C, where the liquid oxide composition is at *g*, and the oxygen-saturated liquid iron phase appears. In this state the equilibrium

$$Fe_{(l)} + \tfrac{1}{2}O_{2(g)} = \text{“FeO”}_{(l)}$$

is established. Temperature increase beyond 1635°C results in the disappearance of the liquid oxide phase and a decrease in the dissolved oxygen content of the liquid iron.

Similarly, isothermal reduction of hematite is achieved by decreasing the partial pressure of oxygen in the system. For example, from Fig. 13.27, at 1300°C hematite is stable until the p_{O_2} has been decreased to 1.34×10^{-2} atm, at which state magnetite of composition b' is in equilibrium with hematite of composition a', magnetite is stable until p_{O_2} has been decreased to 2.15×10^{-8} atm where wustite of composition d' is in equilibrium with magnetite of composition c', and wustite is stable until $p_{O_2} = 1.95 \times 10^{-11}$ where solid iron appears in equilibrium with wustite of composition e'. Further decrease of p_{O_2} results in the disappearance of the oxide phase.

Figure 13.30 shows the phase relationships on a log p_{O_2}-temperature plot, and the points *a*-*g* and a'-e' correspond to those in Fig. 13.27. In that Fig. 13.30 gives no indication of the compositions of coexisting oxide phases, it is less useful than the normal composition-temperature phase diagram containing oxygen isobars. Figure 13.31 shows the equilibria on a plot of log p_{O_2} versus $1/T$. In this figure the slope of any invariant three-phase equilibrium line at any temperature, that is, $d \log p_{O_2}/d(1/T)$, equals $\Delta H/(8.3144 \times 2.303)$, where ΔH is the heat change per mole of oxygen consumed in the oxidation phase change. Over temperature ranges where the compositions of the equilibrated phases are constant, a linear relationship occurs between log p_{O_2} and $1/T$. If the abscissa in Fig. 13.30 is multiplied by $2.303RT$ and is plotted against T, then an Ellingham diagram results. This, for equilibrium in the Fe-O system, is shown in Fig. 13.32, where again the points *a* to *g* and a' to e' correspond to those in Fig. 13.27. Except for the Fe_3O_4-Fe_2O_3 line in Fig. 13.32, the lines drawn are for oxidation reactions, involving the consumption of 1 mole of oxygen, of the type

$$Fe_xO_y + O_2 = Fe_xO_{(2+y)}$$

where the lower oxide of composition Fe_xO_y is in equilibrium with the higher oxide of composition $Fe_xO_{(2+y)}$. The Fe_2O_3-Fe_3O_4 line is hypothetical and applies to the stoichiometric compounds (the stoichiometric Fe_3O_4 contains Fe at a higher activity and oxygen at a lower activity than does the composition in

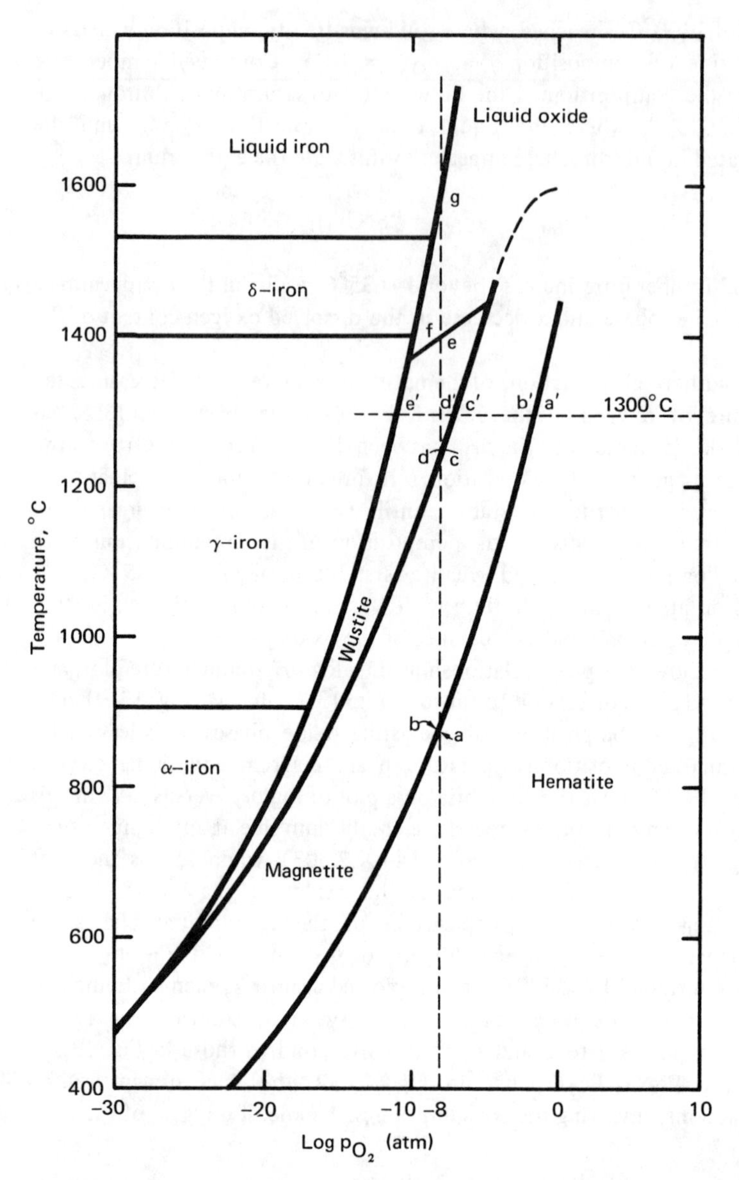

Fig. 13.30. Phase stability in the system Fe–Fe_2O_3 as a function of temperature and log p_{O_2}.

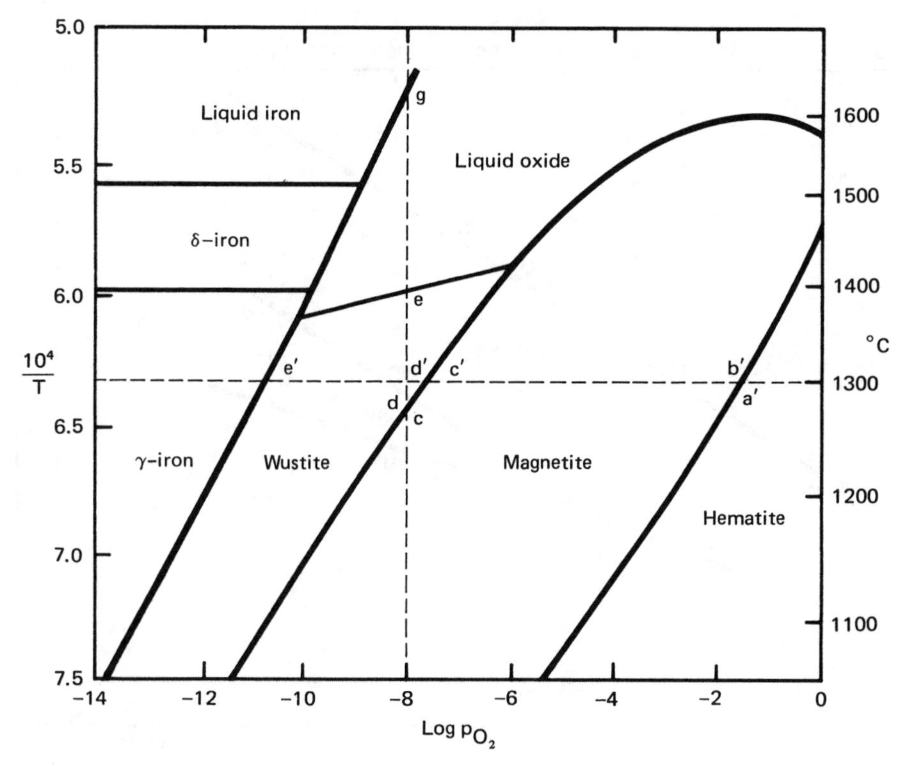

Fig. 13.31. Phase stability in the system $Fe-Fe_2O_3$ as a function of $\log p_{O_2}$ and $1/T$.

equilibrium with hematite). In the Ellingham diagram, lines that radiate from the origin ($\Delta G^0 = 0$, $T = 0$ K), are iso-oxygen activity lines. The distinct advantage of Ellingham-type representation of equilibria is its ability to indicate, at a glance, the relative stabilities of a large number of metal-oxygen systems, as is seen in Fig. 10.13.

13.9 SOLUTIONS CONTAINING SEVERAL DILUTE SOLUTES

Thus far only binary solutions have been discussed, and it is now of interest to examine the thermodynamics of solutions which contain a single solvent and several dilute solutes. In the case of a single dilute solute being present, its behavior is determined by the nature and magnitude of the interactions that occur between the solute and solvent atoms. However, when a second dilute solute is added, three types of interactions occur, namely, solvent-solute I, solvent-solute II, and solute I-solute II, and the thermodynamic behavior of the

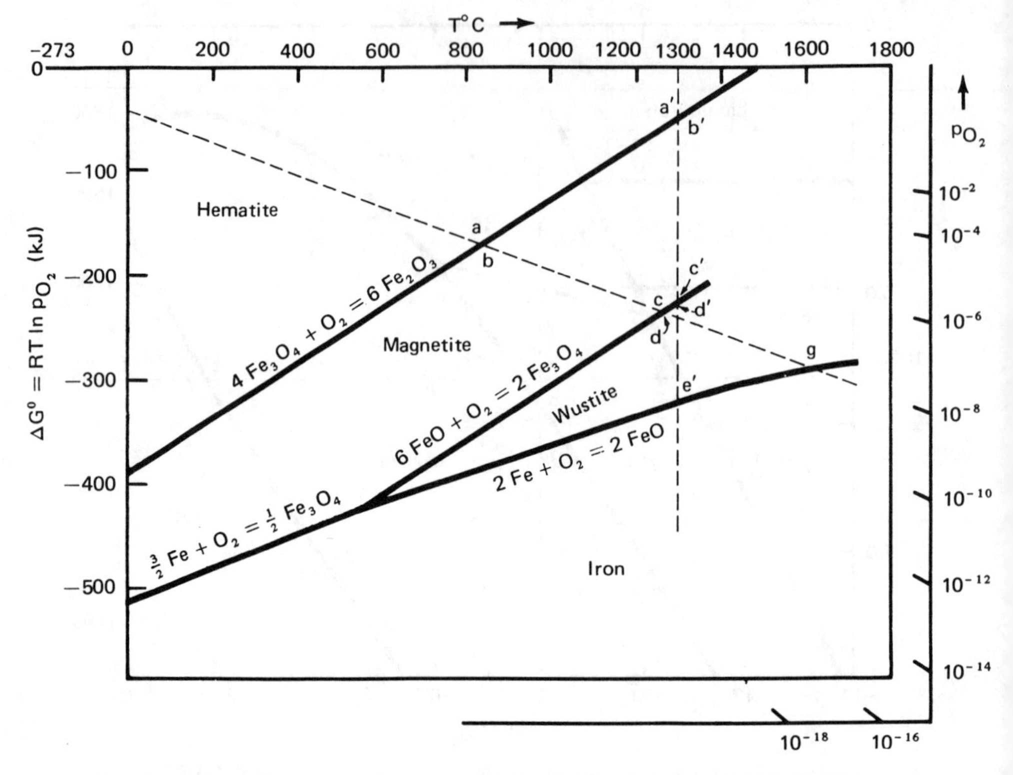

Fig. 13.32 Phase stability in the system Fe–Fe_2O_3 as a function of ΔG° ($= RT \ln p_{O_2}$) and temperature.

system will be determined by the relative magnitudes of the three types of interaction.

Consider the situation where liquid iron is exposed to a gaseous atmosphere of hydrogen and oxygen. In the gas phase, the equilibrium

$$H_{2(g)} + \tfrac{1}{2}O_{2(g)} = H_2O_{(g)} \tag{i}$$

is established, and hence

$$p_{H_2} p_{O_2}^{\frac{1}{2}} = \frac{p_{H_2O}}{K_{(i)}}$$

As both hydrogen and oxygen have some limited solubility in liquid iron, both will dissolve until their respective activities in the iron, with respect to the 1 atm pressure gaseous standard state, equal their respective partial pressures in the gas phase. Alternatively, with respect to the 1 wt% in Fe standard state, as

$$\tfrac{1}{2}O_{2(g)} = O_{(1\ \text{wt\% in Fe})} \tag{ii}$$

and

$$\tfrac{1}{2}H_{2(g)} = H_{(1\ \text{wt\% in Fe})} \tag{iii}$$

for which

$$h_{O(1\ \text{wt\%})} = K_{(ii)} p_{O_2}^{\frac{1}{2}}$$

and

$$h_{H(1\ \text{wt\%})} = K_{(iii)} p_{H_2}^{\frac{1}{2}}$$

equilibrium in the metal phase is given as

$$h^2_{H(1\ \text{wt\%})} h_{O(1\ \text{wt\%})} = \frac{1}{K_{(i)}} K^2_{(iii)} K_{(ii)} p_{H_2O}$$

or

$$f^2_{H(1\ \text{wt\%})} f_{O(1\ \text{wt\%})} [\text{wt\% H}]^2 [\text{wt\% O}] = \frac{K^2_{(iii)} K_{(ii)} p_{H_2O}}{K_{(i)}}$$

The actual solubilities of both H and O (expressed as weight percentages) are thus determined by the values of the activity coefficients of H and O, and hence the questions to be answered are

1. How is the activity coefficient of O in Fe influenced by the presence of H?
2. How is the activity coefficient of H in Fe influenced by the presence of O?

This problem is dealt with by the introduction of interaction coefficients and interaction parameters.

In the binary A-B, the activity of B in dilute solution with respect to the Henrian standard state is given as

$$h_B = f_B^B X_B$$

If, holding the concentration of B constant, the addition of a small amount of solute C alters the activity coefficient of B to f_B, then the difference between f_B and f_B^B is quantified by the expression

$$f_B = f_B^B f_B^C \tag{13.17}$$

where f_B^C is termed the interaction coefficient of C on B and is a measure of the effect on the behavior of B of the presence of a specific concentration of C, at the same concentration of B. Similarly, if a small amount of D is added to the A-B solution as a result of which f_B^B changes to f_B, then

$$f_B = f_B^B f_B^D$$

Now consider the system A-B-C-D. A mathematical analysis of such a system is possible only if f_B^D is independent of the concentration of C and if f_B^C is independent of the concentration of D. If these conditions are not met, then the thermodynamics of the system can only be determined by experiment. Consider that the interaction coefficient of the solute i on the solute j is independent of the other solutes present, in which case the interaction coefficients may be combined by means of a Taylor's expansion of $\ln f_j$ as a function of the concentration of the solutes; e.g., for the binary A-B in which A is the solvent,

$$\ln f_B = \text{some function of the mole fraction of B}$$
$$= \ln f_B^0 + \left.\frac{\partial \ln f_B}{\partial X_B}\right|_{X_B=0} X_B + \frac{1}{2}\left.\frac{\partial^2 \ln f_B}{\partial X_B^2}\right|_{X_B=0} X_B^2 + \cdots$$

and for the multicomponent system A-B-C-D,

$$\ln f_B = \text{some function of the mole fractions of B, C and D}$$
$$= \ln f_B^0 + \left[\left.\frac{\partial \ln f_B}{\partial X_B}\right|_{X_B=0} X_B + \left.\frac{\partial \ln f_B}{\partial X_C}\right|_{X_C=0} X_C + \left.\frac{\partial \ln f_B}{\partial X_D}\right|_{X_D=0} X_D\right]$$
$$+ \left[\frac{1}{2}\left.\frac{\partial^2 \ln f_B}{\partial X_B^2}\right|_{X_B=0} X_B^2 + \left.\frac{\partial^2 \ln f_B}{\partial X_B \partial X_C}\right|_{X_B=X_C=0} X_B X_C + \cdots\right]$$

At very low concentration, terms containing the products of mole fractions are negligible; and also as the Henrian standard state is chosen, that is, $f_B^0 = 1$, then the above expression simplifies to

$$\ln f_B = \left.\frac{\partial \ln f_B}{\partial X_B}\right|_{X_B=0} X_B + \left.\frac{\partial \ln f_B}{\partial X_C}\right|_{X_C=0} X_C + \left.\frac{\partial \ln f_B}{\partial X_D}\right|_{X_D=0} X_D$$
$$= \epsilon_B^B X_B + \epsilon_B^C X_C + \epsilon_B^D X_D \quad (13.18)$$

where

$$\epsilon_j^i = \left.\frac{\partial \ln f_j}{\partial X_i}\right|_{X_i=0}$$

is termed the interaction parameter of i on j and is obtained as the limiting slope of a plot of $\ln f_j$ against X_i at constant X_j.

ϵ_j^i and ϵ_i^j are related as follows. For the general system

$$\frac{\partial^2 G}{\partial n_i \partial n_j} = \frac{\partial \bar{G}_i}{\partial n_j} = \frac{\partial \bar{G}_j}{\partial n_i}$$

and as $\partial \bar{G}_i = RT\,\partial \ln a_i = RT\,\partial \ln f_i$

then
$$\frac{\partial \ln f_j}{\partial n_i} = \frac{\partial \ln f_i}{\partial n_j}$$

and hence
$$\epsilon_i^j = \epsilon_j^i \tag{13.19}$$

It is often more convenient to consider the concentrations of the solutes in terms of weight percentages and to use logarithms to the base ten, in which case Eq. (13.18) becomes

$$\log f_{\mathrm{B}} = \left.\frac{\partial \log f_{\mathrm{B}}}{\partial \,\mathrm{wt\%\,B}}\right|_{\mathrm{wt\%\,B}=0} \mathrm{wt\%\,B} + \left.\frac{\partial \log f_{\mathrm{B}}}{\partial \,\mathrm{wt\%\,C}}\right|_{\mathrm{wt\%\,C}=0} \mathrm{wt\%\,C}$$

$$+ \left.\frac{\partial \log f_{\mathrm{B}}}{\partial \,\mathrm{wt\%\,D}}\right|_{\mathrm{wt\%\,D}=0} \mathrm{wt\%\,D}$$

$$= e_{\mathrm{B}}^{\mathrm{B}}\,\mathrm{wt\%\,B} + e_{\mathrm{B}}^{\mathrm{C}}\,\mathrm{wt\%\,C} + e_{\mathrm{B}}^{\mathrm{D}}\,\mathrm{wt\%\,D} \tag{13.20}$$

Multiplying Eq. (13.20) by 2.303 and comparing, term by term, with Eq. (13.18) gives

$$\epsilon_{\mathrm{B}}^i X_i = 2.303\, e_{\mathrm{B}}^i \,\mathrm{wt\%}\, i$$

and as, at small concentrations of i and B,

$$X_i \doteqdot \frac{\mathrm{wt\%}\, i\, \mathrm{MW_A}}{100\, \mathrm{MW}_i}$$

then
$$e_{\mathrm{B}}^i = \frac{\mathrm{MW_A}}{230.3\, \mathrm{MW}_i} \epsilon_{\mathrm{B}}^i$$

and
$$e_{\mathrm{B}}^i = \frac{\mathrm{MW_B}}{\mathrm{MW}_i} e_i^{\mathrm{B}}$$

As an example, consider the effects of the presence of a second dilute solute on the thermodynamics of nitrogen dissolved in liquid iron. Such systems are particularly amenable to experimental study by virtue of the ease with which the nitrogen activity can be controlled by the gas phase, and several such systems have been studied by Pehlke and Elliott.* The interaction parameters, e_N^i, are determined by maintaining the nitrogen in the melt at constant activity and measuring the variation of nitrogen solubility, at this constant activity, with second solute concentration. As has been seen, nitrogen in liquid iron obeys Henry's law, such that for the Fe-N binary the equilibrium constant for the equilibrium

$$\tfrac{1}{2}N_{2(g)} = [N]_{\text{wt\% in Fe}}$$

can be written as

$$K = \frac{[\text{wt\% N}]}{p_{N_2}^{\frac{1}{2}}}$$

that is, $f_N^N = 1$, and hence e_N^N (and ϵ_N^N) are zero; and so the first terms in Eqs. (13.18) and (13.20) are zero. If the addition of a second solute X to the Fe-N binary (equilibrated with a fixed p_{N_2}) alters the dissolved nitrogen content from $[\text{wt\% N}]_{\text{Fe-N}}$ to $[\text{wt\% N}]_{\text{Fe-N-X}}$, then

$$K = \frac{[\text{wt\% N}]_{\text{in Fe–N}}}{p_{N_2}^{\frac{1}{2}}} = \frac{f_N^X[\text{wt\% N}]_{\text{in Fe–N–X}}}{p_{N_2}^{\frac{1}{2}}}$$

and so f_N^X is obtained experimentally as

$$f_N^X = \left\{ \frac{[\text{wt\% N}]_{\text{in Fe–N}}}{[\text{wt\% N}]_{\text{in Fe–N–X}}} \right\}_{T,P,p_{N_2}}$$

The variation of [wt% N] with [wt% X] is shown for several second solutes in Fig. 13.33, and the corresponding variation of $\log f_N^X$ with wt% X is shown in Fig. 13.34. The values of e_N^X are obtained as the slopes of the linear portions of the lines in Fig. 13.34.

Thus in a multicomponent liquid iron alloy containing several dilute solutes including nitrogen, if the effect of any one solute on f_N is independent of the presence of any other solute, then the total effect of the solute on f_N is the sum of their individual effects, and if $\log f_N^X$ is a linear function of wt% X, then f_N can be calculated from Eq. (13.20). If the concentrations of X exceed their respective concentration limits of linear $\log f_N^X$-wt% X behavior, then a graphical solution is

*R. Pehlke and J. F. Elliott, Solubility of Nitrogen in Liquid Iron Alloys, I: Thermodynamics, *Trans. Met. Soc. AIME*, **218**:1088 (1960).

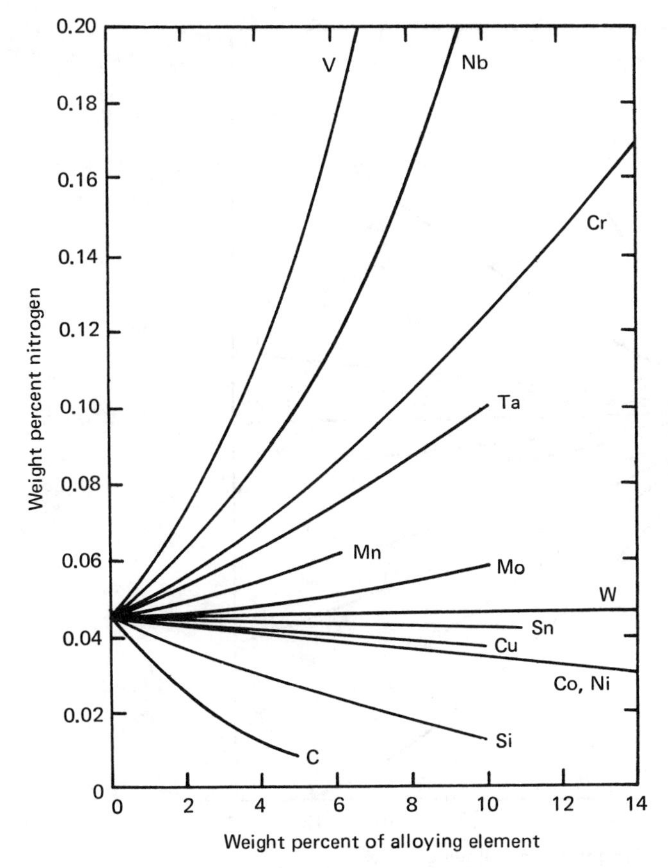

Fig. 13.33. The effect of alloying elements on the solubility of nitrogen at 1 atm pressure in liquid binary iron alloys at 1600°C. *(From R. G. Ward, "Physical Chemistry of Iron and Steelmaking," Arnold, London, 1962.)*

necessary. In such cases the value of $\log f_N^X$ for each wt% X is read from the graph (Fig. 13.34), and $\log f_N$ is obtained as

$$\log f_N = \sum_x \log f_N^X$$

or

$$f_N = \prod^x f_N^X$$

Figure 13.34 indicates that, as a general rule, e_N^X is a negative quantity when X forms a more stable nitride than does iron, and that the order of increasing

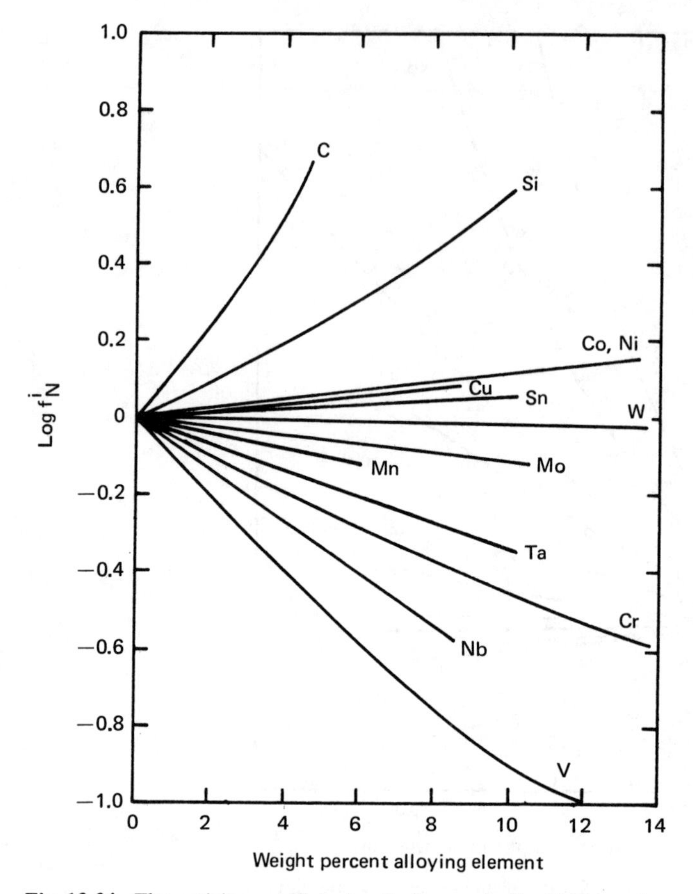

Fig. 13.34. The activity coefficients of nitrogen in liquid binary iron alloys at 1600°C. *(From R. G. Ward, "Physical Chemistry of Iron and Steelmaking," Arnold, London, 1962.)*

magnitude of $|e_N^X|$ follows the order of increasing magnitude of the free energy of formation of the nitride of X. Similarly e_N^X is a positive quantity when X has a greater affinity for iron than either X has for N or iron has for nitrogen.

Example 1

In view of the use of interaction parameters, the example of Si–O equilibrium in liquid iron, discussed in Sec. 13.3, can now be reexamined. In this example it was determined that for

$$Si_{(1\text{ wt\% in Fe})} + O_{2\,(g)} = SiO_{2\,(s)}$$
$$\Delta G^0 = -833{,}400 + 229.5T \text{ joules}$$

For $\frac{1}{2}O_{2\,(g)} = O_{(1\text{ wt\% in Fe})}$

$$\Delta G^0 = -111{,}300 - 6.41T \text{ joules} \tag{i}$$

and hence for

$$Si_{(1\text{ wt\% in Fe})} + 2O_{(1\text{ wt\% in Fe})} = SiO_{2\,(s)}$$

$$\Delta G^0 = -610{,}800 + 242.32T \text{ joules} \tag{ii}$$

From Eq. (i), at 1600°C,

$$\frac{h_{O(1\text{ wt\%})}}{p_{O_2}^{\frac{1}{2}}} = 2.746 \times 10^3 \tag{iii}$$

and from Eq. (ii), at 1600°C,

$$\frac{a_{SiO_2}}{h_{Si(1\text{ wt\%})}h_{O(1\text{ wt\%})}^2} = 2.380 \times 10^4 \tag{iv}$$

Thus for $p_{O_2} = 5.57 \times 10^{-12}$ and $a_{SiO_2} = 1$

Eq. (iii) gives $$h_{O(1\text{ wt\%})} = 6.48 \times 10^{-3} \tag{v}$$

and Eq. (iv) gives $$h_{O(1\text{ wt\%})}^2 h_{Si(1\text{ wt\%})} = 4.20 \times 10^{-5} \tag{vi}$$

Division of Eq. (vi) by Eq. (v) gives

$$h_{Si(1\text{ wt\%})} = 1 \tag{vii}$$

At 1600°C, from Table 13.2,

$$e_O^{Si} = -0.14 \qquad e_O^O = -0.2$$

$$e_{Si}^O = -0.25 \qquad e_{Si}^{Si} = 0.32$$

Thus from Eq. (v),

$$\log f_O + \log[\text{wt\% O}] = \log[6.48 \times 10^{-3}]$$

or $$-0.20 \times [\text{wt\% O}] - 0.14 \times [\text{wt\% Si}] + \log[\text{wt\% O}] = -2.188 \tag{viii}$$

and from Eq. (vii),

$$\log f_{Si} + \log [\text{wt\% Si}] = \log [1]$$

or
$$0.32 [\text{wt\% Si}] - 0.25 [\text{wt\% O}] + \log [\text{wt\% Si}] = 0 \tag{ix}$$

Graphical solution of the simultaneous equations (viii) and (ix) gives

$$[\text{wt\% Si}] = 0.631$$

and
$$[\text{wt\% O}] = 0.00798$$

In the example in Sec. 13.3, where the effect of dissolved oxygen was ignored and it was assumed that $f_{Si} = 1$, the equilibrium weight percentage of Si in the iron when $a_{SiO_2} = 1$ and $p_{O_2} = 5.57 \times 10^{-12}$ was obtained as 1.0. The magnitude of the error incurred due to this assumption is now clearly apparent.

The fact that a significant error will be involved if it is assumed that $f_{Si} = 1$ is immediately apparent when it is seen that $e_{Si}^{Si} = 0.32$, i.e., that the limiting slope of the log f_{Si} versus wt% Si plot as wt% Si $\rightarrow$ 0 is 0.32 and not zero. In view of the tedious method required for the solution of the simultaneous equations (viii) and (ix), it is instructive to see what value of the equilibrium weight percentage of Si would be calculated by taking $e_{Si}^{Si} = 0.32$ and ignoring the effect of oxygen, i.e., assuming $e_{O}^{Si} = e_{Si}^{O} = 0$. From Eq. (ix),

$$0.32 [\text{wt\% Si}] + \log [\text{wt\% Si}] = 0$$

which gives
$$\text{wt\% Si} = 0.629$$

and from Eq. (viii)

$$-0.20 \times [\text{wt\% O}] + \log [\text{wt\% O}] = -2.188$$

which gives
$$[\text{wt\% O}] = 0.00651$$

The error introduced by ignoring the interaction between silicon and oxygen is thus seen to be negligible compared with the error introduced by assuming Si to obey Henry's law over some initial composition range.

Example 2

Calculate the equilibrium oxygen content of an Fe-C-O alloy which, at 1600°C, contains 1 wt% C and is under a pressure of 1 atm of CO.

For $C_{(gr)} + \frac{1}{2}O_{2(g)} = CO_{(g)}$ $\Delta G^0 = -111{,}700 - 87.65T$ joules
For $C_{(gr)} = C_{(1\ wt\%\ in\ Fe)}$ $\Delta G^0 = 22{,}600 - 42.26T$ joules
For $\frac{1}{2}O_{2(g)} = O_{(1\ wt\%\ in\ Fe)}$ $\Delta G^0 = -111{,}300 - 6.41T$ joules

Therefore, for

$$C_{(1\ wt\%)} + O_{(1\ wt\%)} = CO_{(g)} \qquad \Delta G^0 = -23{,}000 - 38.98T \text{ joules}$$

$$\Delta G^0_{1873} = -96{,}010 \text{ joules}$$

and

$$\frac{p_{CO}}{h_C h_O} = 476$$

Thus $$h_C h_O = f_C[\text{wt\% C}] f_O[\text{wt\% O}] = 2.1 \times 10^{-3} p_{CO}$$

At 1600°C

$$e_C^C = 0.22$$

$$e_O^O = -0.2$$

$$e_C^O = -0.097$$

$$e_O^C = -0.13$$

Thus for 1 wt% C and $p_{CO} = 1$,

$$\log \text{wt\% O} - 0.297 \text{ wt\% O} = -2.768$$

solution of which gives wt% O = 0.00171. If all the interaction parameters had been ignored, then

$$\text{wt\% O} = 0.00210$$

Example 3

The partial pressure of hydrogen in the atmosphere is such that an Fe-C-Ti melt containing 1 wt% C and 3 wt% Ti contains 5 parts per million (by weight) of hydrogen at 1600°C. Calculate the vacuum which would be required to decrease the hydrogen content of the melt to 1 ppm. Given that $e_H^{Ti} = -0.08$, that $e_H^C = +0.06$, and that hydrogen in pure iron obeys Henry's law up to a solubility of 0.0027 wt% under a pressure of 1 atm of hydrogen at 1600°C.

For the equilibrium between gaseous hydrogen and dissolved hydrogen, given as $\frac{1}{2}H_{2(g)} = [H]_{(1\text{ wt\% in Fe})}$,

$$K = \frac{f_{(1\text{ wt\%})}[\text{wt\% H}]}{p_{H_2}^{\frac{1}{2}}}$$

In pure iron, as hydrogen obeys Henry's law, $f_{(1\text{ wt\%})} = 1$, and so

$$K_{1873} = 0.0027$$

Thus $$\log f_{(1\text{ wt\%})} + \log[\text{wt\% H}] - \tfrac{1}{2}\log p_{H_2} = \log 0.0027$$

But $$\log f_{(1\text{ wt\%})} = e_H^H[\text{wt\% H}] + e_H^{Ti}[\text{wt\% Ti}] + e_H^C[\text{wt\% C}]$$

As $f_H^H = 1$, $e_H^H = 0$, and hence, at 1600°C,

$$e_H^{Ti}[\text{wt\% Ti}] + e_H^C[\text{wt\% C}] + \log[\text{wt\% H}] - \tfrac{1}{2}\log p_{H_2} = \log 0.0027$$

When $$[\text{wt\% H}] = 5 \times 10^{-4}$$

$$\begin{aligned}\log p_{H_2} &= -(2 \times 0.08 \times 3) + (2 \times 0.06 \times 1)\\ &\quad + 2\log(5 \times 10^{-4}) - 2\log 0.0027\\ &= -1.825\end{aligned}$$

and so $$p_{H_2} = 0.015\text{ atm}$$

Similarly, when $[\text{wt\% H}] = 1 \times 10^{-4}$, $p_{H_2} = 6 \times 10^{-4}$

Thus $$[\text{wt\% H}] = 5\text{ppm when } p_{H_2} = 0.015 \text{ and } P_{total} = 1$$

and so $$[\text{wt\% H}] = 1\text{ppm when } p_{H_2} = 6 \times 10^{-4} \text{ and}$$

$$P_{total} = \frac{6 \times 10^{-4}}{0.015} = 4 \times 10^{-2}\text{ atm}$$

Thus in order to achieve the required decrease in the dissolved hydrogen content, the total pressure must be decreased from 1 atm to 4×10^{-2} atm.

The values of the interaction coefficients for several elements in dilute solution in iron at 1600°C are listed in Table 13.2.

Table 13.2. Some interaction coefficients, $e_i^j \times 10^2$, for dilute solutions of elements dissolved in liquid iron at 1600°C *
Values in parentheses are calculated from $e_i^j = (M_i/M_j)\, e_j^i$

Element (i)	Element (j)											
	Al	C	Co	Cr	H	Mn	N	Ni	O	P	S	Si
Al	4.8	11	...	...	(34)	...	(.5)	...	−160	...	4.9	6
C	(4.8)	22	1.2	−2.4	(72)	...	(11.1)	1.2	(−9.7)	...	9	10
Co	...	(6)	...	...	(11)	...	(4.7)	...	(2.6)	...	...	...
Cr	...	(−10)	...	...	(−11)	...	(−16.6)	...	(−13)	...	(−3.55)	...
H	1.3	6.0	.18	−0.22	0	−.14	...	0	...	1.1	0.8	2.7
Mn	...	...	...	...	(−7.7)	...	(−7.8)	...	(0)	...	(−4.3)	(0)
N	0.3	13	1.1	−4.5	...	−2	0	1	5.0	5.1	1.3	4.7
Ni	...	(5.9)	...	...	(0)	...	(4.2)	0	(2.1)	...	(0)	(1.0)
O	−94	−13	.7	−4.1	...	0	(5.7)	0.6	−20	7.0	−9.1	−14
P	...	...	...	...	(34)	...	(11.3)	...	(13.5)	...	(4.3)	(9.5)
S	5.8	(24)	...	−2.2	(26)	−2.5	(3.0)	0	(−18)	4.5	−2.8	6.6
Si	(6.3)	24	...	...	(76)	0	(9.3)	0.5	(−25)	8.6	(5.7)	32

*Taken from J. F. Elliott, M. Gleiser, and V. Ramakrishna, "Thermochemistry for Steelmaking," vol. 2, Addison-Wesley Publishing Co.. Reading, Mass., 1963.

13.10 TABULAR REPRESENTATION OF THERMODYNAMIC DATA AND THE FREE ENERGY FUNCTION

For precise calculation of chemical reaction equilibria, the use of an analytical expression of the temperature dependence of ΔG^0 for the reaction of interest in the form

$$\Delta G^0 = A + BT \log T + CT$$

is generally inadequate. The inaccuracy arises from the difficulty of curve-fitting an analytical expression, such as the one above, to a series of experimentally determined values of ΔG_T^0. Indeed, even Eq. (10.9) may be less precise than desired, due, again, to the inaccuracy inherent in curve-fitting an expression of the form $c_p = a + bT + cT^{-2}$ to experimentally determined heat capacity data. In order to minimize the error, it would be necessary to use tabulated heat capacity values of the reactants and products and graphically perform the integrations leading to Eqs. (10.8) and (10.9). Although this procedure would be tedious, it would lead to an accurate tabulation of ΔG_T^0 for the reaction of interest, and the temperature interval of tabulation could be made as small as desired. As one such table would be required for each reaction, it is seen that an enormous amount of tabulation would be required to provide a comprehensive presentation of thermodynamic data. In view of this, the following procedure has been developed.

For any reaction the Gibbs-Helmholtz equation can be written in the form

$$d\left(\frac{\Delta G^0}{T}\right) = \Delta H^0 d\left(\frac{1}{T}\right)$$

where ΔG^0 $(= \Delta G_T^0)$ = the standard free energy change for the reaction at the temperature T

ΔH^0 $(= \Delta H_T^0)$ = the standard enthalpy change for the reaction at the temperature T

To this is added the identity

$$-d\left(\frac{\Delta H_0^0}{T}\right) = -\Delta H_0^0\, d\left(\frac{1}{T}\right)$$

where ΔH_0^0 is the standard enthalpy change for the reaction at 0 K. This gives

$$d\left(\frac{\Delta G^0 - \Delta H_0^0}{T}\right) = (\Delta H^0 - \Delta H_0^0) d\left(\frac{1}{T}\right)$$

integration of which from 0 K to T K gives

$$\left.\left(\frac{\Delta G_T^0 - \Delta H_0^0}{T}\right)\right|_T - \left.\left(\frac{\Delta G_T^0 - \Delta H_0^0}{T}\right)\right|_{T=0} = \int_0^T (\Delta H_T^0 - \Delta H_0^0) d\left(\frac{1}{T}\right)$$

As $\Delta G_T^0 - \Delta H_T^0 = -T\Delta S_T^0$, the term $\left.\left(\frac{\Delta G_T^0 - \Delta H_0^0}{T}\right)\right|_{T=0}$ equals $-\Delta S_0^0$,

the standard entropy change at 0 K, which, by virtue of the Third Law of Thermodynamics, is zero. Thus for the given reaction at the temperature T,

$$\frac{\Delta G_T^0 - \Delta H_0^0}{T} = \int_0^T (\Delta H_T^0 - \Delta H_0^0) d\left(\frac{1}{T}\right)$$

For the reaction $a\text{A} + b\text{B} = c\text{C} + d\text{D}$,

$$\Delta H_T^0 - \Delta H_0^0 = [cH_{T(\text{C})}^0 + dH_{T(\text{D})}^0 - aH_{T(\text{A})}^0 - bH_{T(\text{B})}^0] - [cH_{0(\text{C})}^0 + dH_{0(\text{D})}^0 - aH_{0(\text{A})}^0 - bH_{0(\text{B})}^0]$$

which can be rearranged as

$$c[H^0_{T(C)} - H^0_{0(C)}] + d[H^0_{T(D)} - H^0_{0(D)}] - a[H^0_{T(A)} - H^0_{0(A)}] - b[H^0_{T(B)} - H^0_{0(B)}] = \Delta(H^0_T - H^0_0)$$

The term $\Delta H^0_T - \Delta H^0_0 = \Delta(H^0_T - H^0_0)$ is thus the sum of the heats required to raise the temperature of the reaction products from 0 K to T K minus the corresponding sum for the reactants. This can be calculated from heat capacity data alone. The term $(\Delta G^0_T - \Delta H^0_0)/T$ for the reaction can be written as $\Delta(G^0_T - H^0_0)/T$, which is the sum of the terms $(G^0_T - H^0_0)/T$ for the products minus the sum of the terms $(G^0_T - H^0_0)/T$ for the reactants. For each reactant and product

$$\frac{G^0_T - H^0_0}{T} = \int_0^T (H^0_T - H^0_0) d\left(\frac{1}{T}\right)$$

is termed the *free energy function* and can be evaluated if the heat capacity is known from 0 K to T K. Hence from tabulations, for the reactants and products of a reaction, of the free energy function at discrete temperatures, and tabulations of ΔH^0_0, the heat of formation of each reactant and product at 0 K, the value of ΔG^0_T for the reaction is obtained as

$$\frac{\Delta G^0_T}{T} = \frac{\Delta(G^0_T - H^0_0)}{T} + \frac{\Delta H^0_0}{T}$$

Consider the reaction

$$CO_{(g)} + \tfrac{1}{2}O_{2(g)} = CO_{2(g)} \quad \text{at} \quad 1000 \text{ K} \tag{i}$$

Chemical species	$\frac{G^0 - H^0_0}{T}$ at 1000 K joules/degree·mole	ΔH^0_0 kjoules/mole
$CO_{2(g)}$	−226.40	−393.166
$CO_{(g)}$	−204.05	−113.813
$O_{2(g)}$	−212.13	0

$$\frac{\Delta G^0_T}{1000} = \frac{\Delta(G^0 - H^0_0)}{T} + \frac{\Delta H^0_0}{1000}$$

$$= (-226.40) - (-204.05) - \tfrac{1}{2}(-212.13) + \frac{(-393{,}166 + 113{,}813)}{1000}$$

$$= -195.638$$

and thus $$\Delta G^0_{1000} = -195{,}638 \text{ joules}$$

(cf. from Eq. (iii) in Sec. 10.6, $\Delta G^0_{1000} = -195{,}590$ joules). The free energy function may be based on 298 K, in which case

$$\frac{\Delta G^0_T}{T} = \frac{\Delta(G^0 - H^0_{298})}{T} + \frac{\Delta H^0_{298}}{T}$$

where $(G^0 - H^0_{298})/T$ is the 298 K free energy function, and ΔH^0_{298} is the sum of the heats of formation of the reaction products at 298 K minus the sum of the heats of formation of the reactants at 298 K. The advantage of the 298 K free energy function is that it can be used for reactions involving reactants or products for which low temperature heat capacity data are unavailable, provided that the entropy change for the reaction is known (from, for example, an equilibrium measurement). For any reactant or product,

$$S^0_T = -\frac{(G^0_T - H^0_T)}{T}$$

or $$\frac{G^0_T}{T} = \frac{H^0_T}{T} - S^0_T$$

Subtracting H^0_{298} from both sides of this equation and rearranging gives

$$\frac{G^0_T - H^0_{298}}{T} = \frac{H^0_T - H^0_{298}}{T} - S^0_T$$

The 0 K free energy function and the 298 K free energy function are related via

$$\frac{G^0_T - H^0_{298}}{T} = \frac{G^0_T - H^0_0}{T} - \frac{H^0_{298} - H^0_0}{T}$$

Thus for the reaction

$$CO_{(g)} + \tfrac{1}{2}O_{2(g)} = CO_{2(g)}$$

at 1000 K

Chemical species	$\dfrac{G^0 - H_0^0}{T}$ at 1000 K joules/degree·mole	$H_{298}^0 - H_0^0$ kjoules/mole	$\dfrac{G^0 - H_{298}^0}{T}$ at 1000 K joules/degree·mole	ΔH_{298}^0 kjoules/mole
$CO_{2}(g)$	−226.40	9.364	−235.806	−393.514
$CO_{(g)}$	−204.05	8.673	−212.735	−110.525
$O_{2}(g)$	−212.13	8.661	−220.769	0

$$\frac{\Delta G_T^0}{1000} = \frac{\Delta(G^0 - H_{298}^0)}{T} + \frac{\Delta H_{298}^0}{1000}$$

$$= (-235.806 + 212.735 + \tfrac{1}{2} \times 220.769)$$

$$+ \frac{(-393{,}514 + 110{,}525)}{1000}$$

$$= -195.676$$

and thus $\Delta G_{1000}^0 = -195{,}676$ joules.

13.11 ANALYSIS OF EXPERIMENTAL DATA BY THE SECOND LAW AND THIRD LAW METHODS

Combination of the Gibbs-Helmholtz equation,

$$\frac{\partial(\Delta G^0/T)}{\partial(1/T)} = \Delta H^0$$

with

$$\Delta G^0 = -RT \ln K$$

gives

$$\frac{\partial \ln K}{\partial(1/T)} = \frac{-\Delta H^0}{R}$$

and calculation of ΔH^0 for a reaction from the variation of $\ln K$ with $1/T$ is known as the "Second Law" calculation of ΔH^0.

Hildenbrand, Hall, and Potter* have measured the vapor pressure of liquid B_2O_3, and their least-squares analysis of 44 datum points in the temperature range 1410–1590 K gives the variation of $p_{B_2O_3}$ with temperature as

$$\ln p_{B_2O_3} \text{ (atm)} = \frac{-46{,}937}{T} + 20.545$$

For the equilibrium

$$B_2O_{3(l)} = B_2O_{3(v)} \qquad K_p = p_{B_2O_3}$$

and hence
$$\ln K_p = \frac{-46{,}937}{T} + 20.545$$

Thus a Second Law calculation gives $\Delta H^0/R = 46{,}937$ or

$$\Delta H^0 = 8.3144 \times 46{,}937 = 390.3 \text{ kilojoules/mole}$$

This value is for the mean temperature in the experimental range, namely 1500 K, and hence the molar heat of vaporization of B_2O_3 at 1500 K is $\Delta H^0_{1500} = 390.3$ kilojoules/mole. Molar heat capacity data for liquid and vapor B_2O_3 give

$$(H^0_{1500} - H^0_{298})_{B_2O_3(v)} = \int_{298}^{1500} c_{P,B_2O_3(v)}\, dT = 109.3 \text{ kilojoules}$$

and
$$(H^0_{1500} - H^0_{298})_{B_2O_3(l)} = \int_{298}^{1500} c_{P,B_2O_3(l)}\, dT = 142.4 \text{ kilojoules}$$

and thus the Second Law-calculated ΔH^0_{298} is

$$\Delta H^0_{298} = 390.3 + 142.4 - 109.3 = 423.4 \text{ kilojoules/mole}$$

From Eq. (13.20),

$$\frac{\Delta G^0_T}{T} = \frac{\Delta(G^0_T - H^0_{298})}{T} + \frac{\Delta H^0_{298}}{T}$$
$$= -R \ln K$$

*D. L. Hildenbrand, W. F. Hall, and N. D. Potter, Thermodynamics of Vaporization of Lithium Oxide, Boric Oxide and Lithium Metaborate, *J. Chem. Phys.*, **39**:296 (1963).

rearrangement of which gives

$$\frac{\Delta H^0_{298}}{T} = -R \ln K - \frac{\Delta(G^0_T - H^0_{298})}{T} \tag{13.21}$$

and the calculation of ΔH^0_{298} for a reaction from combination of the equilibrium constant, obtained by measurement of the reaction equilibrium, and the 298 K free energy functions for the reactants and products, is known as a "Third Law" calculation.

The measurements of Hildenbrand et al. give $p_{B_2O_3(l)} = 0.231 \times 10^{-5}$, 2.16×10^{-5}, and 15.2×10^{-4} atm at 1400, 1500, and 1600 K respectively, which give the corresponding values of $-R \ln K$ as 107.9, 89.4, and 73.1 joules/degree. The 298 K free energy functions for gaseous and liquid B_2O_3 at 1400, 1500, and 1600 K are listed in columns 1 and 2 of the following table.

T, K	$\frac{G^0 - H^0_{298}}{T}$ joules/degree $B_2O_3(g)$	$\frac{G^0 - H^0_{298}}{T}$ joules/degree $B_2O_3(l)$	$\frac{\Delta(G^0 - H^0_{298})}{T}$ joules/degree	$-R \ln K$ joules/degree·mole	ΔH^0_{298} kjoules/degree·mole
1400	−345.45	−153.33	−192.12	107.9	420.0
1500	−350.41	−159.80	−190.61	89.4	420.0
1600	−355.17	−165.99	−189.18	73.1	419.6

Their difference is listed in column 3, and $-R \ln K$ is listed in column 4. ΔH^0_{298}, calculated by the Third Law method from Eq. (13.21), is listed in column 5.

Generally, analysis of experimental data by the Third Law method is superior to that using the second law, as a value of ΔH^0_{298} is obtained from each datum point.

13.12 NUMERICAL EXAMPLES

Example 1

A piece of iron is to be heat-treated at 1000 K in a CO-CO_2-H_2O-H_2 gas mixture at 1 atm pressure. This gas mixture is obtained by mixing CO_2 and H_2 and allowing the equilibrium $CO_2 + H_2 = CO + H_2O$ to establish. Calculate:

a. The minimum H_2/CO_2 ratio in the inlet gas which can be admitted to the furnace without oxidizing the iron *b*. The activity of carbon (with respect to graphite) in the equilibrated gas of this initial minimum H_2/CO_2 ratio

c. The total pressure to which the equilibrated gas would have to be raised to saturate the iron with graphite at 1000 K

d. The effect, on the partial pressure of oxygen in the equilibrated gas, of this increase in total pressure

Given:

$$C_{(gr)} + \tfrac{1}{2}O_{2(g)} = CO_{(g)} \qquad \Delta G^0 = -111{,}700 - 87.65T \text{ joules}$$
$$C_{(gr)} + O_{2(g)} = CO_{2(g)} \qquad \Delta G^0 = -394{,}100 - 0.84T \text{ joules}$$
$$H_{2(g)} + \tfrac{1}{2}O_{2(g)} = H_2O_{(g)} \qquad \Delta G^0 = -246{,}000 + 54.8T \text{ joules}$$
$$Fe_{(s)} + \tfrac{1}{2}O_{2(g)} = FeO_{(s)} \qquad \Delta G^0 = -259{,}600 + 62.55T \text{ joules}$$

a. Determine the composition of the gas phase which is in equilibrium with Fe and FeO at 1000 K and 1 atm pressure. The system contains four components (Fe, C, H, and O) and exists in three phases. The equilibrium thus has $F = C + 2 - P = 3$ degrees of freedom, two of which are used by fixing the temperature at 1000 K and the pressure at 1 atm. From the stoichiometry of the reaction $CO_2 + H_2 = CO + H_2O$, equal numbers of moles of CO and H_2O are produced, and hence the partial pressures of CO and H_2O in the gas mixture are equal. This uses the third degree of freedom, and hence at 1000 K and $P = 1$ atm, Fe and FeO are in equilibrium with a gas of unique çomposition in which $p_{CO} = p_{H_2O}$. As the system at equilibrium comprises six species (FeO, Fe, CO, CO_2, H_2, and H_2O) and has four components, there exist two independent reaction equilibria, which can be chosen as being

$$CO + FeO = CO_2 + Fe \tag{i}$$

and

$$H_2 + FeO = H_2O + Fe \tag{ii}$$

Summing the standard free energy changes for the reactions $CO + \tfrac{1}{2}O_2 = CO_2$ and $FeO = Fe + \tfrac{1}{2}O_2$ gives

$$\Delta G^0_{(i)} = -22{,}800 + 24.26T \text{ joules}$$

or

$$\Delta G^0_{(i),1000} = 1460 \text{ joules}$$

Thus

$$K_{(i),1000} = 0.839 = \frac{p_{CO_2}}{p_{CO}}$$

Summing the standard free energy changes for the reactions $H_2 + \tfrac{1}{2}O_2 = H_2O$

and $FeO = Fe + \frac{1}{2}O_2$ gives

$$\Delta G^0_{(ii)} = 13{,}600 - 7.75T \text{ joules} \quad \text{or} \quad \Delta G^0_{(ii),1000} = 5850 \text{ joules}$$

Thus

$$K_{(ii),1000} = 0.495 = \frac{p_{H_2O}}{p_{H_2}}$$

But, in the equilibrated gas, $p_{CO_2} + p_{CO} + p_{H_2} + p_{H_2O} = 1$

and so

$$0.839\, p_{CO} + p_{CO} + \frac{p_{H_2O}}{0.495} + p_{H_2O} = 1$$

or, as $p_{CO} = p_{H_2O}$,

$$p_{CO}\left(0.839 + 1 + \frac{1}{0.495} + 1\right) = 1$$

or $p_{CO} = 0.206$ atm. Thus

$$p_{H_2O} = 0.206 \text{ atm}$$

$$p_{CO_2} = 0.206 \times 0.839 = 0.173 \text{ atm}$$

and

$$p_{H_2} = \frac{0.206}{0.495} = 0.416 \text{ atm}$$

From the stoichiometric reaction $CO_2 + H_2 = CO + H_2O$, if the initial mixture contained x moles of CO_2 to each mole of H_2, then the final equilibrated gas mixture would contain $(x - a)$ moles of CO_2, $(1 - a)$ moles of H_2, a moles of H_2O, and a moles of CO, the partial pressures of which would then be

$$p_{CO_2} = \frac{(x-a)}{(1+x)}P$$

$$p_{H_2} = \frac{(1-a)}{(1+x)}P$$

$$p_{CO} = \frac{a}{(1+x)}P$$

and

$$p_{H_2O} = \frac{a}{(1+x)}P$$

For $P = 1$ atm,

$$\frac{p_{CO_2}}{p_{CO}} = \frac{(x-a)}{a} = 0.839 \tag{iii}$$

and

$$\frac{p_{H_2O}}{p_{H_2}} = \frac{a}{(1-a)} = 0.495 \tag{iv}$$

From Eq. (iii), $x = 1.839a$, and from Eq. (iv), $a = 0.331$. Thus $x = 0.609$ = the initial CO_2/H_2 ratio in the inlet gas mixture which, when equilibrated, is in equilibrium with Fe and FeO at 1000 K and 1 atm pressure.

The oxygen pressure in the system is that for equilibrium between Fe and FeO. For $FeO = Fe + \frac{1}{2}O_2$, $\Delta G^0 = 259{,}600 - 62.55T$ joules, and $\Delta G^0_{1000} =$ 197,050 joules.

Thus

$$K_{1000} = p^{1/2}_{O_2\,(\text{eq, Fe/FeO, 1000})} = 5.10 \times 10^{-11}$$

and so

$$p_{O_2\,(\text{eq, Fe/FeO, 1000})} = 2.6 \times 10^{-21} \text{ atm}$$

For the reaction $CO + \frac{1}{2}O_2 = CO_2$, (A)

$$\Delta G^0_A = -282{,}400 + 86.81T \text{ joules}$$

and thus at 1000 K

$$K_A = \frac{p_{CO_2}}{p_{CO}p^{1/2}_{O_2}} = 1.646 \times 10^{10}$$

or

$$p^{1/2}_{O_2}(\text{in the gas mixture}) = \frac{p_{CO_2}}{p_{CO}K_A} = \frac{(x-a)}{a}\frac{1}{K_A} \tag{v}$$

Similarly, for the reaction $H_2 + \frac{1}{2}O_2 = H_2O$ (B)

$$\Delta G^0_B = -246{,}000 + 54.8T \text{ joules}$$

and thus, at 1000 K,

$$K_B = \frac{p_{H_2O}}{p_{H_2}p^{1/2}_{O_2}} = 9.708 \times 10^9$$

or

$$p^{1/2}_{O_2}(\text{in the gas mixture}) = \frac{p_{H_2O}}{p_{H_2}K_B} = \frac{a}{(1-a)K_B} \tag{vi}$$

From Eq. (vi),

$$a = \frac{p_{O_2}^{1/2} K_B}{(1 + K_B p_{O_2}^{1/2})}$$

which, when substituted into Eq. (v) gives

$$p_{O_2}^{\frac{1}{2}} K_A = \left(x - \frac{p_{O_2}^{\frac{1}{2}} K_B}{(1 + p_{O_2}^{\frac{1}{2}} K_B)} \right) \div \left(\frac{p_{O_2}^{\frac{1}{2}} K_B}{1 + p_{O_2}^{\frac{1}{2}} K_B} \right)$$

or

$$p_{O_2}(K_A K_B) + p_{O_2}^{\frac{1}{2}} K_B (1 - x) - x = 0$$

the solution of which is

$$p_{O_2}^{\frac{1}{2}} = \frac{-K_B(1 - x) \mp \sqrt{K_B^2(1 - x)^2 + 4K_A K_B x}}{2K_A K_B}$$

This relates the oxygen pressure in the equilibrated gas to the initial CO_2/H_2 ratio in the inlet gas.

When $x = 0.609$ $p_{O_2} = 2.6 \times 10^{-21}$
when $x > 0.609$ $p_{O_2} > 2.6 \times 10^{-21}$
and when $x < 0.609$ $p_{O_2} < 2.6 \times 10^{-21}$

Thus 0.609 is the maximum allowable CO_2/H_2 ratio in the inlet gas; or, 1.64 is the minimum allowable H_2/CO_2 ratio in the inlet gas.

b. For the reaction $C_{(gr)} + CO_{2(g)} = 2CO_{(g)}$

$$\Delta G^0 = 170{,}700 - 174.5T \text{ joules}$$

$$\Delta G^0_{1000} = -3800 \text{ joules}$$

and

$$K = 1.579$$

Thus

$$1.579 = \frac{p_{CO}^2}{p_{CO_2} a_C} \quad \text{or} \quad a_C = \frac{p_{CO}^2}{p_{CO_2} \times 1.579}$$

$$= \frac{0.206^2}{0.173 \times 1.579} = 0.155$$

Alternatively, as

$$p_{CO_2} = \frac{n_{CO_2}}{n_{\text{total}}} P = \frac{x-a}{1+x} P$$

and

$$p_{CO} = \frac{n_{CO}}{n_{\text{total}}} P = \frac{a}{1+x} P$$

then

$$a_C = \frac{a^2}{(1+x)(x-a)} \frac{P}{1.579}$$

which, for $P = 1$ atm, gives $a_C = 0.155$.

c. To saturate the iron with carbon, the pressure must be raised until $a_C = 1$. As $a_C = 0.155\ P$ (i.e., the activity of carbon in the gas is directly proportional to the total pressure of the gas), then $a_C = 1$ when $P = 1/0.155 = 6.45$ atm.

d. From either Eq. (v) or Eq. (vi), it is seen that the oxygen pressure in the gas mixture is not a function of the total pressure of the gas.

Example 2

Determine the conditions under which an Fe-Cr-O melt is in equilibrium with (i) solid Cr_2O_3 and (ii) solid $FeO{\cdot}Cr_2O_3$ at 1600°C. For

$$2Cr_{(s)} + \tfrac{3}{2}O_{2(g)} = Cr_2O_{3(s)} \qquad \Delta G^0_{(i)} = -1{,}120{,}300 + 259.8T \text{ joules} \qquad \text{(i)}$$

and for $\frac{1}{2}O_{2(g)} = [O]_{(1\text{ wt\%})}$ $\qquad \Delta G^0_{(ii)} = -111{,}070 - 5.87T$ joules $\qquad$ (ii)

At 1600°C, Fe-Cr melts exhibit Raoultian ideality, and the molar heat of fusion of Cr, at its equilibrium melting temperature of 2173 K, is 21,000 joules.

Thus, for $Cr_{(s)} = Cr_{(l)}$,

$$\Delta G^0_m = \Delta H^0_m - T\frac{\Delta H^0_m}{T_m} = 21{,}000 - 9.66T \text{ joules}$$

and for $Cr_{(l)} = [Cr]_{(1\text{ wt\% in Fe})}$,

$$\Delta G^0 = RT \ln \frac{55.85}{100 \times 52.01}$$

$$= -37.70T \text{ joules}$$

Therefore, for $Cr_{(s)} = [Cr]_{(1\text{ wt\%})}$,

$$\Delta G^0_{(iii)} = 21{,}000 - 47.36T \text{ joules} \tag{iii}$$

The standard free energy change for the reaction

$$2[Cr]_{(1\ wt\%)} + 3[O]_{(1\ wt\%)} = Cr_2O_{3(s)} \tag{iv}$$

is thus

$$\Delta G^0_{(iv)} = \Delta G^0_{(i)} - 3\,\Delta G^0_{(ii)} - 2\,\Delta G^0_{(iii)}$$
$$= -829{,}090 + 372.13T \text{ joules}$$
$$= -RT \ln \frac{a_{Cr_2O_3}}{h^2_{Cr(1\ wt\%)} \cdot h^3_{O(1\ wt\%)}}$$

or, at 1873 K,

$$\log \frac{h^2_{Cr(1\ wt\%)} \cdot h^3_{O(1\ wt\%)}}{a_{Cr_2O_3}} = -3.68 \tag{v}$$

Saturation of the melt with respect to solid Cr_2O_3 occurs at $a_{Cr_2O_3} = 1$, and if the interactions between Cr and O in solution are ignored, and it is assumed that oxygen obeys Henry's Law, Eq. (v) can be written as

$$\log\,[\text{wt\% Cr}] = -1.5 \log\,[\text{wt\% O}] - 1.84 \tag{vi}$$

which is thus the variation of [wt% Cr] with [wt% O] in liquid iron required for equilibrium with solid Cr_2O_3 at 1600°C. Equation (v) is drawn as line (vi) in Fig. 13.35.

For

$$Fe_{(l)} + 2Cr_{(s)} + 2O_{2(g)} = FeO{\cdot}Cr_2O_{3(s)} \qquad \Delta G^0_{(vii)} = -1{,}409{,}420 + 318.07T \tag{vii}$$

and thus, for the reaction

$$Fe_{(l)} + 2[Cr]_{(1\ wt\%)} + 4[O]_{(1\ wt\%)} = FeO{\cdot}Cr_2O_{3(s)} \tag{viii}$$

$$\Delta G^0_{(viii)} = \Delta G^0_{(vii)} - 2\Delta G^0_{(iii)} - 4\Delta G^0_{(ii)}$$
$$= -1{,}007{,}140 + 436.27T \text{ joules}$$
$$= -RT \ln \frac{a_{FeO\cdot Cr_2O_3}}{a_{Fe} \cdot h^2_{Cr(1\ wt\%)} \cdot h^4_{O(1\ wt\%)}}$$

or, at 1873 K,

$$\log \frac{a_{Fe} \cdot h^2_{Cr(1\ wt\%)} \cdot h^4_{O(1\ wt\%)}}{a_{FeO \cdot Cr_2O_3}} = -5.30 \qquad \text{(ix)}$$

Saturation of the melt with respect to $FeO{\cdot}Cr_2O_3$ occurs at $a_{FeO \cdot Cr_2O_3} = 1$, and, with the same assumptions as before, and $a_{Fe} = X_{Fe} = 1 - X_{Cr}$, the variation of [wt% Cr] with [wt% O] required for equilibrium with solid $FeO{\cdot}Cr_2O_3$ at 1600°C is

$$\log(1 - X_{Cr}) + 2 \log [\text{wt\% Cr}] + 4 \log [\text{wt\% O}] = -5.30 \qquad \text{(x)}$$

In solutions sufficiently dilute that $X_{Fe} \sim 1$, Eq. (x) can be simplified as

$$\log [\text{wt\% Cr}] = -2 \log [\text{wt\% O}] - 2.65 \qquad \text{(xi)}$$

Equation (x) is drawn as line (x) in Fig. 13.35. Lines (vi) and (x) intersect at the point A, log [wt% Cr] = 0.59, log [wt% O] = −1.62 (wt% O = 0.024, wt% Cr = 3.89), which is the composition of that melt that is in equilibrium with both solid Cr_2O_3 and solid $FeO{\cdot}Cr_2O_3$. From the phase rule, equilibrium in a three-component system (Fe-Cr-O) among four phases (liquid Fe-Cr-O, solid Cr_2O_3, solid $FeO{\cdot}Cr_2O_3$, and a gas phase) has one degree of freedom, which, in the present case, has been used by specifying the temperature to be 1873 K. Thus the activities of Fe, Cr, and O are uniquely fixed, and hence

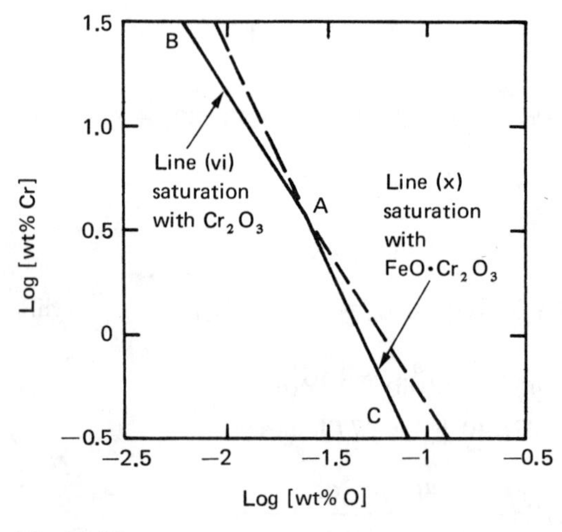

Fig. 13.35.

[wt% Cr] and [wt% O] are uniquely fixed. The equilibrium oxygen pressure in the gas phase is obtained, from Eq. (ii), as

$$\Delta G^0_{(ii),1873\,K} = -122{,}065 \text{ joules} = -8.3144 \times 1873 \ln \frac{[\text{wt\% O}]}{p_{O_2}^{1/2}}$$

which, with [wt% O] $= 0.024$, gives $p_{O_2(eq)} = 8.96 \times 10^{-11}$ atm. The positions of the lines in Fig. 13.35 are such that, in melts of [wt% Cr] > 3.89, Cr_2O_3 is the stable phase in equilibrium with saturated melts along the line AB, and, in melts of [wt% Cr] < 3.89, $FeO \cdot Cr_2O_3$ is the stable phase in equilibrium with saturated melts along the line AC. Alternatively, Cr_2O_3 is the stable phase in equilibrium with saturated melts of [wt% O] < 0.024, and $FeO \cdot Cr_2O_3$ is the stable phase in equilibrium with saturated melts of [wt% O] > 0.024. Consider a melt in which log [wt% Cr] = 1.5. From Fig. 13.35, or Eq. (vi), the oxygen content at this chromium level required for equilibrium with Cr_2O_3 (at the point B in Fig. 13.35) is 5.93×10^{-3} wt%, or log [wt% O] $= -2.25$. From Eq. (v), the activity of Cr_2O_3 in this melt with respect to solid Cr_2O_3 is unity, and hence the melt is saturated with respect to solid Cr_2O_3. However, from Eq. (ix), in the same melt, i.e., $X_{Fe} = 0.668$, [wt% Cr] $= 31.6$, [wt% O] $= 0.00593$, the activity of $FeO \cdot Cr_2O_3$ with respect to solid $FeO \cdot Cr_2O_3$ is only 0.2. Thus the melt is saturated with respect to Cr_2O_3 and is undersaturated with respect to $FeO \cdot Cr_2O_3$. Moving along the line BA from B toward A, $a_{Cr_2O_3} = 1$ and $a_{FeO \cdot Cr_2O_3}$ increases in value from 0.2 at B to unity at A in the doubly saturated melt. Consider a melt in which log [wt% Cr] = −0.5. From Fig. 13.35 the oxygen content required for saturation with $FeO \cdot Cr_2O_3$ is 0.084 wt% (log [wt% O] $= -1.075$ at the point C in Fig. 13.35). From Eq. (ix), the activity of $FeO \cdot Cr_2O_3$ in this melt is unity. However, from Eq. (v), the activity of Cr_2O_3 in the melt, with respect to solid Cr_2O_3, is only 0.285. Thus, this melt is saturated with $FeO \cdot Cr_2O_3$ and is undersaturated with Cr_2O_3. On moving along the line CA from C toward A, $a_{FeO \cdot Cr_2O_3}$ is unity and $a_{Cr_2O_3}$ increases from 0.285 at C to unity at A.

If the various solute-solute interactions had been considered, Eq. (v), with $a_{Cr_2O_3} = 1$, would be written as

$$2 \log h_{Cr(1\,wt\%)} + 3 \log h_{O(1\,wt\%)} = -3.68$$

or

$$2 \log f_{Cr(1\,wt\%)} + 2 \log [\text{wt\% Cr}] + 3 \log f_{O(1\,wt\%)} + 3 \log [\text{wt\% O}] = -3.68$$

or

$$2e_{Cr}^{Cr} \cdot [\text{wt\% Cr}] + 2e_{Cr}^{O} \cdot [\text{wt\% O}] + 2 \log [\text{wt\% Cr}] + 3e_{O}^{O} \cdot [\text{wt\% O}] + 3e_{O}^{Cr} \cdot [\text{wt\% Cr}] + 3 \log [\text{wt\% O}] = -3.68$$

With $e_{\rm Cr}^{\rm Cr} = 0 \quad e_{\rm O}^{\rm O} = -0.2 \quad e_{\rm O}^{\rm Cr} = -0.041$ and $e_{\rm Cr}^{\rm O} = -0.13$

this gives

$$-0.43\,[\text{wt\% O}] + 0.0615\,[\text{wt\% Cr}] + \log\,[\text{wt\% Cr}] + 1.5\log\,[\text{wt\% O}] = -1.84 \qquad \text{(xii)}$$

which is drawn as the line (xii) in Fig. 13.36.

Similarly, with $a_{\rm FeO\cdot Cr_2O_3} = 1$, Eq. (ix) would be written as

$$\log X_{\rm Fe} + 2\log h_{{\rm Cr}(1\,\text{wt\%})} + 4\log h_{{\rm O}(1\,\text{wt\%})} = -5.30$$

or

$$\log X_{\rm Fe} + 2e_{\rm Cr}^{\rm Cr}\cdot[\text{wt\% Cr}] + 2e_{\rm Cr}^{\rm O}\cdot[\text{wt\% O}] + 2\log\,[\text{wt\% Cr}] + 4e_{\rm O}^{\rm O}\cdot[\text{wt\% O}] + 4e_{\rm O}^{\rm Cr}\cdot[\text{wt\% Cr}] + 4\log\,[\text{wt\% O}] = -5.30$$

or

$$\log X_{\rm Fe} - 1.06[\text{wt\% O}] - 0.164[\text{wt\% Cr}] + 2\log\,[\text{wt\% Cr}] + 4\log\,[\text{wt\% O}] = -5.30 \qquad \text{(xiii)}$$

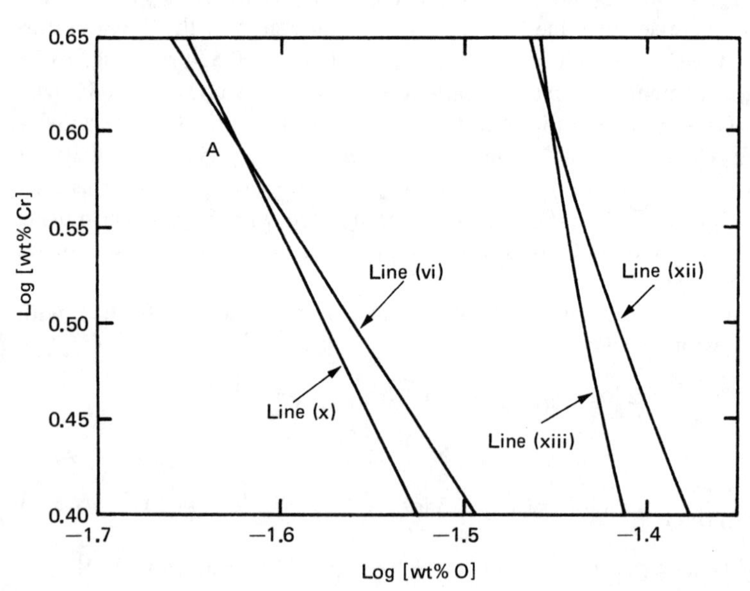

Fig. 13.36.

which is drawn as line (xiii) in Fig. 13.36. Lines (xii) and (xiii) intersect at log [wt% Cr] = 0.615, log [wt% O] = −1.455 ([wt% Cr] = 4.12 [wt% O] = 0.035).

PROBLEMS

13.1 Air at atmospheric pressure is blown over a Cu-rich copper-gold liquid solution at 1500 K. If only the copper is oxidized (to form pure solid Cu_2O), calculate the minimum activity of Cu which can be obtained in the solution.

13.2 Magnesium can be removed from Mg–Al liquid solutions by selectively forming the chloride $MgCl_2$. Calculate the activity of Mg in the liquid Mg–Al system which can be achieved at 800°C by reacting the solution with an H_2-HCl gas mixture containing hydrogen at essentially 1 atm pressure and $p_{HCl} = 10^{-5}$ atm to form pure liquid $MgCl_2$.

13.3 The partial pressure of oxygen in equilibrium with pure liquid lead and pure liquid PbO at 1200 K is 3.72×10^{-9} atm. If SiO_2 is added to the liquid PbO such that equilibrium p_{O_2} for the pure Pb-liquid PbO-SiO_2 solution couple is decreased to 9.29×10^{-10} atm, calculate the activity of PbO in the lead silicate melt.

13.4 Copper, present as an impurity in liquid Pb, can be removed by adding PbS to the Cu-Pb alloy and allowing the exchange reaction

$$2Cu_{(s)} + PbS_{(s)} = Cu_2S_{(s)} + Pb_{(l)}$$

to come to equilibrium.

The solid sulfides are mutually immiscible, Pb is insoluble in solid Cu, and the Cu liquidus, below 850°C, can be represented by

$$\log X_{Cu} = -\frac{3500}{T} + 2.261$$

where X_{Cu} is the solubility of Cu in liquid Pb. If Cu obeys Henry's law in liquid Pb, calculate the extent to which Cu can be removed from liquid Pb by this process at 800°C. Would the extent of the purification be increased by lowering the temperature?

13.5 A CH_4-H_2 gas mixture at 1 atm total pressure, in which $p_{H_2} = 0.955$ atm, is equilibrated with an Fe-C alloy at 1000 K. Calculate the activity of C with respect to solid graphite in the alloy. What would the value of p_{H_2} in the gas mixture (at $P_{total} = 1$ atm) have to be in order to saturate Fe with graphite at 1000 K?

13.6 Calculate the activity of FeO in an FeO–Al_2O_3–SiO_2 melt below which the FeO cannot be reduced to pure liquid iron by a CO-CO_2 mixture of $p_{CO}/p_{CO_2} = 10^5$ at 1600°C.

13.7 The p_{H_2}/p_{H_2O} ratio in an H_2-H_2O gas mixture in equilibrium with pure liquid lead and liquid lead silicate of $X_{PbO} = 0.7$ is 5.66×10^{-4} at 900°C. The corresponding value at 1100°C is 1.2×10^{-3}. Calculate the partial molar heat of solution of liquid PbO in this lead silicate.

13.8 An Fe-Mn solid solution containing $X_{Mn} = 0.001$ is in equilibrium with an FeO-MnO solid solution and an oxygen containing gaseous atmosphere at 1000 K. How many degrees of freedom does the equilibrium have? What is the composition of the equilibrium oxide solution, and what is the oxygen pressure in the gas phase? Assume that both solid solutions are Raoultian.

13.9 The elements A and B, which are both solid at 1000°C, form two stoichiometric compounds A_2B and AB_2, which are also both solid at 1000°C. No solid solutions occur in the A-B system. A has an immeasurably small vapor pressure at 1000°C, and for the change of standard state, $B_{(s)} = B_{(v)}$

$$\Delta G^0 = 187{,}220 - 108.8T \text{ joules}$$

The vapor pressure exerted by an equilibrated AB_2-A_2B mixture is given as

$$\log p(\text{atm}) = -\frac{11{,}242}{T} + 6.53$$

and the vapor pressure exerted by an equilibrated A-A_2B mixture is given as

$$\log p(\text{atm}) = -\frac{12{,}603}{T} + 6.9$$

From these data, calculate the standard free energies of formation of A_2B and AB_2.

13.10 For the change of standard state

$$V_{(s)} = V_{(1\text{ wt\% in Fe})}$$

$$\Delta G^0 = -15{,}480 - 45.61T \text{ joules}$$

Calculate the value of γ_V^0 at 1600°C. If a liquid Fe-V solution is equilibrated with pure solid VO and a gas phase containing $p_{O_2} = 6.45 \times 10^{-11}$ atm, calculate the activity of V in the liquid solution

a. With respect to solid V as the standard state
b. With respect to liquid V as the standard state
c. With respect to the Henrian standard state
d. With respect to the 1 wt% in iron standard state.

13.11 An Fe-Ti liquid solution containing Ti at $h_{(1\ \text{wt\% in Fe})} = 1$ is in equilibrium with pure solid TiO_2 and an H_2O-H_2 gas mixture in which $p_{H_2O}/p_{H_2} = 5.56 \times 10^{-3}$ at 1600°C. Calculate γ^0_{Ti} at 1600°C.

13.12 When an Fe-P liquid solution is equilibrated at 1900 K with solid CaO, solid $3CaO \cdot P_2O_5$, and a gas phase containing $p_{O_2} = 10^{-10}$ atm, the activity of P in the iron, with respect to the 1 wt% in iron standard state, is 20. Given that

$$\Delta G^0_{1900} = -564{,}600 \text{ joules for } 3CaO_{(s)} + P_2O_{5(g)} = 3CaO \cdot P_2O_{5(s)}$$

and $$\Delta G^0 = -122{,}200 - 19.2T \text{ joules}$$

for $$\tfrac{1}{2}P_{2(g)} = P_{(1\ \text{wt\% in Fe})}$$

calculate ΔG^0_{1900} for

$$P_{2(g)} + \tfrac{5}{2}\ O_{2(g)} = P_2O_{5(g)}$$

13.13 Liquid iron, contained in an Al_2O_3 crucible under a gaseous atmosphere of $p_{O_2} = 3 \times 10^{-12}$ atm at 1600°C, contains its equilibrium contents of dissolved oxygen and aluminum. To what value must the p_{O_2} in the equilibrating gaseous atmosphere be raised in order that solid hercynite ($FeO \cdot Al_2O_3$) appears in equilibrium with the melt and solid Al_2O_3? What is the activity of Al (with respect to the 1 wt% in iron standard state) in this state? How many degrees of freedom does this equilibrium have at 1600°C? Given

$$\tfrac{1}{2}O_{2(g)} = O_{(1\ \text{wt\% in Fe})} \qquad \Delta G^0 = -111{,}070 - 5.87T \text{ joules}$$

$$Al_{(l)} = Al_{(1\ \text{wt\% in Fe})} \qquad \Delta G^0 = -43{,}100 - 32.26T \text{ joules}$$

$$FeO \cdot Al_2O_{3(s)} = Fe_{(l)} + O_{(1\ \text{wt\% in Fe})} + Al_2O_{3(s)}$$

$$\Delta G^0 = 146{,}230 - 54.35T \text{ joules}$$

13.14 Pure solid Si, pure solid SiO_2, and pure solid Si_3N_4 are equilibrated with an oxygen-nitrogen gas mixture at 1000 K. How many degrees of freedom does this equilibrium have? What is the equilibrium state? If, maintaining the gas composition constant, the temperature of the system is decreased to 800 K, what happens? If, maintaining the gas composition constant, the temperature of the system is increased to 1200 K, what happens?

13.15 At 1550 K and 1 atm total pressure, can (*a*) Cr_2O_3 or (*b*) MnO be reduced to the respective metals by solid carbon or do the carbides ($Cr_{23}C_6$ and Mn_3C, respectively) form?

13.16 A and B form two compounds, AB and AB_3 for which

$$A + B = AB \qquad \Delta G^0 = -26{,}000 - 4.56T \text{ joules}$$

and

$$A + 3B = AB_3 \qquad \Delta G^0 = -22{,}260 - 8.83T \text{ joules}$$

Solid A and solid B are mutually immiscible, and the two compounds exist at fixed compositions. Calculate the minimum temperatures at which AB_3 and AB are stable.

13.17 In the Pidgeon process for the production of magnesium, dolomite ($MgO \cdot CaO$) is reduced by silicon to form magnesium vapor and Ca_2SiO_4. Calculate the equilibrium pressure of magnesium vapor produced by this reaction at 1200°C. The free energy of formation of dolomite from CaO and MgO is small enough that it can be ignored.

13.18 A mixture of ZnO and graphite is placed in an evacuated vessel that is then heated to 1200 K. Calculate the partial pressures of Zn, CO, and CO_2 that are developed.

13.19 An assemblage of solid CaO, MgO, and $3CaO \cdot Al_2O_3$ and liquid Al exerts an equilibrium vapor pressure of Mg of 0.035 atm at 1300 K. Write the equation for the appropriate reaction equilibrium. Calculate the standard free energy of formation of $3CaO \cdot Al_2O_3$ from CaO and Al_2O_3 and the activity of Al_2O_3 in CaO-saturated $3CaO \cdot Al_2O_3$ at 1300 K.

13.20 An iron–carbon melt containing 0.5 wt% C is prepared in an alumina crucible under an atmosphere of $p_{CO} = 10^{-2}$ atm at 1600°C. Calculate the equilibrium concentration of Al in the melt. Ignore all solute-solute and solvent-solute interactions.

13.21 Calculate the total pressure, i.e., $p_{SO_3} + p_{SO_2} + p_{O_2}$, exerted by equilibrated CoO-$CoSO_4$ at 1223 K.

13.22 It is required that PbO be eliminated from an ore containing PbO, PbS, and $PbSO_4$ by converting it to PbS or $PbSO_4$ by reaction with a SO_2-O_2 gas. Although the pressure of O_2 in the gas can vary within wide limits, the partial pressure of SO_2 in the gas may not be greater than 0.5 atm.

Calculate the maximum temperature at which it can be guaranteed that the PbO phase will be eliminated.

13.23 Cementite, Fe_3C, is metastable with respect to carbon-saturated α-iron and graphite by 2450 joules at 950 K and 1 atm pressure. Given that the molar volumes of α-Fe, $C_{(gr)}$, and Fe_3C at 950 K are 7.32, 5.40, and 23.92 cm^3/mole, calculate the pressure, at 950 K, at which Fe_3C is in equilibrium with carbon-saturated α-Fe and graphite.

13.24 CaO removes sulfur from liquid iron by reacting with the sulfur to produce CaS and oxygen. If the iron contains carbon in solution, it can be considered that the CaO reacts with the S and C in solution to form CaS and CO gas. Calculate the equilibrium sulfur contents of liquid iron-carbon alloys containing 0.2 and 2.0 wt% C that have been treated with CaO under an atmosphere of $p_{CO} = 1$ atm at 1800 K. The interaction coefficients are $e_C^C = 0.22$, $e_S^S = -0.028$, $e_S^C = 0.24$, and $e_C^S = 0.09$.

13.25 An experiment is being conducted on an equilibrated mixture of $CaCO_3$ and CaO contained in a closed vessel at 1200 K. The mixture is contaminated with iron in the form of hematite (Fe_2O_3). The contaminant would not be harmful to the experiment if it occurred either as wustite (FeO) or as cementite (Fe_3C). The necessary changes in the chemical form of the contaminant can be effected by admitting CO gas to the vessel. Calculate the allowable limits of p_{CO} in the vessel for the occurrence of the contaminant (*a*) as wustite and (*b*) as cementite, given that the free energy of formation of Fe_3C from Fe and C is −1550 joules at 1200 K.

13.26 A Cu-Au alloy of $X_{Cu} = 0.5$ is being annealed at 600°C in deoxidized argon. The argon is being deoxidized by being passed over heated pure copper turnings prior to its being admitted to the annealing furnace. The solid Cu-Au system is virtually regular in its solution behavior, with the excess molar free energy of mixing being given by

$$G^{xs} = -28{,}280 X_{Cu} \cdot X_{Au} \text{ joules}$$

Assuming that equilibrium is attained in the deoxidizing furnace, calculate the maximum temperature at which the deoxidizing furnace can be operated without causing oxidation of the copper in the Cu-Au alloy being annealed.

13.27 The variation of γ_{Si}^0 in liquid iron is given by

$$\log \gamma_{Si}^0 = \frac{-6230}{T} + 0.37$$

Calculate the limit to which $\Delta \bar{H}_{Si}^M$ tends as X_{Si} tends to zero.

CHAPTER FOURTEEN

ELECTROCHEMISTRY

14.1 INTRODUCTION

All chemical reactions, in which products form from reactants, involve changes in the state of oxidation or "valency state" of some or all of the participating atoms. By convention, the valency state of an atom in a compound is determined by the number of electrons which surround the nucleus of the atom. The assignment of valency states to the constituent atoms of a compound is, in no way, influenced by the nature of the bonding between the constituent atoms; e.g., although the bonding in the HF molecule is considered to be 50 percent ionic in character and 50 percent covalent in character, where "ionic character" implies complete electron transfer from the H atom to the F atom to form the ions H^+ and F^-, and "covalent character" implies complete electron sharing to give the normal covalent HF molecule, which has zero electric dipole moment, the valency states of H and F in HF are, respectively, +1 and −1.

Changes in the valency state of an element are thus brought about by the addition or removal of electrons, and hence the thermodynamic driving force of any reaction must, in some way, be related to the ease with which the required valency changes of the participating atoms can occur, i.e., to the ease with which the necessary electron transfer can occur. For example, the reaction

$$AO + B = BO + A$$

involves a decrease in the valency state of A from +2 to zero and an increase in the valency state of B from zero to +2; i.e., the reaction involves the transfer of two electrons from B to A, and hence can be written as

$$A^{2+} + B = B^{2+} + A$$

The free energy decrease is thus a manifestation of the energetics of electron transfer. The reaction equation, written as the sum of

$$A^{2+} + 2e^- = A$$

and

$$B - 2e^- = B^{2+}$$

suggests the possibility of conducting the reaction as indicated in Fig. 14.1. A mixture of A+AO and a mixture of B+BO are joined by two connections (*a*) and (*b*), where (*a*) is an electronic conductor through which only electrons can pass, and (*b*) is an ionic conductor through which only oxygen ions can pass. The spontaneous reaction then occurs in the following manner. Two electrons leave a B atom, producing thus a B^{2+} ion, travel from right to left along (*a*), and, on arrival at the A+AO mixture, convert an A^{2+} ion to an A atom. Simultaneously an O^{2-} ion leaves the A+AO mixture and passes through (*b*) to the B+BO mixture. Charge neutrality in the overall system is thus maintained, and the overall reaction can be written as

$$A^{2+}O^{2-} + B = B^{2+}O^{2-} + A$$

As a result of its being conducted in an electrochemical manner, this reaction is termed an electrochemical reaction. The driving force for the electron movement along (*a*) is manifested as an electrical voltage (or electric potential difference) which can be measured by placing in the circuit (*a*) an external opposing voltage which is adjusted until no current flows, i.e., the electrochemical reaction ceases. At this point the external voltage exactly balances the voltage generated by the electrochemical system; i.e., the thermodynamic driving force for the chemical reaction is exactly balanced by the externally applied electric driving force.

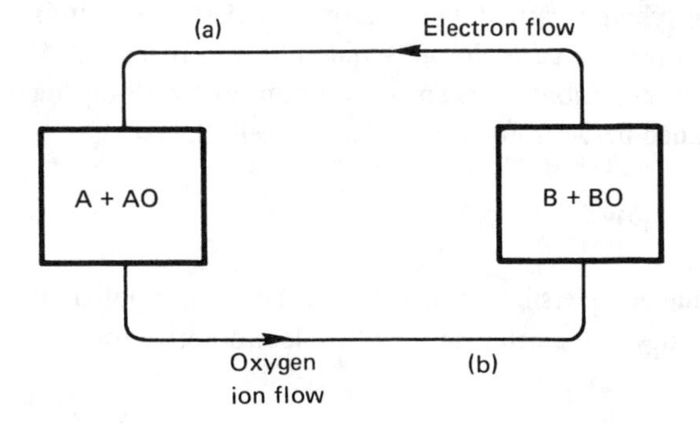

Fig. 14.1. Schematic representation of an electrolytic reaction.

Knowledge of the mathematical relationship between these two types of forces allows the former, (ΔG for the reaction), to be measured. Furthermore, whereas, in Chap. 10, it was seen that pure A, pure AO, pure B, and pure BO are only in thermodynamic equilibrium at the unique temperature at which the two standard Ellingham lines intersect (if, indeed the lines do intersect), it is now seen that the four pure phases can be brought into electrochemical equilibrium at any temperature by balancing the chemical driving force with an opposing electrical driving force. The properties of systems such as this are examined in this chapter.

14.2 THE RELATIONSHIP BETWEEN CHEMICAL AND ELECTRICAL DRIVING FORCES

Equation (5.8) showed that when a system undergoes a reversible process at constant temperature and pressure, the decrease in the free energy of the system equals $w'_{\max}$ the work (other than work of expansion) done by the system. For an increment of such a process,

$$-dG' = \delta w'_{\max}$$

Consider a system which performs electrical work by transporting an electric charge across a voltage difference, i.e., from one electric potential to another. The work performed is obtained as the product of the charge transported, q (coulombs) and the electric potential difference $\Delta\phi$ (volts), and the unit of such work is the joule (= volts $\times$ coulombs). A system which is capable of performing electrical work as a result of the occurrence of a chemical reaction is called a galvanic cell, and the overall chemical reaction is represented by an equation called the cell reaction. The charge carried by 1 gram ion (i.e., Avogadro's number of ions) of unit positive charge is 96,487 coulombs and is termed $\mathfrak{F}$, Faraday's constant. Thus if dn gram ions of valency z are transported through a voltage difference $\Delta\phi$ maintained between the electrodes of a cell, then

$$\delta w' = z\mathfrak{F}\Delta\phi dn$$

If the transportation is conducted reversibly, in which case the electric potential difference between the electrodes of the cell is termed the electromotive force or EMF, $\mathcal{E}$, of the cell, then

$$\delta w'_{\max} = z\mathfrak{F}\mathcal{E}dn = -dG' \tag{14.1}$$

Consider the familiar Daniell cell shown in Fig. 14.2. This cell comprises a zinc electrode dipping into an aqueous solution of $ZnSO_4$ and a copper electrode dipping into an aqueous solution of $CuSO_4$. The two aqueous solutions, which constitute the electrolyte of the galvanic cell (i.e., the medium through which ionic current flows) are prevented from mixing by the insertion between them of a porous diaphragm. The cell reaction is written as

$$Zn + CuSO_4 = Cu + ZnSO_4$$

or as

$$Zn + Cu^{2+} = Cu + Zn^{2+}$$

This can be regarded as being the sum of two half-cell reactions, namely, the half-cell reaction

$$Zn = Zn^{2+} \text{ (in aqueous solution)} + 2e^-$$

occurring at the zinc electrode, and the half-cell reaction

$$Cu^{2+} \text{ (in aqueous solution)} + 2e^- = Cu$$

occurring at the copper electrode.

Consider the Zn-$ZnSO_4$ compartment. When the half-cell reaction occurs, zinc ions enter the solution, and the zinc electrode is left with an excess of electrons. Equilibrium is established when

$$\mu_{Zn^{2+}} \text{ (in the solution)} = \mu_{Zn^{2+}} \text{ (in the zinc electrode)}$$

and

$$\mu_{Zn} = \mu_{Zn^{2+}} + 2\mu_{e^-}^{Zn}$$

where μ_{Zn} is the chemical potential of zinc atoms in the electrode
$\mu_{Zn^{2+}}$ is the chemical potential of zinc ions in the $ZnSO_4$ solution and in the electrode
$\mu_{e^-}^{Zn}$ is the chemical potential of electrons in the zinc electrode

It is seen that the value of $\mu_{Zn^{2+}}$ (in solution), which is determined by the concentration of $ZnSO_4$ in the solution, determines the equilibrium value of $\mu_{e^-}^{Zn}$. Similarly, in the Cu-$CuSO_4$ compartment, when the half-cell reaction occurs, copper comes out of solution to plate out onto the copper electrode and in so doing produces a deficit of electrons in the electrode. Again equilibrium is attained when

$$\mu_{Cu^{2+}} \text{ (in the solution)} = \mu_{Cu^{2+}} \text{ (in the copper electrode)}$$

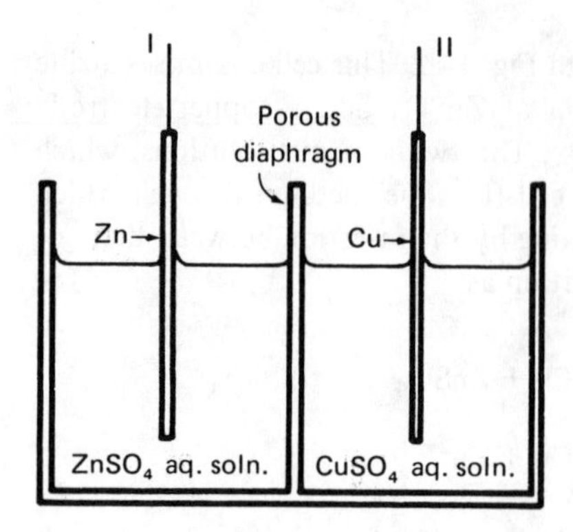

Fig. 14.2. The Daniell cell.

and $$\mu_{Cu^{2+}} + 2\mu_{e^-}^{Cu} = \mu_{Cu}$$

Again it is seen that the equilibrium value of $\mu_{e^-}^{Cu}$ will be determined by the concentration of $CuSO_4$ in the electrolyte. If identical pieces of metal wire are joined to each of the electrodes as extensions, then, as both the wire-zinc electrode and wire-copper electrode are electrical conductors,

$$\mu^{I} = \mu_{e^-}^{Zn} \quad \text{and} \quad \mu^{II} = \mu_{e^-}^{Cu}$$

The reversible transfer of dn moles of electrons from the electric potential ϕ_I to the electric potential ϕ_{II} involves the performance of work $\delta w'_{max}$ according to

$$\delta w'_{max} = z\mathfrak{F}(\phi_{II} - \phi_I)dn$$

and also the reversible transfer of dn moles of electrons from the chemical potential μ^I to the chemical potential μ^{II} at constant temperature and pressure involves the performance of work $\delta w'_{max}$ and a decrease in the free energy, dG', of the system according to

$$\delta w'_{max} = -dG' = -(\mu^{II} - \mu^{I})dn$$

Thus $$(\mu^{II} - \mu^{I}) = z\mathfrak{F}(\phi_{II} - \phi_I) \tag{14.2}$$

where, for electrons, z has the value of minus unity. Equation (14.2) relates chemical potential difference and electric potential difference for electron transfer.

In determining whether the EMF of the cell, $\mathcal{E}$, should be expressed as $(\phi_{II} - \phi_I)$ or as $(\phi_I - \phi_{II})$, the following convention is convenient. The cell reaction is written such that it proceeds spontaneously (i.e., with a decrease in free energy at constant temperature and pressure) from left to right. Thus the Daniell cell reaction would be written as

$$Zn + CuSO_4 = Cu + ZnSO_4$$

In shorthand notation this is written as

$$Zn|Zn^{2+}(\text{aqueous solution}) \vdots Cu^{2+}(\text{aqueous solution})|Cu$$

where the vertical full lines indicate phase boundaries in the cell, and the dashed vertical line represents the porous diaphragm separating the two aqueous solutions. The convention is thus such that the left-hand electrode is negative due to the excess of electrons which builds up in the zinc, and the right-hand electrode is positive due to the deficit of electrons which builds up in the copper. Alternatively, the convention can be remembered as being such that the oxidation reaction, that is, $Zn \rightarrow Zn^{2+}$, occurs at the left-hand electrode (the anode), and the reduction reaction, that is, $Cu^{2+} \rightarrow Cu$, occurs at the right-hand electrode (the cathode). When the cell is represented in this form, the U.S. convention assigns a positive value to the EMF of the cell. The convenience of the convention is that for the passage of dn moles of positive ions of valency z through the cell, which pass through the electrolyte from the anode to the cathode,

$$dG' = \text{a negative quantity}$$

which, from Eq. (14.1), is $-z\mathcal{F}\mathcal{E}dn$, where $\mathcal{E}$ is a positive quantity. Thus for the occurrence of the cell reaction to the extent of 1 mole,

$$\Delta G = -z\mathcal{F}\mathcal{E} \tag{14.3}$$

which is consistent with respect to signs. Equation (14.3) is known as the Nernst equation.

When an opposing electric potential of magnitude $\mathcal{E}$ is externally applied between the electrodes I and II, as the chemical driving force of the cell is exactly balanced by the external opposing voltage, the entire system is at equilibrium. If the magnitude of the external opposing voltage is decreased, equilibrium no longer exists; electronic current flows through the external circuit from I to II, and the equivalent ionic current flows through the cell. In the Daniell cell the ionic current through the cell comprises the passage of SO_4^{2-}

ions from the solution in the $CuSO_4$ compartment (the catholyte) into the $ZnSO_4$ compartment (the anolyte) at a rate equal to the addition of Zn^{2+} ions to the anolyte and hence equal to the rate of removal of Cu^{2+} ions from the catholyte. As the cell reaction proceeds, the concentration of $CuSO_4$ in the catholyte decreases, and the concentration of $ZnSO_4$ in the anolyte increases. As has been seen, the equilibrium values of μ_I and μ_{II}, and hence the value of $\mathcal{E}$ depend on the concentrations of $ZnSO_4$ and $CuSO_4$ in the electrolyte and thus eventually, in the absence of unfavorable kinetic factors, the EMF of the cell decreases to the value of the externally applied opposing voltage, at which point the passage of current ceases and a new equilibrium is established. If, however, by appropriate addition and removal, the concentrations of $ZnSO_4$ and $CuSO_4$ in their respective compartments were maintained constant, the cell reaction would continue indefinitely. When the externally applied voltage is finitely less than the cell EMF, a finite current flows and the cell reaction proceeds irreversibly. In such a situation, less than maximum work is obtained as the electrons in the external circuit are being transported through a smaller voltage difference. In the limit of decrease of the external voltage, i.e., when the external voltage is zero and the cell is short-circuited, the degree of irreversibility of the reaction is maximized, no work is done, and the free energy decrease of the system appears entirely as heat. This situation simply corresponds to that of placing a piece of zinc into an aqueous solution of $CuSO_4$. For the production of maximum work, the cell must be operated reversibly, in which case the externally applied voltage must be only infinitesimally smaller than the cell EMF, giving an infinitesimal current flow in the forward direction. If the cell can be operated reversibly, then an infinitesimal increase in the magnitude of the external voltage reverses the direction of the current flow and the direction of the cell reaction. The cell thus becomes current consuming rather than current producing; i.e., it becomes an electrolysis cell instead of a galvanic cell. This scheme is illustrated in Fig. 14.3.

14.3 THE EFFECT OF CONCENTRATION ON EMF

In the preceding section it was seen that the cell EMF is dependent on the concentrations of $CuSO_4$ and $ZnSO_4$ in the anolyte and catholyte, respectively. The quantitative relationship between concentration, or correctly, activity, and EMF can be introduced as follows. Consider the reaction

$$Zn + CuSO_4 = Cu + ZnSO_4$$

Although it was assumed that the reaction, as written, proceeds spontaneously from left to right, the direction depends on the states of the reactants and products. Consider the reactants and products to occur in their standard states at 298 K. Then

$$Zn_{(s)} + CuSO_{4\,(\text{saturated aqueous solution})}$$
$$= Cu_{(s)} + ZnSO_{4\,(\text{saturated aqueous solution})}$$

which is thermodynamically equivalent to

$$Zn_{(s)} + CuSO_{4\,(s)} = Cu_{(s)} + ZnSO_{4\,(s)}$$

At 298 K, $\Delta G = \Delta G^0_{298} = -213{,}040$ joules, indicating that the spontaneous reaction direction is from left to right, and

$$\mathcal{E}^0 = \frac{-\Delta G^0}{z\mathcal{F}} = \frac{213{,}040}{2 \times 96{,}487} = 1.104 \text{ volts}$$

where $\mathcal{E}^0$, being the EMF of the cell when the reactants and products occur in the designated standard states, is termed the standard EMF of the cell.

Thus when Zn dips into a saturated aqueous solution of $ZnSO_4$ and Cu dips into a saturated aqueous solution of $CuSO_4$, the externally applied voltage which is required to balance the chemical driving force of the cell reaction at 298 K is 1.104 volts.

For the general reaction

$$aA + bB = cC + dD$$

when the reactants and products do not occur in their standard states,

$$\Delta G = \Delta G^0 + RT \ln \frac{a_C^c a_D^d}{a_A^a a_B^b} \tag{13.4}$$

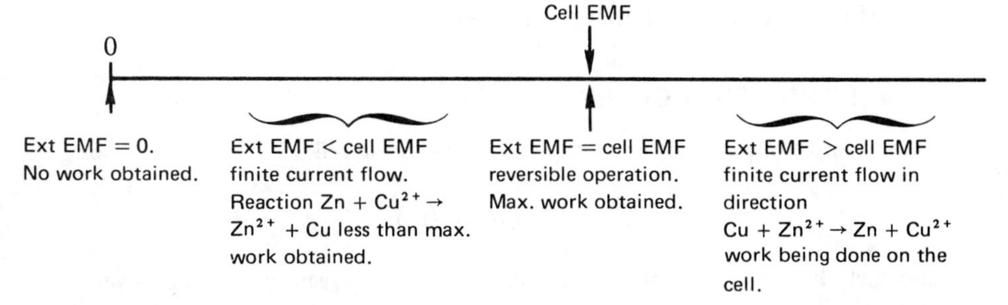

Fig. 14.3. Relationship of a current-producing cell (a galvanic cell) to a current-consuming cell (an electrolysis cell).

and, from Eq. (14.3), the EMF of the cell in which the above reaction is occurring electrochemically is

$$\mathcal{E} = \mathcal{E}^0 - \frac{RT}{z\mathcal{F}} \ln \frac{a_C^c a_D^d}{a_A^a a_B^b} \tag{14.4}$$

Thus with pure Zn and Cu in the Daniell cell, the EMF is given as

$$\mathcal{E} = \mathcal{E}^0 - \frac{RT}{2\mathcal{F}} \ln \frac{a_{ZnSO_4}}{a_{CuSO_4}}$$

which at 25°C gives

$$\mathcal{E} = 1.104 - 0.0296 \log \frac{a_{ZnSO_4}}{a_{CuSO_4}}$$

In order that the cell EMF be zero, the activity quotient must be

$$10^{1.104/0.0296} = 1.98 \times 30^{37}$$

Thus if the $ZnSO_4$ solution is saturated that is, $a_{ZnSO_4} = 1$, then it is necessary that the activity of $CuSO_4$ in the catholyte, with respect to the saturated solution as the standard state, be 5×10^{-38} in order that the occurrence of equilibrium does not require a backing EMF in the external circuit between the electrodes.

The concentration gradient which exists across the porous diaphragm results in ionic diffusion from one compartment to the other, and as diffusion is an irreversible process, the concentration gradient gives rise to a potential known as the liquid junction potential. This potential cannot be measured independently of the electrode potential, and thus the liquid junction either must be minimized by such means as the use of a salt bridge between the anolyte and the catholyte or must be eliminated by altering the cell design.

14.4 FORMATION CELLS

An example of a cell without a liquid junction is the cell

$$Pb_{(l)}|PbO_{(l)}|O_2, (Pt)$$

in which the anode is liquid lead, the cathode is oxygen gas bubbled over a platinum (inert) wire dipping into the electrolyte, and the electrolyte is liquid lead oxide. In this cell the driving force of the cell reaction

$$Pb_{(l)} + \tfrac{1}{2}O_{2\,(g)} = PbO_{(l)}$$

can be balanced by the application of an opposing voltage between the electrodes. Such a cell, which is an example of a formation cell, is illustrated in Fig. 14.4.*

At the molten lead anode,

$$Pb = Pb^{2+} + 2e^-$$

and at the oxygen cathode,

$$\tfrac{1}{2}O_2 + 2e^- = O^{2-}$$

For a pure molten lead anode, pure molten PbO electrolyte, and oxygen gas at 1 atm pressure at the cathode, the standard free energy of formation of lead oxide is obtained as

$$\Delta G^0 = -2\mathcal{F}\mathcal{E}^0$$

Maintaining the oxygen pressure constant at 1 atm and varying the temperature of the cell facilitates determination of the ΔG^0-T relationship for forma-

*R. Sridhar and J. H. E. Jeffes, Thermodynamics of PbO and PbO–SiO_2 Melts, *Trans. Inst. Mining Met.*, 76:C44 (1967).

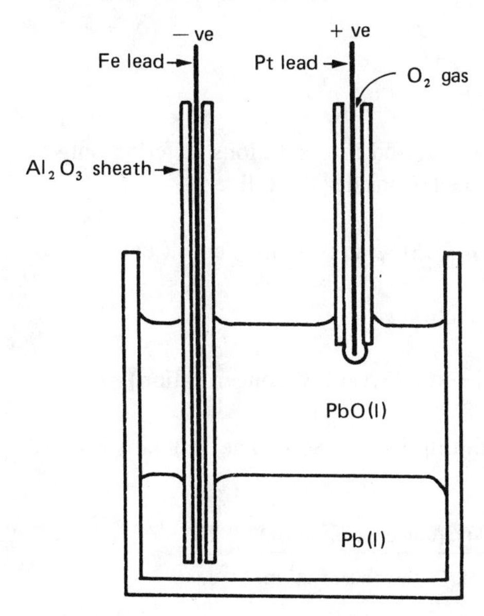

Fig. 14.4. A lead oxide formation cell.

tion of liquid lead oxide from liquid lead and gaseous oxygen. In the temperature range 900 to 1080°C, Sridhar and Jeffes obtained

$$\Delta G^0_{(i)} = -191{,}750 + 79.08T \text{ joules}$$

which can be compared with

$$\Delta G^0_{(i)} = -197{,}280 + 86.23T \text{ joules}$$

obtained from combination of thermochemical data for $Pb_{(l)}$, $PbO_{(s)}$ and $O_{2\,(g)}$ with the heat of fusion of PbO.

The addition to the electrolyte of a second oxide which (1) must be chemically more stable than PbO and (2) which must not introduce any electronic conductivity (for example, SiO_2) influences the cell EMF according to

$$\mathcal{E} = \mathcal{E}^0 - \frac{RT}{2\mathcal{F}} \ln \frac{a_{\mathrm{PbO(in\ PbO-SiO_2)}}}{p_{O_2}^{\frac{1}{2}}}$$

Thus measurement of the cell EMF as a function of PbO concentration in the lead silicate from pure PbO to SiO_2-saturation allows determination of the activity-composition relationship of PbO in the PbO-SiO_2 system and hence determination of the activity-composition relationship of SiO_2 via the Gibbs-Duhem relationship. The results of Sridhar and Jeffes are shown in Figs. 14.5 and 14.6.

14.5 CONCENTRATION CELLS

A cell which has identical electrodes inserted into solutions differing only in concentration is called a concentration cell. Consider the cell

$$\mathrm{Cu|CuSO_4(aq\ soln,\ low\ con)} \,\vdots\, \mathrm{CuSO_4(aq\ soln,\ high\ con)|Cu}$$

The cell EMF is positive, and the cell reaction is

$$\mathrm{CuSO_4(at\ high\ concentration)} \rightarrow \mathrm{CuSO_4(at\ low\ concentration)}$$

i.e., the spontaneous process is the dilution of $CuSO_4$. The standard EMF of such a cell is zero, and the EMF is

$$\mathcal{E} = -\frac{RT}{2\mathcal{F}} \ln \frac{a_{\mathrm{CuSO_4(low\ concentration)}}}{a_{\mathrm{CuSO_4(high\ concentration)}}}$$

If one of the aqueous solutions (say the high-concentration solution) contains $CuSO_4$ in its chosen standard state, then

$$\mathcal{E} = -\frac{RT}{2\mathcal{F}} \ln a_{CuSO_4(\text{low concentration})} = -\frac{\Delta \bar{G}^M{}_{CuSO_4}}{2\mathcal{F}}$$

Thus an electrochemical measurement allows the determination of the partial molar free energy of solution of $CuSO_4$ in water. Such a cell suffers from the disadvantage of having a liquid junction potential.

A form of concentration cell of considerable metallurgical importance is the oxygen concentration cell which uses lime-stabilized zirconia as a solid electrolyte. In certain ranges of oxygen pressure and temperature, this electrolyte is an ionic conductor in which the oxygen ion is the only mobile species. Thus a cell can be constructed, such as

$$O_2\,(\text{g, pressure I}),\ \text{Pt}|\text{CaO-ZrO}_2\,|\text{Pt},\ O_2\,(\text{g, pressure II})$$

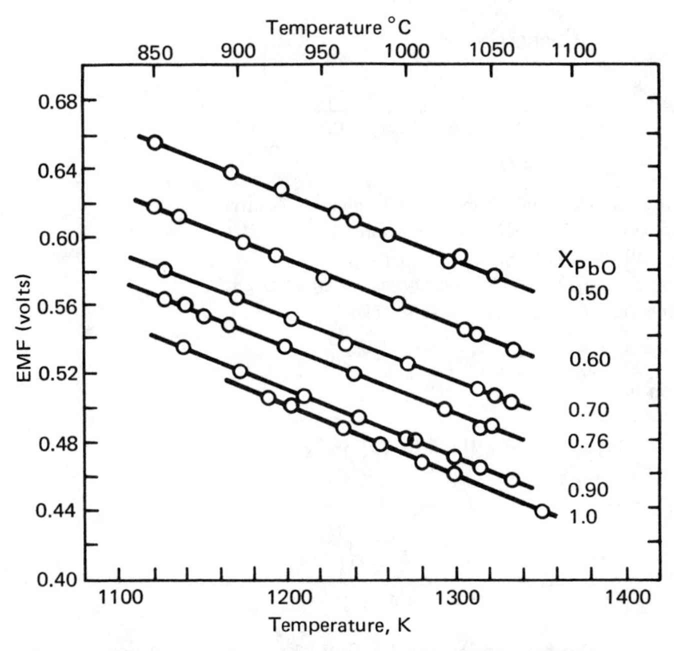

Fig. 14.5. The variation with temperature of the EMF of the cell $Pb_{(l)}|PbO\text{-}SiO_2\,(l)|O_2\,(1\ \text{atm})$ (Pt) with silica content of the electrolyte. *(From R. Sridhar and J. H. E. Jeffes, The Thermodynamics of PbO and PbO-SiO$_2$ Melts, Trans. Inst. Mining Met., 76:C44 [1967]).*

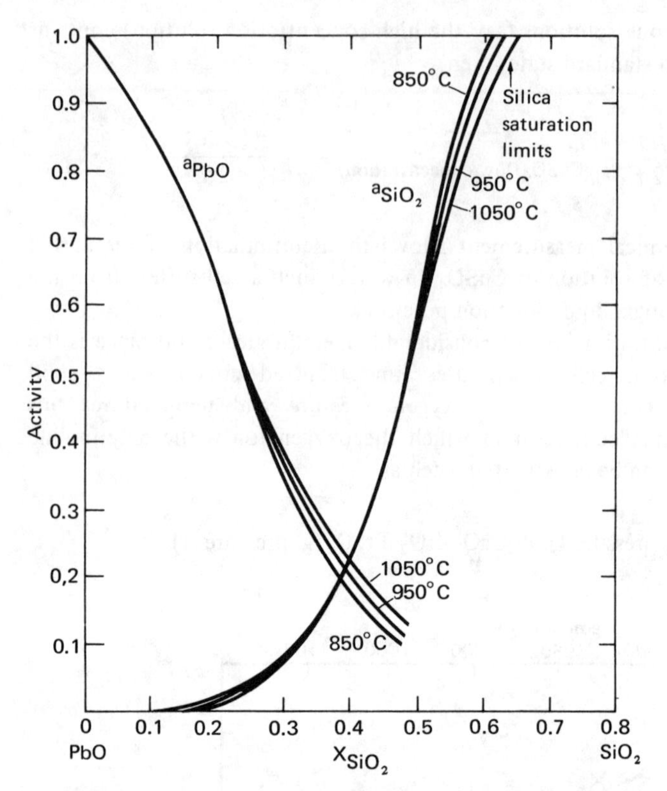

Fig. 14.6. The activities of PbO and Gibbs-Duhem–calculated activities of SiO_2 in the system $PbO\text{-}SiO_2$ as determined from EMF measurements of the cell $Pb_{(l)}|PbO\text{-}SiO_2\,(l)|O_2\,(g,1\text{ atm})$ (Pt). *(From R. Sridhar and J. H. E. Jeffes, The Thermodynamics of PbO and PbO-SiO₂ Melts, Trans. Inst. Mining Met., 76:C44 [1967].)*

and the cell reaction is

$$O_2\,(g, \text{pressure II}) \rightarrow O_2\,(g, \text{pressure I})$$

with a cell EMF of

$$\mathcal{E} = -\frac{RT}{4\mathcal{F}} \ln \frac{p_{O_2(I)}}{p_{O_2(II)}}$$

The oxygen pressures at the electrodes can be fixed by using metal-metal oxide couples; e.g., using the couples X–XO and Y–YO, the cell becomes

$$X|XO|CaO\text{-}ZrO_2\,|YO|Y$$

At the temperature T the anode oxygen pressure $p_{O_2(X,XO)}$ is fixed by the establishment of the chemical equilibrium

$$X_{(s)} + \tfrac{1}{2}O_{2(g)} = XO_{(s)}$$

and the cathode oxygen pressure $p_{O_2(Y,YO)}$ is fixed by the establishment of the chemical equilibrium

$$Y_{(s)} + \tfrac{1}{2}O_{2(g)} = YO_{(s)}$$

Thus the anode half-cell reaction can equivalently be regarded as being

$$X + \tfrac{1}{2}O_{2(\text{at } p_{O_2(\text{eq}, T, X/XO)})} = XO$$

or

$$O^{2-} - 2e^- = \tfrac{1}{2}O_{2(\text{at } p_{O_2(\text{eq}, T, X/XO)})}$$

and similarly the cathode half-cell reaction can equivalently be regarded as being

$$YO = Y + \tfrac{1}{2}O_{2(\text{at } p_{O_2(\text{eq}, T, Y/YO)})}$$

or

$$\tfrac{1}{2}O_{2(\text{at } p_{O_2}(\text{eq}, T, Y/YO))} + 2e^- = O^{2-}$$

The cell reaction is thus

$$YO + X = XO + Y \tag{i}$$

or

$$\tfrac{1}{2}O_{2(\text{at } p_{O_2(\text{eq}, T, Y/YO)})} = \tfrac{1}{2}O_{2(\text{at } p_{O_2(\text{eq}, T, X/XO)})} \tag{ii}$$

Thus, whereas the chemical equilibrium between the pure reactants and products

$$YO + X = XO + Y$$

could be obtained only at the single invariant temperature (for example, T_E in Fig. 10.4) at which

$$p_{O_2(\text{eq}, T, X/XO)} = p_{O_2(\text{eq}, T, Y/YO)}$$

the electrochemical equilibrium

$$YO + X = XO + Y$$

can be obtained at any temperature (within the limits imposed by the electrolyte performance) by placing an external voltage in opposition to the chemical driving force of the cell reaction. For either (i) or (ii) the chemical driving force is

$$\Delta G = RT \ln \frac{p_{O_2(eq,T,X/XO)}^{\frac{1}{2}}}{p_{O_2(eq,T,Y/YO)}^{\frac{1}{2}}}$$

and hence the cell EMF is

$$\mathcal{E} = -\frac{RT}{2\mathcal{F}} \ln \frac{p_{O_2(eq,\, T,\, X/XO)}^{\frac{1}{2}}}{p_{O_2(eq,\, T,\, Y/YO)}^{\frac{1}{2}}}$$

$$\mathcal{E} = -\frac{RT}{4\mathcal{F}} \ln \frac{p_{O_2(eq,\, T,\, X/XO)}}{p_{O_2(eq,\, T,\, Y/YO)}} \qquad \text{(iii)}$$

If one of the metals, say X, is dissolved in an inert solvent, where the requirement for "inertness" is that the equilibrium oxygen pressure for solvent metal-solvent metal-oxide equilibrium is considerably higher than $p_{O_2(eq,T,X/XO)}$, then the activity of X in the alloy can be obtained as follows. If X in solution is denoted $\underline{X}$, then the EMF of the cell

$$\underline{X}|XO|CaO\text{-}ZrO_2|Y|YO$$

is

$$\mathcal{E}' = -\frac{RT}{4\mathcal{F}} \ln \frac{p_{O_2(eq,\, T,\, \underline{X}/XO)}}{p_{O_2(eq,\, T,\, Y/YO)}} \qquad \text{(iv)}$$

As, at the temperature T,

$$a_X p_{O_2(\underline{X},\, XO)}^{\frac{1}{2}} = p_{O_2(X,\, XO)}^{\frac{1}{2}}$$

combination of Eqs. (iii) and (iv) gives

$$\mathcal{E} - \mathcal{E}' = -\frac{RT}{2\mathcal{F}} \ln a_X$$

and so measurement of the variation of $\mathcal{E}'$ with composition of the alloy allows determination of the activity-composition relationship of X in the alloy.

More simply, if the cell is

$$X|XO|CaO\text{-}ZrO_2|\underline{X}|XO$$

then

$$\mathcal{E} = \frac{RT}{2\mathcal{F}} \ln \frac{p_{O_2(eq,\, T,\, \underline{X}/XO)}^{\frac{1}{2}}}{p_{O_2(eq,\, T,\, X/XO)}^{\frac{1}{2}}} = -\frac{RT}{2\mathcal{F}} \ln a_X = -\frac{\Delta \bar{G}_X^M}{2\mathcal{F}}$$

Similarly if the metal oxide XO is dissolved in an inert oxide solvent, then, denoting dissolved XO as $\underline{XO}$, for the cell

$$X|\underline{XO}|CaO\text{-}ZrO_2|X|XO$$

$$\mathcal{E} = -\frac{RT}{2\mathcal{F}} \ln \frac{p^{\frac{1}{2}}_{O_2(eq,\,T,\,X/XO)}}{p^{\frac{1}{2}}_{O_2(eq,\,T,\,X/\underline{XO})}}$$

and as

$$a_{XO} = \frac{p^{\frac{1}{2}}_{O_2(eq,\,T,\,X/\underline{XO})}}{p^{\frac{1}{2}}_{O_2(eq,\,T,\,X/XO)}}$$

then

$$\mathcal{E} = \frac{RT}{2\mathcal{F}} \ln a_{XO}$$

This technique has been used by Kozuka and Samis* to determine the activity-composition relationship of PbO in $PbO\text{-}SiO_2$ melts using the cell

$$Pb_{(l)}|PbO_{(\text{in } PbO\text{-}SiO_2,\,l)}|CaO\text{-}ZrO_2|Pb_{(l)}|PbO_{(l)}$$

Their experimental cell is illustrated in Fig. 14.7, and their results, at 1000°C, are shown, along with the corresponding results obtained by Sridhar and Jeffes, in Fig. 14.8.

14.6 THE TEMPERATURE COEFFICIENT OF A CELL

For any cell reaction at constant temperature and pressure,

$$\Delta G = -z\mathcal{F}\mathcal{E}$$

Differentiation with respect to temperature at constant pressure gives

$$\left(\frac{\partial \Delta G}{\partial T}\right)_P = -z\mathcal{F}\left(\frac{\partial \mathcal{E}}{\partial T}\right)_P = -\Delta S$$

Thus, for the cell reaction,

*Z. Kozuka and C. S. Samis, Thermodynamic Properties of Molten $PbO\text{-}SiO_2$ Systems, *Met. Trans. AIME*, **1**:871 (1970).

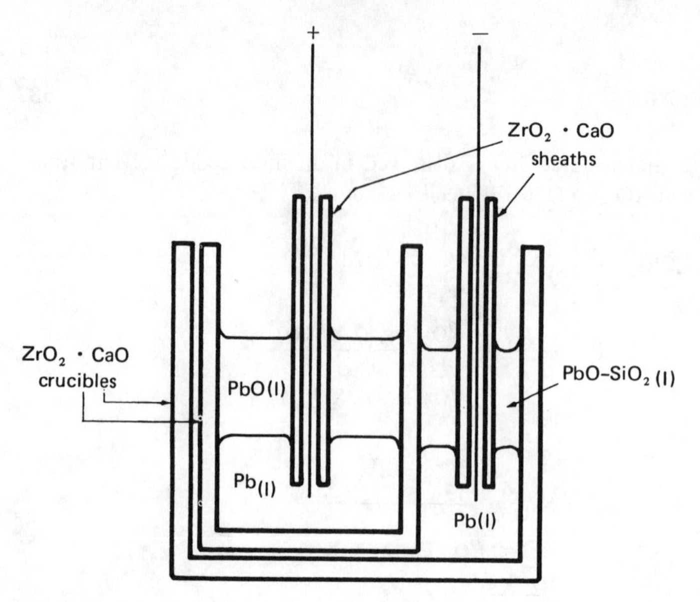

Fig. 14.7. A lead-oxide concentration cell.

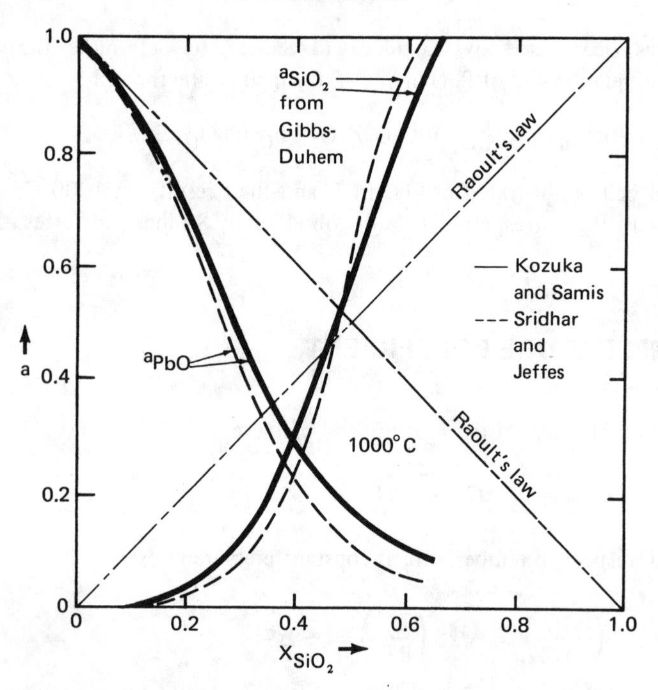

Fig. 14.8. The activities of PbO in liquid $PbO\text{-}SiO_2$ determined from a PbO concentration cell *(from Z. Kozuka and C. S. Samis, Thermodynamic Properties of Molten $PbO\text{-}SiO_2$ Systems, Met. Trans., 1:871 [1970])* and a PbO formation cell *(from R. Sridhar and J. H. E. Jeffes, The Thermodynamics of PbO and $PbO\text{-}SiO_2$ Melts, Trans. Inst. Mining Met., 76:C44 [1967])*.

$$\Delta S = z\mathcal{F}\left(\frac{\partial \mathcal{E}}{\partial T}\right)_P$$

and
$$\Delta H = -z\mathcal{F}\mathcal{E} + z\mathcal{F}T\left(\frac{\partial \mathcal{E}}{\partial T}\right)_P$$

In concentration cells such as

$$A|A^{z+}X^{z-}|A(\text{in A-B alloy})$$

as
$$\mathcal{E} = -\frac{RT}{z\mathcal{F}}\ln a_A$$

the partial molar properties of A in the A–B alloy are obtained as

$$\Delta \bar{S}_A^M = z\mathcal{F}\frac{\partial \mathcal{E}}{\partial T}$$

and
$$\Delta \bar{H}_A^M = -z\mathcal{F}\mathcal{E} + z\mathcal{F}T\frac{\partial \mathcal{E}}{\partial T}$$

Alternatively, from measurements on formation cells of the type

$$A|A^{2+}X^{2-}|X_{2\,(g,\ 1\ \text{atm})} \tag{i}$$

for which

$$\mathcal{E}_{(i)} = \mathcal{E}^0$$

and measurements on cells of the type

$$A(\text{in A-B alloy})|A^{2+}X^{2-}|X_{2\,(g,\ 1\ \text{atm})} \tag{ii}$$

for which

$$\mathcal{E}_{(ii)} = \mathcal{E}^0 - \frac{RT}{2\mathcal{F}}\ln\left(\frac{1}{a_A}\right)$$

the value of a_A is obtained from

$$\mathcal{E}_{(ii)} - \mathcal{E}_{(i)} = \frac{RT}{2\mathcal{F}}\ln a_A \tag{iii}$$

Variation with temperature then gives

$$\Delta \bar{S}_A^M = 2\mathcal{F}\frac{\partial}{\partial T}(\mathcal{E}_{(i)} - \mathcal{E}_{(ii)}) \tag{iv}$$

and

$$\Delta \bar{H}_A^M = 2\mathcal{F}(\mathcal{E}_{(ii)} - \mathcal{E}_{(i)}) - 2\mathcal{F}T\frac{\partial}{\partial T}(\mathcal{E}_{(ii)} - \mathcal{E}_{(i)}) \qquad \text{(v)}$$

Belton and Rao* measured the EMF's of the cells

$$Mg_{(l)}|MgCl_{2\,(l)}|Cl_{2\,(g,\,1\,atm)}$$

and

$$Mg(\text{in } Mg\text{-}Al)_{(l)}|MgCl_{2\,(l)}|Cl_{2\,(g,\,1\,atm)}$$

in the temperature range 700 to 880°C with Mg-Al alloys in the range X_{Mg} = 0.096 to 0.969. For the cell with pure Mg as the anode, they obtained

$$\mathcal{E}^0 = 3.135 - 6.5 \times 10^{-4}T \text{ (volts)}$$

which gives, for the reaction,

$$Mg_{(l)} + Cl_{2\,(g)} = MgCl_{2\,(l)}$$

$$\Delta G^0 = -604{,}970 + 125.4T \text{ joules}$$

Some of their results are listed in Table 14.1. In this table:

Column 1 lists the Mg-Al alloy compositions.
Column 2 lists $\Delta\mathcal{E} = \mathcal{E}_{(i)} - \mathcal{E}_{(ii)}$ at 800°C.

*G. R. Belton and Y. K. Rao, A Galvanic Cell Study of Activities in Mg-Al Liquid Alloys, *Trans. Met. Soc. AIME*, **245**:2189 (1969).

Table 14.1

X_{Mg}	$\Delta\mathcal{E}$ (mv)	$\partial\mathcal{E}/\partial T$ mv/degree $\times 10^2$	a_{Mg}	$\Delta\bar{S}_{Mg}^M$ joules/K	$\Delta\bar{H}_{Mg}^M$ joules
0.096	168.0	9.45	0.027	18.2	−12,890
0.140	137.4	7.77	0.051	15.0	−10,420
0.209	106.6	5.62	0.100	10.8	−8,980
0.340	69.9	4.16	0.22	8.03	−4,870
0.421	52.1	2.83	0.32	5.46	−4,200
0.517	36.3	2.18	0.46	4.21	−2,500
0.668	19.3	1.43	0.66	2.76	−760
0.700	17.7	1.30	0.68	2.51	−720
0.799	10.6	0.88	0.80	1.70	−220
0.844	7.8	0.67	0.84	1.30	−110
0.919	3.9	0.36	0.91	0.69	−10

Column 3 lists the temperature coefficient of $\Delta\mathcal{E}$.
Column 4 lists the values of a_{Mg} calculated from Eq. (iii).
Column 5 lists the values of $\Delta\bar{S}^{M}_{Mg}$ calculated from Eq. (iv).
Column 6 lists the values of $\Delta\bar{H}^{M}_{Mg}$ calculated from Eq. (v).

14.7 HEAT EFFECTS

In examining the properties of the enthalpy, H, in Chap. 5, it was noted that the enthalpy change of a system equals the heat entering or leaving the system in a constant-pressure process only if the work of volume change is the sole form of work performed on or by the system. If the process is performed in a galvanic cell, in which case electrical work is performed, $\Delta H \neq q_p$.

For a change of state at constant temperature and pressure, Eq. (5.6) gave

$$\Delta G = q_p - w + P\Delta V - T\Delta S$$
$$= q_p - w' - T\Delta S$$

If $w' = 0$, then $q_p = \Delta G + T\Delta S = \Delta H$; but if the process, which involves the performance of work w' is carried out reversibly, in which case $-w' = -w'_{max} = \Delta G$, then

$$q_p = T\Delta S$$

Consider the Daniell cell reaction $Zn + CuSO_4 = Cu + ZnSO_4$. When the reactants and products are in their standard states, the free energy change of the cell reaction is

$$\Delta G^0 = -208{,}800 - 13.9T \text{ joules}$$

If the reaction occurs as the result of placing pure solid zinc into saturated copper sulfate solution at 25°C, in which case the reaction proceeds spontaneously and w' is zero, then for the formation of saturated zinc sulfate solution and pure solid copper, per mole of reaction,

$$\Delta H^0 = -208{,}800 \text{ joules}$$

is the heat which *flows from* the system into the thermostating heat reservoir. However if the reaction is performed reversibly in a Daniell cell, in which case

$$w' = -\Delta G^0 = 208{,}800 + 13.9 \times 298 \text{ joules}$$

then $q_p = T\Delta S = 13.9 \times 298 = +4140$ joules is the heat that *flows into* the system from the thermostating heat reservoir.

14.8 THE THERMODYNAMICS OF AQUEOUS SOLUTIONS

The composition of an aqueous solution is usually expressed in terms of the molality, m, or the molarity, M, of the solution, where the molality is the number of moles of solute present in 1000 grams of water and the molarity is the number of moles of solute present in one liter of solution. Mole fraction, molality, and molarity are related as follows. Consider an aqueous solution of m_i moles of solute i in 1000 g H_2O, such that the solution is m_i molal. As the molecular weight of H_2O is 18, 1000 g of H_2O contain 1000/18 gram-moles and hence

$$X_i = \frac{n_i}{n_i + n_{H_2O}} = \frac{m_i}{m_i + 1000/18}$$

Consider an M_i molar solution that contains M_i moles of solute i per liter of solution, i.e., in 1000ρ g of solution, where ρ is the density of the solution in g/cm^3. The number of gram-moles of H_2O in the liter of solution is $(1000\rho - M_i\text{MW}_i)/18$, where MW_i is the molecular weight of i, and hence

$$X_i = \frac{n_i}{n_i + n_{H_2O}} = \frac{M_i}{M_i + (1000\rho - M_i\text{MW}_i)/18}$$

As the solution tends toward infinite dilution,

$$m_i \to \frac{18X_i}{1000} \quad \text{and} \quad M_i \to \frac{18X_i}{1000\rho}$$

In dilute solutions, molality and molarity are essentially equal to one another; e.g., an aqueous solution of NaCl of $X_{NaCl} = 10^{-3}$ is 0.0556 molal and 0.0554 molar.

Just as, in the case of dilute solutes in liquid metals, it was convenient to define the 1 weight percent standard state, and hence the 1 weight percent activity scale, as

$$h_{i(1\,\text{wt\%})} \to [\text{wt\%}\, i] \quad \text{as} \quad [\text{wt\%}\, i] \to 0$$

with the 1 weight percent standard state located on the Henry's Law line at 1 weight percent, it is convenient, in aqueous solutions, to define the analogous unit molality standard state, and hence unit molality activity scale, as

$$a_{i(m)} \to m_i \quad \text{as} \quad m_i \to 0$$

where $a_{i(m)}$ is the activity of the solute with respect to the unit molality standard state, and the unit molality standard state is located on the Henry's Law line at $m_i = 1$. As before, deviations from ideality are accommodated by introducing an activity coefficient defined as

$$\gamma_{i(m)} = \frac{a_{i(m)}}{m_i}$$

Consider the electrolyte, or salt, A_aY_y which, on solution in water, dissociates to form A^{z+} cations and Y^{z-} anions according to

$$A_aY_y = aA^{z+} + yY^{z-}$$

When m moles of A_aY_y are dissolved in n moles of H_2O, the resulting solution can be considered either as

(i) A solution comprising m moles of the component A_aY_y and n moles of H_2O, or
(ii) A solution of am moles of A^{z+} and ym moles of Y^{z-} in n moles of H_2O

In case (i), the variation of the Gibbs free energy of the solution with composition at constant T and P is given by Eq. (11.16) as

$$dG' = \bar{G}_{A_aY_y}\, dm + \bar{G}_{H_2O}\, dn \tag{14.5}$$

In case (ii), the stoichiometry of the dissociation is such that the number of moles of A^{z+}, $m_{A^{z+}}$, is am and the number of moles of Y^{z-}, $m_{Y^{z-}}$, is ym. Thus

$$dm_{A^{z+}} = a\, dm \qquad \text{and} \qquad dm_{Y^{z-}} = z\, dm$$

and, at constant T and P,

$$\begin{aligned} dG' &= \bar{G}_{A^{z+}}\, dm_{A^{z+}} + \bar{G}_{Y^{z-}}\, dm_{Y^{z-}} + \bar{G}_{H_2O}\, dn \\ &= (a\bar{G}_{A^{z+}} + y\bar{G}_{Y^{z-}})\, dm + \bar{G}_{H_2O}\, dn \end{aligned} \tag{14.6}$$

By definition,

$$\bar{G}_{A_aY_y} = \left(\frac{\partial G'}{\partial m}\right)_{T,P,n}$$

$$\bar{G}_{A^{z+}} = \left(\frac{\partial G'}{\partial m_{A^{z+}}}\right)_{T,P,n,m_{Y^{z-}}}$$

and
$$\bar{G}_{Y^{z-}} = \left(\frac{\partial G'}{\partial m_{Y^{z-}}}\right)_{T,P,n,m_{A^{z+}}}$$

As m can be varied at constant n, $\bar{G}_{A_aY_y}$ can be determined experimentally. However, as $m_{A^{z+}}$ and $m_{Y^{z-}}$ cannot be varied independently, neither $\bar{G}_{A^{z+}}$ nor $\bar{G}_{Y^{z-}}$ can be measured. Combination of Eqs. (14.5) and (14.6) gives

$$\bar{G}_{A_aY_y} = a\bar{G}_{A^{z+}} + y\bar{G}_{Y^{z-}} \tag{14.7}$$

which shows that, although neither $\bar{G}_{A^{z+}}$ nor $\bar{G}_{Y^{z-}}$ can be measured, the combination given by Eq. (14.7) can be measured.

If the component A_aY_y occurs in the unit molality standard state, Eq. (14.7) is written as

$$G^0_{A_aY_y} = aG^0_{A^{z+}} + yG^0_{Y^{z-}} \tag{14.8}$$

and subtraction of Eq. (14.8) from Eq. (14.7), noting that

$$\bar{G}_i - G^0_i = RT \ln a_i$$

gives, on rearrangement,

$$a_{A_aY_y} = a^a_{A^{z+}} a^y_{Y^{z-}} \tag{14.9}$$

Thus, again, although neither $a_{A^{z+}}$ nor $a_{Y^{z-}}$ can be measured experimentally, which necessarily means that neither $a_{A^{z+}}$ nor $a_{Y^{z-}}$ have any physical significance, the product given by Eq. (14.9) can be measured and hence does have a physical significance. Equation (14.9) can be written as

$$\begin{aligned} a_{A_aY_y} &= (\gamma_{A^{z+}} m_{A^{z+}})^a (\gamma_{Y^{z-}} m_{Y^{z-}})^y \\ &= \gamma^a_{A^{z+}} \gamma^y_{Y^{z-}} m^a_{A^{z+}} m^y_{Y^{z-}} \end{aligned} \tag{14.10}$$

The mean ionic molality, $m_\pm$, is defined as

$$m_\pm = (m^a_{A^{z+}} m^y_{Y^{z-}})^{1/(a+y)} \tag{14.11}$$

and the mean ion activity coefficient, $\gamma_\pm$, is defined as

$$\gamma_\pm = (\gamma^a_{A^{z+}} \gamma^y_{Y^{z-}})^{1/(a+y)} \tag{14.12}$$

Thus substitution of Eqs. (14.11) and (14.12) in Eq. (14.10) gives

$$a_{A_aY_y} = (\gamma_{\pm} m_{\pm})^{a+y} \qquad (14.13)$$

Consider an m molal solution of NaCl. As $|z^+| = |z^-| = 1$ and $a = y = 1$, Eq. (14.13) gives

$$a_{NaCl(m)} = (\gamma_{\pm} m_{\pm})^2$$

and Eq. (14.11) gives

$$m_{\pm} = (m\,m)^{1/2}$$

Thus

$$a_{NaCl(m)} = (\gamma_{\pm} m_{NaCl})^2$$

and Henrian behavior follows

$$a_{NaCl(m)} = m_{NaCl}^2$$

In an m molal solution of $CaCl_2$, as $|z^+| = 2$, $|z^-| = 1, a = 1$, and $y = 2$,

$$a_{CaCl_2(m)} = (\gamma_{\pm} m_{\pm})^3$$

$$m_{\pm} = [m(2m)^2]^{1/3}$$

and thus

$$a_{CaCl_2(m)} = 4(\gamma_{\pm} m_{CaCl_2})^3$$

Similarly, in an m molal solution of $Fe_2(SO_4)_3$

$$a_{Fe_2(SO_4)_3(m)} = 36(\gamma_{\pm} m_{Fe_2(SO_4)_3})^5$$

The Free Energy of Formation of Ions

Consider the cell

$$Pt, H_{2(g)} | HCl_{(aqueous)} | Hg_2Cl_{2(s)} | Hg_{(l)}$$

set up as illustrated in Fig. 14.9. The half-cell reaction at the anode is

$$\tfrac{1}{2}H_2 = H^+ + e^-$$

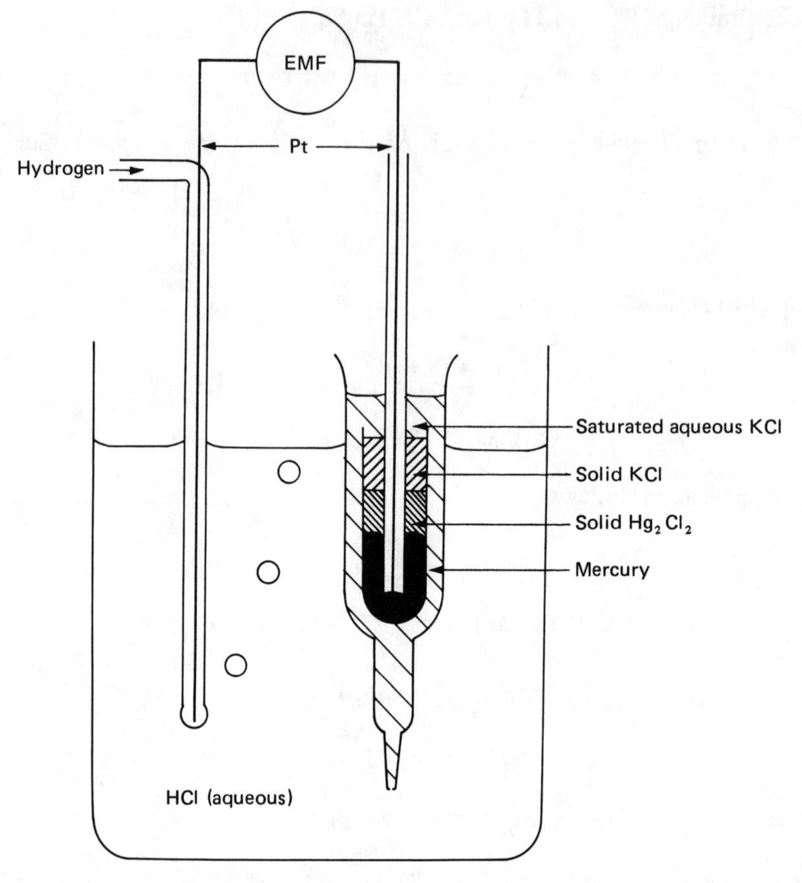

Fig. 14.9. The electrochemical cell Pt, $H_{2(g)}|HCl_{(aq)}|Hg_2Cl_{2(s)}|Hg_{(l)}$.

and the calomel half-cell reaction at the cathode is

$$\tfrac{1}{2}Hg_2Cl_2 + e^- = Hg + Cl^-$$

The overall cell reaction is thus

$$\tfrac{1}{2}H_{2(g)} + \tfrac{1}{2}Hg_2Cl_{2(s)} = Hg_{(l)} + HCl_{(m)} \qquad \text{(i)}$$

and, with Hg and Hg_2Cl_2 occurring at unit activity and $p_{H_2} = 1$ atm, the EMF of the cell is

$$\mathcal{E} = \mathcal{E}^0 - \frac{RT}{\mathcal{F}} \ln a_{HCl_{(m)}}$$

$$= \mathcal{E}^0 - \frac{RT}{\mathcal{F}} \ln (\gamma_\pm m_{\mathrm{HCl}})^2$$

This expression can be rearranged as

$$\mathcal{E} + \frac{2RT}{\mathcal{F}} \ln m_{\mathrm{HCl}} = \mathcal{E}^0 - \frac{2RT}{\mathcal{F}} \ln \gamma_\pm \qquad \text{(ii)}$$

in which the measurable quantities occur on the left-hand side. Extrapolation of the term $\mathcal{E} + (2RT/\mathcal{F}) \ln m_{\mathrm{HCl}}$ to infinite dilution, where $\gamma_\pm \to 1$, allows calculation of $\mathcal{E}^0$ and hence, from Eq. (ii), calculation of the variation of $\gamma_\pm$ with m_{HCl}. The value of $\mathcal{E}^0$ at 298 K has been determined as 0.26796 volts, and the variation of $a_{\mathrm{HCl}(m)}$ with m^2_{HCl} is as shown in Fig. 14.10. As $\mathcal{E}^0 = 0.26796$ volts,

$$\Delta G^0_{(\mathrm{i}),298} = -\mathcal{F}\mathcal{E}^0 = -96{,}487 \times 0.26796 = -25{,}855 \text{ joules}$$

For

$$Hg_{(l)} + \tfrac{1}{2}Cl_{2(g)} = \tfrac{1}{2}Hg_2Cl_{2(s)} \qquad \text{(iii)}$$

$$\Delta G^0_{(\mathrm{iii}),298} = -105{,}320 \text{ joules}$$

and hence, for

$$\tfrac{1}{2}H_{2(g)} + \tfrac{1}{2}Cl_{2(g)} = HCl_{(m)} \qquad \text{(iv)}$$

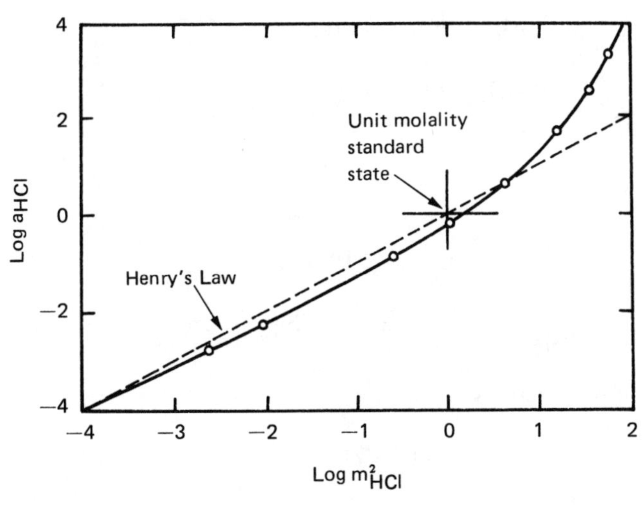

Fig. 14.10. The activity of HCl in aqueous solution.

$$\Delta G^0_{(iv)} = \Delta G^0_{(i)} + \Delta G^0_{(iii)}$$
$$= -25{,}855 - 105{,}320$$
$$= -131{,}175 \text{ joules}$$

Thus the standard molar free energy of formation of HCl in aqueous solution at unit activity from H_2 gas and Cl_2 gas, each at unit pressure, is −131,175 joules at 298 K. For

$$\tfrac{1}{2}H_{2(g)} + \tfrac{1}{2}Cl_{2(g)} = HCl_{(g)} \tag{v}$$

$\Delta G^0_{(v),298} = -94{,}540$ joules, and hence, for the change of state

$$HCl_{(g)} = HCl_{(m)}$$

$$\Delta G^0_{298} = -131{,}175 + 94{,}540$$
$$= -36{,}635 \text{ joules}$$
$$= -8.3144 \times 298 \ln \frac{a_{HCl(m)}}{p_{HCl}}$$

Thus, at 298 K the partial pressure of HCl exerted by an aqueous solution of HCl at unit activity is 3.79×10^{-7} atm.

From Eq. (iv), the standard EMF of the cell

$$Pt,H_{2(g)}|HCl_{(m)}|Cl_{2(g)},Pt$$

is

$$\mathcal{E}^0 = \frac{-\Delta G^0_{(iv)}}{\mathcal{F}} = \frac{131{,}175}{96{,}487} = 1.3595 \text{ volts}$$

and the cell reaction

$$\tfrac{1}{2}H_{2(g)} + \tfrac{1}{2}Cl_{2(g)} = HCl_{(m)}$$

or

$$\tfrac{1}{2}H_{2(g)} + \tfrac{1}{2}Cl_{2(g)} = H^+_{(m)} + Cl^-_{(m)} \tag{vi}$$

is the sum of the half cell reactions

$$\tfrac{1}{2}H_2 = H^+ + e^- \tag{vii}$$

and

$$\tfrac{1}{2}Cl_2 + e^- = Cl^- \tag{viii}$$

It is now convenient to introduce the concept of a half cell, or single electrode potential, the sum of which, in any cell, equals the EMF of the cell. This concept is useful in spite of the fact that it is impossible to construct, and hence to measure the potential of, a cell with a single electrode. In order to give meaning to the concept it is necessary to choose a particular single standard electrode which is arbitrarily assigned a potential of zero. In aqueous solutions this standard single electrode is the Standard Hydrogen Electrode, which is a hydrogen electrode with gas at 1 atm pressure in contact with an aqueous solution containing hydrogen ions at unit activity. Thus, for the cell

$$\mathrm{Pt,H_{2(g)}|H^+_{(m)}Cl^-_{(m)}|Cl_{2(g)},Pt}$$

the cell reaction (vi) is the sum of the anodic half-cell reaction (vii) with a standard single electrode potential $\mathcal{E}^0_a = 0$ (by convention) and the cathodic half-cell reaction (viii) with a standard single electrode potential $\mathcal{E}^0_c$, i.e.,

$$\mathcal{E}^0 = \mathcal{E}^0_a + \mathcal{E}^0_c = \mathcal{E}^0_c = 1.3595 \text{ volts}$$

Thus, by convention, the standard single electrode potential for the reduction of chlorine (the standard reduction potential of chlorine) is

$$\mathcal{E}^0_{\frac{1}{2}\mathrm{Cl_2/Cl^-}} = 1.3595 \text{ volts}$$

or the standard oxidation potential of chlorine is

$$\mathcal{E}^0_{\mathrm{Cl^-/\frac{1}{2}Cl_2}} = -1.3595 \text{ volts}$$

As $\mathcal{E}^0_{\frac{1}{2}\mathrm{H_2/H^+}} = 0$, the standard free energy of formation of the hydrogen ion in aqueous solution is also (by convention) zero and hence the standard free energy of formation of chlorine ions is

$$\Delta G^0_{(\mathrm{viii})} = -\mathcal{F}\mathcal{E}^0_{\frac{1}{2}\mathrm{Cl_2/Cl^-}} = -96{,}487 \times 1.3595 = -131{,}170 \text{ joules}$$

The standard EMF of the cell

$$\mathrm{Zn_{(s)}|ZnCl_{2(m)}\,\vdots\,HCl_{(m)}|H_{2(g)},Pt}$$

is

$$\begin{aligned}\mathcal{E}^0 &= \mathcal{E}^0_c + \mathcal{E}^0_a\\ &= \mathcal{E}^0_{\mathrm{Zn/Zn^{2+}}} + \mathcal{E}^0_{\mathrm{H^+/\frac{1}{2}H_2}}\\ &= 0.763 \text{ volts}\end{aligned}$$

Thus, as, by convention, $\mathcal{E}^0_{H^+/\frac{1}{2}H_2} = 0$, the standard oxidation potential for zinc, $\mathcal{E}^0_{Zn/Zn^{2+}}$, is 0.763 volts, and the standard free energy of formation of zinc ions in aqueous solution is

$$\Delta G^0 = -z\mathcal{F}\mathcal{E}^0 = -2 \times 96{,}487 \times 0.763 = -147{,}200 \text{ joules}$$

The systematic list of standard single electrode potentials, part of which is presented in Table 14.2, is called the Electrochemical Series.

Solubility Products

From Table 14.2, the standard oxidation potential for sodium, $\mathcal{E}^0_{Na/Na^+}$, is 2.714 volts and the standard reduction potential for chlorine, $\mathcal{E}^0_{\frac{1}{2}Cl_2/Cl^-}$, is 1.3595 volts. Thus, for

$$Na = Na^+ + e^-$$

$$\Delta G^0 = -\mathcal{F}\mathcal{E}^0_{Na/Na^+} = -96{,}487 \times 2.714 = -261{,}870 \text{ joules}$$

and for

$$\tfrac{1}{2}Cl_2 + e^- = Cl^-$$

$$\Delta G^0 = -\mathcal{F}\mathcal{E}^0_{\frac{1}{2}Cl_2/Cl} = -96{,}487 \times 1.3595 = -131{,}170 \text{ joules}$$

Summation gives

$$\Delta G^0_{(i)} = -261{,}870 - 131{,}170 = -393{,}040 \text{ joules}$$

for the reaction

$$Na_{(s)} + \tfrac{1}{2}Cl_{2(g)} = Na^+_{(m)} + Cl^-_{(m)} \qquad \text{(i)}$$

For the reaction

$$Na_{(s)} + \tfrac{1}{2}Cl_{2(g)} = NaCl_{(s)} \qquad \text{(ii)}$$

$\Delta G^0_{(ii),298} = -385{,}310$ joules, and combination of (i) and (ii) gives

$$NaCl_{(s)} = Na^+_{(m)} + Cl^-_{(m)} \qquad \text{(iii)}$$

Table 14.2. Standard oxidation potentials*

Electrode Reaction	$\mathcal{E}^0_H$
Acid Solutions	
$Li = Li^+ + e$	3.045
$K = K^+ + e$	2.925
$Cs = Cs^+ + e$	2.923
$Ba = Ba^{++} + 2e$	2.90
$Ca = Ca^{++} + 2e$	2.87
$Na = Na^+ + e$	2.714
$Mg = Mg^{++} + 2e$	2.37
$H^- = \frac{1}{2}H_2 + e$	2.25
$Al = Al^{+++} + 3e$	1.66
$Zn = Zn^{++} + 2e$	0.763
$Fe = Fe^{++} + 2e$	0.440
$Cr^{++} = Cr^{+++} + e$	0.41
$Cd = Cd^{++} + 2e$	0.403
$Sn = Sn^{++} + 2e$	0.136
$Pb = Pb^{++} + 2e$	0.126
$Fe = Fe^{+++} + 3e$	0.036
$D_2 = 2D^+ + 2e$	0.0034
$H_2 = 2H^+ + 2e$	0.000
$H_2S = S + 2\,H^+ + 2e$	−0.141
$Sn^{++} = Sn^{+4} + 2e$	−0.15
$Cu^+ = Cu^{++} + e$	−0.153
$2\,S_2O_3^= = S_4O_6^= + 2e$	−0.17
$Fe(CN)_6^{-4} = Fe(CN)_6^{-3} + e$	−0.36
$Cu = Cu^{++} + 2e$	−0.337
$2\,I^- = I_2 + 2e$	−0.5355
$Fe^{++} = Fe^{+++} + e$	−0.771
$Ag = Ag^+ + e$	−0.7991
$Hg = Hg^{++} + 2e$	−0.854
$Hg_2^{++} = 2\,Hg^{++} + 2e$	−0.92
$2\,Br^- = Br_2\,(l) + 2e$	−1.0652
$Mn^{++} + 2\,H_2O = MnO_2 + 4\,H^+ + 2e$	−1.23
$2\,Cr^{+3} + 7\,H_2O = Cr_2O_7^= + 14\,H^+ + 6e$	−1.33
$Cl^- = \frac{1}{2}Cl_2 + e$	−1.3595
$Ce^{+3} = Ce^{+4} + e$	−1.61
$Co^{++} = Co^{+++} + e$	−1.82
$2\,SO_4^= = S_2O_8^= + 2e$	−1.98
$2\,F^- = F_2 + 2e$	−2.65
Basic Solutions	
$2\,OH^- + Ca = Ca(OH)_2 + 2e$	3.03
$3\,OH^- + Cr = Cr(OH)_3 + 3e$	1.3
$4\,OH^- + Zn = ZnO_2^= + 2\,H_2O + 2e$	1.216
$2\,OH^- + Cn^- = CNO^- + H_2O + 2e$	0.97
$2\,OH^- + SO_3^= = SO_4^= + H_2O + 2e$	0.93
$H_2 + 2\,OH^- = 2\,H_2O + 2e$	0.828
$2\,OH^- + Ni = Ni(OH)_2 + 2e$	0.72
$OH^- + Fe(OH)_2 = Fe(OH)_3 + e$	0.56
$O_2 + 2\,OH^- = O_3 + H_2O + 2e$	−1.24

*W. M. Latimer, *The Oxidation States of the Elements and Their Potentials in Aqueous Solutions* (New York: Prentice-Hall, 2nd ed., 1952).

for which
$$\Delta G^0_{(iii)} = \Delta G^0_{(i)} - \Delta G^0_{(ii)}$$
$$= -393{,}040 + 385{,}310$$
$$= -7730 \text{ joules}$$
$$= -8.3144 \times 298 \ln \frac{(\gamma_\pm m_{NaCl})^2}{a_{NaCl}}$$

In Eq. (iii), the standard state on the left-hand side is pure solid NaCl and the standard state on the right-hand side is the unit molality standard state. Saturation of the aqueous solution occurs when NaCl has dissolved to the extent that the activity of NaCl in the solution, with respect to solid NaCl as the standard state, is unity. At this state

$$(\gamma_\pm m_{NaCl})^2 = 22.6$$

or
$$\gamma_\pm m_{NaCl} = 4.76$$

Thus, if the ions in solution are behaving ideally, the saturated aqueous solution of NaCl at 298 K is 4.76 molal. When the activity of NaCl in the solution is unity with respect to solid NaCl as the standard state, the term $(\gamma_\pm m_{NaCl})^2$ is called the *solubility product,* K_{sp}. Thus, generally, for the salt A_aY_y,

$$K_{sp} = (\gamma_\pm m_\pm)^{a+y} = \exp \frac{-\Delta G^0}{298R}$$

where ΔG^0 is the standard free energy change for the change of state

$$A_aY_{y(\text{Raoultian})} = aA^{z+}_{(m)} + yY^{z-}_{(m)}$$

Example 1

Calculate the solubility of AgBr in water at 298 K. From Table 14.2, $\mathcal{E}^0_{Ag/Ag^+} = -0.7991$ and $\mathcal{E}^0_{\frac{1}{2}Br_2/Br^-} = 1.0652$. Therefore, for the reaction

$$Ag_{(s)} + \tfrac{1}{2}Br_{2(l)} = Ag^+_{(m)} + Br^-_{(m)}$$

$$\Delta G^0_{298} = -\mathcal{F}(-0.7991 + 1.0652) = 25{,}675 \text{ joules}$$

For
$$Ag_{(s)} + \tfrac{1}{2}Br_{2(l)} = AgBr_{(s)}$$

$$\Delta G^0_{298} = -95{,}670 \text{ joules}$$

and hence for

$$AgBr_{(s)} = Ag^+_{(m)} + Br^-_{(m)}$$

$$\begin{aligned}\Delta G^0 &= 25{,}675 + 95{,}670 \\ &= 121{,}345 \text{ joules} \\ &= -8.3144 \times 298 \ln K_{sp}\end{aligned}$$

Thus $K_{sp} = (\gamma_{\pm} m_{AgBr})^2 = 5.4 \times 10^{-22}$ or m_{AgBr} at saturation is 2.3×10^{-11}, which indicates that AgBr is virtually insoluble in water.

Example 2

Calculate the molalities of H^+ and OH^- in water at 298 K. From Table 14.2, the standard reduction potential for the half cell reaction

$$H_2O_{(l)} + e^- = \tfrac{1}{2}H_{2(g)} + OH^-_{(m)}$$

is −0.828 volts and the standard potential for the reaction

$$\tfrac{1}{2}H_{2(g)} = H^+_{(m)} + e^-$$

is zero. Summing gives

$$H_2O_{(l)} = H^+_{(m)} + OH^-_{(m)}$$

for which

$$\begin{aligned}\Delta G^0_{298} = -\mathcal{F}\mathcal{E}^0 &= -96{,}487 \times (-0.828) \\ &= 79{,}900 \text{ joules} \\ &= -8.3144 \times 298 \ln \gamma_{\pm}^2 m_{H^+} m_{OH^-}\end{aligned}$$

Thus, ignoring the activity coefficient,

$$m_{H^+} m_{OH^-} = 9.92 \times 10^{-15}$$

or, from the stoichiometry of the dissociation, as $m_{H^+} = m_{OH^-}$,

$$m_{H^+} = m_{OH^-} = 10^{-7}$$

The Influence of Acidity

The single electrode potential for the half cell reaction

$$\tfrac{1}{2}H_2 = H^+ + e^-$$

is

$$\mathcal{E} = \mathcal{E}^0_H - \frac{RT}{\mathcal{F}} \ln \frac{a_{H^+}}{p^{1/2}_{H_2}}$$

where $\mathcal{E}^0_H = 0$. Thus, if the hydrogen ions are behaving ideally,

$$\mathcal{E} = -\frac{RT}{\mathcal{F}} \ln \frac{m_{H^+}}{p^{1/2}_{H_2}} \qquad \text{(i)}$$

i.e., for a fixed pressure of hydrogen gas, $\mathcal{E}$ is a linear function of the logarithm of the molality of the hydrogen ions. The concentration of hydrogen ions in an aqueous solution determines the acidity of the solution and, conventionally, this acidity is quantified by the definition of pH, where

$$\text{pH} = -\log\,[H^+] \qquad \text{(ii)}$$

and $[H^+]$ is the molarity of the hydrogen ions, i.e., the number of moles of H^+ per liter of solution. Substitution of (ii) into (i) would require that either the half-cell potentials be determined with reference to the unit molarity standard state, or that pH be redefined as $-\log\,(m_{H^+})$. As, in dilute solutions, molality and molarity are virtually identical, this theoretical difficulty is of no practical significance. In the following discussions, the unit molarity standard state will be used for ions in solution. This standard state is defined as

$$a_{A^{z+}} \rightarrow [A^{z+}] \quad \text{as} \quad [A^{z+}] \rightarrow 0$$

Furthermore, in the following discussions it will be assumed that all ions in aqueous solution behave ideally such that

$$a_{A^{z+}} = [A^{z+}] \doteqdot m_{A^{z+}}$$

With this understanding, Eq. (i) becomes

$$\mathcal{E} = -\frac{RT}{\mathcal{F}} \ln \frac{[H^+]}{p^{1/2}_{H_2}}$$

which, with Eq. (ii), becomes

$$\mathcal{E} = -\frac{2.303 \times 8.3144 \times 298}{96{,}487} \log [H^+] + \frac{2.303 \times 8.3144 \times 298}{2 \times 96{,}487} \log p_{H_2}$$

$$= 0.0591\,(\text{pH}) + 0.0296 \log p_{H_2}$$

or, with $p_{H_2} = 1$ atm,

$$\mathcal{E} = 0.0591\,(\text{pH})$$

In Example 2 of the previous section, it was found that the molality (and hence the molarity) of H^+ in H_2O is 10^{-7}. Thus the pH of H_2O is 7, and hence the hydrogen electrode potential in water is $0.0591 \times 7 = 0.414$ volts. The hydrogen electrode has its standard value of zero at pH = 0, i.e., at $[H^+] = 1$.

14.9 POURBAIX DIAGRAMS

Pourbaix diagrams, or "potential-pH" diagrams, are graphical representations of thermodynamic and electrochemical equilibria occurring in aqueous systems. They are thus the electrochemical analogues of the chemical stability diagrams discussed in Sec. 13.4. This approach to the representation of equilibria has been developed by M. Pourbaix in Belgium* and hence uses the European EMF sign convention. The European convention is simply the opposite of the U.S. convention, i.e.,

$$\text{with the U.S. convention} \qquad \Delta G^0 = -z\mathcal{F}\mathcal{E}^0$$

$$\text{and with the European convention} \qquad \Delta G^0 = +z\mathcal{F}\mathcal{E}^0$$

To distinguish between the two conventions, an EMF written in conformity with the U.S. convention will be denoted $\mathcal{E}$, and the EMF written in conformity with the European convention will be denoted E. Thus, always, $E = -\mathcal{E}$.

As an example, consider the Daniell cell

$$\text{Zn}|\text{Zn}^{2+} \vdots \text{Cu}^{2+}|\text{Cu}$$

With the U.S. convention,

$$\mathcal{E}^0 = \mathcal{E}^0_{\text{Zn/Zn}^{2+}} + \mathcal{E}^0_{\text{Cu}^{2+}/\text{Cu}} = 0.763 + 0.337 = 1.1 \text{ volts}$$

*M. Pourbaix, "Atlas of Electrochemical Equilibria in Aqueous Solutions," National Association of Corrosion Engineers, Houston, Texas, 1974.

and the free energy change for the cell reaction is

$$\Delta G^0 = -2\mathscr{F}\mathscr{E}^0 = -212{,}300 \text{ joules}$$

With the European convention

$$E^0 = E^0_{Zn/Zn^{2+}} + E^0_{Cu^{2+}/Cu}$$

$$= -0.763 - 0.337$$

$$= -1.1 \text{ volts}$$

and the free energy change for the cell reaction is

$$\Delta G^0 = 2\mathscr{F}E^0 = -212{,}300 \text{ joules}$$

For any Daniell cell

$$\mathscr{E} = \mathscr{E}^0 - \frac{RT}{2\mathscr{F}} \ln \frac{[Zn^{2+}]a_{Cu}}{[Cu^{2+}]a_{Zn}}$$

or

$$E = E^0 + \frac{RT}{2\mathscr{F}} \ln \frac{[Zn^{2+}]a_{Cu}}{[Cu^{2+}]a_{Zn}}$$

Consider the cell

$$Pt,H_{2(g)}|H_2O|O_{2(g)},Pt$$

The galvanic cell reaction

$$H_{2(g)} + \tfrac{1}{2}O_{2(g)} = H_2O_{(l)} \tag{i}$$

is the sum of the half cell reactions

$$H_2 = 2H^+ + 2e^- \tag{ii}$$

and

$$\tfrac{1}{2}O_2 + 2H^+ + 2e^- = H_2O \tag{iii}$$

Thus

$$\Delta G^0_{(i)} = \Delta G^0_{(ii)} + \Delta G^0_{(iii)}$$

$\Delta G^0_{(i),298} = -237{,}190$ joules and, by convention, $\Delta G^0_{(ii),298} = 0$. Thus the standard reduction potential for reaction (iii) is

$$E^0_{(iii)} = \frac{\Delta G^0_{(iii)}}{2\mathcal{F}}$$

$$= \frac{-237{,}190}{2\mathcal{F}}$$

$$= -1.229 \text{ volts}$$

or the standard oxidation potential is +1.229 volts. With all of the reactants and products occurring in their standard states electrochemical equilibrium is established when the backing EMF applied to the cell is 1.229 volts. If the backing EMF is less than this value the cell is a current-producing galvanic cell and the cell reaction occurring is

$$H_{2(g)} + \tfrac{1}{2}O_{2(g)} \rightarrow H_2O_{(l)}$$

with the oxidation $2H_2 \rightarrow 2H^+ + 2e^-$ occurring at the anode and the reduction $\frac{1}{2}O_2 + 2H^+ + 2e^- \rightarrow H_2O$ occurring at the cathode. However, if the backing EMF is greater than 1.229 volts, the cell is a current-consuming electrolysis cell and the cell reaction occurring is

$$H_2O_{(l)} \rightarrow H_{2(g)} + \tfrac{1}{2}O_{2(g)}$$

with the oxidation reaction $H_2O \rightarrow \frac{1}{2}O_2 + 2H^+ + 2e^-$ occurring at the anode and the reduction $2H^+ + 2e^- \rightarrow H_2$ occurring at the cathode. Thus, with hydrogen gas and hydrogen ions at unit activity, electrochemical equilibrium is established at the hydrogen electrode when the electrode is at zero potential. If the potential of the electrode is increased above zero the anodic reaction $H_2 \rightarrow 2H^+ + 2e^-$ occurs and if the potential of the electrode is decreased below zero the cathodic reaction $2H^+ + 2e^- \rightarrow H_2$ occurs. Similarly, with oxygen at unit pressure and hydrogen ions at unit activity, electrochemical equilibrium at the oxygen electrode is established when the electrode potential is 1.229 volts. If the potential of the electrode is increased above 1.229 volts the anodic reaction $H_2O \rightarrow \frac{1}{2}O_2 + 2H^+ + 2e^-$ occurs, and if the potential of the electrode is decreased below 1.229 volts the cathodic reaction $\frac{1}{2}O_2 + 2H^+ + 2e^- \rightarrow H_2O$ occurs. Thus, generally, if the potential, E, of an electrode at equilibrium is increased an anodic oxidation reaction occurs, and if it is decreased a cathodic reduction reaction occurs.

As both half-cell reactions (ii) and (iii) involve hydrogen ions, the half cell potentials are functions of the pH of the aqueous solution. At the hydrogen electrode, the electrochemical equilibrium

$$\tfrac{1}{2}H_2 = H^+ + e^- \tag{a}$$

is established at

$$E_{(a)} = \frac{RT}{\mathcal{F}} \ln \frac{[H^+]}{p_{H_2}^{1/2}}$$

$$= -0.0591 \text{ (pH)} - 0.0298 \log p_{H_2}$$

which, with $p_{H_2} = 1$ atm, is drawn as line (a) in Fig. 14.11. At the oxygen electrode, the electrochemical equilibrium

$$H_2O = \tfrac{1}{2}O_2 + 2H^+ + 2e^- \qquad \text{(b)}$$

is established at

$$E_{(b)} = 1.229 + \frac{RT}{2\mathcal{F}} \ln [H^+]^2 p_{O_2}^{1/2}$$

$$= 1.229 - 0.0591 \text{ (pH)} + 0.0148 \log p_{O_2}$$

which, with $p_{O_2} = 1$ atm, is drawn as line (b) in Fig. 14.11. Lines (a) and (b) in Fig. 14.11 define the *domain of thermodynamic stability of water* in aqueous solutions under a pressure of 1 atm of H_2 and 1 atm of O_2. Below line (a), the equilibrium pressure of hydrogen gas is greater than 1 atm, and hence hydrogen gas is cathodically evolved from an aqueous solution at an electrode, the potential of which is moved below line (a) when the pressure of hydrogen gas at the electrode is 1 atm. Similarly, above line (b) the equilibrium pressure of oxygen gas is greater than 1 atm and hence oxygen gas is anodically evolved from an aqueous solution at an electrode, the potential of which is moved above line (b)

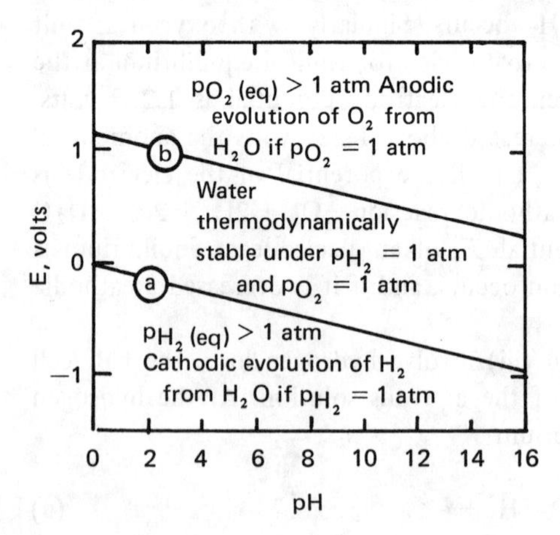

Fig. 14.11. The domain of thermodynamic stability of water.

when the oxygen pressure at the electrode is 1 atm. Between lines (a) and (b), water is thermodynamically stable with oxygen and hydrogen pressures of 1 atm.

The Pourbaix Diagram for Copper

The species participating in the various chemical and electrochemical equilibria are the solids Cu, Cu_2O, and CuO, and the ions Cu^+, Cu^{2+}, $HCuO_2^-$, and CuO_2^{2-}, and the pertinent standard free energies are

	ΔG^0_{298}, joules
$2Cu_{(s)} + \frac{1}{2}O_{2(g)} = Cu_2O_{(s)}$	−146,360
$Cu_{(s)} + \frac{1}{2}O_{2(g)} = CuO_{(s)}$	−127,190
$Cu_{(s)} = Cu^+_{(m)} + e^-$	50,210
$Cu_{(s)} = Cu^{2+}_{(m)} + 2e^-$	64,980
$\frac{1}{2}H_{2(g)} + Cu_{(s)} + O_{2(g)} + e^- = HCuO^-_{2(m)}$	−256,980
$Cu_{(s)} + O_{2(g)} + 2e^- = CuO^{2-}_{2(m)}$	−182,000
$H_{2(g)} + \frac{1}{2}O_{2(g)} = H_2O_{(l)}$	−237,190

Consider first the various equilibria that occur among the ions in solution. The equations for the electrochemical equilibria are always set up with the lower oxidation state on the left and the higher oxidation state on the right, i.e., with the balancing electronic charge on the right hand side of the equation.

1. The Equilibrium between Cu^+ and Cu^{2+}

$$Cu^+ = Cu^{2+} + e^- \tag{1}$$

$$\Delta G_1^0 = 64{,}980 - 50{,}210 = 14{,}770 \text{ joules} = \mathcal{F}E_1^0$$

Thus

$$E_1 = E_1^0 + \frac{RT}{\mathcal{F}} \ln \frac{[Cu^{2+}]}{[Cu^+]}$$

$$= \frac{14{,}770}{96{,}487} + \frac{8.3144 \times 298 \times 2.303}{96{,}487} \log \frac{[Cu^{2+}]}{[Cu^+]}$$

$$= 0.153 + 0.0591 \log \frac{[Cu^{2+}]}{[Cu^+]}$$

and hence $[Cu^{2+}] = [Cu^{+}]$ at $E = 0.153$ volts. $E = 0.153$ volts is drawn as line 1 in Fig. 14.12*a*. At potentials greater than 0.153 volts, $[Cu^{2+}] > [Cu^{+}]$, and at potentials less than 0.153 volts, $[Cu^{+}] > [Cu^{2+}]$.

2. The Equilibrium between Cu^{2+} and $HCuO_2^-$

$$Cu^{2+} \leftrightarrow HCuO_2^-$$

The procedure for deriving the expression for the equilibrium is as follows:

(i) Balance the oxygen with H_2O, i.e.,

$$Cu^{2+} + 2H_2O \leftrightarrow HCuO_2^-$$

(ii) Balance the hydrogen with H^+, i.e.,

$$Cu^{2+} + 2H_2O \leftrightarrow HCuO_2^- + 3H^+$$

(iii) Balance the charge with e^-.

As the equilibrium between Cu^{2+} and $HCuO_2^-$ is not electrochemical, i.e., does not involve a change in the oxidation state of the copper, step (iii) is not necessary. The desired expression is thus

$$Cu^{2+} + 2H_2O = HCuO_2^- + 3H^+ \tag{2}$$

for which

$$\Delta G_2^0 = -256{,}980 + (2 \times 237{,}190) - 64{,}980$$

$$= 152{,}420 \text{ joules}$$

$$= -8.3144 \times 298 \times 2.303 \log \frac{[H^+]^3[HCuO_2^-]}{a^2_{H_2O}[Cu^{2+}]}$$

or, with $a_{H_2O} = 1$,

$$\log \frac{[HCuO_2^-]}{[Cu^{2+}]} = 3\,(\text{pH}) - 26.71$$

Thus $[HCuO_2^-] = [Cu^{2+}]$ at pH = 8.9, which is drawn as line 2 in Fig. 14.12*a*. In solutions more acidic than pH = 8.9 $[Cu^{2+}] > [HCuO_2^-]$, and in less acidic solutions $[HCuO_2^-] > [Cu^{2+}]$. At the point of intersection of lines 1 and 2,

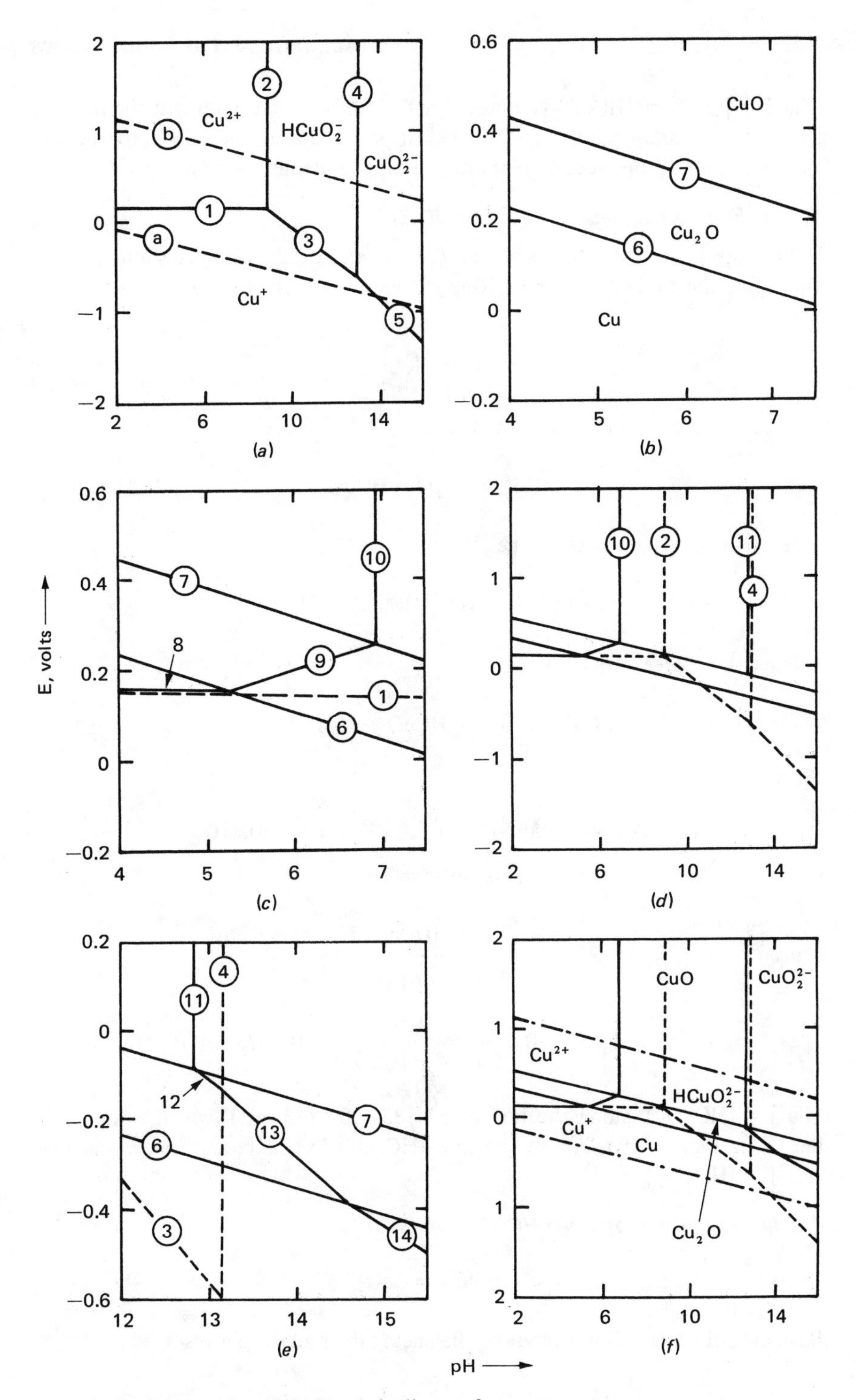

Fig. 14.12. Construction of the Pourbaix diagram for copper.

$[Cu^+] = [Cu^{2+}] = [HCuO_2^-]$. Lines 1 and 2 define what is termed the *domain of relative predominance* of Cu^{2+}, i.e., in states above line 1 and to the left of line 2, Cu^{2+} is the predominant copper ion in aqueous solutions.

3. The Equilibrium between Cu^+ and $HCuO_2^-$

With the lower oxidation state on the left and the higher oxidation state on the right, the procedure for deriving the expression for the equilibrium begins with

$$Cu^+ \leftrightarrow HCuO_2^-$$

Then, as before, balance the oxygen with H_2O, i.e.,

$$Cu^+ + 2H_2O \leftrightarrow HCuO_2^-$$

balance the hydrogen with H^+, i.e.,

$$Cu^+ + 2H_2O \leftrightarrow HCuO_2^- + 3H^+$$

balance the charge with e^-, i.e.,

$$Cu^+ + 2H_2O = HCuO_2^- + 3H^+ + e^- \qquad (3)$$

for which

$$\Delta G_3^0 = -256{,}980 + (2 \times 237{,}190) - 50{,}210$$
$$= 167{,}190 \text{ joules} = \mathcal{F}E_3^0$$

Thus
$$E_3 = E_3^0 + \frac{RT}{\mathcal{F}} \ln \frac{[H^+]^3 [HCuO_2^-]}{[Cu^+] a_{H_2O}}$$
$$= 1.733 + 0.0591 \log \frac{[HCuO_2^-]}{[Cu^+]} - 0.1773 \text{ (pH)}$$

$[Cu^+] = [HCuO_2^-]$ along the line $E = 1.733 - 0.1773$ (pH), which is drawn as line 3 in Fig. 14.12*a*. Above the line $[HCuO_2^-] > [Cu^+]$, and below the line $[Cu^+] > [HCuO_2^-]$.

4. The Equilibrium between $HCuO_2^-$ and CuO_2^{2-}

$$HCuO_2^- \leftrightarrow CuO_2^{2-}$$

Balancing of oxygen is not required. Balancing the hydrogen gives

$$HCuO_2^- = CuO_2^{2-} + H^+ \tag{4}$$

which does not require balancing of the charge.

$$\Delta G_4^0 = -182{,}000 + 256{,}980$$
$$= 74{,}980 \text{ joules}$$
$$= -RT \ln \frac{[H^+][CuO_2^{2-}]}{[HCuO_2^-]}$$

or

$$\log \frac{[CuO_2^{2-}]}{[HCuO_2^-]} = -13.14 + \text{pH}$$

Thus $[HCuO_2^-] = [CuO_2^{2-}]$ at pH = 13.14, which is drawn as line 4 in Fig. 14.12*a*. Lines 2, 3, and 4 define the limits of the domain of relative predominance of the $HCuO_2^-$ ion, and lines 3 and 4 intersect at the state in which $[Cu^+] = [HCuO_2^-] = [CuO_2^{2-}]$.

5. The Equilibrium between Cu^+ and CuO_2^{2-}

$$Cu^+ \leftrightarrow CuO_2^{2-}$$
$$Cu^+ + 2H_2O \leftrightarrow CuO_2^{2-}$$
$$Cu^+ + 2H_2O \leftrightarrow CuO_2^{2-} + 4H^+$$

and

$$Cu^+ + 2H_2O = CuO_2^{2-} + 4H^+ + e^- \tag{5}$$

for which

$$\Delta G_5^0 = -182{,}000 + (2 \times 237{,}190) - 50{,}310$$
$$= 242{,}170 \text{ joules}$$
$$= \mathcal{F}E_5^0$$

Thus

$$E_5 = E_5^0 + \frac{RT}{\mathcal{F}} \ln \frac{[H^+]^4 [CuO_2^{2-}]}{[Cu^+] a_{H_2O}}$$
$$= 2.510 + 0.0591 \log \frac{[CuO_2^{2-}]}{[Cu^+]} - 0.2364 \text{ (pH)}$$

$[CuO_2^{2-}] = [Cu^+]$ along the line $E = 2.510 - 0.2364$ (pH), which is drawn as line 5 in Fig. 14.12*a*. Lines 1, 3, and 5 define the limits of the domain of relative predominance of Cu^+, and lines 4 and 5 define the limits of the domain of relative predominance of CuO_2^{2-}. Figure 14.12*a*, which includes the lines (a) and (b) defining the domain of thermodynamic stability of water at 1 atm, shows that

i. Acidic ions predominate at low pH
ii. Basic ions predominate at high pH
iii. Higher valent states predominate at more positive values of E, and
iv. Lower valent states predominate at more negative values of E.

Consider, now, the equilibria occurring among the pure solid phases.

6. The Equilibrium between Solid Cu and Solid Cu₂O

$$2Cu \leftrightarrow Cu_2O$$

Balancing the oxygen:

$$2Cu + H_2O \leftrightarrow Cu_2O$$

Balancing the hydrogen:

$$2Cu + H_2O \leftrightarrow Cu_2O + 2H^+$$

and balancing the charge gives

$$2Cu + H_2O = Cu_2O + 2H^+ + 2e^- \qquad (6)$$

for which

$$\Delta G_6^0 = -146{,}360 + 237{,}190$$
$$= 90{,}830 \text{ joules} = 2\mathcal{F}E_6^0$$

Thus, with $a_{Cu} = a_{Cu_2O} = a_{H_2O} = 1$,

$$E_6 = E_6^0 + \frac{RT}{2\mathcal{F}} \ln [H^+]^2$$
$$= 0.471 - 0.0591 \text{ (pH)}$$

which is drawn as line 6 in Fig. 14.12*b*. An increase in potential above the equilibrium value for any pH causes reaction to proceed in the anodic direction, i.e., from left to right. Thus above line 6 solid Cu_2O is stable relative to solid Cu, and below the line Cu is stable relative to Cu_2O.

7. The Equilibrium between Solid Cu_2O and Solid CuO

For the reaction

$$Cu_2O + H_2O = 2CuO + 2H^+ + 2e^- \tag{7}$$

$$\Delta G_7^0 = -(2 \times 129{,}170) + 237{,}190 + 146{,}360$$
$$= 129{,}170 \text{ joules} = 2\mathcal{F}E_7^0$$

Thus, with $a_{Cu_2O} = a_{CuO} = a_{H_2O} = 1$,

$$E_7 = E_7^0 + \frac{RT}{2\mathcal{F}} \ln [H^+]^2$$
$$= 0.669 - 0.0591 \text{ (pH)}$$

which is drawn as line 7 in Fig. 14.12*b*. Again, if the potential is increased above the equilibrium value for any pH, the reaction proceeds in the anodic direction, i.e., from left to right. Thus CuO is stable relative to Cu_2O above line 7, and Cu_2O is stable relative to CuO below the line. Figure 14.12*b* shows the fields of stability of the three solid phases.

Now consider the various chemical and electrochemical equilibria which occur among the ionic species in the aqueous solution and the solid phases.

8. The Equilibrium between Solid Cu and Cu^{2+} in Solution

For the reaction

$$Cu = Cu^{2+} + 2e^- \tag{8}$$

$$\Delta G_8^0 = 64{,}980 \text{ joules} = 2\mathcal{F}E_8^0$$

Thus, with $a_{Cu} = 1$,

$$E_8 = E_8^0 + \frac{RT}{2\mathcal{F}} \ln [Cu^{2+}]$$
$$= 0.337 + 0.0296 \log [Cu^{2+}] \text{ volts}$$

which is the variation of the single electrode potential at a copper electrode with the equilibrium concentration of Cu^{2+} ions in an aqueous solution. Pourbaix considered that, if the concentration of ions in electrochemical equilibrium with the solid phase is equal to or less than 10^{-6} moles per liter, the solid is "immune" from electrochemical attack (corrosion) by the aqueous solution. With $[Cu^{2+}] = 10^{-6}$ moles per liter, $E_8 = 0.159$ volts, which is drawn as line 8 in Fig. 14.12*c*. Line 8 lies just within the domain of relative predominance of Cu^{2+}, and, from Eq. (1), the concentration of Cu^+ ions in equilibrium with a Cu electrode at 0.159 volts is 8×10^{-7} moles per liter. Line 8 intersects line 6 at pH = 5.28, which is thus the state at which solid Cu, solid Cu_2O, and an aqueous solution of $[Cu^{2+}] = 10^{-6}$ moles/liter are in equilibrium.

9. The Equilibrium between Solid Cu_2O and Cu^{2+} in Solution

For the reaction

$$Cu_2O + 2H^+ = 2Cu^{2+} + H_2O + 2e^- \tag{9}$$

$$\Delta G_9^0 = 39{,}130 \text{ joules} = 2\mathfrak{F}E_9^0$$

Thus

$$E_9 = E_9^0 + \frac{RT}{2\mathfrak{F}} \ln \frac{[H^+]^2}{[Cu^{2+}]^2}$$

$$= 0.203 + 0.0591\,(\text{pH}) + 0.0591 \log\,[Cu^{2+}]$$

With $[Cu^{2+}] = 10^{-6}$, moles/liter,

$$E_9 = -0.1516 + 0.0591\,(\text{pH})$$

which is drawn as line 9 in Fig. 14.12*c*. At the point of intersection of line 9 with line 7 ($E = 0.259$, pH = 6.94), an aqueous solution of $[Cu^{2+}] = 10^{-6}$ moles/liter is in equilibrium with a Cu_2O-CuO electrode.

10. The Equilibrium between Solid CuO and Cu^{2+} in Solution

For the reaction

$$Cu^{2+} + H_2O = CuO + 2H^+ \tag{10}$$

$$\Delta G_{10}^0 = 45{,}020 \text{ joules}$$

$$= -8.3144 \times 298 \ln \frac{[H^+]^2}{[Cu^{2+}]^2}$$

or

$$\log\,[Cu^{2+}] = 7.89 - 2\,(\text{pH})$$

which is independent of electrode potential. Thus, with $[Cu^{2+}] = 10^{-6}$ moles/

liter, the pH required for equilibrium with CuO is 6.94. This is drawn as line 10 in Fig. 14.12*c*. The concentration of Cu^{2+} in equilibrium with CuO increases with decreasing pH.

11. The Equilibrium between Solid CuO and $HCuO_2^-$ in Solution

For the reaction

$$CuO + H_2O = HCuO_2^- + H^+ \tag{11}$$

$$\Delta G_{11}^0 = 107{,}400 \text{ joules} = -8.3144 \times 298 \ln [H^+][HCuO_2^-]$$

or

$$\log [HCuO_2^-] = -18.82 + \text{pH}$$

Thus, with $[HCuO_2^-] = 10^{-6}$ moles/liter, the required pH is 12.82. This is drawn as line 11 in Fig. 14.12*d*. Line 11 lies just within the domain of relative predominance of $HCuO_2^-$ and, from Eq. (4), lies at a concentration of 4.8×10^{-7} moles/liter of CuO_2^{2-}, which is thus the concentration of CuO_2^{2-} required for equilibrium with CuO at pH = 12.82. The concentrations of both $HCuO_2^-$ and CuO_2^{2-} in equilibrium with CuO increase with increasing pH, with the equilibrium concentration of CuO_2^{2-} exceeding that of $HCuO_2^-$ at pH greater than that given by line 4, i.e., at pH greater than 13.14.

12. The Equilibrium between Solid Cu_2O and $HCuO_2^-$ in Solution

For the reaction

$$Cu_2O + 3H_2O = 2HCuO_2^- + 4H^+ + 2e^- \tag{12}$$

$$\Delta G_{12}^0 = 343{,}970 \text{ joules} = 2\mathfrak{F}E_{12}^0$$

Thus

$$E_{12} = E_{12}^0 + \frac{RT}{2\mathfrak{F}} \ln [H^+]^4 [HCuO_2^-]^2$$

$$= 1.782 - 0.1182\,(\text{pH}) + 0.0591 \log [HCuO_2^-]$$

Thus, with $[HCuO_2^-] = 10^{-6}$ moles/liter,

$$E_{12} = 1.427 - 0.1182\,(\text{pH})$$

which is drawn as line 12 in Fig. 14.12*e*. Line 12 intersects line 4 at the state at which the concentrations of $HCuO_2^-$ and CuO_2^{2-} in equilibrium with a Cu_2O electrode are both 10^{-6} moles/liter.

13. The Equilibrium between Solid Cu_2O and CuO_2^{2-} in Solution

For the reaction

$$Cu_2O + 3H_2O = 2CuO_2^{2-} + 6H^+ + 2e^- \tag{13}$$

$$\Delta G_{13}^0 = 493{,}930 \text{ joules} = 2\mathcal{F}E_{13}^0$$

Thus
$$E_{13} = E_{13}^0 + \frac{RT}{2\mathcal{F}} \ln [H^+]^6 [CuO_2^{2-}]^2$$

$$= 2.560 - 0.1773 \text{ (pH)} + 0.0591 \log [CuO_2^{2-}]$$

With $[CuO_2^{2-}] = 10^{-6}$ moles/liter,

$$E_{13} = 2.205 - 0.1773 \text{ (pH)}$$

which is drawn as line 13 in Fig. 14.12*e*. Line 13 intersects line 6 at $E = -0.396$, pH = 14.67, which is the state in which a solution containing 10^{-6} moles/liter of CuO_2^{2-} is in electrochemical equilibrium with a Cu-Cu_2O electrode.

14. The Equilibrium between Solid Cu and CuO_2^{2-} in Solution

For the reaction

$$Cu + 2H_2O = CuO_2^{2-} = 4H^+ + 2e^- \tag{14}$$

$$\Delta G_{14}^0 = 292{,}380 \text{ joules} = 2\mathcal{F}E_{14}^0$$

Thus
$$E_{14} = E_{14}^0 + \frac{RT}{2\mathcal{F}} \ln [H^+]^4 [CuO_2^{2-}]$$

$$= 1.515 - 0.1182 \text{ (pH)} + 0.0296 \log [CuO_2^{2-}]$$

With $[CuO_2^{2-}] = 10^{-6}$ moles/liter,

$$E_{14} = 1.337 - 0.1182 \text{ (pH)}$$

which is drawn as line 14 in Fig. 14.12*e*. The concentrations of $HCuO_2^-$ and CuO_2^{2-} in equilibrium with each of Cu, Cu_2O, and CuO increase with increasing pH.

The complete Pourbaix diagram for Cu-H_2O is shown in Fig. 14.13. This figure includes lines, obtained from Eqs. (8), (9), and (10), along which the concentrations of Cu^{2+} in equilibrium with each of the three solid phases are 1, 10^{-2}, and 10^{-4} moles/liter. Also included are lines along which the concen-

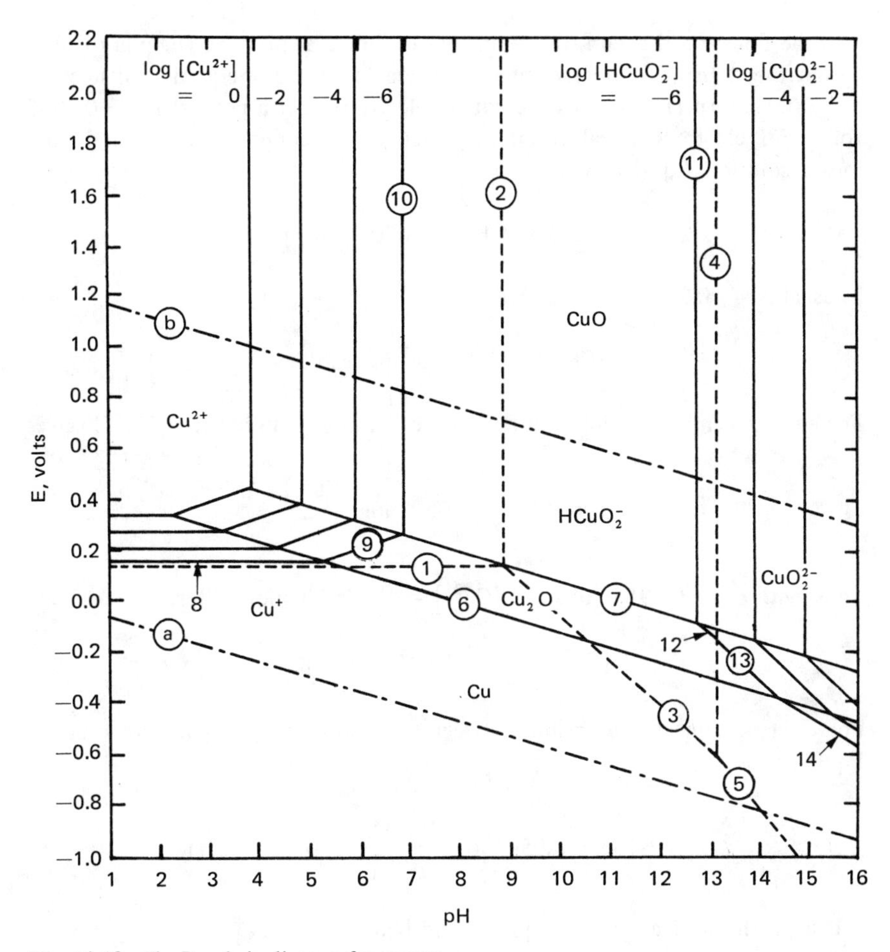

Fig. 14.13. The Pourbaix diagram for copper.

trations of CuO_2^{2-} in equilibrium with each of the three solid phases are 10^{-2} and 10^{-4} moles/liter.

Examples

Calculate the concentrations of the various copper ions in an aqueous solution of pH = 5 under a gaseous atmosphere of $p_{H_2} = 1$ atm that is in equilibrium with a copper electrode at which the cathodic evolution of hydrogen gas is imminent.*

*This is an equilibrium calculation and hence does not consider any hydrogen overvoltage at the copper electrode.

In the Pourbaix diagram, this point lies on line *a* at pH = 5. The point is thus within the domain of relative predominance of Cu^+. The equation of line a is $E = -0.0591$ (pH), and thus the copper electrode is at a potential of -0.2955 volts. For electrochemical equilibrium between a copper electrode and Cu^{2+} ions in solution, Eq. (8) gives

$$E = 0.337 + 0.0296 \log [Cu^{2+}]$$

Thus, at $E = -0.2955$,

$$[Cu^{2+}] = 4 \times 10^{-22} \text{ moles/liter}$$

For electrochemical equilibrium between Cu^+ and Cu^{2+} in solution, Eq. (1) gives

$$E = 0.153 + 0.0591 \log \frac{[Cu^{2+}]}{[Cu^+]}$$

Thus, with $E = -0.2955$ volts and $[Cu^{2+}] = 4 \times 10^{-22}$ moles/liter,

$$[Cu^+] = 1.7 \times 10^{-14} \text{ moles/liter}$$

For electrochemical equilibrium between Cu^+ and $HCuO_2^-$ in aqueous solution, Eq. (3) gives

$$E = 1.733 + 0.0591 \log \frac{[HCuO_2^-]}{[Cu^+]} - 0.1773 \text{ (pH)}$$

Thus, with the given values of pH, E, and $[Cu^+]$,

$$[HCuO_2^-] = 8 \times 10^{-34} \text{ moles/liter}$$

A similar calculation using Eq. (5) gives $[CuO_2^{2-}] = 6 \times 10^{-42}$ moles/liter.

If the potential of the Cu is increased at constant pH, at what potential does Cu_2O begin to form?

In the Pourbaix diagram the required state lies on line 6 at pH = 5, where electrochemical equilibrium is established between Cu and Cu_2O. The equation of line 6 is

$$E = 0.471 - 0.0591 \text{ (pH)}$$

and thus, with pH = 5, $E = 0.1755$ volts.

What adjustments must be made to the potential of the Cu–Cu_2O electrode and the pH of the solution in order that the equilibrium concentration of Cu^{2+} ions in the solution be 0.1 moles/liter?

The state must be moved along line 6 until $Cu^{2+} = 0.1$. For equilibrium between solid Cu and Cu^{2+}, Eq. (8) gives

$$E = 0.337 + 0.0296 \log [Cu^{2+}]$$

Thus, for $[Cu^{2+}] = 0.1$, E must be 0.307 volts, and for this point to lie on line 6, given by

$$E = 0.471 - 0.0591 \text{ (pH)}$$

pH = 2.77. The required state is thus $E = 0.307$, pH = 2.77.

The solution is made more basic and the potential is maintained constant. At what pH does CuO form on the Cu_2O?

The required state lies on line 7 at $E = 0.307$ volts. The equation for line 7 is

$$E = 0.669 - 0.0591 \text{ (pH)}$$

Thus, with $E = 0.307$ volts, pH = 6.11.

To what value of pH must the solution be adjusted in order that the concentration of CuO_2^{2-} in equilibrium with CuO be 0.1 moles/liter?

The concentration of $HCuO_2^-$ in equilibrium with CuO is given by Eq. (11) as

$$\log [HCuO_2^-] = -18.82 + \text{pH}$$

and the concentration of CuO_2^{2-} in equilibrium with $HCuO_2^-$ is given by Eq. (4) as

$$\log [CuO_2^{2-}] = \log HCuO_2^- - 13.14 + \text{pH}$$

Combination gives the concentration of CuO_2^{2-} in equilibrium with CuO as

$$\log [CuO_2^{2-}] = -31.96 + 2 \text{ (pH)}$$

Thus $[CuO_2^{2-}] = 0.1$ moles/liter at pH = 15.48.

The Solubility of Copper and Its Oxides in Aqueous Solutions

Consider the solubilities of the three solids Cu, Cu_2O, and CuO at a constant potential of 0.2 volts. From Fig. 14.13 [or Eq. (6)], with $E = 0.2$ volts, Cu and

Cu_2O are in equilibrium with one another at pH = 4.585, and, from Eq. (7), Cu_2O and CuO are in equilibrium with one another at pH = 7.94. Also, with $E = 0.2$ volts, Cu^{2+} predominates at pH less than 8.9, CuO_2^{2-} predominates at pH greater than 13.14, and $HCuO_2^-$ predominates between these values.

Consider the concentrations of Cu^{2+} in equilibrium with each of the three solid phases. For equilibrium with Cu, Eq. (8) gives [log Cu^{2+}] = −4.63, which is drawn as line *ab* in Fig. 14.14; for equilibrium with Cu_2O, Eq. (9) gives log [Cu^{2+}] = −0.051 − pH, which is drawn as line *bc*; and for equilibrium with CuO, Eq. (10) gives log [Cu^{2+}] = 7.89 − 2 (pH), which is drawn as line *cd*. From Eq. (1), with E = 0.2 volts, [Cu^+] = 0.16 [Cu^{2+}]. The concentrations of Cu^+ in equilibrium with Cu, Cu_2O, and CuO are thus as given by lines *ef*, *fg*, and *gh*, respectively, in Fig. 14.14.

Consider the concentrations of $HCuO_2^-$ in equilibrium with each of the three phases. $HCuO_2^-$ is the predominant ion between pH = 8.9 and pH = 13.14, which, with E = 0.2 volts, is within the field of stability of CuO. From Eq. (11),

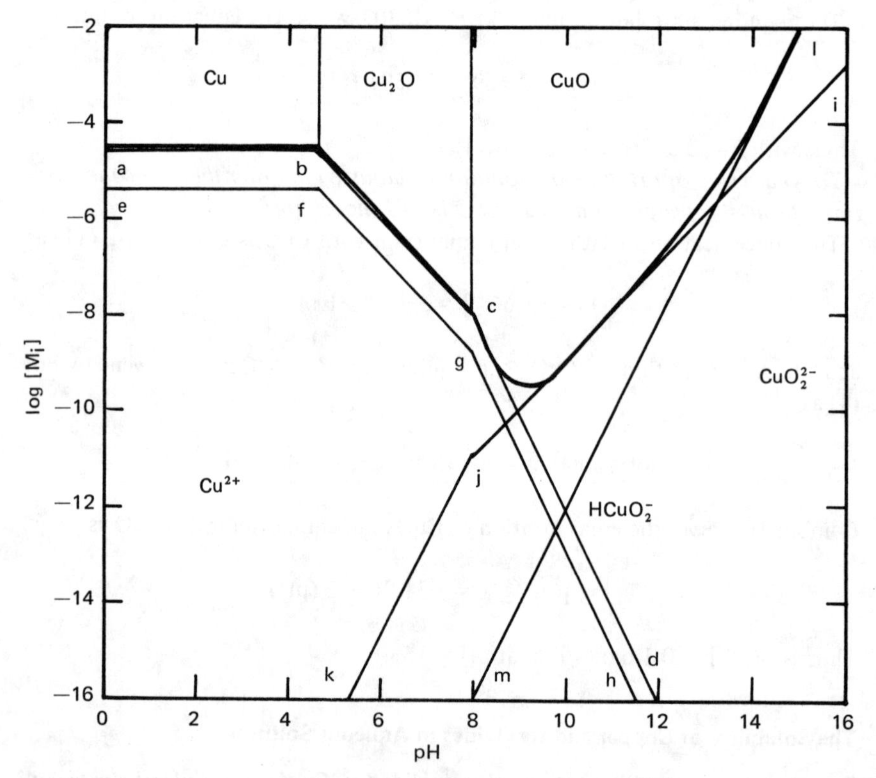

Fig. 14.14. The solubility of copper and copper oxides in aqueous solutions as a function of pH at E = 0.2 volts.

the concentration of $HCuO_2^-$ in equilibrium with CuO is log $[HCuO_2^-]$ = -18.82 + pH, which is drawn as line *ij* in Fig. 14.14. For equilibrium with Cu_2O at $E = 0.2$ volts, Eq. (12) gives log $[HCuO_2^-] = -26.77 + 2$ (pH), which is drawn as line *jk*. The concentration of $HCuO_2^-$ in equilibrium with Cu at $E =$ 0.2 volts, given by log $[HCuO_2^-] = -31.34 + 3$ (pH), is small enough to be ignored.

Combination of Eq. (11) and Eq. (4) to eliminate $[HCuO_2^-]$ gives the concentration of CuO_2^{2-} in equilibrium with CuO, as log $[CuO_2^{2-}] = -31.96 + 2$ (pH), which is drawn as line *lm* in Fig. 14.14. The concentrations of CuO_2^{2-} in equilibrium with Cu_2O and Cu, given, respectively, as log $[CuO_2^{2-}] = -39.91 +$ 3 (pH) and log $[CuO_2^{2-}] = -44.48 + 4$ (pH), are small enough to be ignored.

Lines *cd* and *ij* intersect at pH = 8.9, which is the limit of the domains of relative predominance of Cu^{2+} and $HCuO_2^-$, and lines *ij* and *lm* intersect at pH = 13.14, which is the limit of the domains of relative predominance of $HCuO_2^-$ and CuO_2^{2-}. The solubilities of the three phases are obtained by summing the equilibrium concentrations of the four ionic species, to give the full line in Fig. 14.14.

14.10 SUMMARY

1. The electromotive force $\mathcal{E}$ produced by a galvanic cell is related to the free energy decrease due to the electrochemical reaction in the cell, ΔG, by the equation

$$\Delta G = -z\mathcal{F}\mathcal{E}$$

where $\mathcal{F}$ is Faraday's constant (= 96,487 coulombs/mole) and z is the number of moles of electrons transported through the external circuit. If the cell reaction is

$$Cu_2O + Ni = NiO + 2Cu$$

the half-cell reactions are

$$2Cu^+ + 2e^- = 2Cu$$

and

$$Ni - 2e^- = Ni^{2+}$$

In such a case, if ΔG is per mole of Ni consumed, then $z = 2$; or if G is per mole of Cu produced, then $z = 1$.

Any electrochemical cell reaction can be written as the sum of two half-cell reactions–an oxidation reaction involving the release of electrons which occurs at the anode, and a reduction reaction involving the consumption of electrons which occurs at the cathode.

2. The U.S. convention concerning the sign of the EMF of a galvanic cell is as follows. The cell reaction is written such that it proceeds spontaneously from left to right, e.g.,

$$A + BX = B + AX \qquad \Delta G \text{ is negative}$$

In shorthand notation the cell reaction is written with the anode to the left, i.e.,

$$A|A^{z+}X^{z-}\,\vdots\, B^{z+}X^{z-}|B$$

and the EMF of the cell, as written, is positive.

3. The EMF of a cell is measured as the opposing external voltage which must be applied between the electrodes of the cell such that no current flows in the external circuit.

4. As

$$\Delta G = \Delta G^0 + RT \ln Q$$

and

$$\Delta G = -z\mathcal{F}\mathcal{E}$$

then

$$\mathcal{E} = \mathcal{E}^0 - \frac{RT}{z\mathcal{F}} \ln Q$$

where Q is the activity quotient of the species participating in the reaction. When the reactants and products occur in their chosen standard states, the EMF of the cell is termed the standard EMF, $\mathcal{E}^0$ (which occurs when $Q = 1$). When $Q = K$ (the equilibrium constant), then $\mathcal{E} = 0$. If the system is such that all but one of the activities of the participating species are known, then the unknown activity can be determined electrochemically.

5. If the cell reaction proceeds reversibly, i.e., the externally applied opposing voltage is only infinitesimally smaller than the EMF of the cell, such that an infinitesimal current flows in the external circuit, then the work done by the cell, being the reversible work, equals the free energy decrease. In such a case the heat transfer between the cell and the thermostat is $T\Delta S$. If the cell is operated irreversibly, i.e., such that a finite current flows in the external circuit, then less than the maximum work is performed. In the limit of irreversibility, i.e., short-circuiting of the cell, the cell performs no work, and the free energy decrease is manifested as a heat release, that is, $q = \Delta H$.

6. For a variation in temperature at constant pressure,

$$\left(\frac{d\Delta G}{dT}\right)_P = -z\mathfrak{F}\left(\frac{d\mathcal{E}}{dT}\right)_P = -\Delta S$$

and thus for the cell reaction

$$\Delta S = z\mathfrak{F}\left(\frac{d\mathcal{E}}{dT}\right)_P$$

and

$$\Delta H = -z\mathfrak{F}\mathcal{E} + z\mathfrak{F}T\left(\frac{d\mathcal{E}}{dT}\right)_P$$

7. The molality, m, and the molarity, M, of an aqueous solution are, respectively, the number of moles of solute per kilogram of solvent and the number of moles of solute per liter of solution. They are related to the mole fraction, X_i, of the solute via

$$X_i = \frac{m_i}{m_i + 1000/18} = \frac{M_i}{M_i + (1000\rho - M_i \cdot \text{MW}_i)/18}$$

As the solution tends toward infinite dilution, $m_i \rightarrow 18X_i/1000$ and $M_i \rightarrow 18X_i/1000\rho$.

8. If a salt or electrolyte A_aY_y dissociates in aqueous solution according to

$$A_aY_y = aA^{z+} + yY^{z-}$$

The activity of the component A_aY_y can be written as

$$a_{A_aY_y} = a^a_{A^{z+}} a^y_{Y^{z-}}$$

where $a_{A^{z+}}$ and $a_{Y^{z-}}$ are the individual, unmeasureable activities of the ions. The mean ionic molality, $m_\pm$, is defined as

$$m_\pm = (m^a_{A^{z+}} m^y_{Y^{z-}})^{1/(a+y)}$$

and the mean ion activity coefficient, $\gamma_\pm$, is defined as

$$\gamma_\pm = (\gamma^a_{A^{z+}} \gamma^y_{Y^{z-}})^{1/(a+y)}$$

Thus

$$a_{A_aY_y} = (\gamma_\pm m_\pm)^{a+y}$$

9. By convention, the standard single electrode potential for the half cell reaction $\frac{1}{2}H_{2(g)} = H^+_{(\text{aqueous})} + e^-$ is assigned a value of zero. Thus, by con-

vention, the standard free energy of formation of the aqueous hydrogen ion is also zero. The standard reduction potential for the element Y, $\mathcal{E}^0_{Y/Y^-}$, is obtained as the standard EMF of the cell

$$H_{2(g)}|H^+Y^-|Y$$

and the standard free energy of formation of the ion Y^- is $-\mathcal{F}\mathcal{E}^0_{Y/Y^-}$. Similarly, the standard oxidation potential of the element A, $\mathcal{E}^0_{A/A^+}$, is obtained as the standard EMF of the cell

$$A|A^+ \vdots H^+|H_{2(g)}$$

and the standard free energy of formation of the A^+ ion is $-\mathcal{F}\mathcal{E}^0_{A/A^+}$. The standard EMF of the cell

$$A|A^+Y^-|Y$$

is thus the sum of the standard oxidation potential of A and the standard reduction potential of Y.

$$\mathcal{E}^0_{A/A^+} = -\mathcal{E}^0_{A^+/A}$$

The electrochemical series is a list of the standard oxidation potentials of the elements based on $\mathcal{E}^0_{\frac{1}{2}H_2/H^+} = 0$.

10. The solubility product for an electrolyte, K_{sp}, is given by

$$K_{sp} = (\gamma_\pm m_\pm)^{a+y} = \exp \frac{-\Delta G^0}{298R}$$

where ΔG^0 is the standard free energy change for the change of state

$$A_aY_{y\,(\text{Raoultian})} = aA^{z+}_{(m)} + yY^{z-}_{(m)}$$

11. The European EMF sign convention is simply the opposite of U.S. sign convention, i.e.,

in the U.S. convention $\quad \Delta G^0 = -z\mathcal{F}\mathcal{E}^0$

in the European convention $\quad \Delta G^0 = +z\mathcal{F}E^0$

Thus, in the U.S. convention,

$$\mathcal{E} = \mathcal{E}^0 - \frac{RT}{z\mathcal{F}} \ln Q$$

and in the European convention,

$$E = E^0 + \frac{RT}{z\mathcal{F}} \ln Q$$

14.11 NUMERICAL EXAMPLES

Example 1

The standard free energy of formation of MnO in the temperature range 298 to 1500 K is given as $\Delta G^0 = -384{,}700 + 72.8T$ joules, and the standard free energy formation of H_2O in the temperature range 298 to 2500 K is given as $\Delta G^0 = -246{,}000 + 54.8T$ joules.

a. Calculate the H_2/H_2O ratio in equilibrium with pure Mn and MnO at 1400 K.

b. The reaction $Mn + H_2O = MnO + H_2$ is set up in a galvanic cell. Calculate the standard EMF of the cell, the maximum work obtainable from this cell, and the heat transfer between the cell and its constant-temperature heat reservoir when it is operated reversibly.

c. Calculate the EMF of the cell when the ratio of the pressures of H_2 and H_2O is maintained at 1000/1.

d. Manganese is alloyed with a metal X in the solid state. The composition of the alloy is 50 atom percent, and the partial molar enthalpy of mixing of Mn in this alloy is known to be −5000 joules. If the solution behaves regularly and the alloy is placed in the galvanic cell in place of the pure manganese, calculate the EMF of the cell when $p_{H_2}/p_{H_2O} = 1000$.

e. If the MnO is dissolved in an oxide solution and a new EMF measurement is made for any of the above cells, it is found that EMF has become more positive by 0.05 volt. Calculate the activity of MnO in the oxide solution.

a. For $Mn_{(s)} + H_2O_{(g)} = MnO_{(s)} + H_{2\,(g)}$

$$\Delta G^0 = -138{,}700 + 18.0T \text{ joules}$$

Thus

$$\Delta G^0_{1400} = -113{,}500 \text{ joules}$$

and

$$\left(\frac{p_{H_2}}{p_{H_2O}}\right)_{eq,1400} = 1.717 \times 10^4$$

b. As $\Delta G^0 = -z\mathcal{F}\mathcal{E}^0$, then the standard EMF of the cell,

$$\mathcal{E}^0 = -\frac{\Delta G^0}{z\mathcal{F}} = \frac{113{,}500}{2 \times 96{,}487} = 0.5882 \text{ volts}$$

$$w'_{\max} = -\Delta G^0 = 113{,}500 \text{ joules}$$

which is thus the work done by the standard cell per mole of reaction when it is operated reversibly.

The heat transferred $= q = T\Delta S^0 = 1400 \times -18.0 = -25{,}200$ joules. Thus 25,200 joules of heat flow from the cell into the constant-temperature heat reservoir per mole of reaction when the standard cell is operating reversibly.

c. Generally for the cell

$$\mathcal{E} = \mathcal{E}^0 - \frac{RT}{z\mathcal{F}} \ln \frac{a_{MnO} p_{H_2}}{a_{Mn} p_{H_2O}} \qquad \text{(i)}$$

Thus when $a_{Mn} = a_{MnO} = 1$ and $p_{H_2}/p_{H_2O} = 1000$,

$$\mathcal{E} = 0.5882 - \frac{8.3144 \times 1400}{2 \times 96{,}487} \ln 1000$$

$$= 0.1715 \text{ volts}$$

d. In the regular system Mn-X,

$$\Delta \bar{H}^M_{Mn} = \bar{G}^{xs}_{Mn} = RT \ln \gamma_{Mn} = -5000 \text{ joules}$$

Thus $$\ln \gamma_{Mn} = -\frac{5000}{8.3144 \times 1400} = -0.4295$$

and $\gamma_{Mn} = 0.651$. As $X_{Mn} = 0.5$, then

$$a_{Mn} = \gamma_{Mn} X_{Mn} = 0.651 \times 0.5 = 0.325$$

Substituting into Eq. (i) gives

$$\mathcal{E} = 0.5882 - 0.0603 \ln \frac{1000}{0.325}$$

$$= 0.1038 \text{ volts}$$

e. If, as the result of the solution of MnO, the EMF of any cell becomes 0.05

volt more positive, then,

$$0.05 = -0.0603 \ln a_{MnO} \qquad \text{at } 1400 \text{ K}$$

Thus $\qquad \ln a_{MnO} = -0.829 \qquad \text{and} \qquad a_{MnO} = 0.436$

Example 2

Fayalite, $2FeO \cdot SiO_2$, is the only iron silicate compound formed by reaction of FeO with SiO_2 at a total pressure of 1 atm, and the standard free energy change for the reaction

$$2FeO_{(s)} + SiO_{2(s)} = 2FeO \cdot SiO_{2(s)}$$

is −11,070 joules at 1200 K. Calculate the EMF of the cell

$$Fe|SiO_2|2FeO \cdot SiO_2|CaO\text{-}ZrO_2|FeO|Fe$$

at 1200 K.

This is an oxygen concentration cell in which the cell reaction can be written as

$$O_{2(\text{higher pressure at the cathode})} \rightarrow O_{2(\text{lower pressure at the anode})} \qquad \text{(i)}$$

and for which the EMF is

$$\mathcal{E} = -\frac{RT}{4\mathcal{F}} \ln \frac{p_{O_2}\,(\text{at the anode})}{p_{O_2}\,(\text{at the cathode})} \qquad \text{(ii)}$$

The oxygen pressures at the electrodes are fixed by the chemical equilibrium

$$Fe + \tfrac{1}{2}O_2 = FeO \qquad \text{(iii)}$$

for which

$$K_{(iii)} = \frac{a_{FeO}}{a_{Fe} p_{O_2}^{1/2}}$$

At the cathode the activity of FeO, with respect to Fe-saturated pure FeO, is unity and, at the anode, the activity of FeO is that occurring in $2FeO \cdot SiO_2$ saturated with Fe and SiO_2. As

$$K_{(iii)} = \frac{a_{FeO\,(cathode)}}{a_{Fe\,(cathode)}\,p^{1/2}_{O_2\,(cathode)}} = \frac{a_{FeO\,(anode)}}{a_{Fe\,(anode)}\,p^{1/2}_{O_2\,(cathode)}}$$

and

$$a_{Fe\,(anode)} = a_{Fe\,(cathode)}$$
$$= a_{FeO(cathode)} = 1$$

$$\frac{p_{O_2\,(anode)}}{p_{O_2\,(cathode)}} = a^2_{FeO\,(anode)}$$

and, hence, in Eq. (ii),

$$\mathcal{E} = -\frac{RT}{4\mathcal{F}} \ln a^2_{FeO\,(anode)}$$

For $2FeO + SiO_2 = 2FeO{\cdot}SiO_2$,

$$\Delta G^0_{1200} = -11{,}070 \text{ joules}$$
$$= -8.3144 \times 1200 \ln \frac{a_{2FeO \circ SiO_2}}{a^2_{FeO}a_{SiO_2}}$$

Thus, at the anode, with $a_{2FeO \cdot SiO_2} = a_{SiO_2} = 1$,

$$a_{FeO\,(anode)} = 0.574$$

and hence

$$\mathcal{E} = -\frac{8.3144 \times 1200}{4 \times 96{,}487} \ln (0.574)^2$$
$$= 0.0287 \text{ volts}$$

Alternatively, the anode half cell reaction can be written as

$$O^{2-} = \tfrac{1}{2} O_{2(eq.Fe/\underline{FeO})} + 2e^-$$

and the cathode half cell reaction can be written as

$$\tfrac{1}{2}O_{2(eq.\ Fe/FeO)} + 2e^- = O^{2-}$$

or, at the anode,

$$2Fe + 2O^{2-} + SiO_2 = 2FeO{\cdot}SiO_2 + 4e^-$$

and, at the cathode,

$$2FeO + 4e^- = 2Fe + 2O^{2-}$$

summation of which gives the cell reaction as

$$2FeO + SiO_2 = 2FeO \cdot SiO_2$$

The free energy change for the cell reaction is

$$\Delta G^0 = -z\mathcal{F}\mathcal{E}^0 = -11{,}070 \text{ joules}$$

and thus $$\mathcal{E} = \frac{-\Delta G}{z\mathcal{F}} = \frac{11{,}070}{4 \times 96{,}487} = 0.0287 \text{ volts}$$

PROBLEMS

14.1 The EMF of the galvanic cell

$$Pb_{(s)}|PbCl_{2\,(s)}|HCl_{(aqueous)}|AgCl_{(s)}|Ag_{(s)}$$

where all components are present as pure solids in contact with an HCl electrolyte, is 0.490 volt at 25°C and, at that temperature, the temperature coefficient of the EMF is $-$ 1.84 $\times$ 10^{-4} volt/degree. Write the cell reaction and calculate the free energy and entropy changes for this reaction at 298 K.

14.2 At 25°C, the EMF of the cell

$$Pb|PbCl_2 \,\vdots\, Hg_2Cl_2|Hg$$

is + 0.5357 volt and the temperature coefficient is 1.45 $\times$ 10^{-4} volt/degree. Calculate:

a. The maximum work available from the cell at 25°C per mole of Pb reacted

b. The entropy change of the cell reaction

c. The heat absorbed by the cell at 25°C per mole of Pb reacted when the cell is operating reversibly

The Hg electrode in the cell is replaced by an Hg–X alloy in which X_{Hg} = 0.3 and where X is inert. The cell EMF at 25°C is found to increase by 0.0089 volt. Calculate the activity of Hg in the alloy at 25°C.

14.3 The solid-state electrochemical cell

$$(\mathrm{Pt}),\ \mathrm{O_2}\ \text{gas at}\ p_{\mathrm{O_2}}\,|\mathrm{CaO\text{-}ZrO_2}|\mathrm{Fe}|\mathrm{FeO},\ (\mathrm{Pt})$$

is built to measure the partial pressure of oxygen in gases. Write an equation relating the oxygen pressure in and temperature of the gas to the cell EMF.

14.4 The EMF of the cell

$$\mathrm{Ag}_{(s)}|\mathrm{AgCl}_{(s)}|\mathrm{Cl_2}\ (1\ \text{atm}),\ \mathrm{Pt}$$

is found to be

$$\mathcal{E}(\text{volts}) = 0.977 + 5.7 \times 10^{-4}(350 - t) - 4.8 \times 10^{-7}(350 - t)^2$$

in the temperature range $t = 100°\mathrm{C}$ to $t = 450°\mathrm{C}$. Calculate the value of Δc_p for the cell reaction.

14.5 A galvanic cell is set up with electrodes of solid aluminum and solid aluminum-zinc alloy and an electrolyte of a fused $AlCl_3$-NaCl mixture. When the mole fraction of Al in the alloy electrode is 0.38, the EMF of the cell is 7.43 millivolts at 380°C, and the temperature coefficient of the EMF is 2.9×10^{-5} volt/degree. Calculate:

a. The activity of Al in the alloy
b. The partial molar free energy of mixing of Al in the alloy
c. The partial molar enthalpy of mixing of Al in the alloy

14.6 By measuring the EMFs of cells of the type

$$\mathrm{Ni}_{(s)}|\mathrm{NiO}_{(s)}|\mathrm{CaO\text{-}ZrO_2}|\mathrm{Cu}_{(l)}\ \text{containing dissolved oxygen}$$

it has been established that e_O^O in liquid copper is -0.16 and that the standard free energy change for

$$\tfrac{1}{2}\mathrm{O}_{2(g)} = [\mathrm{O}]_{1\ \text{wt\%}\ \text{in Cu}}$$

is $\Delta G^0 = 86{,}680 - 18.67T$ joules. If the EMF of such a cell is 0.166 volts at 1373 K, calculate:

a. The activity of oxygen in the liquid copper cathode with respect to a standard state of oxygen gas at 1 atm pressure
b. The activity of Cu_2O in the cathode metal with respect to Cu-saturated pure solid Cu_2O
c. The weight percentage of oxygen dissolved in the copper cathode

14.7 Calculate the conditions under which an aqueous solution of $[Pb^{2+}] = 1$ mole/liter is in equilibrium with metallic Pb and solid PbO at 25°C. Is any other lead ion present in significant concentration in this solution? Given:

Species	ΔG^0_{298} joules
$PbO_{(s)}$	−189,300
$Pb^{2+}_{(m)}$	−24,310
$Pb^{4+}_{(m)}$	+302,500
$HPbO_2^-(m)$	−339,000
$PbO^{2-}_{3(m)}$	−277,570
$PbO^{4-}_{4(m)}$	−28,210
$H_2O_{(l)}$	−237,190

14.8 Aluminum can be produced by electrolysis of Al_2O_3 dissolved in molten cryolite, $3NaF \cdot AlF_3$. If inert electrodes are used in an electrolysis cell and and the cryolite electrolyte is saturated with Al_2O_3 at 1000°C, what is decomposition voltage of Al_2O_3? The Hall-Heroult process for the electrolysis of Al_2O_3 uses carbon as the anode material, such that the gas evolved at the anode is essentially pure CO_2 at 1 atm pressure. Calculate the decomposition voltage of Al_2O_3 in an Al_2O_3-saturated $3NaF \cdot AlF_3$ electrolyte at 1000°C in the Hall-Heroult cell.

APPENDIX A

THERMODYNAMIC TABLES

Table A-1. The standard free energy changes for several reactions (standard states are noted by subscript)

Reaction	ΔG^0, joules	Range, K
$2Al_{(l)} + \frac{3}{2}O_{2\,(g)} = Al_2O_{3\,(s)}$	$-1{,}676{,}000 + 320T$	923-1800
$Al_{(l)} = [Al]_{1\,wt\%\,in\,Fe}$	$-43{,}100 - 32.26T$	
$\frac{23}{6}Cr_{(s)} + C_{(s)} = \frac{1}{6}Cr_{23}C_{6\,(s)}$	$-68{,}530 - 6.44T$	298-1673
$2Cr_{(s)} + \frac{3}{2}O_{2\,(g)} = Cr_2O_{3\,(s)}$	$-1{,}120{,}300 + 260T$	298-2100
$C_{(s)} + 2H_{2\,(g)} = CH_{4\,(g)}$	$-69{,}120 + 22.25T \ln T - 65.34T$	298-1200
$C_{(s)} + \frac{1}{2}O_{2\,(g)} = CO_{(g)}$	$-111{,}700 - 87.65T$	298-2500
$C_{(s)} + O_{2(g)} = CO_{2\,(g)}$	$-394{,}100 - 0.84T$	298-2000
$C_{(s)} = [C]_{1\,wt\%\,in\,Fe}$	$22{,}600 - 42.26T$	
$2Ca_{(v)} + O_{2\,(g)} = 2CaO_{(s)}$	$-1{,}590{,}800 + 390.1T$	1760-2500
$2Ca_{(v)} + S_{2\,(g)} = 2CaS_{(s)}$	$-1{,}408{,}800 + 382.6T$	1760-2500
$2CaO_{(s)} + SiO_{2\,(s)} = Ca_2SiO_{4\,(s)}$	$-126{,}400 - 5.02T$	298-1700
$CoO_{(s)} + SO_{3\,(g)} = CoSO_{4\,(s)}$	$-227{,}860 + 165.3T$	298-1449
$CaO_{(s)} + CO_{2\,(g)} = CaCO_{3\,(s)}$	$-168{,}400 + 144T$	449-1500
$2Cu_{(s)} + \frac{1}{2}O_{2\,(g)} = Cu_2O_{(s)}$	$-169{,}000 - 7.12T \ln T + 123T$	298-1356
$2Cu_{(l)} + \frac{1}{2}O_{2\,(g)} = Cu_2O_{(s)}$	$-195{,}000 - 7.12T \ln T + 143T$	1356-1503
$2Cu_{(s)} + \frac{1}{2}S_{2\,(g)} = Cu_2S_{(s)}$	$-142{,}900 - 11.3T \ln T + 120.2T$	623-1360
$Fe_{(s)} + \frac{1}{2}O_{2\,(g)} = FeO_{(s)}$	$-259{,}600 + 62.55T$	298-1642
$Fe_{(l)} + \frac{1}{2}O_{2\,(g)} = FeO_{(l)}$	$-232{,}700 + 45.31T$	1808-2000

Table A-1. The standard free energy changes for several reactions (standard states are noted by subscript) (*Continued*)

Reaction	ΔG^0, joules	Range, K
$3Fe_{(s)} + 2O_2(g) = Fe_3O_4(s)$	$-1,091,060 + 312.75T$	298-1642
$2Fe_{(s)} + \frac{3}{2}O_2(g) = Fe_2O_3(s)$	$-810,520 + 254.0T$	298-1460
$H_2(g) + \frac{1}{2}O_2(g) = H_2O_{(g)}$	$-246,000 + 54.8T$	298-2500
$H_2(g) + Cl_2(g) = 2HCl_{(g)}$	$-182,200 + 3.60T \ln T - 43.68T$	298-2100
$H_2(g) + \frac{1}{2}S_2(g) = H_2S_{(g)}$	$-90,290 + 49.39T$	298-2000
$Mn_{(l)} + \frac{1}{2}S_2(g) = MnS_{(l)}$	$-262,600 + 64.4T$	1803-2000
$Mn_{(s)} + \frac{1}{2}O_2(g) = MnO_{(s)}$	$-384,700 + 72.8T$	298-1500
$3Mn_{(s)} + C_{(s)} = Mn_3C_{(s)}$	$-13,900 - 1.09T$	298-1010
$2MgO_{(s)} + SiO_2(s) = Mg_2SiO_4(s)$	$-63,260$	298-1700
$Mg_{(g)} + \frac{1}{2}O_2(g) = MgO_{(s)}$	$-759,800 - 13.39T \ln T + 316.7T$	1380-2500
$MgO_{(s)} + CO_2(g) = MgCO_3(s)$	$-117,600 + 170T$	298-1000
$Mg_{(l)} + Cl_2(g) = MgCl_2(l)$	$-605,000 + 125.4T$	973-1133
$Mg_{(g)} + Cl_2(g) = MgCl_2(l)$	$-770,300 - 37.63T \ln T + 508.4T$	714-1710
$2Ni_{(s)} + O_2(g) = 2NiO_{(s)}$	$-489,100 + 197T$	298-1725
$2Ni_{(l)} + O_2(g) = 2NiO_{(s)}$	$-524,300 + 217.4T$	1725-2200
$\frac{1}{2}O_2(g) = [O]_{1\ wt\%\ in\ Fe}$	$-111,070 - 5.87T$	
$2Pb_{(l)} + S_2(g) = 2PbS_{(s)}$	$-314,500 + 160T$	600-1380
$Pb_{(l)} + \frac{1}{2}O_2(g) = PbO_{(l)}$	$-191,750 + 79.1T$	1200-1400
$PbO_{(s)} + SO_2(g) + \frac{1}{2}O_2(g) = PbSO_4(s)$	$-402,290 + 269.24T$	298-1159
$PCl_3(g) + Cl_2(g) = PCl_5(g)$	$-95,600 - 7.94T \ln T + 235.2T$	298-1000
$\frac{1}{2}S_2(g) = [S]_{1\ wt\%\ in\ Fe}$	$-131,460 + 22.03T$	
$SO_2(g) + \frac{1}{2}O_2(g) = SO_3(g)$	$-94,560 + 89.37T$	318-1800
$\frac{1}{2}S_2(g) + O_2(g) = SO_2(g)$	$-362,400 + 72.43T$	718-2273
$Si_{(s)} + O_2(g) = SiO_2(s)$	$-902,000 + 174T$	700-1700
$3Si_{(s)} + 2N_2(g) = Si_3N_4(s)$	$-741,000 - 10.5T \ln T + 403T$	298-1686
$Sn_{(l)} + Cl_2(g) = SnCl_2(l)$	$-333,000 + 118T$	520-925
$Ti_{(s)} + O_2(g) = TiO_2(s)$	$-910,000 + 173T$	298-2080
$2V_{(s)} + O_2(g) = 2VO_{(s)}$	$-861,500 + 150T$	900-1800
$Zn_{(g)} + \frac{1}{2}O_2(g) = ZnO_{(s)}$	$-482,920 - 18.80T \ln T + 344.7T$	1170-2000
$Zn_{(g)} + \frac{1}{2}S_2(g) = ZnS_{(s)}$	$-397,400 - 14.62T \ln T + 313.5T$	1120-2000

Table A-2. The constant-pressure molar heat capacities of various substances

Substance	$c_p = a + bT + cT^{-2}$, joules/degree·mole			
	a	$b \times 10^3$	$c \times 10^{-5}$	Range, K
$Al_{(s)}$	20.7	12.4		298–932
$Al_{(l)}$	29			932–1273
$Al_2O_{3(s)}$	106.6	17.8	–28.5	298–1800
$Au_{(s)}$	23.7	5.19		298–T_m
$Bi_{(s)}$	18.8	23		298–T_m
$Bi_{(l)}$	20	6.15	21.1	T_m–820
$C_{(diamond)}$	9.12	13.2	–6.19	298–1200
$C_{(graphite)}$	17.2	4.27	–8.79	298–2300
$Co_{(\alpha)}$	21.4	14.3	–0.88	298–650
$Co_{(\beta)}$	13.8	24.5		715–1400
$Co_{(\gamma)}$	40			1400–T_m
$CoO_{(s)}$	48.28	8.54	1.7	298–1800
$Cr_{(s)}$	24.4	9.87	–3.7	298–T_m
$Cr_2O_{3(s)}$	119.4	9.2	–15.6	350–1800
$H_{2(g)}$	27.3	3.3	0.5	298–3000
$Mn_{(\alpha)}$	21.6	15.9		298–993
$Mn_{(\beta)}$	34.9	2.8		993–1373
$Mn_{(\gamma)}$	44.8			1373–1409
$Mn_{(\delta)}$	47.3			1409–1517
$Mn_{(l)}$	46.0			1517–T_b
$MnO_{(s)}$	46.48	8.12	–3.7	298–1800
$Ni_{(\alpha)}$	32.6	–1.97	–5.586	298–630
$Ni_{(\beta)}$	29.7	4.18	–9.33	630–T_m
$NiO_{(\alpha)}$	–20.9	157.2	16.3	298–525
$O_{2(g)}$	30	4.18	–1.7	298–3000
$Pb_{(s)}$	23.6	9.75		298–600
$Pb_{(l)}$	32.4	–3.1		600–1200
$PbO_{(s)}$	37.9	26.8		298–1000

Table A-3. The standard molar heats of formation and molar entropies of various substances at 298 K

Substance	ΔH^0_{298}, joules	S^0_{298}, joules/degree
Ag	0	42.68
Ag_2O	–30,500	122
Al_2O_3	–1,674,000	50.6
$C_{(graphite)}$	0	5.694
$C_{(diamond)}$	1,900	2.44
Ca	0	41.6
CaO	–634,300	40
Ca_2SiO_4	–126,400*	128
Co	0	30
CoO	–239,000	52.93
Cr_2O_3	–1,120,300	81.2
MnO	–384,700	59.12
Ni	0	29.8
NiO	–244,600	38
O_2	0	205
Pb	0	64.9
PbO	–219,000	67.4
Si	0	19
SiO_2(quartz)	–910,900	41.5

*From $2CaO + SiO_2$.

Table A-4. The vapor pressures of various elements

Element	$\ln p(\text{atm}) = -A/T + B \ln T + CT + D$				Range, K
	A	B	$C \times 10^3$	D	
$Ag_{(s)}$	34,300	−0.85		21.46	298–T_m
$Ag_{(l)}$	33,200	−0.85		20.31	T_m–T_b
$Au_{(s)}$	45,650	−0.306	−0.37	18.26	298–T_m
$Au_{(l)}$	44,400	−1.01		21.88	T_m–T_b
$Cu_{(s)}$	40,930	−0.86		21.67	298–T_m
$Cu_{(l)}$	40,350	−1.21		23.79	T_m–T_b
$Hg_{(l)}$	7,610	−0.795		17.214	298–T_b
$I_2(s)$	8,240	−2.51		34.164	298–T_m
$I_2(l)$	7,380	−5.18		47.83	T_m–T_b
$Ni_{(l)}$	51,590	−2.01		32.40	T_m–T_b
$Pb_{(l)}$	23,330	−0.985		19.07	T_m–T_b
$Zn_{(s)}$	15,780	−0.755		19.25	273–T_m
$Zn_{(l)}$	15,250	−1.255		21.79	693–1180

Table A-5. Molar heats of fusion and transformation of various elements and compounds

Element	Trans	ΔH_{trans} (joules)	T_{trans}, K
Al	$s \rightarrow l$	10,500	932
Au	$s \rightarrow l$	12,760	1336
Bi	$s \rightarrow l$	10,900	544
Cr	$s \rightarrow l$	21,000	2173
Cs	$s \rightarrow l$	2,090	302.8
Cu	$s \rightarrow l$	12,970	1356
FeO	$s \rightarrow l$	30,960	1643
I_2	$s \rightarrow l$	15,800	387
Mg	$l \rightarrow v$	129,580	1363
Mn	$\alpha \rightarrow \beta$	2,010	993
Mn	$\beta \rightarrow \gamma$	2,300	1373
Mn	$\gamma \rightarrow \delta$	1,800	1409
Mn	$\delta \rightarrow l$	13,400	1517
MnO	$s \rightarrow l$	54,400	2148
Ni	$s \rightarrow l$	17,150	1728
Pb	$s \rightarrow l$	4,810	600
PbO	$s \rightarrow l$	27,490	1158
Rb	$s \rightarrow l$	2,197	312
Si	$s \rightarrow l$	50,630	1683
Ti	$s \rightarrow l$	18,800	1940
V	$s \rightarrow l$	17,600	2188

APPENDIX B

SOURCES OF THERMODYNAMIC AND THERMOCHEMICAL DATA

Thermochemistry for Steelmaking, vol. I
J. F. Elliott and M. Gleiser
Addison-Wesley Publishing Co., Reading, Mass., 1960

Physical properties of selected elements. Tabulations in 100° temperature intervals of c_p, $H_T^0 - H_{298}^0$, S_T^0, $-(G_T^0 - H_{298}^0)/T$, ΔH_f^0, ΔG_f^0, and log K_p for selected elements. Tabulations in 100° temperature intervals of standard heats and free energies of formation of selected carbides, nitrides, oxides, phosphides, silicides, and sulfides. Tabulations in 100° intervals of vapor pressures of selected elements and compounds.

Thermochemistry for Steelmaking, vol. II
J. F. Elliott, M. Gleiser, and V. Ramakrishna
Addison-Wesley Publishing Co., Reading, Mass., 1963

Tabulations in 100° temperature intervals of standard free energies and heats of formation of selected complex oxide compounds. Binary phase diagrams for selected elements and oxides. Effect of alloying elements on the solubility of graphite in iron. Thermodynamic properties of important solutions. Physical properties of liquid solutions.

Thermodynamics (2nd ed.)
G. N. Lewis and M. Randall (rev. by K. S. Pitzer, and L. Brewer)
McGraw-Hill Book Company, New York, 1961

Tabulations of $-(G_T^0 - H_{298}^0)/T$ and $H_{298}^0 - H_0^0$ for selected solid, liquid, and gaseous elements. Tabulations of $-(G_T^0 - H_{298}^0)/T$ and ΔH_{298}^0 for selected halides, oxides, sulphides, carbides, nitrides, and related compounds.

Metallurgical Thermochemistry (5th ed.)
O. Kubaschewski and C. B. Alcock
Pergamon Press, New York, 1979

Data for an extensive list of elements and compounds including ΔH^0_{298}, S^0_{298}, transition temperatures and structures. Heats of transformation, fusion, and evaporation. Heat capacities presented as $c_p = a + bT + cT^{-2}$. Vapor pressures presented as $\log p = AT^{-1} + B \log T + CT + D$. Standard free energies of reactions presented as $\Delta G^0_T = A + BT \log T + CT$. Thermochemical data for solutions, $(\Delta \bar{H}^M_i, \Delta \bar{S}^M_i, \Delta \bar{G}^M_i, \Delta H^M, \Delta S^M, \Delta G^M)$ tabulated against composition.

Selected Values of Thermodynamic Properties of Metals and Alloys
R. Hultgren, R. L. Orr, P. D. Anderson, and K. K. Kelley
John Wiley, New York, 1963

Tabulations of the properties of condensed and gaseous metallic elements: c_p, $H^0_T - H^0_{298}$, $S^0_T - S^0_{298}$, $(G^0_T - H^0_{298})/T$, vapor pressure, heats and free energies of evaporation. Tabulations of properties of binary alloys: ΔG^M, ΔH^M, ΔS^M, G^{xs}, S^{xs}, a_i, and γ_i tabulated against composition.

JANAF Thermochemical Data
Army-Navy-Air Force Thermochemical Panel
The Dow Chemical Co., Midland, Mich., 1962-63

Exhaustive tabulations of c_p, S^0_T, $-(G^0_T - H^0_{298})/T$, $H^0_T - H^0_{298}$, ΔH^0_f, ΔG^0_f and $\log K_p$ in temperature intervals of 100° for elements and compounds.

Contributions to the Data on Theoretical Metallurgy XIII (U.S. Bureau of Mines, Bulletin 584)
K. K. Kelley
U.S. Government Printing Office, Washington, D.C., 1960

Tabulations of $H^0_T - H^0_{298}$ and $S^0_T - S^0_{298}$ along with corresponding analytical expressions, and analytical expressions for c_p for an extensive list of elements and compounds.

Thermodynamic Properties of Minerals and Related Substances at 298.15 K and One Atmosphere Pressure and at Higher Temperatures
R. A. Robie and D. R. Waldbaum
Geological Survey Bulletin 1259. U.S. Dept. of the Interior, U.S. Government Printing Office, Washington, D.C., 1968.

Tabulations of S^0_{298}, V^0_{298}, ΔH^0_{298}, ΔG^0_{298} for elements, compounds, and aqueous ions. Tabulations of $H^0_T - H^0_{298}$, $(G^0_T - H^0_{298})/T$, S^0_T, ΔG^0_T, ΔH^0_T, $\log K_p$ at

100 K intervals up to 2000 K for an extensive list of compounds. Values of T_m, T_b, ΔH_m^0, ΔH_b^0, $H_{298} - H_0$, V^0.

Constitution of Binary Alloys (2nd ed.)
M. Hansen
McGraw-Hill Book Company, New York, 1958
See also first supplement by R. P. Elliott, 1965
and second supplement by F. A. Shunk, 1969

An exhaustive collection of phase diagrams of binary elemental systems.

Phase Diagrams for Ceramists
E. M. Levin, C. R. Robbins, and H. F. McMurdie
American Ceramic Society, Inc., Columbus, Ohio, 1964
Also 1969 supplement

An exhaustive collection of binary, ternary, and quaternary phase diagrams of oxide, halide, and sulfate systems.

RECOMMENDED READING

The Scientific Papers of J. Willard Gibbs, vol. I, *Thermodynamics*
Dover Publications, New York, 1961 (originally published by Longmans, Green and Company, 1906)

Thermodynamics of Alloys
C. Wagner
Addison-Wesley Publishing Co., Reading, Mass., 1952

Thermodynamics (2nd ed.)
G. N. Lewis and M. Randall (rev. by K. S. Pitzer and L. Brewer)
McGraw-Hill Book Company, New York, 1961

Physical Chemistry of Metals
L. S. Darken and R. W. Gurry
McGraw-Hill Book Company, New York, 1953

The Principles of Chemical Equilibrium (3rd ed.)
K. Denbigh
Cambridge University Press, Cambridge, England, 1971

Alloy Phase Equilibria
A. Prince
Elsevier Publishing Company, New York, 1966

Ternary Systems
G. Masing (trans. by B. A. Rogers)
Dover Publications, New York, 1960 (originally published by Reinhold Publishing Corp., 1944)

Thermodynamics
H. B. Callen
John Wiley & Sons, New York, 1960

Metallurgical Thermochemistry (5th ed.)
O. Kubaschewski and C. B. Alcock
Pergamon Press, New York, 1979

Thermodynamics (5th ed.)
E. A. Guggenheim
North Holland Publishing Co., Amsterdam, Holland, 1967

Entropy
J. D. Fast
McGraw-Hill Book Company, New York, 1962

Thermodynamics
N. A. Gokcen
Techscience Inc., 1977

An Introduction to the Physical Chemistry of Iron and Steel Making
R. G. Ward
Edward Arnold (Publishers), Ltd., London, 1962

Physical Chemistry of Iron and Steel Manufacture
C. Bodsworth
Longmans, Green and Co., London, 1963

Principles of Extractive Metallurgy
T. Rosenqvist
McGraw-Hill Book Company, New York, 1974

Problems in Applied Thermodynamics
C. Bodsworth and A. S. Appleton
Longmans, Green and Co., London, 1965

Problems in Metallurgical Thermodynamics and Kinetics
G. S. Upadhyaya and R. K. Dube
Pergamon Press, New York, 1977

Problem Manual for Metallurgical Thermodynamics
A. E. Morris
Department of Metallurgical Engineering, University of Missouri-Rolla, 1973

ANSWERS

CHAPTER TWO

2.1 (*a*) $V = 100$ liters, $w = +23.34$ kilojoules, $q = +23.34$ kilojoules, $\Delta U' = 0$, $\Delta H' = 0$; (*b*) $V = 39.81$ liters, $w = 9.151$ kilojoules, $q = 0$, $\Delta U' = -9.151$ kilojoules, $\Delta H' = -15.25$ kilojoules.

2.2 (*a*) $P = 1$ atm, $V = 8$ liters, $T = 273$ K; (*b*) $P = 1$ atm, $V = 16.0$ liters, $T = 546$ K; (*c*) $\Delta U' = 1215$ joules, $\Delta H' = q_p = 2025$ joules; (*d*) $C_p \doteq 20.78$ joules/degree, $C_v = 12.47$ joules/degree.

2.3 Work done in (*a*) = 5226 joules; work done in (*b*) = −18,071 joules; work done in (*c*) = 4609 joules; total work done = −8236 joules. Heat change in (*a*) = 5226 joules; heat change in (*b*) = 0; heat change in (*c*) = −13,462 joules; total heat change = −8236 joules.

2.4 $P = 0.3$ atm.

2.5 Work done = 7958 joules, $T_{max} = 1116$ K, $T_{min} = 255$ K.

CHAPTER THREE

3.1 Minimum heat withdrawn = 3525 joules; efficiency in (*a*) = $(T_2 - T_1 + \Delta T)/(T_2 + \Delta T)$; efficiency in (*b*) = $(T_2 - T_1 + \Delta T)/T_2$; thus the efficiency in (*b*) is greater than that in (*a*).

3.2 (*a*) $\Delta S = 19.14$ joules/degree; (*b*) $\Delta S = 0$; (*c*) $\Delta S = -28.7$ joules/degree.

3.3 $\Delta U = 0$, $\Delta H = 0$, $q = w = 1462$ joules, $\Delta S = 9.12$ joules/degree.

3.4 Final temperature = 321.7 K; $\Delta S'_{Pb} = -28.63$ joules/degree, $\Delta S'_{H_2O} = 32.07$ joules/degree, thus entropy created = 3.44 joules/degree.

3.5 108.5 joules/degree.

3.6 As the process is reversible, no entropy is produced, i.e.,

$$\Delta S_{\text{heat source}} = \Delta S_{\text{heat sink}}$$

or
$$\int_{T_2}^{T_f} C_2 \frac{dT_2}{T_2} = -\int_{T_1}^{T_f} C_1 \frac{dT_1}{T_1}$$

Integrating between the final and initial temperature gives

$$\left(\frac{T_f}{T_2}\right)^{C_2} = \left(\frac{T_1}{T_f}\right)^{C_1}$$

and hence

$$T_f = (T_1^{C_1} \cdot T_2^{C_2})^{1/(C_1 + C_2)}$$

w = high-temperature heat entering engine − low-temperature heat rejected by engine

$= [-C_2(T_f - T_2)] - [C_1(T_f - T_1)]$

CHAPTER FOUR

4.1 $R \ln 4; R \ln 8;$ zero; $R \ln (32/27)$.
4.2 1.022 joules/degree.

CHAPTER FIVE

5.1 $dS = dU/T + (P/T)dV$. For an isothermal process, dU for an ideal gas is zero. Thus $(dS/dV)_T = P/T = R/V$.

5.2
$$\left(\frac{dG}{dP}\right)_T = V \quad \text{and} \quad \left(\frac{d^2G}{dP^2}\right)_T = \left(\frac{dV}{dP}\right)_T$$

$$\left(\frac{dA}{dV}\right)_T = -P \quad \text{and} \quad \left(\frac{d^2A}{dV^2}\right)_T = -\left(\frac{dP}{dV}\right)_T$$

so
$$\left(\frac{d^2G}{dP^2}\right)_T = -\frac{1}{(d^2A/dV^2)_T}$$

5.3
$$c_p = (dH/dT)_P, \text{ so}$$

$$(dc_p/dP)_T = \left[\frac{d}{dP}\left(\frac{dH}{dT}\right)_P\right]_T = \left[\frac{d}{dT}\left(\frac{dH}{dP}\right)_T\right]_P$$

But $(dH/dP)_T = T(dS/dP)_T + V$ and $\left(\frac{dS}{dP}\right)_T = -\left(\frac{dV}{dT}\right)_P$

so $$\left(\frac{dc_p}{dP}\right)_T = \frac{d}{dT}\left[-T\left(\frac{dV}{dT}\right)_P + V\right]$$

$$= -T\left(\frac{d^2V}{dT^2}\right)_P - \left(\frac{dV}{dT}\right)_P + \left(\frac{dV}{dT}\right)_P$$

$$= -T(d^2V/dT^2)_P$$

5.4 $$dS = \frac{\delta q}{T} = \frac{c_P dT}{T} \quad \text{so } (dS/dT)_P = c_P/T$$

Thus $$\frac{T}{c_P}\left(\frac{dV}{dT}\right)_P = \left(\frac{dT}{dS}\right)_P\left(\frac{dV}{dT}\right)_P$$

But $$\left(\frac{dT}{dS}\right)_P = -\left(\frac{dT}{dP}\right)_S\left(\frac{dP}{dS}\right)_T \quad \text{and} \quad \left(\frac{dS}{dP}\right)_T = -\left(\frac{dV}{dT}\right)_P$$

so $$\frac{T}{c_P}\left(\frac{dV}{dT}\right)_P = \left(\frac{dT}{dP}\right)_S\left(\frac{dT}{dV}\right)_P\left(\frac{dV}{dT}\right)_P = \left(\frac{dT}{dP}\right)_S$$

5.5 $$c_P - c_V = \left(\frac{dH}{dT}\right)_P - \left(\frac{dU}{dT}\right)_V$$

$$= \left(\frac{dH}{dT}\right)_P - \left(\frac{dH}{dT}\right)_V + V\left(\frac{dP}{dT}\right)_V$$

but $$\left(\frac{dH}{dT}\right)_P = \left(\frac{dH}{dT}\right)_V + \left(\frac{dH}{dV}\right)_T\left(\frac{dV}{dT}\right)_P$$

so $$c_P - c_V = \left(\frac{dH}{dT}\right)_V + \left(\frac{dH}{dV}\right)_T\left(\frac{dV}{dT}\right)_P - \left(\frac{dH}{dT}\right)_V + V\left(\frac{dP}{dT}\right)_V$$

$$= V\left(\frac{dP}{dT}\right)_V + \left(\frac{dH}{dV}\right)_T\left(\frac{dV}{dT}\right)_P$$

Now $$\left(\frac{dV}{dT}\right)_P = -\left(\frac{dV}{dP}\right)_T\left(\frac{dP}{dT}\right)_V \quad \text{so}$$

$$c_P - c_V = -\left(\frac{dH}{dV}\right)_T\left(\frac{dV}{dP}\right)_T\left(\frac{dP}{dT}\right)_V + V\left(\frac{dP}{dT}\right)_V$$

$$= \left(\frac{dP}{dT}\right)_V\left[V - \left(\frac{dH}{dP}\right)_T\right]$$

Now $\left(\frac{dH}{dP}\right)_T = T\left(\frac{dS}{dP}\right)_T + V$ and $(dS/dP)_T = -(dV/dT)_P$

$$\text{so}\quad c_p - c_v = \left(\frac{dP}{dT}\right)_V\left[V + T\left(\frac{dV}{dT}\right)_P - V\right]$$

$$= T\left(\frac{dP}{dT}\right)_V\left(\frac{dV}{dT}\right)_P$$

5.6 $(dP/dV)_S = -(dP/dS)_V(dS/dV)_P$

$$dS = \frac{c_v dT}{T} \text{ at constant } V \text{ and } = \frac{c_p dT}{T} \text{ at constant } P$$

$$\text{So}\quad \left(\frac{dS}{dP}\right)_V = \frac{c_v}{T}\left(\frac{dT}{dP}\right)_V \text{ and } \left(\frac{dS}{dV}\right)_P = \frac{c_p}{T}\left(\frac{dT}{dV}\right)_P$$

$$\text{Thus}\quad \left(\frac{dP}{dV}\right)_S = -\frac{T}{c_v}\left(\frac{dP}{dT}\right)_V \frac{c_p}{T}\left(\frac{dT}{dV}\right)_P$$

$$\text{but}\quad \left(\frac{dP}{dT}\right)_V\left(\frac{dT}{dV}\right)_P = -\left(\frac{dP}{dV}\right)_T = \frac{1}{\beta V}$$

$$\text{so}\quad \left(\frac{dP}{dV}\right)_S = -\frac{c_p}{c_v V\beta}$$

5.7 The process is adiabatic, and so $q = 0$. The work done by the gas $= w = P_2V_2 - P_1V_1$. Thus $\Delta H = q - w + (P_2V_2 - P_1V_1) = 0$.

$$\mu_{J\text{-}T} = \left(\frac{dT}{dP}\right)_H$$

$$\text{but}\quad \left(\frac{dT}{dP}\right)_H = -\left(\frac{dT}{dH}\right)_P\left(\frac{dH}{dP}\right)_T = -\frac{1}{c_p}\left(\frac{dH}{dP}\right)_T$$

$$\text{and}\quad \left(\frac{dH}{dP}\right)_T = T\left(\frac{dS}{dP}\right)_T + V = -T\left(\frac{dV}{dT}\right)_P + V = -T\alpha V + V$$

$$\text{so}\quad \mu_{J\text{-}T} = -\frac{1}{c_p}(-\alpha TV + V)$$

$$= \frac{V}{c_p}(\alpha T - 1)$$

For an ideal gas $\alpha = 1/T$, so $\alpha T = 1$, and $\mu_{J\text{-}T} = 0$.

5.8 (*a*) (1) $\Delta U = \Delta H = 0 \quad \Delta S = R \ln 4 \quad \Delta A = \Delta G = -RT \ln 4$

(2) $\Delta U = \Delta H = 0 \quad \Delta S = R \ln 8 \quad \Delta A = \Delta G = -RT \ln 8$

(3) $\Delta U = \Delta H = \Delta S = \Delta A = \Delta G = 0$

(4) $\Delta U = \Delta H = 0 \quad \Delta S = R \ln (32/27) \quad \Delta A = \Delta G = -RT \ln (32/27)$

(*b*) $\Delta U = \Delta H = 0 \quad \Delta S = R \ln (V_2/V_1) = R \ln 2 \quad \Delta A = \Delta G = -RT \ln 2$

(*c*) $\Delta U = c_v(T_2 - T_1) \quad \Delta H = c_p(T_2 - T_1) \quad \Delta S = 0$

$\Delta A = \Delta U - S(T_2 - T_1) = (c_v - S)(T_2 - T_1)$

$\Delta G = \Delta H - S(T_2 - T_1) = (c_p - S)(T_2 - T_1)$

(*d*) $\Delta U = c_v(T_2 - T_1) \quad \Delta H = c_p(T_2 - T_1)$

$\Delta S = c_v \ln (T_2/T_1) + R \ln(V_2/V_1) = c_p \ln(T_2/T_1)$

$\Delta A = \Delta U - (T_2 S_2 - T_1 S_1)$

$= \Delta U - (T_2(S_1 + \Delta S) + T_1 S_1)$

$= \Delta U - (T_2 - T_1)S_1 - T_2 \Delta S$

$\Delta G = \Delta H - (T_2 - T_1)S_1 - T_2 \Delta S$

(*e*) $\Delta U = c_v(T_2 - T_1) \quad \Delta H = c_p(T_2 - T_1) \quad \Delta S = c_v \ln (T_2/T_1)$

$\Delta A = \Delta U - (T_2 - T_1)S_1 - T_2 \Delta S$

$\Delta G = \Delta H - (T_2 - T_1)S_1 - T_2 \Delta S$

CHAPTER SIX

6.1 $\Delta H_{1000} = -216{,}600$ joules; $\Delta S_{1000} = -97.1$ joules/degree.

6.2 $\Delta H_{m,800} = 9550$ joules; $\Delta S_{m,800} = 18.10$ joules/degree; $\Delta G_{m,800} = -4930$ joules.

6.3 $H_{\text{diamond},1000} - H_{\text{graphite},1000} = 906$ joules. Thus the oxidation of diamond is 906 joules/mole more exothermic at 1000 K than the oxidation of graphite.

6.4 13 kilograms.

6.5 $\Delta H_{600} = 4250$ joules, $\Delta S_{600} = 12.99$ joules/degree. Thus $\Delta G_{600} = -3544$ joules. $\Delta G_{600} = \Delta H_{298} - 600 \times \Delta S_{298} = -3238$ joules.

6.6 $P_3 = 494$ atm; $\Delta S(1 \rightarrow 2) = -1.780$ joules/degree, $\Delta S(2 \rightarrow 3) = -0.022$ joules/degree. Thus $S_3 - S_1 = -1.802$ joules/degree.

6.7 $\Delta H_{298} = 231.3$ kilojoules, $\Delta S_{298} = 33.2$ joules/degree, $\Delta G_{298} = 221.4$ kilojoules.

6.8 $c_{p,\text{Al}} = 21 + 0.0116T$ joules/degree·mole.

6.9 $\Delta H_{1800} = -404.9$ kilojoules.

6.10 (*a*) $\Delta H_{298} = -710.8$ kilojoules, $\Delta S_{298} = -346$ joules/degree, $\Delta G_{298} = -607.7$ kilojoules. (*b*) $\Delta H_{298} = -417.2$ kilojoules, $\Delta S_{298} = -20$ joules/

degree, $\Delta G_{298} = -411.2$ kilojoules. (*c*) $\Delta H_{298} = -244.8$ kilojoules, $\Delta S_{298} = -146$ joules/degree, $\Delta G_{298} = -201.3$ kilojoules.

CHAPTER SEVEN

7.1 $\Delta H_{evap, Ag, 2147} = 260.9$ kilojoules/mole, $c_{p, Ag(v)} - c_{p, Ag(l)} = -7.07$ joules/degree·mole.

7.2 $p^0_{Hg,373} = 3.72 \times 10^{-4}$ atm.

7.3 $\ln p^0_{Fe(l)}$ (atm) $= -45{,}434/T - 1.269 \ln T + 23.94$.

7.4 2995 atm.

7.5 At 150°C, $p^0_{I_2(l)} = 0.39$ atm, and $p^0_{I_2(s)} = 0.61$ atm. Thus, isothermal compression at 150°C from $P = 0.04$ atm causes condensation of the liquid state at $P = 0.39$ atm. For the cooling at constant volume, $P = 9.453 \times 10^{-5} T$, which intersects the solid–vapor equilibrium line first at $P = 0.031$ atm, $T = 362$ K.

7.6 $p^0_{CO_2(l),298} = 73.4$ atm. As the triple-point pressure is 5.14 atm, the 1-atm isobar does not pass through the liquid phase field.

7.7 At $P = 5000$ kg/cm^2, the tangent to the variation of P with T, $(dP/dT) = \Delta H/T\ \Delta V = 100$ kg/cm^2·degree ($= 100 \times 0.969$ atm/degree). Thus $\Delta H_{m,Na} = 2298$ joules/mole.

CHAPTER EIGHT

8.1 $(P_R - 3/V_R^2)(V_R - \frac{1}{3}) = 8T_R/3$, which, in containing no constants, is applicable to all van der Waals gases. $Z_{cr} = 0.375$; $(\partial U/\partial V)_T = a/V^2$, for a van der Waals gas.

8.2 $n_A/n_B = 1$; $P = 1.414$ atm.

8.3 (*a*) $a = 6.78$ liters^2·atm·mole^{-2}, $b = 0.0568$ liters·mole^{-1}. (*b*) $V_{cr} = 0.1704$ liters·mole^{-1}. (*c*) $P = 65.5$ atm, $P_{ideal} = 82.1$ atm.

8.4 $w = -301.1$ kilojoules; (*a*) w for a van der Waals gas $= -309.4$ kilojoules. (*b*) w for an ideal gas $= -272.2$ kilojoules.

8.5 (*a*) $f = 689$ atm; (*b*) $P = 1083$ atm; (*c*) $\Delta G = 16{,}188$ joules with a nonideal contribution of 791 joules.

CHAPTER NINE

9.1 $\Delta G^0_{1000} = -27{,}387$ joules; $K_p = 1$ at 892 K; in both (*a*) and (*b*), the equilibrium shifts toward the $CO + H_2$ side.

9.2 18.33% CO_2, 18.33% H_2, 56.67% CO, 6.67% H_2O.

9.3 See Sec. 9.6. $\Delta H^0 = -94{,}600$ joules and $x = 0.463$. Thus, the heat evolved $= 0.463 \times 94{,}600 = 43{,}800$ joules.

9.4 $CO_2/H_2 = 1.285$; $p_{O_2} = 1.08 \times 10^{-6}$ atm.

9.5 $PCl_5/PCl_3 = 0.3715$.

9.6 $p_H = 1.67 \times 10^{-3}$ atm.

CHAPTER TEN

10.1 $T = 565$ K.

10.2 $T_{m,\mathrm{Ni}} = 1725$ K, $\Delta H_{m,\mathrm{Ni}} = 17{,}600$ joules, $\Delta S_{m,\mathrm{Ni}} = 10.2$ joules/degree.

10.3 $p_{O_2(\mathrm{eq},923,\mathrm{Cu/Cu_2O})} = 3.86 \times 10^{-12}$ atm. Thus the poorest vacuum that could be tolerated is $3.86 \times 10^{-12}/0.21 = 1.84 \times 10^{-11}$ atm. The temperature at which $p_{O_2(\mathrm{eq,Cu/Cu_2O})} = 10^{-4}$ atm is higher than the melting temperature of copper, and hence solid copper could not be annealed in this atmosphere without oxidation occurring.

10.4 (*a*) 463 K; (*b*) 422 K.

10.5 For $Cr_2O_{3(s)} + 3H_{2(g)} = 2Cr_{(s)} + 3H_2O_{(g)}$, $p_{H_2O}/p_{H_2(\mathrm{eq},1500)} = 1.69 \times 10^{-3}$ and thus, with $P_{\mathrm{total}} = 1$ atm, the maximum tolerable water vapor pressure is 1.69×10^{-3} atm.

10.6 If equilibrium had been attained, the exit gas would have contained 11.4% HCl. As the exit gas contains only 7% HCl, equilibrium has not been attained.

10.7 $p_{CO_2(\max)} = 0.988$ atm.

10.8 $\Delta G^0 = 172{,}274 - 15T$ joules.

10.9 $p_{\mathrm{Mg},1673} = 0.0146$ atm.

10.10 (*a*) The system contains 0.01 mole of $CaCO_3$ and hence 0.01 mole of CO_2. The variation of the pressure of CO_2 in equilibrium with CaO and $CaCO_3$ is

$$\ln p_{CO_2(\mathrm{eq},T)} = \frac{-20{,}254}{T} + 17.32$$

and if all of the CO_2 in the system exists in the gas phase, the pressure of CO_2 varies as

$$P = 8.206 \times 10^{-4}\, T$$

These lines intersect at $T = 1166$ K, at which temperature decomposition of the $CaCO_3$ is complete. Below this temperature the pressure of CO_2 is that for equilibrium with CaO and $CaCO_3$, and above this temperature the pressure of CO_2 is given by the ideal gas law. (*b*) 0.053 atm. (*c*) 1.23 atm.

10.11 $\Delta G^0 = 283{,}690 - 124.3T$ joules.

10.12 For $2Fe + Fe_3O_4 = 4FeO + Fe$, $\Delta G^0 = 52{,}600 - 62.6T$ joules. $\Delta G^0 = 0$ at $T = 840$ K, at which temperature Fe, FeO, and Fe_3O_4 exist in equilibrium. At $T < 840$, ΔG^0 is positive and hence Fe_3O_4 is the stable oxide in equilibrium with Fe. At $T > 840$, ΔG^0 is negative and hence FeO is the stable oxide in equilibrium with Fe.

10.13 $ZrCl_4$ reacts with the gas to form solid ZrO_2, which decreases the partial pressure of $ZrCl_4$ in the gas to 1.72×10^{-8} atm. $HfCl_4$ does not react with the gas.

10.14 (I) is for oxidation of solid Mg, (II) is for oxidation of gaseous Mg, and (III) is for oxidation of liquid Mg.

$$T_{m,\mathrm{Mg}} = 928 \text{ K} \qquad T_{b,\mathrm{Mg}} = 1372 \text{ K}$$

10.15 $p_{O_2} = 1.058 \times 10^{-5}$ atm, $p_{H_2S} = 0.0151$ atm, $p_{S_2} = 0.356$ atm, and $p_{Zn} = 2.608$ atm.

10.16 4.62×10^{-7} atm is the poorest vacuum allowable.

10.17 $p_{H_2} = 0.995$ atm, $p_{Mg} = 0.00167$ atm, $p_{HCl} = 0.00344$ atm. The maximum tolerable water vapor pressure is 1.36×10^{-7} atm.

10.18 $P = 5428$ atm.

CHAPTER ELEVEN

11.1 Heat change = heat of fusion of Cr + $\Delta\bar{H}^M_{Cr}$ = 21,000 + 0 = 21,000 joules. Entropy change = entropy of fusion of Cr + $\Delta\bar{S}^M_{Cr}$ = 21,000/2173 − 8.3144 × ln 0.2 = 23.05 joules/degree.

11.2

X_{Sn}	0	.1	.2	.3	.4	.5	.6	.7	.8	.9	1
a_{Sn}	0	.015	.077	.201	.333	.457	.577	.692	.799	.900	1
a_{Cu}	1	.853	.644	.472	.362	.279	.210	.150	.097	.048	0

11.3 (*a*) $\bar{G}^{xs}_{Au} = -2166$ joules, $\bar{G}^{xs}_{Cu} = -11{,}791$ joules. (*b*) $\Delta G^M = -12{,}925$ joules. (*c*) $p_{Au} = 4.1 \times 10^{-7}$ atm, $p_{Cu} = 1.76 \times 10^{-6}$ atm.

11.4 (*a*) Up to $X_A = 0.4$; (*b*) 0.4; (*c*) $\Delta\bar{H}^M_A = -2093$ joules; (*d*) $\Delta H^M = -2093X_A$.

11.5 See Figs. 11.12, 11.13, and 11.15.

11.6 See Figs. 11.10, 11.11, and 11.14.

11.7 $\log \gamma_{Sn} = -0.32X^2_{Pb}$; (*a*) 16,541 joules; (*b*) 17.67 joules/degree; (*c*) a_{Pb} at 746 K = 0.416, a_{Pb} at 1000 K = 0.436.

11.9 $\log \gamma_{Cd} = 0.185X^2_{Zn} + 0.13X^3_{Zn}$; $a_{Cd} = 0.577$.

CHAPTER TWELVE

12.1 (*a*) $\Delta G^0_{m,\mathrm{Fe}} = 13{,}010 - 1.3T \ln T + 2.554T$ joules. (*b*) $a_{Fe} = 0.795$; (*c*) $a_{Fe} = 0.76$; (*d*) $G^{xs} = 4979$ joules; (*e*) $X_{Fe} = 0.39$, $X_{FeS} = 0.61$.

12.2 (*a*) −11,310 joules; (*b*) zero.

12.3 −1447 joules.

12.4 $T = 1889$ K; $X_{FeO} = 0.671$ (the liquidus composition at 1889 K); $T = 1973$ K; $X_{FeO} = 0.342$ (the solidus composition at 1973 K).

CHAPTER THIRTEEN

13.1 $a_{Cu} = 0.141$.

13.2 $a_{Mg} = 0.000866$.

13.3 $a_{PbO} = 0.5$.

13.4 $X_{Pb(eq, 1073)} = 0.9784$. The extent to which Cu is removed is increased by decreasing the temperature.

13.5 $a_{C(gr)} = 0.5$; for $a_{C(gr)} = 1$, $p_{H_2} = 0.917$ atm.

13.6 $a_{FeO} = 6.05 \times 10^{-5}$.

13.7 $\Delta \bar{H}^M_{PbO} = -3937$ joules.

13.8 The equilibrium has two degrees of freedom which may be chosen from T, p_{O_2}, one activity in the metal phase, and one activity in the oxide phase. Fixing T and X_{Mn} (which fixes a_{Mn}) determines the values of the other variables. Oxide is of $X_{FeO} = 9.98 \times 10^{-4}$ and $p_{O_2(eq)} = 2.6 \times 10^{-27}$ atm.

13.9 For $2A + B = A_2B$, $\Delta G^0 = -24{,}371$ joules; and for $A + 2B = AB_2$, $\Delta G^0 = -23{,}189$ joules.

13.10 $\gamma^0_V = 0.14$; (*a*) 10^{-3}; (*b*) 8.5×10^{-4}; (*c*) 7.4×10^{-3}; (*d*) 0.65.

13.11 $\gamma^0_{Ti} = 0.012$.

13.12 $\Delta G^0 = -567{,}440$ joules.

13.13 $p_{O_2(eq, 1873, Fe(l)/FeO \cdot Al_2O_3/Al_2O_3)} = 5.18 \times 10^{-10}$ atm. $h_{Al(1\ wt\%)} = 6.9 \times 10^{-6}$. The four-phase, three-component equilibrium has one degree of freedom that is used by specifying the temperature. Thus the values of all the other variables are fixed.

13.14 The four-phase, three-component equilibrium has one degree of freedom that is used by specifying the temperature. Thus p_{O_2} and p_{N_2} are uniquely fixed by the equilibria $Si + O_2 = SiO_2$ and $3Si + 2N_2 = Si_3N_4$, respectively. The equilibrium state is $p_{O_2} = 9.4 \times 10^{-39}$ atm, $p_{N_2} = 1.9 \times 10^{-11}$ atm. At higher temperatures, the system exists as Si + gas, and at lower temperatures it exists as SiO_2 + gas.

13.15 (*a*) $Cr_{23}C_6$ is stable with respect to Cr and Cr_2O_3. (*b*) MnO is stable with respect to Mn and Mn_3C.

13.16 AB_3 decomposes to AB and B at temperatures below 876 K. AB is a stable solid at all temperatures up to its melting point.

13.17 $p_{Mg} = 0.021$ atm.

13.18 $p_{Zn} = 0.544$ atm; $p_{CO} = 0.532$ atm; $p_{CO_2} = 5.9 \times 10^{-3}$ atm.

13.19 $3CaO_{(s)} + 2Al_{(l)} + 3MgO_{(s)} = 3Mg_{(v)} + 3CaO \cdot Al_2O_{3(s)}$
$\Delta G^0_{3CaO \cdot Al_2O_3, 1300} = -49{,}995$ joules; $a_{Al_2O_3} = 0.0098$.

13.20 [Al] = 0.396 wt%.

13.21 $p_{SO_3} = 0.0799$ atm, $p_{SO_2} = 0.614$ atm, and $p_{O_2} = 0.307$. Thus $P_{total} = 1$ atm.

13.22 1050 K.

13.23 $P = 7030$ atm.

13.24 With [C] = 0.2 wt%, [S] = 0.27 wt%, and with [C] = 2.0 wt%, [S] = 0.00416 wt%.

13.25 (*a*) For wustite, $0.31 < p_{CO} < 2.9$ atm. (*b*) For cementite, $8.0 < p_{CO} \leqslant 8.65$ atm.

13.26 1026 K.

13.27 −119,290 joules.

CHAPTER FOURTEEN

14.1 $Pb_{(s)} + 2AgCl_{(s)} = PbCl_{2(s)} + 2Ag_{(s)}$
$\Delta G^0_{298} = -94{,}560$ joules, $\Delta S^0_{298} = -35.5$ joules/degree.

14.2 (*a*) 103,380 joules; (*b*) 27.98 joules/degree; (*c*) 8338 joules, $a_{Hg} = 0.707$.

14.3 $\ln p_{O_2} = -4.642 \times 10^4 \mathcal{E}/T - 6.243 \times 10^4/T + 15.07$.

14.4 $\Delta c_p = -0.093$ joules/degree.

14.5 (*a*) 0.673; (*b*) −2150 joules; (*c*) 3329 joules.

14.6 (*a*) 1.32×10^{-6}; (*b*) 0.5; (*c*) 0.266 wt%.

14.7 $E = -0.126$ volts, pH = 6.33. In this state, $[Pb^{4+}] = 2.76 \times 10^{-62}$ moles/liter, $[HPbO_2^-] = 1.12 \times 10^{-9}$ moles/liter, $[PbO_3^{2-}] = 3.13 \times 10^{-47}$ moles/liter, and $[PbO_4^{4-}] \sim 0(\ln [PbO_4^{4-}] = -274)$.

14.8 Decomposition voltage of Al_2O_3 at 1000°C = 2.19 volts. In the Hall-Heroult cell, decomposition voltage at 1000°C is 1.17 volts.

INDEX